D1285976

THE COMPLETE ENCYCLOPEDIA

Antarctica

and the Arctic

Antarctica
and the Arctic

David McGonigal   Dr Lynn Woodworth

FIREFLY BOOKS

KALAMAZOO PUBLIC LIBRARY

A FIREFLY BOOK

Published by Firefly Books Ltd., 2001

© Global Book Publishing Pty Ltd 2001

All rights reserved. No part of this publication
may be reproduced, stored in a retrieval system,
or transmitted in any form or by any means,
electronic, mechanical, photocopying, recording
or otherwise, without the prior written permission
of the publisher.

The moral rights of all contributors have been
asserted.

First printing

National Library of Canada Cataloguing in
Publication Data

Antarctica and the arctic : the complete
encyclopedia

Includes index.
ISBN 1-55297-545-2

1. Antarctica.   2. Arctic regions.   I. McGonigal, David

G587.A57 2001   998   C2001-900551-2

Published in Canada in 2001 by
Firefly Books Ltd.
3680 Victoria Park Avenue
Willowdale, Ontario M2H 3K1

Printed in Hong Kong by Sing Cheong Printing Co. Ltd
Film separation Pica Digital Pte Ltd, Singapore

Captions for images in the preliminary pages
Page 1: Ice mask
Pages 2–3: Emperor penguins and chicks
Page 4–5: Emperor penguins, Ross Ice Shelf
Pages 6–7: Mount Paget, South Georgia
Page 8–9: Ice cave, Erebus Ice Tongue
Page 12: Vanderford Glacier, East Antarctica
Page 16: Eroded icebergs, Antarctic Peninsula

Details of cover images and part openers appear
on page 608.

Photographers are acknowledged on page 608.

Publisher	Gordon Cheers
Associate publisher	Margaret Olds
Art director	Stan Lamond
Managing editor	Tracy Tucker
Chief text editor	Janet Healey
Senior text editor	Judith Simpson
Editors	Maggie Aldhamland
	Derek Barton
	Anna Cheifetz
	Dannielle Doggett
	Fiona Doig
	Kate Etherington
	Vanessa Finney
	Denise Imwold
	Margaret McPhee
Additional text	Mike Craven
	Peter Gill
	Mark Hindell
	Luke Saffigna
	Kim Westerskov
Book and cover design	Stan Lamond
Cartographer	John Frith
Additional cartography	Robert Taylor
Map contour artwork	Oliver Rennert
Managing map editor	Valerie Marlborough
Map editors	Vanessa Finney
	Janet Parker
	Judith Simpson
	Jan Watson
Cartographic consultants	Vivienne Allan
	Henk Brolsma
Wildlife distribution maps	Lynn Woodworth
Principal photographer	David McGonigal
Photo library	Alan Edwards
Photo and illustration research	Gordon Cheers
	Joanna Collard
	Heather McNamara
	Tracy Tucker
Illustrations	Glen Vause
Diagrams	Mike Gorman
Index	Di Harriman
Typesetting	Dee Rogers
Publishing assistant	Erin King
Production	Rosemary Barry
	Bernard Roberts

Contributors

Don Adamson

Dr Don Adamson is a Senior Research Fellow in biology at Macquarie University in Sydney, Australia. Over the past 23 years he has undertaken summer fieldwork in the Vestfold Hills, the Prince Charles Mountains, and near Mawson in continental Antarctica, as well as on sub-Antarctic Macquarie Island. His research interests include geomorphology, vegetation–landscape interactions, and environmental history.

Thomas Bauer

Thomas Bauer is an Assistant Professor in the Department of Hotel and Tourism Management at the Hong Kong Polytechnic University. During a 10-year affiliation with Antarctica, he has participated in tourism research, lectured aboard cruise ships, and driven Zodiacs. He is the photographer and co-producer of the *Voyage to Antarctica* CD-ROM and video, produced with the Computer Aided Learning Centre at Victoria University of Technology, Melbourne. Thomas's writings include *Tourism in the Antarctic: Opportunities, Constraints, and Future Prospects.*

Gary Burns

Dr Gary Burns is a Principal Research Scientist at the Australian Antarctic Division. He has wintered at Casey and on Macquarie Island, and spent a summer at Davis overseeing the equipment used to study aurora and the upper atmosphere for Australian National Antarctic Research Expeditions (ANARE). Gary is presently monitoring the Antarctic upper atmosphere for indicators of climate change. He has published articles on pulsating aurora, the auroral green line, and auroral linkages between the hemispheres.

Robert Clancy

In his professional life, Robert Clancy is Professor of Pathology at Australia's Royal Newcastle Hospital, a clinical immunologist, and Director of a biotechnology company related to mucosal immunology. His recreational passion is Antarctica. Robert first visited Macquarie Island in 1959 as an Australian Scout, and then returned there in 1998 to research the effect of Antarctic living on the immune system. He collects antique maps of *Terra Australis*, documents of Antarctic exploration, and published records of discovery.

Louise Crossley

Louise Crossley—scientist, historian, university lecturer, museum director, broadcaster, futurist, Antarctic expeditioner, and writer—has spent two winters and three summers in Antarctica with ANARE. She has also been a guide/lecturer on icebreaker cruises to the Antarctic Peninsula, the Weddell Sea, the Ross Sea, and on three semi-circumnavigations. Louise is author of *Explore Antarctica* (1995), a general introduction to the continent, and editor of *Trial by Ice: The Antarctic Journals of Captain John King Davis* (1997).

Arthur Ford

For nearly three decades Dr Arthur Ford led or participated in numerous United States Geological Survey expeditions into many areas of Antarctica, and has also undertaken geological research in Alaska. Since retiring in 1995 he has lectured on Antarctic cruise ships. The chapter on Antarctica in *Encyclopædia Britannica* is among his more than 200 publications. A recipient of the United States Antarctica Service Medal, Arthur is a Fellow of the Royal Geographical Society, the Explorers Club, and the Geological Society of America.

Julia Green

Dr Julia Green has been studying, researching, writing, and lecturing on Antarctic law and policy for more than 10 years at the University of Tasmania. She recently visited Casey Station with ANARE. Julia has a B.A. degree in politics, philosophy, and sociology, and a Graduate Diploma (honors) in Antarctic and Southern Ocean studies. Her Ph.D. research investigated the changing nature of sovereignty in the Arctic and Antarctic in response to global environmental interdependence.

Robert Headland

Robert Headland, Curator of the Scott Polar Research Institute, Cambridge, United Kingdom, is particularly interested in the historical geography of polar regions. Besides being familiar with the archival resources of the Institute and other organizations, he has served with the British Antarctic Survey and similar expeditions in Antarctica, and has lectured aboard tourist vessels. Currently, Robert is updating his "Antarctic Chronology," an extensive work detailing all known expeditions to the region from the earliest days until 2000.

Robert Hill

Professor Robert Hill spent 20 years at the University of Tasmania researching the plant macrofossil record in Tasmania and Antarctica over the past 40 million years. Now a senior Research Fellow at the University of Adelaide, he is investigating the impact of Australia's drying climate on vegetation over the past 30 million years. Robert is Professor Emeritus at the University of Tasmania and Professor at the University of Adelaide.

Peter Hillary

Peter Hillary's extensive mountaineering experience includes more than 30 alpine expeditions in the Himalayas, the Karakoram, New Zealand's Southern Alps, the Andes, Antarctica, and North America. He forged a new route on foot up the Shackleton Glacier to the South Pole in 1999, 42 years after his father, Sir Edmund Hillary's epic Antarctic journey. In between expeditions, Peter writes, lectures, designs and markets outdoor equipment, and is an adventure travel operator.

Bernadette Hince

Bernadette Hince loves words and is lost without a dictionary. She wrote the natural history definitions in the *Australian National Dictionary* and advised on natural history for the *Dictionary of New Zealand English*. Like her hero, Dr Samuel Johnson, she believes that humor has a place in dictionaries. As part of 11 years researching and writing for her Antarctic dictionary, Bernadette spent the summer of 1995–96 in Antarctica with innumerable Weddell seals.

Paul Holper

Paul Holper is Communication Manager for Australia's Commonwealth Scientific and Industrial Research Organisation (CSIRO) Atmospheric Research division, which conducts research into climate, weather, and pollution. He has a first class honors degree in chemistry from the University of Melbourne, a Diploma of Education, and a Graduate Diploma in Science Communication. Paul writes extensively on science and the environment, and is the author of numerous articles, five science textbooks, and a series of science books for children.

Paul Lehmann

Dr Paul Lehmann received a Ph.D. in upper atmospheric physics at the University of Melbourne, followed by research fellowships in atmospheric physics at the University of Illinois and the Max Planck Institute, Germany. In 1980 he was appointed to the Physics Department at the University of Melbourne. In 1989 Paul became senior physicist in the area of stratospheric ozone depletion at the Bureau of Meteorology, Melbourne. He has represented Australia at United Nations meetings on ozone depletion.

Peter Lemon

Peter Lemon has worked with Australia's Peregrine Adventures for 17 years and has undertaken 3 trips to the Antarctic, 12 to Nepal and over 30 to southern Africa. He is especially interested in wildlife and landscape photography. Peter is also fascinated by vintage and veteran aircraft and Australian commercial aviation history, particularly that between the two World Wars. He enjoys flying in aircraft—the older the better, within reason.

Harvey Marchant

Professor Harvey Marchant is Leader of the Australian Antarctic Biology Program at the Australian Antarctic Division and a Professorial Fellow of the University of Melbourne. He has an honors degree in science from the University of Adelaide, a Ph.D. from the Australian National University, and received a Senior Fulbright Award in 1987. Author of more than 100 publications and editor of the international journal, *Polar Biology*, Harvey has visited Antarctica 10 times with Australian, Japanese, and United States programs.

David McGonigal

David McGonigal is an award-winning travel writer and author/editor of more than a dozen books. In 1997 he rode his motorcycle in Antarctica (then to Alaska and across Siberia) on the world's first seven-continent motoring journey. A graduate in Arts/Law and a Royal Geographical Society Fellow, David convenes university courses on Antarctica and publishes and broadcasts widely. He has made 25 journeys to the Antarctic Peninsula, Ross Sea, and Northeast and Northwest passages as leader, lecturer, and photographer.

Gary Miller

Dr Gary Miller is a Research Assistant Professor of Biology at the University of New Mexico. More than 25 years ago, he began studying the behavior of Polar bears and the distribution of Bowhead whales in the Arctic, but was soon drawn to Antarctica to study the behavior, ecology, genetics, and diseases of penguins and skuas. Gary has participated in United States, New Zealand, and Australian Antarctic programs, as well as lecturing on ships visiting Antarctica.

Stephen Nicol

Dr Stephen Nicol has worked on krill since 1979. He joined the Australian Antarctic Division (AAD) in 1987 to lead the krill research team, and has made six voyages to the Antarctic. Since 1999 Stephen has been Program Leader for the AAD's Antarctic Marine Living Resources Program. He has been a member of Australia's delegation to the Commission for the Conservation of Antarctic Marine Living Resources (CCAMLR) since 1987, and is currently vice chair of CCAMLR's Scientific Committee.

Hugh Pennington

Professor Hugh Pennington graduated in medicine and gained his Ph.D. from St Thomas's Hospital Medical School, London. After posts there, at the University of Wisconsin, and the University of Glasgow, in 1979 he was appointed to the Chair of Bacteriology at Aberdeen University. His research—sometimes involving British Antarctic Survey medical officers—focuses on developing molecular typing methods for bacteria. Hugh is a Fellow of the Royal Society of Edinburgh and the Academy of Medical Sciences.

Kim Pitt

Kim F. Pitt, A.M., joined the Australian Antarctic Division in October 1997 as Assistant Director of Expedition Operations. He is responsible for providing the infrastructure (station and field operations, shipping and air operations, and engineering and construction programs) that support Australia's interests in the Antarctic and sub-Antarctic. Before this, he spent 32 years in the Royal Australian Navy serving in submarines, the intelligence community, and the Navy's Support Command.

Martin Riddle

Dr Martin Riddle has spent his entire career researching the effects of human activity on the environment. His studies have encompassed the North Sea oil industry, coral reefs, and the effects of ocean disposal of sewage effluent. In 1994 Martin became the first program leader of the Australian Antarctic Division's Human Impacts Research Program, which aims to provide sound information on which to base guidelines and procedures to reduce the effect of humans on the Antarctic environment.

Stephen Rintoul

Dr Stephen R. Rintoul studies the impact of Southern Ocean currents on the earth's climate. For the past 10 years, he has led the Southern Ocean program at CSIRO Marine Research, where he is a Principal Research Scientist. He received his Ph.D. from the Woods Hole Oceanographic Institution and Massachusetts Institute of Technology, and in 2000 was awarded the Australian Meteorology and Oceanography Society's Priestley Medal. Stephen has published numerous papers on oceanography and climate.

Annie Rushton

Annie Rushton has published widely and worked in Australian ABC radio and television. She visited Macquarie Island in 1994, and more recently spent an Antarctic summer at Casey station. She was researcher, writer, and associate producer for ABC-TV's documentary *South of No North*, which celebrated Hobart's role as an Antarctic gateway. Annie has developed content for videos and CD-ROMs on Antarctica and is currently setting up an interactive website, "Expeditioner," for the Australian Antarctic Division's Operations Branch.

Pat Selkirk

Dr Pat Selkirk is a Senior Lecturer in biology at Macquarie University in Sydney, Australia. During the past 22 years she has spent many summers undertaking Antarctic fieldwork on Ross Island, in Victoria Land, and near Casey, and sub-Antarctic fieldwork on Macquarie and Heard islands. Her research interests include plant–environment interactions, environmental history, biogeography, and evolutionary biology.

Robert B. Stephenson

Robert B. Stephenson spent many years as a town planner but his avocation has always been Antarctica. This interest was sparked through book collecting in the 1970s. He now visits the Southern Continent as a lecturer on tourist ships, and coordinates "The Antarctic Circle"—an informal international group of scholars and knowledgeable amateurs actively involved in non-scientific Antarctic studies. Robert also zealously hunts down sites of Antarctic interest beyond Antarctica.

Michael Stoddart

Professor Michael Stoddart is ANARE Chief Scientist. After 16 years as Lecturer and Reader in Zoology at the University of London, he took the Chair of Zoology at the University of Tasmania in 1985. In 1994 he became Deputy Vice Chancellor at the University of New England, Armidale, New South Wales, before moving back to Hobart in 1998. Michael who holds a Professorial Fellowship at Melbourne University, has written three textbooks on vertebrate olfactory biology, and more than 100 scientific papers.

Bernard Stonehouse

Polar biologist Dr Bernard Stonehouse is especially interested in whales, seals, and penguins. He has spent four winters and innumerable summers in Antarctica, where he first worked as a meteorologist and pilot, and ran a dog team on sledging surveys. Bernard has taught in universities in Britain, New Zealand, Canada, and the United States. Now retired, he is a Senior Associate of the Scott Polar Research Institute, and is currently researching the ecological impacts of tourism on polar regions.

Patrick Toomey

Retired from the Canadian Coast Guard, Captain Patrick Toomey is presently a consultant ice navigation specialist and an assessor for the Canadian Federal Court. He has piloted Russian icebreakers in Arctic waters, made transits of the Northwest Passage, and reached the North Pole. He has also been an ice pilot and lecturer on Antarctic voyages, and has circumnavigated the Southern Continent. Patrick specializes in nautical journalism, the development of international regulations concerning ice navigation, and logistics studies for polar transportation.

Kim Westerskov

Dr Kim Westerskov is an award-winning freelance photographer and writer with a Ph.D. in Marine Sciences. Based in Tauranga, New Zealand, he specializes in Antarctica, and in the seas and wildlife from Antarctica to the tropics. Kim has written and illustrated many books, including ones on Emperor penguins, Weddell seals, and Antarctica's huskies. His awards include five first prizes in the British "Wildlife Photographer of the Year" competition.

Lynn Woodworth

Dr Lynn Woodworth completed her Ph.D. on maximizing genetic diversity in endangered species. Her honors work was in mammalian infectious diseases and genetic causes of multiple births. She has managed a genetics research laboratory and coordinated conferences for the Society for Conservation Biology. Lynn's first Antarctic voyage was in 1995, and she has been there every year since, spending several months working on polar vessels as an assistant expedition leader, Zodiac driver, and wildlife lecturer.

 C o n t

e n t s

Foreword

Not very long ago a young woman asked me whether there are any adventures left in the world for young people to strive for. I replied that there are many outdoor quests that qualify as true adventures—they are just harder to find than they were half a century ago. The world's polar regions, in particular the barren wasteland of the Antarctic ice sheet, are fertile grounds for such adventures, as they have been for the past 100 years.

Antarctica is accurately described as the "last great wilderness on earth." When my fellow New Zealanders and I arrived at the South Pole on 4 January 1958 (in Ferguson tractors as part of the Commonwealth Transantarctic Expedition) we were the first overland party to do so since Robert Scott and his ill-fated party in 1912. Before Scott, the only other party to reach the Pole was Roald Amundsen's, which arrived six weeks before Scott and claimed the South Pole for Norway. What a contrast with today! Now, several expeditions each southern summer take the same journey as ours—or even longer ones, along more difficult routes—to a destination that for so long proved elusive, and sometimes tragic.

In recent years, the Antarctic coastline, too, has become increasingly accessible. When we built Scott Base on Ross Island in 1957, we could not possibly have imagined that tourist icebreakers would be regularly visiting it by the end of the second millennium. It is remarkable to reflect that some contributors to this book have been to Antarctica 20, even 50, times or more on scheduled voyages aboard ice-strengthened ships.

Antarctica—the myth and the reality—has fascinated the world for more than 2,000 years. At the opening of the third millennium, it is gratifying to introduce a book that brings together early speculations about Antarctica and our present hard-won knowledge. In 1999, I had the pleasure of meeting David McGonigal and Lynn Woodworth,

the principle authors of this book, at an Antarctic seminar convened by David. Their passion for Antarctica shines out from the pages that follow.

Antarctica is a continent where the unusual prevails. As a New Zealander I look toward Antarctica to see what our weather has in store, but Antarctica's weather patterns are the driving force for the entire world's weather. Like every expeditioner, I have reveled in the antics of penguins and seals and watched the soaring albatross with pure joy; the wildlife section brings us up to date on what scientists have learned about these remarkably adapted creatures. And within these pages we find out about the research that is going on across the continent that acts as the largest cold storage center on earth.

Anyone with a sense of adventure and an active imagination can't fail to be moved by the tales of early polar exploration, such as Nansen deliberately setting the *Fram* into the ice in the hope that he would drift closer to the North Pole, or Mawson starving and alone on an epic trek of survival after his companions perished exploring the Antarctic interior. For much of my life I've been fascinated by Ernest Shackleton, one of the greatest expedition leaders of all time. He is the polar explorer I would have most enjoyed traveling with. Here, I read of his exploits and those of all the others, and enjoy them once again.

The year after I arrived at the South Pole, the governments of the world reached rare accord: that Antarctica should be maintained as a place of peaceful scientific research and cooperation. That cause is best served if people understand how special Antarctica is. Now that it is more readily accessible, I recommend an Antarctic trip to anyone. But, whether you set foot on the "seventh continent" or not, this book will provide an insight into the unique nature of what lies at the ends of the earth.

Sir Edmund Hillary

Antarctica

Key to Antarctic Peninsula stations
1. Comandante Ferraz (Brazil)
2. Arctowski (Poland)
   Teniente Jubany (Argentina)
3. King Sejong (South Korea)
   Artigas (Uruguay)
   Presidente Eduardo Frei (Chile)
   Bellingshausen (Russian Fed)
   Great Wall (China)
4. Capitán Arturo Prat (Chile)
5. General Bernardo O'Higgins (Chile)
6. Almirante Brown (Argentina)
   González Videla (Chile)

60°W · 31 · 50°W · 32 · 40°W · 33 · 30°W · 34

30
Península Valdés
Trelew
Comodoro Rivadavia
Golfo de San Jorge
Puerto Deseado
70°W
ARGENTINA
29
Río Gallegos
Bahía Grande
Cabo Vírgenes
Strait of Magellan
Tierra del Fuego
Ushuaia
CHILE
Puerto Natales
Punta Arenas
Isla de los Estados
Cabo de Hornos (Cape Horn)
Is Diego Ramírez

Falkland Islands (UK)
Stanley
Saunders I
Westpoint I
West Falkland
East Falkland
Port Stephens

DRAKE PASSAGE

South Orkney Is (UK)
Powell I
Coronation I
Laurie I
Orcadas (Argentina)
Signy (UK)

Clarence I
Elephant I
Joinville I
Moody Point
Dundee I
King George I
South Shetland Is (UK)
Esperanza (Argentina)
Vicecomodoro Marambio (Argentina)
James Ross I
Livingston I
Snow I
Smith I
Palmer Archipelago
Brabant I
Anvers I
Palmer (USA)
Academician Vernadsky (Ukraine)
Renaud I
Biscoe Islands
Lavoisier I
Adelaide I
Rothera (UK)
Marguerite Bay
Wordie Ice Shelf

Bransfield Strait
Robertson I
Cape Framnes
Larsen Ice Shelf
Cape Agassiz
Hearst I
Ewing I
Dolleman I
Cape Knowles
Kemp Peninsula
New Bedford Inlet
General San Martin (Argentina)
Palmer Land
Smith Peninsula
Mt Coman 3,655 m (11,991 ft)

WEDDELL SEA

Berkner Island
Ronne Ice She
He Ice
Korff Ice Rise
Skytrain Ice Rise

SOUTH PACIFIC OCEAN

Rothschild I
Alexander Island
George VI Sound
Wilkins Sound
Charcot I
Latady I
Spaatz I
Smyley I
Rydberg Peninsula

Bellingshausen Sea
Eltanin Bay
Allison Pen
Venable Ice Shelf
Fletcher Pen
Dendtler I
Peter I Øy
Abbot Ice Shelf
Dustin I
Jones Mts
Evans Pen
Noville Pen
Thurston I
Cape Flying Fish
Burke I
King Pen
Pine I Bay
Thwaites Glacier
Mt Takahe 3,398 m (11,148 ft)
Thwaites Iceberg Tongue
Bear Peninsula
Cape Herlacher
Martin Pen
Mt Frakes 3,677 m (12,064
Wright I
Getz Ice Shelf
Mt Sidley 4,285 m (14,058 f
Carney I
Siple I
Mt Siple 3,110 m (10,203 ft)
Dean I
Grant I
Hull Bay
Cruzen I

Ellsworth Land
Mts
Ellsworth
Vinson Mass 4,897 m (16,0

WEST ANTARCTIC

Mar
Byr
Lan
Flood Rang

Amundsen Sea

Polar Front (Antarctic Convergence)
Antarctic Circle

J · H · G · F · E · D · C
90°W

28
27
26
25

0 200 400 600 kilometers
0 200 400 miles
Scale at latitude 70°S

■ Major scientific station
▲ Summit, height on rock (not on ice)

100°W
110°W
120°W · 24 · 130°W · 23 · 140°W · 22 · 150°W · 21 · 160°W

Introduction

"Glittering white, shining blue, raven black, in the light of the sun the land looks like a fairy tale. Pinnacle after pinnacle, peak after peak—crevassed, wild as any land on our globe, it lies, unseen and untrodden." So wrote Roald Amundsen after discovering Antarctica's Queen Maud Range in 1911. Every visitor to Antarctica falls under the spell of the continent at the bottom of the world—a region of endless ice, strange and endearing creatures, and no indigenous inhabitants. Antarctica makes visitors feel like privileged strangers, as they would if they landed on the moon.

For every visitor to the polar regions, there must be a thousand enthusiasts at home who are in thrall to the magic of these extreme environments. This book preserves that sense of awe, but also offers a new and contemporary view of Antarctica and the Arctic. Collectively, those who have contributed have spent several lifetimes "on the ice"—yet all maintain that their passion for the polar regions remains undimmed by any difficult experience of cold and challenging conditions. Indeed, some of the text was written on icebreakers carving a passage to the shores of the Antarctic Continent, or in tents overlooking frolicking penguins. It was an all-encompassing project, spanning the globe from pole to pole, and contributions have come from an international galaxy of experts.

PART I, THE ENDS OF THE EARTH, describes how the laws of nature and physics gave rise to these remote regions of ice and snow, and have populated them with flora and fauna perfectly adapted to survive these most hostile of conditions.

PART II, POLAR REGIONS, gazettes the most important places in Antarctica and its surrounding oceans, from towering peaks to scientific bases, from historic sites to crowded penguin colonies and odiferous seal wallows. It also describes the settlements and major features of the lands and islands surrounding the North Pole.

PART III, POLAR WILDLIFE, includes up-to-the-minute information about penguins, albatrosses, whales, seals, Walruses, and Polar bears; indeed, about all the extraordinary creatures that live and thrive at the poles. Antarctica is a continent dedicated to scientific study that transcends national boundaries, and recent research has added greatly to the international fund of knowledge about the unique wildlife of the poles.

PART IV, EXPLORATION OF ANTARCTICA, tells a story that defies superlatives. It relates the almost unbelievable feats of the brave men who ventured to the dangerous ends of the earth, first in frail wooden sailing ships, then in flimsy aircraft. The story of polar exploration reflects the development of technology, although in the savage ice many innovations proved inadequate.

PART V, LIFE AT THE POLES, tells how the human history of Antarctica did not end with wooden ships and iron men, and how both scientists and visitors are learning important lessons from the world's last great wilderness. It also discusses the issues facing the polar regions today and the challenges for the future.

PART VI, RESOURCES, directs the reader to research sources, including books, video material, and websites, that can provide a key to further knowledge of contemporary Antarctic research and reading.

Antarctica then, now, and in the future

The Heroic Age of Antarctic exploration coincided with the development of photography. As the true nature of Antarctica has been discovered by indefatigable human efforts, it has been invariably captured on film. Within these pages, polar landscapes and wildlife come to life through photographs spanning more than a century. Many images reproduced here were commissioned specifically for this book, and each shows the unique nature of high latitudes that few have seen—or will ever see.

A printed page cannot capture the experience of an Antarctic blizzard's icy blast or the sensual overload generated by the cacophony of thousands of penguins. However, understanding the nature of the polar regions illuminates parts of the world that, within living memory, were blanks on world maps. As the icy ramparts of Terra [once] Incognita fall to the assault of satellite transmitters and icebreakers, the world is discovering regions even more marvelous than the conjectures of ancient imaginations. The points around which the earth rotates form the axis of the harsh yet inspiring wildernesses at the ends of the earth.

David McGonigal   Lynn Woodworth

Part I

THE ENDS OF THE EARTH

In polar regions, the rules by which people normally live break down. These are areas of everlasting ice, days lasting six months, large populations of a few wildlife species, erratic compass readings, and the meeting of all longitudes. They are places of dazzling beauty: ice reflects rainbow shards of light, snow petrels are black beaks and eyes against the snow, and the night sky is a screen for the scintillating aurora. Here, the sciences are not dull texts but guidelines for human survival. Frozen in time and ice, Antarctica holds secrets of the earth's past—and clues to its future.

The Geology
of the Poles

About the polar regions

From space, the earth is a blue ball capped with white at the poles, but few pause to wonder why the poles are cold enough to maintain permanent ice fields while the tropics are always hot. The energy generated by the sun is capable of heating the whole planet to 14°C (57°F), and the sun is 150 million kilometers (93 million miles) away—so why should it heat one part of the earth more than another?

The main reason is that the earth is spherical, with the equator pointing directly at the sun but the poles tilting away so that sunlight strikes the surface at an oblique angle. At 30° north or south of the equator (roughly the latitude of Florida or Sydney) this cuts the amount of sunlight received to about 86 percent of that falling on the equator. At 60° (about the latitude of Oslo or the South Sandwich Islands) the intensity of sunlight is reduced to 50 percent, and by 80° (about the latitude of the northern coast of Greenland or the edge of the Ross Ice Shelf, in the Antarctic) it has fallen to 17.4 percent. From the poles the sun is lower in the sky and its rays less warming. If the earth were not tilted, the poles themselves would receive no sunlight at all, and the sun would travel around the horizon, never rising above it.

Two other factors contribute to depriving the poles of the sun's warmth. First, the greater angle means that the sun's rays must penetrate more of the earth's atmosphere before reaching the surface. More importantly, of the solar radiation that does reach the earth's surface, the ice and snow that cover the poles reflect back at least 85 percent of it into the atmosphere.

Each 24 hours the earth rotates through 360°, so that alternate heating and cooling moderate the impact of the energy emanating from the sun; if it were not so, the side facing the sun would be too hot to sustain life, and the area in shadow too cold. The poles form the axis of this rotation, so they move very little, whereas points on the equator move at 1,670 kilometers per hour (1,037 mph) to complete a rotation in 24 hours.

The earth also travels around the sun. If the poles were perpendicular to the sun's rays, the whole earth would have 12 hours of daylight and 12 hours of darkness every day; there would be no seasons—and no easily definable year. The patterns of night and day and the seasons come about because the earth's axis of rotation inclines somewhat from the perpendicular, so that for half the year the North Pole leans toward the sun and experiences summer, while for the other half it is the South Pole that is inclined toward the sun. At the two solstices—June and December—the earth is at the points in its orbit where the North Pole or the South Pole, respectively, are most inclined toward the sun. At these times one pole is bathed in 24-hour daylight and gets more sunlight than anywhere else on earth.

Previous pages

◄ **COLD DESERTS**
The Dry Valleys in the Transantarctic Mountains, seen from the top of the Olympus Range. The largest snow- and ice-free area in Antarctica, these "cold deserts" are among the driest places on earth—even drier than the Sahara Desert.

◄ **MAGNIFICENT SUNRISE**
Its image distorted by the laws of atmospheric physics, the sun rises over the Ross Ice Shelf in early summer. South of the Antarctic Circle there are some winter days when the sun never rises above the horizon and an equal number of summer days when the sun never sets below it.

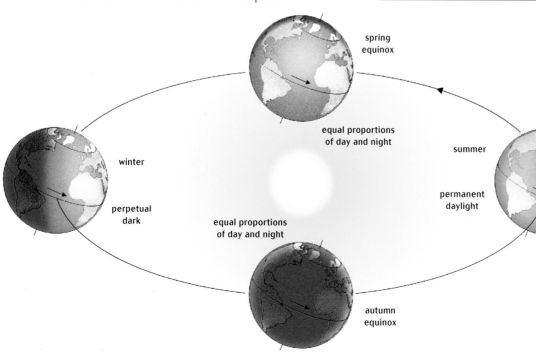

spring equinox

equal proportions of day and night

winter

summer

perpetual dark

permanent daylight

equal proportions of day and night

autumn equinox

The polar seasons

The planet's seasons are caused by the tilt of the earth. The closer a location is to either of the two poles, the more extreme the variation between summer and winter. As the illustration shows, during winter in the Antarctic the South Pole is tilted away from the sun, and at midwinter it lies in perpetual darkness. At midsummer in the southern hemisphere, the South Pole is tilted more toward the sun, and thus is bathed in perpetual daylight. The North Pole, being tilted in the opposite direction, experiences the opposite season, so in summertime in Antarctica it is winter at the North Pole.

> RESTING ON ICE
Midnight—but the sun is still high in the summer sky above the snow-free surface of a frozen tarn on Ross Island. Ice of such Antarctic ponds and lakes is frequently beautifully patterned. At this latitude the sun never sets between late October and late February. Continuous sunlight every day tempts many Antarctic visitors to push themselves to their physical limits.

Midway between the solstices are the September and March equinoxes, when radiation from the sun falls vertically at the equator. The name suggests that everywhere on earth has equal hours of light and darkness on that day. However, sunrise and sunset are judged not from the angle of the sun but from the appearance of its first and last rays, so different locations experience equal hours of light and darkness several days on either side of the true equinox. (The time also varies, defined by established time zones, so midday may not be midway between sunrise and sunset.) The equinoxes also mark the point where the poles graduate from six months of sunshine to six months of darkness, and vice versa. The earth does not orbit the sun in a perfect circle: when the South Pole is leaning toward the sun, the earth is about three percent closer to the sun. Nevertheless, Antarctica is much colder than the Arctic, mainly because of the dominant effect of its high polar ice sheet, but also because it is a landmass that blocks the moderating influence of the ocean.

The earth's poles

In reference to the earth, a pole is defined as either end of an axis around which the planet revolves. The earth has no fewer than seven poles. Some move over time; others are fixed, although the markers left by humans move as the Antarctic ice sheet slides toward the sea and the Arctic ice shifts with ocean currents.

Geographic poles: Fixed poles at latitudes 90°N and 90°S. Halfway between them—almost 10,000 kilometers (6,220 miles) from each—lies the equator. This is despite an attempt in the late eighteenth century to define the meter as one ten-millionth of this distance.

Poles of rotation: Movable poles forming the axis of the earth's rotation. They are within about 20 meters (66 ft) of the geographic poles, around which they rotate over 435 days. The exact locations of the geographic poles are calculated by taking the average measurement of the rotational poles.

Celestial poles: Movable poles at the positions in the sky occupied by an imaginary line extended infinitely through the geographic poles. They move because the earth wobbles on its axis over a variety of cycles that last 100,000, 41,000, or 23,000 years, due to the way in which the earth orbits around the sun.

Magnetic poles: Movable poles positioned where the magnetic field is at right angles to the earth's surface. They fluctuate daily in an oval under the influence of solar winds, moving about 2,000 kilometers (1,240 miles) in the last 400 years. Since 1841, the South Magnetic Pole has moved northwest at an average of 9 kilometers (6 miles) a year; in 2001 it was at 65°S 139°E and the North Magnetic Pole was at 79°01′N 105°08′W.

Geomagnetic poles: Calculations of where the magnetic poles would be if the earth's magnetic field worked like a simple bar magnet, they are situated at 78°39′N 69°W and 78°30′S 111°E. These poles are also the outer limits of the earth's geomagnetic field and the centers of auroral activity.

Poles of inaccessibility: In the Arctic, the position equidistant from the surrounding landmasses—84°03′N 174°51′W; in Antarctica the position, on average, that is furthest from the sea—85°50′S 65°47′E.

While not technically a pole, the point of the coldest place on earth is currently defined as Russia's base Vostok, high on the ice sheet. A temperature of −89.2°C (−128.6°F) was recorded there in July 1983.

▲ THE SOUTH POLE
It is said that an Archbishop came to Antarctica to say Midnight Mass at the South Pole, after having first said it at McMurdo Sound. Going to Byrd Station, he crossed the date line to say Mass again— the third time within 24 hours, always at the correct time.

> **THE BIRTH OF ICEBERGS**
Andvord Bay on the western side of the Antarctic Peninsula, near Anvers Island, has the spectacularly rugged landscape of mountains and glaciers that is typical of the Peninsula. These alpine glaciers occur in the high mountains and flow all the way down to the coast. When the ice flow reaches the coastline, the glaciers break off in cliffs and spawn myriad icebergs.

Polar contrasts

While the two polar regions share extremes of daylight and darkness, in other respects they differ widely.

ANTARCTIC	ARCTIC
Continent surrounded by ocean.	Ocean surrounded by continents.
South Pole: 2,836 m (9,300 ft) above sea level; bedrock 30 m (100 ft) above sea level.	North Pole: 1 m (3 ft) of sea ice; bedrock 427 m (1,400 ft) below sea level.
Deep, narrow continental shelf; restricted ice-free frozen ground; no tree line; no tundra; no native population.	Shallow, extensive continental shelf; extensive frozen ground; clear tree line; well defined tundra; circumpolar native populations.
South Polar ice sheet covering 98% of land.	Limited land ice.
Icebergs from glaciers and shelf ice measured in cubic kilometers.	Icebergs from glaciers measured in cubic meters.
Sea ice mainly annual, salty, and less than 2 meters (6 ½ ft) thick.	Sea ice mainly multi-year, low in salinity, and more than 2 meters thick (6 ½ ft).
Mean annual temperature at South Pole: −50°C (−58°F).	Mean annual temperature at North Pole: −18°C (0°F).
Marine mammals (whales and seals); no terrestrial mammals.	Terrestrial mammals (reindeer, wolf, musk ox, hare, lemming, fox); marine mammals (whales, seals, polar bears).
Less than 20 bird species between latitudes 70° and 80°.	More than 100 bird species between latitudes 75° and 80°.
Lichens at latitude 82°.	About 90 flowering plant species at latitude 82°.

Measuring Antarctica

The rock and permanent ice of the Antarctic landmass cover about 14 million square kilometres (5½ million sq. miles), making it the fifth largest continent, and considerably larger than Europe. If the ice melted, Antarctica would consist of the East Antarctic continent and the archipelago of West Antarctica leading northward to the Antarctic Peninsula; it would then be the smallest continent, at about half its present size. The winter sea ice roughly doubles the effective area of Antarctica; if it were permanent, Antarctica would be third in size after Asia and Africa. At its deepest point, the dome of the polar ice sheet is 4,800 meters (15,800 ft) thick, and the South Pole stands on 2.8 kilometers (1¾ miles) of ice. The average elevation of Antarctica is 2,160 meters (7,100 ft)—Asia, the next highest, is about 1,000 meters (3,300 ft).

Defining the polar regions

The Arctic and Antarctic Circles, at about 66°33′, mark the furthest latitudes from the poles where there is at least one day when the sun does not dip below the horizon at midsummer or rise above it at midwinter. Because the earth wobbles on its axis—making its tilt from the perpendicular vary—these circles slowly shift back and forth latitudinally. This occurs over one of the periods known as the Milankovitch cycles. They can last 100,000, 41,000, and 23,000 years. At present the circles are moving closer to the poles, and, being about halfway to the turning point of a 41,000-year cycle (where the angle of the earth's axis is between 22° and 24.5°) they are moving at close to the maximum rate of about 14.5 meters (47½ ft) per year. In 1996 the polar circles were at 66°33′37″; now they lie at about 66°33′39″; they will reach 68° in about 10,000 years.

It would be convenient to define the polar regions as those lying within the polar circles, but conditions at the circles differ so much that definition on this basis is meaningless. In fact each region has several different definitions, and the criteria also differ between the south

and the north polar regions. For example, the Antarctic Peninsula extends beyond the Antarctic Circle and almost all the winter sea ice develops north of the circle.

Antarctica is defined as the Antarctic Continent, and "the Antarctic" has a variety of definitions but is loosely defined as the area south of 60°S. An alternative definition of the Antarctic is that it lies south of the natural boundary formed by the Polar Front—the convergence of oceanic waters that encircles the continent roughly between latitudes 50°S and 60°S. There is also a legal definition, embodied in the Antarctic Treaty, which defines Antarctica as everywhere south of 60°S.

The Arctic, on the other hand, is not a landmass but frozen sea surrounded by land; even in winter the sea ice stops forming well north of the Arctic Circle. One defining criterion for the Arctic region is the northern limit of the tree line: in Scandinavia there are farms and towns that lie on the Arctic Circle, and trees will survive even further north. A definition that effectively covers the same area is based on temperature: the Arctic is defined as a northern region where the average temperature during the warmest month is under 10°C (50°F).

◄ **VOLCANIC PLUG**

A black plug of volcanic rock off the north coast of Siberia rises above the snow and ice in the Kara Sea in the high Arctic. The definition of the Arctic is based on quite different parameters to those used to define the Antarctic. The Arctic is, essentially, a frozen sea, while the Antarctic includes the large continental landmass of Antarctica.

▼ SOUTH POLE STATION
The United States operates Amundsen–Scott station, a year-round facility located at the South Pole. Because all longitudes and time zones meet here, one can zip around the world's time zones in mere minutes, and by crossing the international date line, step from today into tomorrow.

Time at the ends of the world

At the South Pole's Amundsen–Scott Base, a popular exercise is to take a quick walk around the world. This is possible because the base is located at the South Geographic Pole, where all the meridians of longitude converge, so that you can cross all the world's time zones as fast as you can run in a circle.

Until the end of the nineteenth century, towns and cities operated on the time for their specific longitude: at noon in New York City, it was 12.12 pm in Boston. But the rapid development of transport and communications made some standardization of time essential; in October 1884 the meridian of longitude through Greenwich, England, became the Prime Meridian, or 0 degrees, and Greenwich Mean Time (GMT), now known as Universal Time (UT), became the world's fixed time reference.

The Prime Meridian marks the international date line, running through the Bering Sea between eastern Russia and Alaska and then across the largely landless Pacific Ocean, with some deviations so that the few populated areas do not have to manage the problem of neighboring towns operating a day apart. At the poles, the 0° meridian is also the 180° meridian, dividing Antarctica into East Antarctica and West Antarctica and generating some confusion in one of the two parts of the world where "east" and "west" are largely meaningless concepts: at the South Geographic Pole every direction is north.

In polar regions, where the sun provides few temporal clues, clocks are important for human routine. At the equator, each degree of longitude occupies 111.36 kilometers (69.17 miles) and each time zone covers 15 degrees of longitude (1,670 kilometers/ 1,037 miles), but at the polar circles each time zone occupies only 665 kilometers (413 miles). Bases on the Antarctic Continent, where longitude means little during 24 hours of daylight or darkness, often use the same time as their home country for convenience. DM

Time zones

Each time zone occupies 15° of longitude. The first figure for each zone indicates the number of hours each time zone is ahead or behind Universal Time (UT). The figures in brackets show the time in each zone when it is 12:00 noon UT.

▼ MEASURING BY THE MOON
The full moon shines on Castle Rock, a small crater of volcanic breccia on Hut Point Peninsula, Ross Island. Scientists monitoring ozone levels can measure them with reasonable accuracy from the sunlight reflected off the moon—if the moon is at least half full, and high enough in the sky.

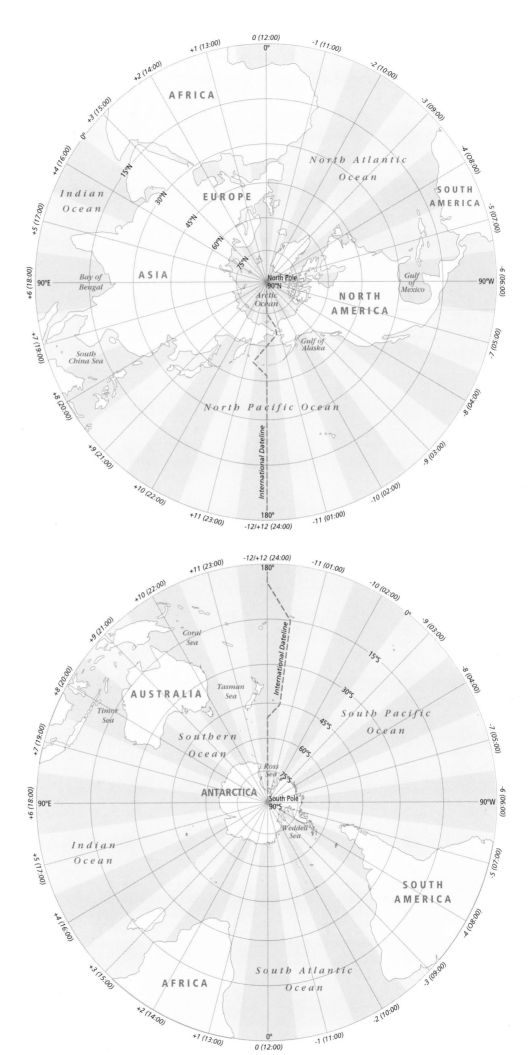

The polar landscape

Antarctica, sometimes called the Crystal Desert, has two faces today: a visible one largely of ice, and a masked one of bedrock. Ice averaging 2.3 kilometers (1½ miles) thick covers over 98 percent of the continent. In other continents, "land" is those parts lying above water; however, in Antarctica, "land" is predominantly ancient bedrock clothed in water that is crystalline except rarely in midsummer. Snow equivalent to only about 2 centimeters (¾ in) of water accumulates over much of the interior. In places, major mountain ranges 3,000 or more meters high (9,850 ft), such as the Gamburtsev Subglacial Mountains of East Antarctica, lie entirely hidden. An Antarctic "island"—the bedrock of a mountain peak or hill encircled by ice—is called by the Inuit name *nunatak*.

▲ FIRST WINTER LANDING
Cape Adare, at the northern end of Victoria Land, is a complex of overlapping, small volcanic cones. Named by Captain James Ross in 1841 for his friend Viscount Adare, MP for Glamorganshire, it was also the site of Carsten Borchgrevink's landing.

The masked face of Antarctica

The geological framework of Antarctica's approximately 14,200,000 square kilometers (5,500,000 sq. miles) can be inferred from scattered nunataks and ice-bathed mountain ranges. Extensive geophysical surveys using seismic soundings, rock magnetism, radio echo-sounding, and gravity have established the features of the buried face. Comparisons with continents to which Antarctica was joined until 130 million years ago as the hub of a giant southern continent, Gondwana, also provide useful clues.

Antarctica is roughly pear-shaped: the large, bulbous East Antarctica lies mostly in eastern longitudes; the smaller West Antarctica, with the spine of the Antarctic Peninsula stretching toward South America, is in western longitudes. The Ross Sea at the southern end of the Pacific Ocean and the Weddell Sea at the south end of the Atlantic Ocean deeply indent the continent's outline. In the nineteenth and early twentieth centuries, geologists speculated that the Ross and Weddell seas might be connected beneath the ice of the interior, but surveys during the 1957–58 International Geophysical Year (IGY) clearly showed that an ice-buried ridge above sea level links the Ellsworth Mountains of West Antarctica to the Transantarctic Mountains of East Antarctica.

A mountainous continent

Antarctica contains several of earth's major mountain belts, the largest being the great belt of the Transantarctic Mountains, a chain of nunataks, mountains, and ranges 3,200 kilometers (1,990 miles) long, extending with some interruptions, from Cape Adare on southernmost Pacific shores to Coats Land on the Atlantic side. These mountains reach to within about 500 kilometers (310 miles) of the pole, and form a topographic marker between East and West Antarctica.

The Antarctic Peninsula forms a second major mountain system that stretches about 1,600 kilometers (990 miles) south from Drake Passage to disappear as a chain of nunataks beneath the ice sheets of Ellsworth Land. An early ship's captain noted that there was a great similarity between rocks in the Peninsula and those of South America's Andes, and called the belt the Antarctandes. For example, the Andes in Chile and numerous Antarctic Peninsula rocks show evidence of copper deposits. Where this belt disappears in Ellsworth Land, its structures begin bending westward and seem to extend, largely beneath ice, to reappear along the coasts of the Bellingshausen and Amundsen

◄ ROCKS OF ANCIENT LINEAGE
Nunataks of the Framnes Mountains in the vicinity of Amery Inlet occupy an area geologically a part of the Precambrian shield of East Antarctica. They were first sighted by the BANZARE (British, Australian and New Zealand Antarctic Research Expedition) party led by Douglas Mawson in 1931.

seas. In the Cretaceous Period (144 to 65 million years ago), when ice-free Antarctica was connected to Australia and dinosaurs and other reptiles roamed, these Antarctandean mountains apparently extended the Andes around the southern Pacific to the eastern rim of Australia. Much later, this rim broke off and moved eastward during the opening and widening of the Tasman Sea to form New Zealand.

The spectacular Ellsworth Mountains at the head of the Weddell Sea, where Antarctica's highest point—Vinson Massif—rises to 5,140 meters (16,864 ft), are a geological enigma. Sandstones, shale, limestone, and ancient glacial deposits of Paleozoic age reveal a geological history more like that of East than West Antarctica. These mountains in West Antarctica contain the long-

extinct plant *Glossopteris,* a late Paleozoic tree and a fossil characteristic not only of East Antarctica but of all Gondwana. Geologists think that these mountains probably originally formed somewhere in the Transantarctic Mountains and later became displaced.

Geophysically, the average thickness of the earth's crust in Antarctica matches that of other continents, roughly 30–32 kilometers (19–20 miles) thick in West Antarctica and about 40 kilometers (25 miles) thick in East Antarctica. A sharp change in thickness along the Transantarctic Mountains front may indicate a deep crustal fault system. The overall crustal stability of present-day Antarctica is confirmed by the absence of any significant earthquake activity: earthquake activity is recorded only sporadically in a few volcanoes.

▲ **PEAK LANDSCAPE**
Spring sunshine bathes the ice-locked coastline and Transantarctic Mountains north of Terra Nova Bay on the western shoreline of the Ross Sea. Mount Melbourne, one of Antarctica's few volcanoes considered to be "active" is near the coastline (right). The Deep Freeze Range occupies the middle distance, with the Eisenhower Range beyond it.

The formation of Antarctica

Geologists use the term tectonics (as in "plate tectonics") for movements of the earth's large, rigid, crustal plates, and for their deformation by folding and breaking (faulting) of rocks under compression, or by tension at places where, moving differently, the plates touch. Lavas—melted rocks—erupt along mid-ocean ridges as the ocean floors widen and continents separate. Along collision zones, surface rocks can be carried to great depths and mountain ranges squeezed up. At great depths, the rocks change under high pressures and temperatures, and in places melt. Lavas rise along conduits to form volcanic arcs where plates collide; the Andes, for instance, resulted from the collision of Pacific and South American plates. The geological map of Antarctica reveals a long history of plate collisions. Some geologists hypothesize that the northern extremity of Victoria Land, near Cape Adare, was an island mass that collided with Antarctica by plate-tectonic movements. Because it originated away from the continent, it is termed an "exotic terrane" of rocks.

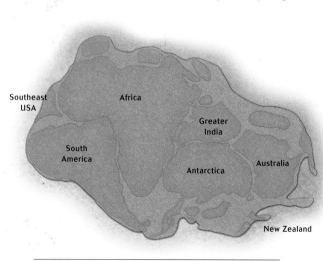

▲ A CRUCIAL QUESTION
Was Antarctica a single continent or really two, as depicted in this vintage map? That was one major question that geophysicists of the 1957–58 IGY (International Geophysical Year) wanted to answer while they were exploring the land beneath the ice by seismology.

Making mountains

An episode of mountain building by folding under compression, faulting, or upthrusting is known as an orogeny. Fossils in the rocks and radioactive dating of materials such as volcanic rock and granites formed during such events are used to determine when ancient, long-eroded mountains formed.

Like other continents, Antarctica has a tectonic framework consisting of a long-stable core of rocks of Precambrian age (older than 543 million years) called a shield, adjoined by orogenic belts of younger rocks. East and West Antarctica are geological provinces of greatly different character, with West Antarctica having a record of much younger crustal movement. The Antarctic Peninsula formed as a volcanic arc during the Mesozoic Era, and was an active arc up to the beginning of the Cenozoic Era, about 60 million years ago. By 30 million years ago, in the mid-Cenozoic era, South America stretched northward from Antarctica, became disrupted, and formed the small, new plate of the Scotia Sea. The very active volcanoes of the South Sandwich Islands at the easternmost end of the Scotia Plate mark its collision with an oceanic plate to the east.

Continents adrift

The geological history of Antarctica is largely the record of ancient plate movements. Pangaea, a supercontinent that encompassed most of earth's continental crust from more than 300 million to about 200 million years ago, rifted apart to form Laurasia, a northern landmass encompassing today's Europe, North America, Asia, and Gondwana, which later broke up into the southern continents. These ancient lands fractured along faults and split apart, perhaps many times. The Atlantic seems to have opened once, then closed, and then reopened again into today's ocean.

The 1950s controversy in some northern hemisphere universities over "continental drift" resolved into the theory of plate tectonics

GONDWANA

The southern supercontinent Gondwana, as it was about 180 million years ago. It comprised all the present-day southern continents and started to show signs of rifting at this time. A landmass consisting of Africa, South America, and India was first to separate; Australasia split off later.

Southeast USA
Africa
Greater India
South America
Antarctica
Australia
New Zealand

ERA	PERIOD	EPOCH	YEARS AGO (millions)
Cenozoic	Quaternary	Holocene	The present / 0.01
		Pleistocene	1.8
	Tertiary	Pliocene	5.3
		Miocene	
		Oligocene	24 / 34
		Eocene	55
		Paleocene	65
Mesozoic	Cretaceous		
	Jurassic		144
	Triassic		206
Paleozoic	Permian		248
	Carboniferous		290
	Devonian		354
	Silurian		417
	Ordovician		443
	Cambrian		490
	Precambrian		543
			3800

by the late 1960s; the lateral motion of continents became scientifically acceptable once a plausible mechanism was identified. Some geologists had long known how continents could be refitted, as by matching the Brazilian bulge with Africa's great Ivory Coast indentation by the closing of the Atlantic Ocean. Other geophysicists had calculated that this was a mathematical impossibility, and many geologists believed them, but no arguments are now heard against oceans opening and closing, nor against Antarctica's key geological role in the configuration of Gondwana. It is accepted that Gondwana broke apart along a number of rifts that developed into mid-ocean ridge systems, and that new oceans carrying fragments of continents spread ever wider.

Antarctica forms one of a small number of the earth's rigid crustal plates. Today's plates formed as they broke and rifted apart along faults of some earlier plate, under crustal stresses produced by thermal plumes rising from the earth's interior. Early stages in the isolation of

BLUE SKY, WHITE PEAKS
Towering, snowy peaks rise high above Cape Renard, which separates the Danco and Graham coasts of the west side of the Antarctic Peninsula. They were first sighted in 1898 by the men on board Adrien de Gerlache's Belgian Antarctic Expedition on the *Belgica*.

A LANDSCAPE SHAPED BY ICE
The rugged western coast of the Antarctic Peninsula vividly encapsulates what a powerful force ice can be in sculpting these mountains into a landscape of valleys and hills. Valley glaciers spill over the vast cliffs into the sea and, in doing so, they spawn huge icebergs.

Antarctica can be witnessed in the rifts that are beginning to tear East Africa from West Africa, a process that is proceeding southward from the Red Sea into the Rift Valleys. Numerous active volcanoes mark the zones of heat from the interior along these rifts in Kenya and Tanzania. The Red Sea is a more advanced stage in the plate tectonics of the region, in which rifting has developed into an incipient oceanic-ridge system, with seafloor spreading and separation of Arabia from Africa. Antarctica, too, seems to be at an early stage of interior splitting. The Ross Sea region and Marie Byrd Land contain active volcanoes (for example, Mount Erebus, Mount Terror, and Mount Melbourne) of the same unusual alkaline chemistry as those of the East African rifts. Geologists find evidence in the area for a major zone of rifting, called the West Antarctic Rift System.

East Antarctica's Precambrian shield was part of a much larger rock shield that included much of western Australia, southern Africa, Madagascar, and India. Some of the oldest rocks yet known—about 3,800 million years old—are reported from such places. Original sedimentary and igneous rocks became deeply buried in the crust, and tectonic processes again and again transformed them by metamorphism into many types of gneiss and schist. Eventually, late in the Precambrian Era, a plate of an ancestral Pacific Ocean developed by volcanism associated with crustal rifting, and by seafloor spreading.

As it opened, this ancestral plate pushed against and collided with the adjoining continental plate of Pangaea, and was dragged beneath the continent in the process geologists call subduction. Where subducted crust was dragged sufficiently deep, parts melted under the high temperatures, forming molten silicate rock material, called magma, that was injected upward to crystallize as granite, or extruded in volcanoes. The Pacific crust collided more than once along the edge of "paleo-Antarctica," marked by today's Transantarctic Mountains. Ancient mountains formed and eroded. The processes successively added to the continent, which grew toward the Pacific Ocean. Sediments were deposited in marine or terrestrial basins formed ahead of and behind

▲ **LIFE AT THE SUMMIT**
Mount Melbourne, a little-dissected stratovolcanic cone rises 2,732 meters (8,963 ft) above the western shore of the Ross Sea. It was named by James Clark Ross in 1841 after the British Prime Minister at the time. At the summit is a rare Antarctic habitat: a few hundred square meters of steam-warmed, ice-free ground where thin coverings of algae, mosses, and liverworts can grow.

◄ **MOUNT EREBUS**
Its summit hidden by cloud, Mount Erebus is the youngest of several volcanoes that form Ross Island, and the southernmost active volcano. Castle Rock, the snow-free knob (left) is Hut Point Peninsula's tallest point. In the foreground and middle distance, pressure ridges of jumbled sea ice push against the shoreline.

▼ FOSSIL EVIDENCE
Fossil leaves of *Glossopteris*. Fossils of this extinct plant are found in all the fragments of the supercontinent Gondwana. These fossils were the key to proving Gondwana's former connections. When explorer Scott was found dead in his tent, *Glossopteris* fossils were found with him—having abandoned all unnecessary equipment, his group had retained the fossils, recognizing this plant's scientific significance.

▼ ANCIENT SANDSTONE
Found throughout the Transantarctic Mountain range is a thick sequence of light yellow-brown sandstones called the Beacon Sandstone. Apart from being uplifted, they have barely been disturbed since their deposition over a vast period of time—from 190 to 390 million years ago.

volcanic arcs. Granites crystallized from materials melted at depth, and lavas were forced out where the magmas coursed up fractures to the surface. The youngest belt thus formed is the mountain system of the Antarctic Peninsula and its continuation along coastal West Antarctica. Ancient New Zealand and the Andes in Chile were extensions of this system. New Zealand, once part of West Antarctica, rifted away about 85 million years ago.

The Beacon Sandstone

Antarctica's geological evolution followed a course that was generally similar to that of the other Gondwanan continents. Most of the record through the early Mesozoic Era, up to about 248 million years ago, is well displayed in the rock sequence of the Transantarctic Mountains, called the Beacon Sandstone by early British explorers. To the excitement of paleontologists at the British Museum, pioneer geologists of the British Scott and Shackleton expeditions in 1904–12 collected leaf fossils and coal samples of late Paleozoic age from the Beardmore Glacier at the edge of the polar plateau, and fossils of Devonian Age fish (about 380 million years old) from the quartzitic sandstones near McMurdo Sound. The coals were found to be extensive and were considered a potential mineral resource. Fossils of late Paleozoic amphibians and dinosaurs have now been found, and also of the late Paleozoic leaf, *Glossopteris*, a hallmark of Gondwana in the Permian Period. These are compelling evidence that the southern lands were once connected, and separate from the northern continents.

The Beacon Sandstone (now called the Beacon Supergroup), has since been mapped throughout the Transantarctic Mountains and at places far within East Antarctica. It generally forms flat beds of Devonian or older to Jurassic age, overlying an eroded surface on strongly folded, trilobite-bearing Cambrian limestone and other bedrock. Such structural relations, combined with an absence of sedimentary rocks aged between

Cambrian and Devonian, indicate that strong deformation and mountain building occurred after Cambrian and before Devonian time. Granites also formed during this event, which is known as the Ross Orogeny. Effects of that orogeny and of an older one of the late Precambrian Era, called the Beardmore Orogeny, are known throughout the Transantarctic Mountains. They are believed to result from collisions between an ancestral Pacific plate and the East Antarctic shield. During the late Mesozoic and early Cenozoic Eras, the continent continued to enlarge by subduction of the Pacific Ocean crust, accretion (the addition of new material), and formation of volcanic arcs along the Antarctic Peninsula and its extensions, in an event called the Andean Orogeny.

Gondwanan folding

In *Our Wandering Continents*, published in 1937, South African Alex du Toit predicted that folded rocks similar to those of South Africa's Cape Fold Belt near Cape Town must once have once extended into Antarctica at the head of the Weddell Sea. Lincoln Ellsworth, on his 1935 inaugural transcontinental flight, had seen high mountains in that area. The later-named Ellsworth Mountains were geologically wholly unknown until a United States exploration party visited them in 1958. Du Toit had predicted correctly, on the basis of how he thought the continents had to refit. That tectonic event, in which rocks as young as Permian (290–248 million years) were strongly folded during the earliest Mesozoic Era, affected Antarctic regions only around the margin

of the Weddell Sea, including the Antarctic Peninsula, the Pensacola Mountains, and the Ellsworth Mountains. The Cape Fold Belt extended west into ancestral Argentina. When the Atlantic rifted open, one small fragment, today's Falkland Islands, was dragged out from Africa and left on Argentina's continental shelf. Unlike the ocean–continent collisions, the Gondwanan orogeny seemingly occurred *within* a continent by some little-understood mechanism.

In the middle and late Mesozoic Era, heat currents ("plumes") rising through the earth's mantle led to successive crustal rifting accompanied by volcanism on an immense scale. Gondwana was dismembered, and the new, smaller plates of today's continents formed. In the mountains of Victoria Land, these events are marked by

▽ ROCK GLACIERS

A light dusting of snow highlights the rock glaciers at the head of the Beacon Valley, a Dry Valley in the Transantarctic Mountains. Composed of rock debris held together with interstitial ice, they move very slowly down the valley—at only 0.5 to 2 meters (1 ½ to 6 ½ ft) per year. The dark rocks are Jurassic basalt magma that was forced sideways, between layers of lighter sandstone, as thick horizontal sheets.

▲ LAYERS OF HISTORY
Large boulders at the Upper
Wright Glacier, in the Dry
Valleys. The lighter rocks are
sandstone; the darker rocks
are dolerite. Bands of these
totally different rock types
are visible in the mountains
behind. Sandstone was
deposited by rivers and lakes
during a geologically quiet
period, while dolerite is formed
by intrusion of molten rock.

the conspicuous black dolerite, also known as diabase,
that intrudes into the horizontally layered, white,
quartz-rich rocks of the Beacon Sandstone. Some of
these intrusions, called sills, parallel the sandstone beds,
and others, called dykes, cut across them. Rifting began
early in the Jurassic Period, about 190 million years ago.
Vast piles of flood-lavas across the Transantarctic
Mountains also mark this as one of the most dramatic
volcanic events in the earth's history. An exceptionally
large body of igneous rock, the Dufek layered igneous
complex near the southeastern Weddell Sea, may mark a
mantle plume associated with the rifting of Africa from
Antarctica in an early phase of the breakup of Gondwana.

Isolation sets in

Plant and animal migration routes that had apparently
freely interconnected all the southern continents were
cut off at various times. Africa rifted off first. At Hope
Bay, at the north end of the Antarctic Peninsula, Mount
Flora's spectacularly plant-rich beds of Jurassic age have
no counterparts in southern Africa. Similarly, the vast
early Cenozoic forests of southern beech (*Nothofagus*),
about 100 million years old, fossils of which are found
in Antarctica and living forms in most southern lands,
are not represented in Africa. Antarctica was becoming
isolated at a time when land mammals were diversifying
and flourishing elsewhere, populating all the other

continents. It had long been thought that Antarctica was a migration path for marsupials moving between South America and Australia during earliest Cenozoic time, and proof of this was found in the 1982 discovery of a marsupial fossil on Seymour Island near the Antarctic Peninsula.

The final phase in Antarctica's isolation was the development of a rift-fault system with Australia near the end of the Cretaceous Period. Growth of the new Southeast Indian Ocean by seafloor spreading had important consequences for climates and biology; it started new ocean circulation patterns such as the West Wind Drift around Antarctica, and cut off migration routes of animals such as marsupials. Collisions of the Pacific and other oceanic plates against the growing continental crust of West Antarctica and the Antarctic Peninsula continued from mid-Mesozoic time through the Cenozoic Era. Subduction of the oceanic crust produced basins in which sandstone and shale were formed, accompanied by extensive intrusion of granite bodies and extrusion of lavas of various types. Such materials are widely exposed in the South Shetland Islands and the Antarctic Peninsula.

Volcanic activity indicates the continuation of the plate-tectonic processes, as in the West Antarctic Rift System, in northern parts of the Antarctic Peninsula, and in the highly active volcanoes of the South Sandwich Islands of the Scotia volcanic arc at the eastern end of the Scotia Sea. Continuing eruptions of the volcanic caldera of Deception Island, with major activity in 1967–70 that destroyed two research stations, mark the development of a new ocean-ridge spreading system: the South Shetland Islands are moving westward from the Antarctic Peninsula and widening Bransfield Strait. Modern global positioning system (GPS) instruments are now directly measuring crustal movements and movements of continents to test past speculations by geologists and geophysicists.

> RECORD-KEEPING ROCKS
Sedimentary rocks of Seymour Island, in the northwestern Weddell Sea, contain an unusual abundance of fossils and a unique geological record of earth's history across what geologists call the "K-T" boundary, which marks the end of the Mesozoic Era, and the extinction of the dinosaurs.

◄ MOUNTAIN SENTINELS
Mountain spires of massive black rocks rise precipitously above the scenic Gerlache Strait, a channel separating the island-dotted Palmer Archipelago from the west side of the Antarctic Peninsula. The area was explored in 1898 by the Belgian expedition led by Adrien de Gerlache.

> PENGUIN PLAYGROUND, HUMAN CHALLENGE
Chinstraps frolic on the black, basaltic-sand beach of Baily Head, on the north side of Deception Island's volcanic caldera. Heavy sea surges test the skills of Zodiac drivers trying to land on this beach, which is flanked by layered rocks called "tuffs," consisting of basaltic ash and other volcanic materials.

The formation of the Arctic

The Arctic is named for the north polar constellation *Arktos*—Greek for "bear." It contains areas north of a somewhat irregular zone marking the northern limit of stands of trees, and delineated by a mean July temperature of approximately 10°C (50°F). All of Greenland and Spitsbergen, and the northern parts of Alaska, Canada, Norway, and Russia, lie within the Arctic. The Arctic Ocean, with earth's north rotational pole is central to this region, but was not discovered until 1893, when Norwegian explorer Fridtjof Nansen locked his ship, *Fram*, into pack ice off Siberia for a three-year drift across the top of the world to test his theory of the origin of North Atlantic currents. The Arctic, at 14.5 million square kilometers (5½ million sq. miles), is almost exactly the same size as its polar opposite, Antarctica, but popular images of the Arctic feature lush tundra and large mammals such as bears, wolves, caribou, and walruses, as well as a human occupation dating back 10,000 years or more, whereas images of Antarctica include thick ice sheets, scattered outcrops of bare rock, and penguins.

▲ FRANZ JOSEF LAND
Flat-lying rocks form a long cliffline above the Arctic Ocean off Franz Josef Land. This is an uninhabited group of over 60 ice-covered islands. Russian territory, they are the northernmost islands in Eurasia. Franz Josef Land was discovered by an Austrian expedition in 1873.

Polar differences

The two polar regions are fundamentally different: one is an ocean surrounded by land, the other a continent surrounded by ocean. Ice sheets and glaciers form only on land. Antarctica developed ice sheets as early as 40 million years ago, long before the Pleistocene Epoch (1.8 million to 10,000 years ago), when ice sheets spread across much of the polar north. Oceanographers believe that warmer waters, circulating from the Atlantic, probably prevented much of the Arctic Ocean from developing permanent pack ice until about 850,000 years ago. During the Ice Ages, great volumes of water froze, sea levels fell, large expanses of the Arctic continental shelves became exposed, and the English Channel became dry land. A land bridge across the Bering Strait allowed people to cross from Asia into North America.

Off Siberia, the Arctic Ocean has an unusually large area underlain by a continental shelf, its width approaching 1,600 kilometers (1,000 miles). The surrounding lands are generally low-lying, with one of the flattest regions at the head of the Siberian Shelf, at the deltas of several of the world's largest rivers: the Ob', the Lena, and the Yenisey. The immense submarine Lomonosov Ridge traverses the Arctic from north of Greenland to the Siberian continental shelf (135°E) via the North Pole, and separates two major basins: the Eurasia Basin, north of Russia, and the Amerasia Basin, north of Alaska and Canada. Other transoceanic submarine ridges, Nansen Ridge and Alpha Ridge (also called Alpha Cordillera), subdivide the basins into four smaller ones.

Geology and plate tectonics

Great differences in geological history and plate-tectonic activity have made the Arctic an ocean and the Antarctic a continent. More than 300 million to about 200 million years ago, both were parts of the Paleozoic continent of Pangaea, in which ancient mountain belts formed and eroded many times. Three large Precambrian shields 543 million years old or more form principal structural elements in lands around the Arctic Basin: the Canadian–Greenland, the Baltic, and the Russian Angara shields. Remnant mountain belts of Paleozoic age and younger wrap around the shields. Such old mountain belts of Greenland and Scandinavia were interconnected before the Atlantic Ocean opened in the Cenozoic Era. The Appalachian Mountains, farther south, also once continued into northern Europe. Similarly, ancient structures, as seen in Alaska's Brooks Range and the Seward Peninsula, continued west across the Bering Strait into Siberia. Rift opening and ocean widening interrupted mountain belts across the Atlantic; rising sea levels submerged land across the Bering Strait as ice sheets 20,000 years old melted.

Latitudinal rifting along an ancestral Mediterranean Sea at the end of the Paleozoic Era separated northern Laurasia from southern Gondwana. The lands around the Arctic Ocean have a long and diverse geological record, ranging from the early Precambrian Period to

▶ NORTHWEST TERRITORIES
Spectacular coastal scenery of Elmerson Peninsula and Greely Fiord, Ellesmere Island, in Canada's Northwest Territories. This region is in the high Arctic, well north of the Arctic Circle, and the closest land to the North Pole. It thus has been important as a stepping-off point for expeditions to the north.

the present. Southwest Greenland contains some of the world's oldest known rocks, about 3,500 million years old. In Antarctica, continental rifts formed about 180–120 million years ago in the mid-Mesozoic Era and developed into mid-ocean-ridge systems, and spreading of ocean floors from ridges around the continent progressively isolated it. The plate-tectonic history of the Arctic Basin, known largely from marine geophysical surveys, was quite different.

The Eurasia Basin appears to be a northern extension of the Atlantic Ocean formed by seafloor spreading along the mid-ocean ridge axis of Nansen Ridge, between Lomonosov Ridge and the Siberian mainland. Nansen Ridge seems to be an extension of the Mid-Atlantic Ridge, although it shows dislocation by faulting around northern Greenland. Geologists propose that a narrow splinter of the Asian continental margin ripped off during the start of rifting that led to the development of Nansen Ridge. Lomonosov Ridge is considered to consist of continental-type rock in the middle of the Arctic Basin. Seafloor spreading from

Nansen Ridge created two smaller basins: Nansen Basin on the Siberian side of the axis, and Fram Basin on the Alaskan side. The opening of the Arctic Basin, however, seems to have involved more tectonic activity than just ocean-floor spreading from Nansen Ridge.

The origin of the Amerasia Basin, north of Alaska, has been much debated. An early thought was that Alpha Ridge might be a relict, non-active, mid-ocean ridge system, from which spreading took place in the Mesozoic Era (248–65 million years ago). Others proposed that the ridge is an incipient volcanic island arc reflecting plate movements and compression in the North Atlantic. Currently, researchers favor an opening of the basin by anticlockwise rotation of the Arctic–Alaskan crustal plate away from North America in latest Mesozoic time. Marine geophysical studies and core drilling continue, contributing to a better understanding of the geological framework of the petroleum provinces along Alaska's north coast.   AF

▲ FOSSILS TELL A STORY
Marine invertebrate fossils from the Qutinirtaaq National Park, Ellesmere Island, show that the area was once part of a shallow-water marine environment and that marine life flourished here in the past. Such fossils give clues to the age, deposition environment, and similarities with life elsewhere at the time.

Evolution in Antarctica

A ntarctic wildlife is usually perceived in terms of the animals that live in the region, either never visiting land, or using it as a temporary base for breeding. Land plants are mainly mosses and lichens, and only two flowering plants—a grass and a cushion plant that cling to the Antarctic Peninsula. However, probably as recently as 5 million years ago, woody flowering plants were present well inland, suggesting complex and diverse vegetation, and there is evidence of much more complex plant and animal communities even further back in time.

The continent now called Antarctica has occupied its present polar position for tens of millions of years, but long ago, it was nowhere near the South Pole, and even when it was, it supported life forms as complex as those of any landmass in low latitudes: for example, the *Glossopteris* flora that dominated Gondwana 248–290 million years ago. Scientists once believed that Antarctica's high latitude prevented the development of complex vegetation, because plants could not survive the extremely harsh conditions of long polar winters. However, experiments on living plants related to those that thrived in Antarctica in the remote past have proved that plants can survive such conditions, especially if temperatures are not too high. The extreme latitudes of

▼ CLINGING TO LIFE
Plant life in Antarctica today has a tenuous existence. This whale skull is encrusted with lichen, one of the major plant forms now present in Antarctica—a far cry from the polar forests that once covered much of the continent.

Antarctica are no longer seen as an impediment to the development of complex plant and animal communities.

The reduction of Antarctica's plant life to extremely sparse and simple vegetation is one of the greatest natural extinction events in the history of the earth. The cause is clear—extreme climate change. The history of Antarctic life over the past 60 million years clearly shows the impact of climate change, and the potential impact of human activity on life on earth.

Climate change

The climate of the southern hemisphere is dominated

by massive oceans, their currents spanning large latitudinal belts and exerting a profound influence on climate. Water warmed by the sun in equatorial regions can transfer energy to high latitudes, but the Antarctic Circumpolar Current, which circulates vast amounts of water around Antarctica, never leaves very high latitudes, and so draws very little energy from incoming sunlight. This body of extremely cold water is the main reason for the freezing conditions that exist in Antarctica today.

When the southern continents were part of the supercontinent Gondwana, no circum-Antarctic current could form because there was no Antarctic continent; the major southern currents flowed from equatorial to polar latitudes and back again, so that the water reaching high latitudes remained relatively warm, and produced much milder conditions there; indeed, it was sometimes so warm that there was no polar ice cap. Warm seawater leads to high evaporation, and, consequently, high rainfall. In those times, land at very high latitudes was both warm and wet—perfect conditions for the development of complex forests.

The rifting of Gondwana, the separation of Australia, and then the opening of the Drake Passage between South America

▶ SPARSE COVERAGE
Once diverse and widespread In Antarctica, mosses now grow only sparsely on the Peninsula and along the coast of the Continent. Vigorous moss beds occasionally form, often along drainage lines, where conditions are favorable.

and Antarctica allowed full development of the Antarctic Circumpolar Current. Circulation of this cool water forced the formation of the Antarctic ice sheet (although prior to this, higher areas may have had ice caps), which reflected much of the incoming solar radiation, making it even cooler so that the ice sheet extended even further, cooling the surrounding ocean. This feedback system is complex and often unpredictable, but the overall outcome was the modern Antarctic environment, where the ancient conditions suitable for complex forest growth are long gone.

Evidence for past Antarctic life

Fossils are usually quite common, but Antarctica is a special case, where the fossil record is particularly difficult to assemble. Today, the continent lies mostly beneath a thick ice sheet that either conceals fossils or has scoured them from the surface and deposited them in the Southern Ocean. Fortunately, the fossil records of other Gondwanan landmasses that were once connected to Antarctica—for example, South America, Australia, and New Zealand—are much more complete. They provide many ideas about Antarctic species during times when the ice cover was greatly reduced, or even absent.

When the Antarctic climate was able to support complex ecosystems, day length was a major factor affecting life. Generally, observation of similar living organisms casts light upon plant and animal communities of the past. However, all present-day high-latitude areas, with very long days in summer and continuous darkness during at least part of the winter, have extremely cold climates that prevent complex forest vegetation from developing. There are no modern forest communities growing under these light conditions, which makes reconstruction of the past more difficult.

▲ FOSSIL RECORD
Around 40 million years ago the Southern beech, *Nothofagus*, was dominant in parts of Antarctica. At least some of the species were winter deciduous and shed leaves that were trapped in underwater sediments.

Ancient Antarctic forests

▼ STILL STANDING
This 100-million-year-old petrified tree is still in its growth position. The roots were preserved in the fossil soil, and the trunk was entombed in flood-borne volcanic-rich sands.

▼ LIVING FOSSIL
Fossil leaves similar to living *Ginkgo biloba* are common in Cretaceous sediments in Antarctica. These plants may have been winter deciduous, like the living species, in response to the long periods of darkness in winter.

◄ CLEAR RECORD
The woods of conifers preserve well, and thin sections of fossils show clearly the individual cells and their wall patterning.

▲ MODERN RECONSTRUCTION
Fossil trunks found in their growth positions suggest that a Cretaceous conifer forest may have looked like this. The wide-spaced trees with their cascading canopies would have allowed light penetration from the low-angled sun.

One very lucky fossil find in Antarctica was a forest of conifer trunks in their growing positions. The trunks are 3–5 meters (10–16 ft) apart, and the tallest one preserved is 7 meters (23 ft) high. These relatively widely spaced trees would have captured light for photosynthesis very efficiently. Trees in very high latitudes must not shade each other too much because the sun moves around close to the horizon during the day, so that incoming solar radiation is at a very low angle and comes from different directions throughout the day. To capture the optimum amount of light, the Antarctic conifers would probably have had their foliage cascading down the sides of the trunks, rather than spreading overhead like the tree canopies of tropical forests, where the sun is predominantly high in the sky. Ancient Antarctic forests had open canopies to allow adequate light uptake. The fossil tree rings are very clear and large, suggesting seasonal growth under good growing conditions.

These open forests provided habitats for many animals. One of the most interesting, from the Early Cretaceous period (144–99 million years ago), is the dinosaur *Leaellynasaura*. These quite small dinosaurs had unusually prominent optic lobes, suggesting that very large eyes allowed them to remain active during prolonged winter darkness. If this was so, then they may have been warm-blooded to cope with the relative cold of the dark winters. These dinosaurs are usually reconstructed as being smooth-skinned, but they may have been feathered to provide greater insulation. There were also several different carnivorous dinosaurs present at this time.

Before flowering plants arrived, conifers dominated the forests, with an understory of now extinct seed plants, ferns, mosses, liverworts, and lichens. Sometimes the seed-bearing species *Ginkgo* (which now grows naturally only in China) occurred, and its deciduousness probably allowed it to survive the long winters.

Flowering plants evolved 130–160 million years ago, probably in northern South America and Africa. These pioneering early flowering plants had generalized pollination and seed dispersal, which enabled them to spread over a wide range, and they soon began to disperse along coastlines of the rift valleys that formed as the supercontinent of Gondwana began to break up.

Fossil evidence for Antarctic animals is sparse, but some fossils from nearby lands are helpful. For example, platypus-like bones around 61–63 million years old from Patagonia suggest that there were monotremes in South America at that time, and later across Antarctica, and similar evidence suggests that marsupials lived in Antarctica at least 60 million years ago. Other land animals have been recorded as rare fossils, including a giant, probably flight-less bird, 40 million years old, on the Antarctic Peninsula.

> **PAST LIVES**
Monotremes, including the living platypus, are now restricted to Australia and New Guinea, but have been found as fossils in southern South America. Monotremes must have once been part of the Antarctic fauna.

> **ANTARCTIC DINOSAUR**
Cretaceous dinosaur remains occur on the southern Australian shoreline, which was once adja-cent to Antarctica. The dinosaur *Leaellynasaura* was a small animal with relatively large eyes, possibly an adaptation to low winter light levels. These dinosaurs may have been bare-skinned, or perhaps had feathers for insulation.

◄ WINTER DECIDUOUS
Nothofagus was prominent in Antarctica until relatively recently, and many of the species found in fossils shed their leaves in winter. *Nothofagus pumilio* in South America, shown here during leaf turn, is a living winter deciduous species.

▽ AUSTRALIAN ENDEMIC
Nothofagus gunnii, a Tasmanian endemic species, is the only winter deciduous species in Australia today, and the only living winter deciduous *Nothofagus* species outside South America.

◄ ARCTIC EQUIVALENT
This fossil *Nothofagus* wood, probably five to seven million years old from the Sirius Formation, shows very small, asymmetrical growth rings, suggesting horizontal growth in harsh conditions. The growth pattern is similar to that in Dwarf Arctic willow that survive today on the tundra.

Nothofagus

▲ PERFECT FORM
A single, beautifully preserved *Nothofagus* leaf from 40 million-year-old sediments on King George Island.

◄ ▲ GROUND HUGGER
A single mat of fossil leaves of *Nothofagus beardmorensis* was discovered in Sirius Formation sediments. The plant was probably ground hugging, as shown in the reconstruction on the left, with relatively large and fragile leaves, as seen above.

After the break up

Some prominent plants in the early Antarctic flora still exist elsewhere today, and their ecology is well understood. A good example is *Nothofagus*, the southern beech, which has been studied extensively across its modern range. In South America, New Zealand, and New Guinea, this species tends not to regenerate continuously where climatic conditions favor complex forests. In these environments, its seeds germinate and establish in fresh, cleared subsoil after catastrophic disturbance, which is common in the unstable mountain chains. This behavior is probably similar to regeneration early in the history of *Nothofagus* in Antarctica, when developing rift valleys provided unstable habitats. Until about 40 million years ago, unstable habitats at high southern latitudes provided a migration pathway for early flowering plants. They were also an opportunity for incidental populations of widespread species to become isolated, which may have been a critical factor for evolution.

The first Antarctic flowering plants—herbaceous plants or shrubby trees—did not appear until later. All these flowering plants migrated eastwards across Antarctica, from South America towards Australia and New Zealand.

In the Late Cretaceous period (99–65 million years ago), there were strong similarities between vegetation within the same latitudes of Australasia, Antarctica, and southern South America. Conifers and *Nothofagus* were the major species of the tall, open forests. There were also some Proteaceae and Myrtaceae, although both were less diverse than in Australia, where they are still prominent. By 65 million years ago, vegetation dominated by woody flowering plants was well established in Antarctica.

Major extinctions

The end of the Cretaceous period, about 65 million years ago, is noted as a time of major extinction, probably resulting from the impact of a massive extraterrestrial body. There is scant evidence of this event in Antarctica. Scientists have suggested that the collision occurred in June (mid-winter in high southern latitudes), when animals and plants, for example, winter-deciduous trees, would have been dormant. This scenario explains the paucity of evidence for the effect of the impact on Antarctic life. However, information is difficult to obtain and it is too early to conclude too much from current data.

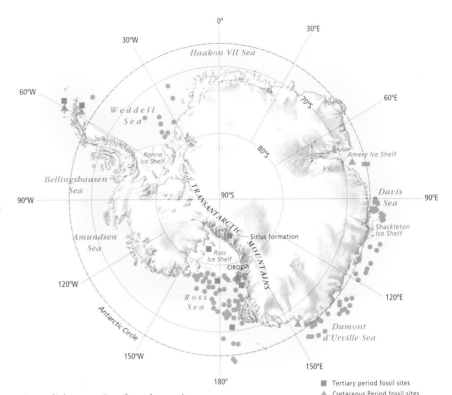

Australia's separation from Antarctica began in the Late Cretaceous period, although opportunities for plants and animals to cross water gaps may have persisted for a long time thereafter. The Drake Passage probably began to open about 23–25 million years ago, but details are lacking concerning how much dry land there was between the Antarctic Peninsula and South America before then. There are few Antarctic fossils from the past 65 million years, but some are very instructive.

Cenozoic plant macrofossils (less than 65 million years old) are known only from the Antarctic Peninsula, a drill hole in McMurdo Sound, and the Sirius Formation in the Transantarctic Mountains. Controversy surrounding these fossils centers on the accuracy of the identifications, and the age of the sediments containing them. Most are poorly preserved, which makes classifying them difficult, especially for *Nothofagus* leaves, which are relatively common.

■ Tertiary period fossil sites
▲ Cretaceous Period fossil sites
● Recycled fossils (no longer located in original sediments)

▲ **FOSSIL SITES**
Cretaceous and Cenozoic fossils are primarily located in offshore sediments. The fossils are mainly pollen and though originally laid into sediments on the Continent have been relocated (recycled) by ice scouring. For this reason, the fossil record is often difficult to interpret.

➤ **LONG-TIME SURVIVOR**
The waratah belongs to the ancient Gondwanan plant family, Proteaceae. This family occurred in Antarctica for tens of millions of years. Today, the species occurs only in Australia.

▲ SIMILAR SITUATION
Mountain beech (*Nothofagus solandri*) occurs in New Zealand and forms part of a cool temperate ecosystem which is probably similar in structure to forests that occurred in Antarctica prior to the separation of Australia and South America. The very wet conditions are conducive to fern and moss growth on the trunks.

Pollen grains fossilize more readily than other plant parts. The pollen record, however, is often hard to interpret because much of it is not preserved where it was deposited, but lies in offshore sediments where it has been carried by ice scouring. Nevertheless, there is evidence of quite diverse and endemic species of flowering plants in the early Cenozoic era, and this vegetation persisted without much change until about 35 million years ago. Glaciers had probably reached sea level by then, so this vegetation persisted into the early phases of chilling.

The pollen record does not reveal when increasing cold and ice cover eliminated the Antarctic Cenozoic vegetation, largely because of confusion caused by the movement of the pollen once glacial erosion and sedimentation had begun. This was complicated by the discovery of probable Pliocene fossils from the Sirius Formation.

There is evidence for several glaciations on the Antarctic Peninsula between about 20 and 100 million years ago, separated by non-glacial periods when macrofossils were deposited. During the Late Cretaceous period and the Early Cenozoic era, there were enough refugia (havens during climatic change) in Antarctica or nearby to allow woody plants to retreat to them during a glaciation, and subsequently to recolonize West Antarctica during the climatic improvements that

followed. But the formation of an Antarctic-wide ice cap about 32–33 million years ago may have brought about a major change in plant communities, and probably removed all woody plants from the Antarctic Peninsula for several million years. Antarctica's growing isolation and lengthening plant migration routes meant that many species could not return when the glaciation waned.

Thus, the Antarctic Peninsula's macrofossil record suggests a flora in decline through time. Presumably land fauna also declined, but there is no direct evidence for this. Away from the Antarctic Peninsula, a single *Nothofagus* leaf fragment, 24–29 million years old, from the CIROS-1 borehole demonstrates the presence of at least that genus at a much higher latitude, and well away from the Antarctic Peninsula. The record from the Antarctic Peninsula ends about 24 million years ago, but can be put in context with the Sirius Formation flora, which is probably Pliocene, 1.8–5.3 million years old (although some scientists believe it is much older).

Sirius Formation flora

The Sirius Formation sediments contain beautifully preserved pollen, *in situ* roots, leaves, and wood; the plants that produced these fossils must have been growing on site when the fossils were deposited. The fossil pollen is dominated by *Nothofagus*, although pollen from other flowering plants and conifers is recorded.

The fossil wood and leaves are also *Nothofagus*, and it can be assumed that they all came from the same species.

The stems are up to 1.3 centimeters (½ in) in diameter, and several are gnarled and contorted and contain branching junctions. Extremely narrow growth rings indicate extremely slow growth. In one specimen, more than 60 rings were measured along a radius of only 5 millimeters (less than ¼ in). The stems are distinctly asymmetrical, in a pattern known as "reaction wood." In flowering plants, this forms on the upper side of branches and in horizontal trunk wood. The dominance of reaction wood among Sirius samples suggests that these stems grew horizontally, like living Dwarf Arctic willows, which grow close to the ground, thus protected from freezing winds. Asymmetric growth rings are also present in prostrate forms of living *Nothofagus gunnii* from alpine Tasmania. Some fossil branches show evidence of traumatic events and scarring, and that is a common feature in living Dwarf Arctic willows.

Many hundreds of fossil leaves have been recovered, each with a very strong ribbed and creased pattern where it was folded like a fan in the bud. In living *Nothofagus* species, this indicates deciduousness and, along with the dense accumulation of leaves within a single thin layer, suggests that the fossil leaves are the result of a single, seasonal leaf fall.

Since the fossil wood and leaves from the Sirius group are found among glacial sediments, an environment similar to the present high Arctic is envisaged for the Antarctic Pliocene epoch. Pulses of glacial melt and rapid erosion, like those that occur in the Arctic spring and summer, probably damaged the tree stems and periodically retarded their growth. The vegetation probably resembled living Tasmanian alpine communities but was less diverse, with winter deciduous *Nothofagus*, other flowering plants, and a variety of conifers.

Controversy surrounding the Sirius sediments centers on interpretations of prevailing climate, particularly temperatures, based on the presence of the *Nothofagus* fossils and their nearest living relatives, and the implications for the Pliocene climate. The dwarfed growth forms of the fossil *Nothofagus* suggest that, although summer temperatures may have been around 5°C (41°F), these conditions would have lasted for only about three months during summer. For the rest of the year, temperatures would have been below freezing, and the plants would have remained dormant. The lower temperature limit would probably have been at least −15°C (5°F) to −22°C (−8°F), and possibly much colder, if snow cover had protected the dormant plants. This gives a mean annual temperature well below freezing, and perhaps in the region of −16°C (3°F).

Nothofagus-dominated vegetation must have been present in Antarctica for about 80 million years, until the time that the Sirius *Nothofagus* was deposited. This is consistent with a progressive simplification of Antarctic vegetation, possibly with some form of tundra vegetation towards the end of the process. It should be noted that the Sirius Formation fossils are well inland in Antarctica, suggesting that there may have been other, more diverse vegetation at lower sites closer to the coast.

No simple conclusions

Scarceness is the most obvious feature of the vegetation of Antarctica today. Many scientists did not believe it possible that Antarctica was once thickly vegetated. However, a combination of plate tectonics and other factors can be used to explain a very different climate at high latitudes in the past, and it is now known that diverse life was possible in Antarctica without other physical changes. It is particularly interesting that, while Antarctica is now almost barren, the rest of the southern hemisphere is covered in wildlife that owes much to this enigmatic region.

The Cenozoic decline in Antarctic plants and animals was undoubtedly climatically induced, but there is little agreement about other details. Evidence from the Antarctic Peninsula demonstrates a gradual impoverishment of the flora in response to glacial cycles, coupled with the increasing isolation of the continent and thus a strengthening of migration barriers during succeeding warmer phases. However, the probable Pliocene Sirius Formation fossils suggest that woody vegetation remained in Antarctica long after most paleoclimatologists had previously suggested that it was far too cold. The fact that one fossil find can cause so much uncertainty shows how much there is still to learn about the history of life on this most mysterious of continents.

▲ GROWTH RINGS
This flowering plant wood fossil shows well defined growth rings, indicative of a strongly seasonal climate. The cells in the wood are characteristic of flowering plants and are for water transport through the plant.

Evolutionary novelties

Evolutionary novelties are the result of major evolutionary changes in plants and animals that occur over relatively short periods of time. These changes often produce quite distinct life forms, distinguished as genera or even families of species that have many characteristics in common. High latitude areas were, and still are, an important source of evolutionary novelties. The reason for this is not well understood, although several theories have been proposed. However, these theories are difficult to test.

The Weddellian Biogeographic Province, extending from southern South America along the Antarctic Peninsula and West Antarctica to Tasmania, southeastern Australia, and New Zealand, was the center of origin and diversification for many organisms during the past 100 million years. The development of marsupials is an obvious example. Many new plant species arose and survived in this region, including important groups like *Nothofagus*, the casuarinas, and the Proteaceae. Much of Antarctica's endemic plant and animal life remained isolated because of geographical, climatic, or biological barriers. However, many plants and animals spread northward from the Weddellian Province to mid- and low-latitude regions, where their descendants still occur today, long after their place of origin has become too cold to support such life forms.

▼ Fire oak (*Casuarina cunninghamiana*).

The Polar Environment

The Polar Front

Previous pages

◀ **PEARLY CLOUDS**
Nacreous ("mother of pearl")
clouds are occasionally seen
in polar regions during the
coldest months, when the sun
is below the horizon but illuminates high-altitude wave
clouds—stationary clouds
formed downwind of mountain
ranges, in this case the
Transantarctic Mountains.
Nacreous clouds are a type
of polar stratospheric cloud
that is important in the breakdown of ozone.

Early explorers seeking a great southern continent found that both the air and the sea became cooler as they sailed south across the Southern Ocean. The whalers and sealers who followed them in quest of quick and easy profits noticed this too. These intrepid travelers also observed that strong currents set their ships to the east where the temperature change occurred. This transition between warm subtropical waters and cold polar waters became known as the Antarctic Convergence. Today, it is called the Polar Front.

Defining the Polar Front

The Front encircles Antarctica between latitudes 40°S and 60°S. Changes in sea temperature across the Front occur in a few sharp jumps—fronts—rather than as a gradual decrease across the width of the Southern Ocean. The two most important fronts are the Subantarctic and Polar fronts. The rapid temperature changes across the fronts are linked to the strong eastward flow of the Antarctic Circumpolar Current.

These fronts act as boundaries that define zones with different temperatures, salinities, and nutrient concentrations. The different zones also tend to be populated by distinct communities of plants and animals; early oceanographers could determine which side of the Polar Front they were on by the presence or absence of particular species of krill.

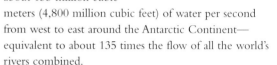

Antarctica's Circumpolar Current stretches for more than 20,000 kilometers (12,400 miles) around Antarctica. The surface speed of the current is modest, but its great depth and width make it the largest of all currents in the world's oceans. It carries about 135 million cubic meters (4,800 million cubic feet) of water per second from west to east around the Antarctic Continent—equivalent to about 135 times the flow of all the world's rivers combined.

The massive flow of the Circumpolar Current is driven by some of the strongest winds on earth. The persistent westerly winds, which are punctuated by frequent violent storms, prompted sailors to dub the southern latitudes the Roaring Forties and the Furious Fifties. The strong winds acting over the circumpolar extent of the Southern Ocean also create the largest waves on the planet.

Ocean conveyor belt

The Antarctic Circumpolar Current plays a unique role in the earth's climate system. Each of the world's major ocean basins is enclosed by land except at its southern boundary, and the Circumpolar Current functions as a pipe connecting these basins, smoothing out variations in water properties between the basins, and—more importantly—permitting a truly global ocean circulation pattern. Water, made cold and saline at high latitudes, becomes heavy enough to sink into the deep ocean;

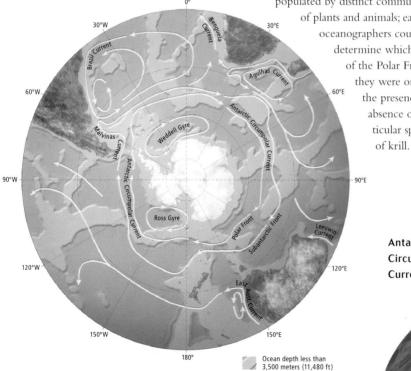

0°
30°W
30°E
Benguela Current
Brazil Current
Agulhas Current
60°W
60°E
Malvinas Current
Weddell Gyre
Antarctic Circumpolar Current
90°W
90°E
Antarctic Circumpolar Current
Ross Gyre
Polar Front
Subantarctic Front
Leeuwin Current
120°W
120°E
East Aust. Current
150°W
150°E
180°

▨ Ocean depth less than 3,500 meters (11,480 ft)

▲ **SOUTHERN OCEAN CURRENTS**
The currents of the Southern Ocean circulate west to east, unimpeded by land. This is the only place on the planet where the oceans can circulate around the globe, uninterrupted by any continental landmass. These currents loosely follow a route tracing deeper waters, and allow water transfer between oceans.

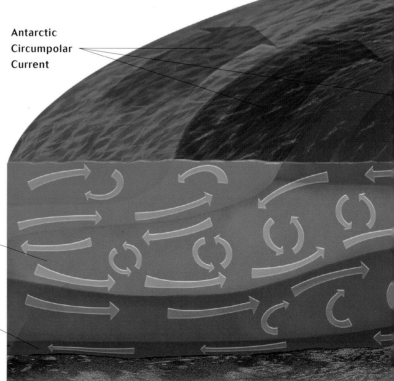

Antarctic
Circumpolar
Current

Antarctic
Intermediate
Water

Antarctic
Bottom Water

warm water flows into the high latitude regions to replace the sinking dense water. The exchange of warm and cold water carries heat from low latitudes to high latitudes, cooling the former and warming the latter, and so maintaining the earth's climate. The Circumpolar Current, by allowing the free exchange of water between the ocean basins, is a key link in this so-called "overturning circulation," or "ocean conveyor belt."

Polar differences

The southern polar region differs in important ways from the northern. Antarctica is a continent surrounded by the Southern Ocean, whereas the Arctic is virtually a landlocked sea; only relatively small amounts of water are exchanged through the narrow straits between the Arctic seas and the northern Pacific and Atlantic oceans.

The two polar regions also differ in the character of the sea ice. Antarctic sea ice tends to be carried away from the continent by winds and currents, whereas Arctic sea ice is trapped by surrounding landmasses and has a tendency to pile up, becoming much thicker than the sea ice of Antarctica. Freezing of the ocean surface in the Antarctic effectively doubles the size of the continent in winter, but in summer most of this ice melts, whereas the Arctic is ice-covered all year round.

As a result of these differences, there is no Arctic equivalent of the Antarctic Polar Front. In the northern hemisphere the transition between polar and subtropical waters occurs further south, outside of the Arctic, and is not circumpolar due to the presence of landmasses. SR

▲ FORCE 11 GALE
The gale force winds and huge seas battled by the early explorers are the same as those that confront visitors to the Southern Ocean today. The ships may have changed, but the Southern Ocean's power to generate the wildest conditions on earth still poses a great challenge to modern-day sailors.

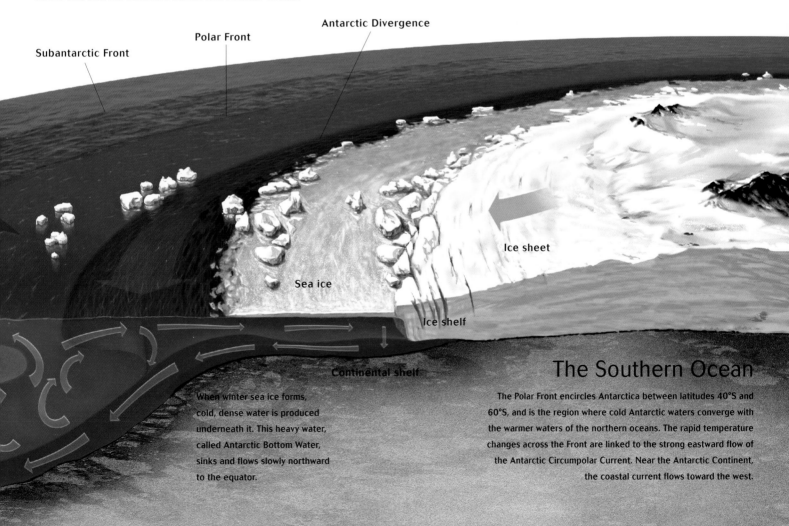

Subantarctic Front

Polar Front

Antarctic Divergence

Sea ice

Ice sheet

Ice shelf

Continental shelf

When winter sea ice forms, cold, dense water is produced underneath it. This heavy water, called Antarctic Bottom Water, sinks and flows slowly northward to the equator.

The Southern Ocean

The Polar Front encircles Antarctica between latitudes 40°S and 60°S, and is the region where cold Antarctic waters converge with the warmer waters of the northern oceans. The rapid temperature changes across the Front are linked to the strong eastward flow of the Antarctic Circumpolar Current. Near the Antarctic Continent, the coastal current flows toward the west.

Cold, dry, and windy

The north and south polar regions are "heat sinks" that influence the whole world's climate. A complex meteorological system is created by the different levels of energy received by the poles and the tropics, and is greatly affected by the spinning of the globe. Air heated at the equator rises and flows toward the poles, where it cools and sinks; this dense, cool polar air flows back to the low-pressure area created at the equator by the rising warm air.

Although the Antarctic polar ice sheet is crowned by constant high pressure, it is surrounded by a region of low pressure. Since Antarctica and South America separated, probably about 25–30 million years ago, the Southern Ocean has completely encircled Antarctica, allowing the winds to flow unimpeded. Here, eastward-bound low-pressure systems are generated in never-ending succession, circling Antarctica like a procession of spinning tops. Sailors call these latitudes the Roaring Forties, the Furious Fifties, and the Screaming Sixties.

Weather observations, first by explorers and then by scientists, have been invaluable sources of meteorological information about Antarctica. The International Geophysical Year (IGY) of 1957–58 led to the establishment of many Antarctic research bases, and a few years later the development of weather satellite technology provided a wealth of information. Today, about a hundred ground stations, some attended (mainly coastal) and some automatic (mainly across the interior), provide weather information for distribution around the world.

More than 97 percent of Antarctica is covered with snow, and this has a significant impact on climate all over the world. Solar energy is reflected from the earth in an effect known as albedo. The Antarctic surface absorbs radiation for a short midsummer period only—for the rest of the year it re-radiates more energy than it receives, the balance being restored by heat transfer from the tropics.

World's coldest

The lowest temperature recorded on earth was taken by A. Budretski on 21 July 1983 near Russia's Vostok station: it was –89.2°C (–128.6°F), lower than anything experienced at the South Pole because of Vostok's higher altitude—3,500 meters (11,475 ft), as opposed to the South Pole's 2,836 meters (9,300 ft). Vostok's record is likely to be broken only if a base is established at the top of the polar ice sheet—an altitude of 4,000 meters (13,115 ft). Antarctica rapidly becomes colder in autumn, reaching its extreme in the lightless depths of winter. Inland, winter temper-atures can range from –40°C to –70°C (–40°F to –94°F); coastal temperatures in winter range from –15°C to –35°C (5°F to –31°F). The Antarctic Peninsula is by far the warmest part of Antarctica; in the middle of summer, temperatures there can reach 15°C (59°F), while in East Antarctica they range from 0°C (32°F) on the coast to –25°C (–13°F) inland.

Antarctic temperatures are lower than Arctic temperatures, which range from about 0°C (32°F) in summer to –35°C (–31°F) in winter. Antarctic latitudes experience about the same temperatures as Arctic latitudes 500 kilometers (310 miles) closer to the North Pole, because of the height of the polar ice sheet, and because Arctic temper-atures are moderated by the ocean: the Southern Ocean generates little warmth, because it is much colder than the Arctic Ocean and does not receive the warm currents of the northern hemisphere. As in Antarctica, the lowest Arctic temperatures occur far from the ocean. Verkhoyansk, on the Yana River in Siberia, has recorded –67.8°C (–90°F), but has peaked in summer to 37°C (almost 100°F). Temperatures across the Arctic range widely, depending on proximity to warm or cold ocean currents.

World's driest

The air over Antarctica is generally too cold to hold water vapor, so there is very little precipitation: snowfall on the Polar Plateau equates to less than 50 millimeters (2 in) of rain per year. Antarctica, the world's driest

> SCULPTURES IN SNOW
Light winds deposit snow in the lee of obstacles and undulations across the polar landscape, and then blizzards erode the snow into sculptured shapes known as sastrugi. These polished forms can stand several meters high and be as hard as concrete. A vast field of sastrugi, like the one pictured here, appears as a snap-frozen, storm-tossed sea.

▼ COLD AND WINDY
Temperature and wind speed vary greatly over the Antarctic Continent. In general, the heart of the Continent (Dome C) is very cold, especially during the dark of winter. Antarctica's coastal regions (Cape Denison, Butler Island) are milder, warmed by the oceans, but are often the windiest areas, buffeted by freezing blasts of katabatic winds from the icy slopes.

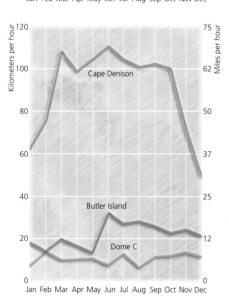

desert, ranks with the Sahara as the world's largest desert. But Antarctica retains what moisture it receives; about 75 percent of the world's fresh water is stored as ice, and 90 percent of that ice is in Antarctica.

The Arctic, too, is a desert, where annual precipitation ranges from 100 millimeters to 400 millimeters (4–15¾ in). As in Antarctica, the northern polar air is too cold to hold much moisture, but what there is falls on ocean—there are ice sheets only on Greenland and nearby islands. The Greenland ice sheet is about one-tenth the size of the Antarctic ice sheet.

➤ **READING THE PULSE**

A field surveyor aligns an automatic weather station mast north, as a baseline for wind direction measurements and to maximize power output from a solar panel. Hourly meteorological readings are automatically transmitted around the world via satellite.

"The merciless blast"

At Cape Denison, Mawson measured wind speeds over a year on a 24-hour average. The wind nearly always came from the south–southeast. On only one day was the wind speed less than 4.5 meters per second (15 fps), and for only about a third of the year was the average wind speed less than 18 meters per second (59 fps). There were six days when the average wind speed was greater than 27 meters per second (88 fps). Over the whole year, the average wind speed was 19.4 meters per second ($63\frac{1}{2}$ fps), or about 70 kilometers per hour (44 mph). In the stormiest hour of that year, on 6 July 1913, the wind speed was 43 meters per second (141 fps), or 154 kilometers per hour (96 mph). The mean wind speed for the quietest month was 11.7 meters per second ($38\frac{1}{2}$ fps), or 42 kilometers per hour (26 mph).

A French station later established nearby discovered that Mawson's anemometer may have been registering under-readings. This area certainly records the world's strongest winds at sea level—only a few mountain peaks may have recorded higher wind speeds.

World's windiest

Antarctica is the windiest place on earth. A variety of different winds blow there, from the inversion winds at the Pole to winds funneled between islands along the coast, and violent blizzards can develop with incredible speed. The wind creates its own landforms: sastrugi, irregular ridges up to 1 meter (3 ft) high, carved by blowing snow. Like sand dunes, their undulating surfaces indicate the prevailing wind direction. The strongest winds, however, are on the coast, where wind speeds of up to 300 kilometers per hour (185 mph)—twice the velocity of hurricane-force winds—have been recorded.

On 8 January 1911, Douglas Mawson's Australasian Antarctic Expedition arrived at Cape Denison. That afternoon, the wind rose as they unloaded their supplies. They waited for it to drop, but it never really did. Mawson called his account of the 1911–14 expedition *The Home of the Blizzard*—not without good reason. He had built his hut where blasts of katabatic wind flow down from the Polar Plateau. Simply put, in Antarctica, katabatic wind is cold and dense air that pours down the ice slope to the sea, becoming denser and picking up speed as it goes. On average, the wind speed at Cape Denison reaches hurricane force every three days. Mawson's own words sum up the human response to Antarctic weather: "… the drift is hurled, screaming through space at a hundred miles an hour, and the temperature is below zero Fahrenheit. Shroud the infuriated elements in the darkness of a polar night and … a plunge into the writhing snow-whirl stamps upon the senses an indelible and awful impression seldom equaled in the whole gamut of natural experience … We stumble and struggle through the Stygian gloom; the merciless blast—an incubus of vengeance—stabs, buffets and freezes, the stinging drift blinds and chokes … We had found an accursed country." DM

▲ SHAPED BY THE WIND
Sculpted by sunlight and wind, a piece of broken sea ice rests on the frozen surface of the Ross Sea. By summer it will be gone. Wind shapes everything in Antarctica: snow, ice, rocks —and people's lives. Antarctica abounds in stories of being trapped inside tents for days by raging blizzards.

▼ THE COLD SHOULDER
Adélie penguins turn their backs to the wind. Conserving energy is one of the keys to survival in Antarctic blizzards. Adélie penguins spend their winters on the pack ice, where air temperatures are always higher than on land.

➤ DOWN-FLOWING WINDS
Katabatic winds sweep down the coastal slopes of Terra Nova Bay. Unlike winds in other parts of the world, Antarctica's katabatic winds are caused by the shape of the land: cold, dense air on the high ice sheet flows down the coastal slopes under the influence of gravity.

The Antarctic ice sheet

Antarctica's ice sheet is immense. It covers about 14 million square kilometers (5,500,000 sq. miles), and averages about 2,300 meters (7,500 ft) thick (from bedrock), with local thicknesses up to nearly 5,000 meters (16,400 ft). Antarctica's ice potentially locks up about 75 percent of the world's fresh water. Surprisingly, seismic soundings have recently confirmed the presence of many large lakes under East Antarctica, but little is known about them, especially about their age. Just above one of the largest of these lakes, Lake Vostok, the drilling of an ice core more than 2 kilometers (1¼ miles) long has been discontinued until techniques can be developed for sampling its waters without running the risk of contaminating them. It is believed that the lake's waters may contain extremely ancient organisms (perhaps now extinct elsewhere), as well as a record of climatic conditions in the world tens or hundreds of thousands of years ago.

▲ DRILLING RIG
Law Dome is an isolated ice cap near Casey station. Due to the underlying bedrock, the ice of the dome is moving radially out from its center. Separated from the main ice sheet and circular in shape, the dome is a miniature of the Antarctic ice sheet. At the top of the dome, scientists drill ice core samples to learn about climate change, especially about the makeup of the earth's past atmosphere.

Antarctica's ice sheet

The East Antarctic Ice Sheet contains a great volume of ice, and much, though not all, of it is grounded on bedrock near or well above sea level. In contrast, much of the far smaller West Antarctic Ice Sheet lies on rock below sea level, which will be an important factor if climates warm sufficiently for the polar ice sheets to melt. Glaciologists debate whether the Antarctic ice sheets are expanding or decaying, but they generally believe that the continental ice sheets, especially that of East Antarctica, are more or less in a state of equilibrium, in contrast to some of Antarctica's ice shelves, which seem to be breaking up. The ice sheets are like settling tanks, trapping and accumulating materials from the atmos-

phere. They contain a record of ancient climates, and of global volcanic activity extending back for thousands of years, as well as trapped meteorite fragments.

Storm tracks seldom reach the interior of East Antarctica, and much of this region is true desert, where snow accumulates at very low rates, generally about 2 centimeters (¾ in) of equivalent water per year. As snow layers become buried, they transform into ice, brittle at upper levels, but becoming more plastic under pressure at deeper levels. Under the thickest areas, the ice can reach its pressure melting point and water can form, allowing basal sliding—the rapid movement of ice over bedrock under the pull of gravity. The highest part of the ice sheet, about 4,000 meters (13,000 ft) above sea level, lies not at the South Pole but in the middle of the East Antarctic Ice Sheet at about 80°S latitude and 75°E longitude, south of the Amery Ice Shelf. From near there, ice flows outward more or less radially by flow under gravity. Where it reaches the sea, much of the ice calves off as icebergs, and flows onto and adds to nearby masses of floating ice shelves that fringe the coasts. In places, ice movement is channeled into rapidly flowing ice streams. Regional ice "domes" occur at various places.

The East Antarctic Ice Sheet also flows toward and becomes largely dammed by the Transantarctic Mountains. In many places the dam is breached by outlet glaciers that squeeze through the mountains and plunge down to feed the Ross Ice Shelf, or flow into the Ross Sea, farther north. The giant, highly crevassed

A cross section of Antarctica

The shape of Antarctica, with the exception of the Transantarctic Mountains and a few locations around the coast, is lost under a blanket of ice. In some locations in East Antarctica, the ice is up to several kilometers thick.

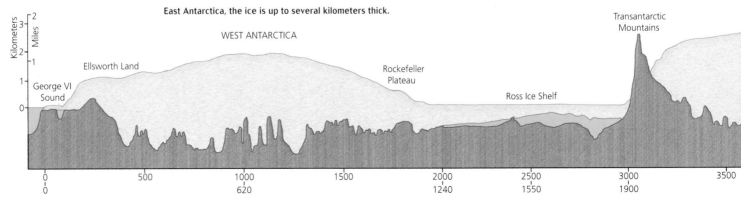

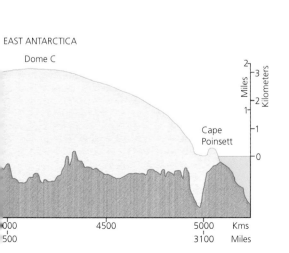

EAST ANTARCTICA

Dome C

Cape
Poinsett

Miles
Kilometers

2
3
2
1
1
0
0

000   4500   5000   Kms
500   3100   Miles

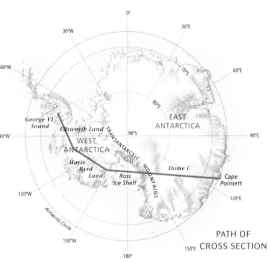

30°W   0°   30°E

60°W   60°E

70°S

80°S

George VI
Sound   EAST
Ellsworth Land   TRANSANTARCTIC   ANTARCTICA

90°W   90°S   90°E

WEST
ANTARCTICA

Marie
Byrd   Dome C
Land   Ross
Ice Shelf   Cape
Poinsett

120°W   120°E

Antarctic Circle

150°W   180°   150°E

PATH OF
CROSS SECTION

▲ WATERFALLS OF ICE
At the head of the Taylor
Glacier in the Dry Valleys,
the ice of the Polar Plateau
pours over the hidden lip of
the Transantarctic Mountains,
(here buried under the ice).
Looking like a giant waterfall
in slow motion, the ice breaks
up into heavily crevassed
"icefalls," the tops of which
are shown here.

Beardmore Glacier was used in the early 1900s by Ernest Shackleton and Robert Scott's expeditions to the Polar Plateau on their way to the South Pole. The Axel Heiberg Glacier gave Roald Amundsen a more direct route to reach the Pole first—in November 1911.

The West Antarctic Ice Sheet forms several major ice streams that feed the head area of the great Ross Ice Shelf, and are its main source. Its center is at only half the elevation of that of the East Antarctic Ice Sheet. The accumulation rate is much higher than that of East Antarctica because many maritime storm tracks cross the area. The Antarctic Peninsula lies outside the area of the ice sheets, and it contains mostly glaciers of alpine type, as well as larger ice fields.

Theories of origin

Today we live in an ice age world (often referred to as an "interglacial" stage), with large ice sheets covering Antarctica and Greenland. Antarctica's ice sheet gives the best picture of how northern North America must have looked 20,000 years ago, at the end of the last glacial maximum, under the Laurentide Ice Sheet, which had largely disappeared by 10,000 years ago. In Antarctica, ice seems to have begun forming between 35 and 50 million years ago and the ice sheet probably remains in much the same form as it has throughout the ice age. Major glaciations seem not to be synchronous between hemispheres (though there is some debate about this), whereas the advances and retreats of smaller valley glaciers tend to be more sensitive to both global and regional climate changes.

The development of Antarctica's ice sheet seems to have commenced about 10 million years after the continent "drifted" southward into the polar region as part of Gondwana about 200 million years ago. The reason for the long delay is unknown, but oceanographers suggest that it may be due to a decline in ocean temperatures about 50 million years ago, at a time when glaciers seem to have started forming in Antarctica, probably in high country such as the Ellsworth Mountains and the Transantarctic Mountains. As climates cooled, valley glaciers advanced and coalesced into piedmont glaciers in plains at the foot of mountains, which in turn expanded, and eventually built up into ice sheets that covered the continent. The record of ice build-up is found in surrounding ocean sediments. Cores drilled from the Ross Sea in the 1970s recovered boulders and pebbles that must have been carried along by icebergs and dropped into seafloor mud about 30–40 million years ago. Such rocks, deposited far from their place of origin, are called glacial erratics. The volume of south polar ice must have fluctuated greatly since the birth of the Antarctic ice sheets, and the continent has probably been continuously glaciated since their formation, but there is some evidence of extensive deglaciation about 5 million years ago.

Evidence and opinions that try to explain the formation of glaciers and their oscillations often conflict. In 1920 the Serbian mathematician Milutin Milankovitch found evidence for cyclic variations (of 23,000, 41,000, and 100,000 years) in the earth's orbit around the sun, and in movements of the earth's axis that might explain variations in solar radiation absorbed to warm or cool the earth. Scientists of many specialties are refining the Milankovitch Cycles in ever-increasing detail in order to improve correlations with geologists' evidence from field work. Others see the sun as the prime mover of climatic change, and search for variations in solar cycles (such as the 11-year sunspot cycle), and in radiation from a "turbulent sun." These sorts of possibilities for explaining continental glaciations are not yet well understood. AF

▼ CLOUD AT LAW DOME
Low, medium, and high level clouds streak the sky above this featureless plateau. Close to the South Pole, clouds tend to form at lower altitudes than elsewhere. At 70° south, mid-level cloud may form at 2,000 meters (6,560 ft), whereas at 40° south, mid-level cloud forms at no lower than 3,000 meters (9,840 ft).

▶ GLACIAL ERRATICS
Millions of years ago, these banded sandstone boulders were picked up and carried many miles in the ice before being deposited by the Vanderford Glacier on this smooth granite surface in East Antarctica.

▼ FROST FLOWERS

Sublimation, the process of condensation directly from the gaseous phase to the solid state, allows water vapor to deposit delicate hoar crystals on the cold sea ice surface of Erebus Bay in spring. Governed by the hexagonal (six-sided) symmetry of crystal geometry, these formations sprout intricate fronds in the chill air of a calm, clear Antarctic night.

➤ ICY POLES

Smooth mounds appear where wind-blown snow has been plastered onto blocks of ice on the walls of an open Ross Ice Shelf crevasse. The warmth of the summer sun generates considerable melt causing water drops to refreeze into long stalactite spears; stalagmite spikes jut skywards (top left) beneath dripping overhangs above.

◄ BLUE HEAVEN

Expeditioners gaze in awe at an exquisite ice cave in the coastal cliffs near the Vestfold Hills. Snow looks white due to scattering of light in the many air spaces between grains. Glacier ice, which forms when the air spaces are compressed, absorbs light preferentially at the red end of the spectrum, allowing a brilliant blue to radiate out.

▲ FROZEN VEIL

Pressure cracks splinter the frigid surface of Lake Vanda in the Dry Valleys region of Victoria Land. As fresh water in the upper layers of the lake freezes, the ice expands, causing myriad fractures that fizz and crackle as they appear.

Ice shelves and glaciers

▼ WAVES OF ICE

As the vast Ross Ice Shelf approaches immovable Hut Point Peninsula on Ross Island, it buckles, looking like huge waves about to break on land. The flat ice in the foreground is sea ice just a few meters thick that later in summer will probably break out, leaving open sea before lower autumn temperatures freeze it again. The long, straight line in the distance is a service road.

Australian geologist Douglas Mawson may have been the first person (in 1912) to use the term "ice shelves" to describe sheets of very thick, mostly floating ice. Ice shelves make up nearly 50 percent of the Antarctic coastline. They are fed by glaciers or ice streams from Antarctica's continental ice sheets, and are additionally nourished by snow accumulating at their surface, and probably, in some cases, by seawater freezing at their base. Some areas may be aground. The Ross Ice Shelf is the world's largest ice shelf, averaging about 330 meters (1,100 ft) thick and increasing to a maximum of about 700 meters (2,300 ft) toward its southern boundary.

Ice shelves

Ice shelves generally fill embayments, and so are landlocked on three sides. They have level or gently undulating surfaces, and their seaward side spawns the tabular bergs characteristic of the Southern Ocean. Tabular bergs are usually easily distinguished from bergs formed by a calving glacier; the typical shape of a newly calved glacier berg is irregular.

Where ice shelves are joined to land, immense cracks or crevasses can develop due to ocean tides and currents. The Grand Chasm, at the head of the Filchner Ice Shelf, is such a zone of almost impenetrable crevasses. It was the first challenge for English geologist Vivian Fuchs and his companions on their way to the continent itself in the first successful continental crossing from the Weddell Sea to the Ross Sea in 1957–58. There had been two previous attempts, by German Wilhelm Filchner in 1911–12, and by Englishman Ernest Shackleton in 1914–16. Both had their ships imprisoned by the pack ice.

Most Antarctic ice shelves are small, but the three principal ones, in decreasing order of size, are the Ross,

the Ronne, and the Amery. The Ross Ice Shelf lies at the head of the Ross Sea and covers an area about the size of France. Another, the Larsen Ice Shelf of the Antarctic Peninsula, has decreased greatly in recent years and seems to be breaking up. The Larsen Ice Shelf is the probable source of many tabular bergs seen by visitors on cruise ships.

The first humans to reach the Ross Sea were awed by the impressive cliff front of the Ross Ice Shelf, which to the south rises up to 50 meters (165 ft) above the sea—it seemed such an impediment to travel inland that it became known as the Ross (or Great) Barrier.

Visitors to the Ross Sea and McMurdo Station are likely to see some of the most developed pressure ridges in Antarctica. At this point, the Ross Ice Shelf, moving northward, collides with Ross Island and is thrown into spectacular giant folds.

Antarctica's ice shelves

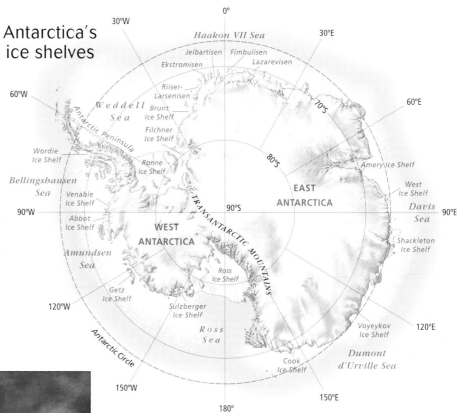

◄ **PRESSURE RIDGES**
Storms can put pack ice under immense pressure, crushing and upending floes, rafting one on top of another, and forming pressure ridges (as shown here) where the floe buckles. In some places, the resulting sea ice can be up to several meters thick.

▼ **DIRTY ICE**
Ice shelves are typically much "cleaner" than this area, called the Dirty Ice, which lies at the edge of the ice shelf at McMurdo Sound in midsummer when the surrounding sea ice has broken out. Each of the ponds in this area has a unique chemistry and is full of cyano-bacterial microbial mats.

Glaciers

In Antarctica, rivers of ice, called valley glaciers, form at the outlets of the polar ice sheets that flow down through the Transantarctic Mountains to feed the Ross Ice Shelf. The Lambert Glacier, 40 kilometers (25 miles) wide, and probably the world's largest glacier by volume, drains a major area of the East Antarctic Ice Sheet to feed the Amery Ice Shelf. The fastest-flowing Antarctic glacier known, the Shirase in Queen Maud Land, flows at a rate of 2 kilometers (1¼ miles) per year. On the Antarctic Peninsula there is no ice sheet, so glaciers are defined by the area's mountain topography. Only glaciers of alpine type are seen, including cirque glaciers that occupy high mountain amphitheaters, and valley glaciers, which flow down mountain chasms. Many of these glaciers show evidence of retreat in recent years. Glaciers can carry rocks on their surfaces and deposit them along their path, where they end up as ridges of bouldery material called moraines. Such ridges show the former extents of glaciers long after the ice has melted back.   AF

▼ SEA CAVES
When the ocean washes up against icebergs, it exploits any areas of weakness. Eventually the weakest section of ice gives way, leaving a cave-like hollow. An average berg has a life of four years. As they age, bergs split along lines of natural weakness and roll over when they become unstable.

▲ VALLEY GLACIERS

Antarctica is drained by vast systems of glaciers and ice streams. Valley glaciers, the most spectacular of them, pass through exposed mountain ranges, flowing around large, exposed mountain tops or hills called nunataks.

◄ GLACIER TONGUE

Vast floating glacier tongues such that of the Vanderford Glacier (seen here), can spawn large tabular icebergs not easily distinguished from true ice shelf bergs. Glacier tongues project out into the ocean, landlocked at the rear, where the feed glacier comes from; ice shelves are usually landlocked on three sides.

◄ FROM THE MOUNTAINS TO THE SEA

The lower slopes of Ross Island are shaped by heavily crevassed glacial ice, formed by the compaction of accumulated snow. Here, at the sea edge, flat sea ice meets the glacial ice. Unbroken sea ice attached to a coastline is called fast ice, which is strong enough to support ice vehicles, and even to land large aircraft on.

Icebergs

And now there came both mist and snow,
And it grew wondrous cold:
And ice, mast-high, came floating by,
As green as emerald.

And through the drifts the snowy clifts
Did send a dismal sheen:
Nor shapes of men nor beasts we ken—
The ice was all between.

SAMUEL TAYLOR COLERIDGE (1772–1834)
THE RIME OF THE ANCIENT MARINER

▽ TABULAR BERG

Icebergs are categorized according to their shapes and sizes. This iceberg is known as a tabular berg, with its table-like top and sheer sides. Tabular bergs can be enormous, sometimes hundreds of square kilometers in surface area. What appears above the water is just a fraction of the overall size of the iceberg. About four-fifths of an average iceberg is submerged.

◁ ADRIFT IN ICE

These small bergs have run aground in shallow waters and are now locked in place by sea ice. Icebergs can remain grounded for many years. Jade bergs (foreground) form under special conditions at the base of floating ice shelves. The impact of large icebergs hitting the ocean floor may be nothing less than catastrophic for the dwellers on the ocean floor.

△ FANTASTIC SHAPES

The tall ice sculpture at the center of the picture is a weathered berg, which has eroded (probably from an old cave) to form a pillar on the left, and tilted so that an old waterline is exposed at an angle across the berg. Bergy bits bob up and down in the foreground.

There are now websites analyzing Samuel Taylor Coleridge's *Rime of the Ancient Mariner* for the geographical information it contains. Coleridge never saw an albatross, far less an iceberg, but his description is remarkably accurate. Although an iceberg is unlikely to be "as green as emerald"—for such bergs occur only when the ice holds a large quantity of organic matter—the relationship Coleridge describes between icebergs and the "snowy clifts" of an ice shelf rings very true.

Birth of an iceberg

An iceberg adrift is the archetypal image of ice in Antarctic waters. However, icebergs are not composed of frozen salt water; they are born on land and made of fresh water. They begin as snow falling on the continental ice sheet; these accumulate and are compressed over many years, as the ice sheet flows toward the sea. Upon reaching the sea, the ice spills into the water when a glacier calves, or floats on the ocean surface as an ice tongue or a massive ice shelf. The Ross Ice Shelf is just a floating ice block, abutting the fast ice along its seaward front so that it rises and falls with the tides.

Every year, thousands of icebergs break off Antarctica, scattering across the Southern Ocean, mainly below the Polar Front, where the water is cold enough to slow their melting. There are fewer icebergs in the Arctic because there is less glaciated land—Greenland alone provides a comparable source of icebergs, and it is only a fraction the size of the Antarctic Continent.

Huge tabular bergs sometimes break off ice shelves, and drift northward under the influence of winds and

▼ MELTING MOMENTS

▼ **MELTING MOMENTS**
In the warmer months, the unstable edges of this huge grounded tabular berg melt, the water gathering to create an ice pond between the berg and the fast ice. During the night, icicles form at the base of the berg, only to melt away again during the day.

currents, to melt eventually in the warmer water north of the Polar Front. On rare occasions, they have been found as far north as 45°S in the South Pacific, and at 35°S in the Indian and South Atlantic oceans.

The giant Ronne–Filchner ice shelf of the Weddell Sea spawns giant flat-topped or tabular bergs. These are trapped in the field of sea ice that rotates clockwise and almost fills that sea. It was such a rotation that eventually carried Shackleton's party to open water long after their ship, the *Endurance*, was crushed in the pack ice.

As well as the huge icebergs that regularly make the headlines, there are innumerable smaller ones. One survey carried out in 1985 found 30,000 icebergs in an area of 4,000 square kilometers (1,500 sq. miles) of ocean. Today, satellite imaging and satellite tracking gather a more accurate record. The United States National Ice Center maintains a database of the world's large icebergs (more than 10 kilometers/6 miles long). It can be accessed at the website: http://www.natice.noaa.gov/, which also carries satellite imagery of the larger, tabular bergs.

Profiles of bergs

Icebergs are named by their quadrant of origin. Those designated "A" are from 0° to 90°W (the Bellingshausen Sea and the Weddell Sea); "B" bergs are from 90°W to 180°W (the Amundsen Sea and eastern Ross Sea); "C" bergs are from 180° to 90°E (the western Ross Sea and Wilkes Land), and "D" bergs are from 90°E to 0° (the Amery Ice Shelf and the eastern Weddell Sea). A large tabular iceberg can be 300 meters (980 ft) thick, may weigh hundreds of million tonnes, and may contain enough fresh water to supply a city of a million people for three years.

A vast hunk of the Amery Ice Shelf broke off in 1963, and, four years later, collided with the Fimbul Ice Shelf to produce two tabular bergs, 110 by 75 kilometers (68 by 47 miles) and 104 by 53 kilometers (65 by 33 miles). The last recognizable chunk of these was seen off the Antarctic Peninsula in 1976. Toward the end of the twentieth century, there was considerable interest in berg B-10A, which broke off in 1992, and in 1998 entered shipping lanes between the Antarctic Peninsula and South America. It measured about 100 by 50 kilometers (62 by 30 miles), and in March 1999 headed toward South Georgia.

But these giants were dwarfed in 2000, when the Ross Ice Shelf calved the biggest iceberg ever seen. The larger of the two segments that broke away from the shelf in March 2000 was about 300 kilometers long and 40 kilometers wide (185 miles by 25 miles) and had an area of more than 10,915 square kilometers (4,214 sq. miles). Soon afterward, on 4 May and 6 May, two massive icebergs calved from the Ronne Ice Shelf. The first one soon broke in two to become A-43A, measuring 168 by 33 kilometers (86 by 20½ miles), and the smaller A-43B, which was 84 by 35 kilometers (52 by 22 miles). The other was named A-44, and measured 60 by 32 kilometers (37 by 20 miles).

Whereas ice shelves give birth to tabular icebergs, most glacial tongues

▲ BLUE ICE

As snow falls, its weight compresses the snow beneath, and it turns to ice. At first, the ice is air-filled and granular, but by the time it sinks to 50 meters (165 ft), it has become solid ice clouded with bubbles of air. As it descends deeper, the bubbles are compressed out, and the ice becomes clear blue. When a glacier, containing this blue ice breaks off and floats away, it becomes a "blue berg"—one of Antarctica's most attractive features.

▲ JADE GARDEN

Snow-blasted winter travelers stop alongside an upturned iceberg trapped in fast ice near Mawson. The iceberg originally formed under the Amery Ice Shelf, frozen onto its base. The minute aquatic organisms captured within the ice layer impart a stunning translucent jade color to the now upturned berg.

(unless they are huge) create icebergs of irregular shapes that can inspire flights of fantasy. Frank Worsley, captain of Shackleton's *Endurance*, described the ice that threatened their lifeboat, *James Caird*, as they struggled to reach South Georgia: "Swans of weird shape pecked at our planks, a gondola steered by a giraffe ran foul of us, which much amused a duck sitting on a crocodile's head … All the strange, fantastic shapes rose and fell in stately cadence with a rustling, whispering sound and hollow echoes to the thudding seas."

Icebergs range from crystal clear and pure white to green, brown, and blue—even pink from algae. Most have fissures of bright electric blue and many are fringed by icicles. The shapes are unimaginable—from the hewn straight sides of a recently calved tabular iceberg to the grottoes, pinnacles, and arches of a berg that has rolled to reveal its partially melted underside.

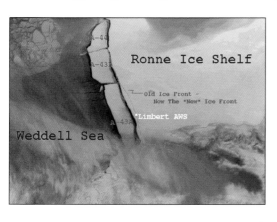

◄ ICEBERG ARMADAS

In recent years, gigantic icebergs have calved from almost the entire extent of the Ross and Ronne ice shelves. These episodic events are not climate-related, but the result of natural dynamic cycles as the shelves thin and rapidly spread seaward. The Amery Ice Shelf is currently undergoing fracture-forming huge rifts—sites of future iceberg production.

▲ FINDING A BALANCE
With erosion by wind and waves, sections of tabular bergs break away. The iceberg slumps into the ocean, but finds its balance moments later, revealing a magnificent ice cave. The broken chunks of ice float up to the surface.

Dangers and difficulties

The common expression—"just the tip of the iceberg"—takes on new significance in polar regions. Less than 10 percent of an iceberg is visible above the water, but the actual percentage can change depending on such factors as impurities in the ice and the salinity and temperature of the water. Although they look huge and unchanging, icebergs are inherently unstable, as they are constantly melting from both the top and the bottom.

Smaller icebergs (relative terminology if they are seen from a ship) become top-heavy as they melt underwater, and can roll quickly when their equilibrium shifts. Large bergs may seem to explode violently, as they collapse into many pieces, creating large waves and dangerous vortices.

Ice shelves spawn tabular bergs that may tip over to become capsized bergs, or they may create castellated bergs after a lot of the ice has melted from underneath. Glacial ice breaks into the ocean as smaller icebergs that decay to become "bergy bits," then "growlers," when they have melted enough so that very little is visible above the water. Growlers are very dangerous because they may float unnoticed by radar or helmsman until a ship collides with one. The most notable shipping disaster involving an iceberg was the *Titanic*, which sank at a latitude of 41°46´N, on 15 April 1912, on its maiden voyage from Southampton to New York—1,503 people died that night. The *Titanic* did not hit a growler but came to grief on the extended underwater "foot" of an iceberg.

While most icebergs simply drift toward warmer, temperate waters, where they quickly melt, some run aground on shallow shoals and stay intact for years. In the 1970s, there was considerable interest in the idea of towing icebergs towards the desert coastlines of Australia, South America, and Africa, where their fresh water could be used to irrigate barren land. This bold plan was not completely impractical, although the cost would have been enormous, but it was abandoned after consideration of all the details. One difficulty was that an iceberg breaking up while under tow in warmer water would have created safety difficulties for the operation. The main problem, however, was the underwater bulk of the iceberg; it would almost certainly have become grounded on the continental shelf at a distance too far to pipe the water ashore.   DM

A TYPICAL MIXTURE

A typical scene off the coast of the Antarctic: a mix of sea ice and land ice. Ice floes are sections of sea ice and can be up to 2 kilometers (1 1/4 miles) long. In between the floes are smaller tabular bergs, capsized bergs, and the beginnings of a castellated berg.

JELLY MOLD

The bulbous mound of an overturned iceberg is scarred by parallel melt channels, which formed while the berg sat upright in the water. Smooth scallop features, the size of a cupped hand, pockmark the once submerged surface. The old crown of the berg juts skyward in the background.

The frozen seas

As winter approaches, the shores of Antarctica go through a transformation. By March, after weeks of 24-hour sunlight, most of the sea ice that normally surrounds the continent has melted. All that is left is a rim of ice in some places, most notably the Weddell Sea and the Bellingshausen Sea.

The formation of sea ice

Each evening as winter draws closer, the sun dips further below the horizon and the air temperature falls enough to create a thin sheet of ice on the water's surface, seen first as ice needles and tiny ice plates known as frazil. It develops into a thin sludge that sailors call grease ice. At first the ice crystals are about 2.5 centimeters (1 in) wide and about 0.15 centimeters (less than $\frac{1}{8}$ in) thick, but when the sea is calm, they can grow quickly to form sheets, especially when snow falls on top of this skin of ice. The plastic ice, up to 10 centimeters (4 in) thick, which bends readily with the waves, is called nilas. Wind and wave motion, however, may break the ice into plates, and then bang them together to create pancake ice—a clear sign of substantial freezing, and a warning to polar mariners that winter is certainly on the way.

If temperatures stay low for a few days, the pancakes meld and thicken to form floes of new ice. In a month, the ice may thaw, freeze, and be covered by snow, forming a sheet of sea ice 15–60 centimeters (6–25 in) thick. Sea ice that is attached to the edge of the continent is called fast ice; ice that drifts with the currents offshore is called pack ice. Fast ice thickens mostly as the result of new ice forming on the underside of the surface ice (this is also particularly true of the way Arctic sea ice forms), whereas 50 percent or more of the thickening of pack ice is due to the pancakes or small floes piling on top of one another (rafting) during storms.

This expanse of new ice grows at an incredible rate: up to 60 square kilometers (23 sq. miles) per minute. Throughout the winter, it advances about 4 kilometers ($2\frac{1}{2}$ miles) each day, adding almost 100,000 square kilometers (40,000 sq. miles) of new ice. In October, at the end of an average winter, it has effectively more than doubled the size of Antarctica. The 14 million square kilometers (5,400,000 sq. miles) of land abut 20 million

▼ STRONG ICE
By midsummer, there will be open water around the Cape Royds coastline (with Orcas and minke whales feeding) but in spring the fast ice is strong enough to travel safely on. Broken floes are here trapped in the sea ice, whose uneven surface and cracks bear testimony to some of the pressures acting on it.

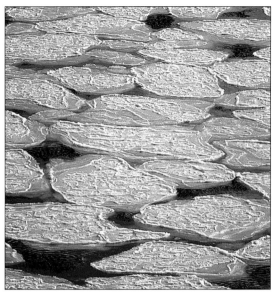

◀ PANCAKE ICE
Pancake ice looks very much
like pale waterlily leaves—
each flat plate has a raised
edge around it from collisions
with its neighbors.

Far left
◀ NEW ICE FOR OLD
Beyond the edge of last
winter's sea ice in McMurdo
Sound, new sea ice is forming.
The appearance of new sea
ice depends on conditions at
the time. On calm seas a thin
slick forms, which gradually
thickens into glossy sheets
just a few centimeters thick.
Typically, disturbances from
sea or wind break these
fragile sheets up and they ride
over each other, creating
amazing patterns.

square kilometers (7,700,000 sq. miles) of frozen ocean, reaching out more than 2,000 kilometers (1,240 miles) from the coast. The ice is generally about 1 meter (3 ft) thick, but can be up to 10 meters (33 ft) thick in places. By October, the Pacific pack ice extends northward to about 62°S, while the Atlantic pack extends much farther north, to approximately 52°S. The northern range of icebergs is from 6° to 10° beyond the pack ice.

New ice does not accumulate uniformly around the whole continent. It starts to form in the Weddell Sea, and then in the Ross Sea and the Bellingshausen Sea. As Ernest Shackleton had ample time to observe, sea ice moves continually. In the Weddell Sea, where his *Endurance* was trapped and finally crushed, a clockwise drift is more defined than the drift in either the Ross Sea or the Bellingshausen Sea.

▲ WIDE WATERS
WIDE WATERS

Even at the coldest time of
year there is a polynya (a
large area of open water) at
McMurdo Sound and Ross
Island. The nearest part of
Ross Island is Cape Bird, the
site of large Adélie penguin
colonies in summer. The high
points of Ross Island are
Mount Terror (center) and
Mount Erebus (right).

Nor is the spring melt a simple reversal of the
freezing process. Ice that once formed near the coast
can be found at huge distances from the shore by
December, and in late summer, there can be a thick
ring of pack ice far from land, while the coastal waters
are relatively ice-free and navigable.

Polar dissimilarities

Sea ice of the two polar regions differs in many ways.
Arctic sea ice may have a life span of up to eight years,
but most Antarctic sea ice is less than a year old—only
within the Weddell and Bellingshausen seas is sea ice
found that is up to three years old. About 90 percent of
Arctic sea ice is older than a year, more than 2 meters
(6 ½ ft) thick, very strong, and low in salinity. A similar
proportion of Antarctic sea ice is less than a year old, less
than 2 meters (6 ½ ft) thick, structurally weak, and quite
saline. Old sea ice is much stronger than first-year ice,
and is much more difficult for ships to break through.

In winter, the area of Antarctic pack ice is much
larger: 20 million square kilometers (7,700,000 sq. miles),
compared with 14 million square kilometers (5,400,000
sq. miles) in the Arctic. In summer, the sea ice in both
regions is much reduced: 4 million square kilometers

(1,500,000 sq. miles) in Antarctica, as against 7 million
square kilometers (2,700,000 sq. miles) in the Arctic.

Sea ice nomenclature

Sea ice is named by size. Fragments of ice less than
2 meters (6 ½ ft) across—smaller than a grand piano—
are called "brash ice." Ice floes are "small" if they are
less than 100 meters (330 ft) wide, "medium" if less
than 300 meters (990 ft) wide, and "large" if less than
2 kilometers (1 ¼ miles) wide. A "vast floe" measures
up to 10 kilometers (6 ¼ miles) across; anything larger
than that achieves "giant floe" status.

When wind and ocean currents push the ice apart,
the open water is called a lead, and the heat released
from the water below can create frost smoke. A really
large area of open water is known by the Russian
name: *polynya*. In shipping terms, "open pack" is water
containing up to six-tenths ice, and "close pack" is
seven- or eight-tenths ice, when some water is still
visible. An icebreaker or ice-strengthened vessel is
needed for movement in "very close pack"—nine-
tenths ice. "Compact" pack is all ice, no leads; no water
at all is visible—even most icebreakers would have difficulty passing through it.

Winter–summer sea ice

The continent always retains some sea ice in the sheltered areas around its coast, even in summer. In winter, when the sea freezes, the effective size of the Antarctic Continent increases dramatically, and the continent itself becomes locked away, largely unreachable by either human or animal life, for several months.

How sea ice behaves

In some ways, it is miraculous that sea ice forms at all. Only a few substances are less dense as solids than they are as liquids; water is one of them. The density of pure ice is 91.7 percent that of pure water—so it floats with 8.3 percent of its mass above the water. Further, because of its salt content, seawater freezes at about −1.8°C (29°F) and, unlike fresh water, increases in density as it approaches freezing point. So when it cools, it sinks. Only when the cold air above has cooled a top layer of water around 3 meters (10 ft) thick to freezing point can sea ice form.

The dynamics beneath sea ice are very different from those in a lake or river covered in ice. As fresh water cools, it increases in density until about 4°C (39° F), then it slightly decreases in density until 0°C (32°F). The density of salt water increases between 0°C and its sub-zero freezing point. So, in an ice-covered lake the coldest water is lighter than the warmer water, and the water forms bands: ice on top, colder water below it, and warmer water on the bottom. In the sea, the ice cools the water directly below it, some of which freezes to the bottom of the ice, while the rest falls (when its density increases as it approaches freezing point), to be replaced by warmer, lighter water. So the water is constantly circulating below the sea ice.

The extent of sea ice varies considerably from year to year, and climatologists are looking for patterns that may reveal evidence of global warming. The presence of so much ice certainly influences the climate of Antarctica. The sea ice "skin" restricts the normal heat exchange between the ocean and the atmosphere. In spring, the sea ice reflects back solar energy that would otherwise warm the ocean and start summer earlier. At the start of winter, new sea ice reflects the last sunlight away, and isolates the (relatively) warm ocean from the air, so winter comes much sooner.

Is sea ice salty?

Sea ice always contains some salt, but its salinity is generally about one-tenth that of seawater. As seawater freezes, the salt is forced out but gets caught up in tiny brine pockets within the ice. The faster the freezing, the more salt is captured. Over weeks and months, the trapped salt migrates downward through the ice in response to gravity. So, gradually, the sea ice becomes less saline at the top than at the bottom. In the Arctic, the Inuit often use surface ice for their drinking water; sea ice that is more than a year old has a salt content of less than 0.1 percent.

Freezing and melting of polar sea ice greatly affect surface-water salinities of the sea. As sea ice forms, the exclusion of salts from the growing crystals of ice concentrates the salts into the seawater, increasing the sea's salinity. Through spring and summer the surface seawater loses salinity as this sea ice melts.   DM

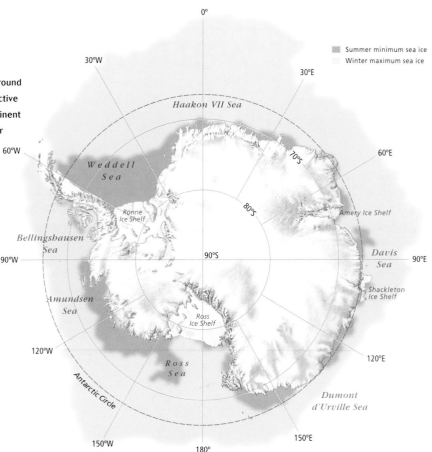

▲ SHATTERED ICE
Angular in shape, the ice forms collide with one another. These floes may freeze together and break up several times before forming a solid cover. As the ice thickens, ocean swells break it into larger pieces.

▼ BEWARE OF CRACKS!
A recurring crack in Erebus Bay shows that even the fast ice is subject to pressures such as tides, currents, and weather. Fast ice is a convenient traveling surface between winter and early summer—but care is necessary.

Global warming

▼ NATURAL MELT?

According to glacial geologists there have been four or more major advances and then retreats of continental ice sheets over the past two million years. The major question scientists face is how the man-made "greenhouse" effect modifies these natural cycles.

Today, the earth supports six times more people than it did 200 years ago. Between 1760 and 1820, the population of Britain jumped from 6 million to 14 million, its agrarian, handicraft economy becoming one dominated by industry and machines. The Industrial Revolution rapidly spread across the world, increasing the demand for energy. The burning of wood, coal, oil, and natural gas generated carbon dioxide.

Greenhouse gases

Carbon dioxide is one of the greenhouse gases that occur naturally in the atmosphere, absorbing heat released by the land and oceans in a process known as the greenhouse effect. Without the heat-trapping properties of these gases, the earth would be too cold for human habitation.

However, atmospheric carbon dioxide concentrations are now more than 30 per cent higher than they were 200 years ago. A typical car, for example, releases tonnes of carbon dioxide each year. Methane concentrations are 140 percent greater; cows burp as much as 100 kilograms (220 lb) of methane per year. New greenhouse gases, such as chlorofluorocarbons (CFCs), have been added to the air. This extra greenhouse effect is warming the lower atmosphere and changing earth's climate.

Likely climatic changes

Average surface temperature has already risen by between 0.3°C (0.5°F) and 0.6°C (1°F) during the past 100 years; the twentieth century was the warmest of the past millennium. In the northern hemisphere, the 1990s was the warmest decade, and 1998 was the warmest year. World rainfall patterns are changing, sea level is rising, and glaciers are retreating.

By the end of the twenty-first century, global temperatures are likely to be between 1.4°C (2.5°F) and 5.8°C (10°F) higher than now. This anticipated warming

▲ LESS ICE, MORE VEGETATION

It is anticipated that global temperatures will rise significantly by the end of the twenty-first century. Land will warm more than the sea, and the greatest warming is likely to occur near the Poles. Luxuriant vegetation, such as the above, is now found on only a few sub-Antarctic islands but may well become more widespread.

will be greater than any recent natural fluctuations, and will happen faster than any other known changes experienced since the last Ice Age.

Land will warm more than the sea, with the greatest warming expected to occur in regions close to the Poles. The temperature difference between night and day is likely to decrease. Arid and semi-arid parts of Africa, southern Europe, and the Middle East, and areas of Latin America and Australia will become drier as evaporation increases.

A warmer atmosphere will heat the oceans' upper layers, and the expanding water will raise sea level. Land-based ice in temperate regions such as South and North America and Greenland will melt more rapidly, adding to this rise. Conversely, increased precipitation over Antarctica and Greenland will lock water away in the ice sheets. Scientists estimate that by 2100, sea level will be 9–88 centimeters (3–35 in) higher.

Almost everywhere, these weather and climate differences will require people to adapt significantly. The changes will be even more serious for the natural environment, as climate change will exacerbate pressures on ecosystems from population growth and human exploitation.   PH

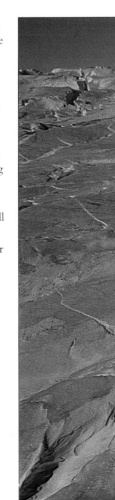

When the ice melts

The huge weight of the polar ice sheets depresses the earth's crust—remove the ice and the crust will rise again to its original level. The shoreline terraces of eastern Canada's Great Lakes formed by this process of "glacial rebound." They tilt gradually northward from near Chicago to a maximum of about 300 meters (990 ft) near Hudson Bay, and reflect the disappearance of the 10,000-year-old continental ice sheets of North America. This model indicates that the rock surface of Antarctica would rise by glacial rebound when the ice melts. The uplifted crust would displace seawater and the sea level would rise, but in small amounts and extremely slowly. Under the thickest ice of East Antarctica the crust is estimated to be depressed as much as 950 meters (3,100 ft). How much sea levels would rise from that amount of uplift is unknown.

Melting ice would return water formerly held on land to the world's oceans and, again, the sea levels would rise. Doubtless, this would begin slowly, as climates warmed sufficiently to begin melting the polar ice sheets. Warming and thermal expansion of the world's oceans might alone result in a rise of some 0.5–4 meters (1½–13 ft), but it might take a thousand years for atmospheric warming to reach oceanic bottom waters. Already, however, sea levels are rising almost imperceptibly. A quickening rate would have serious consequences for the world's coastal cities.

Antarctica contains 90 percent of the world's ice, enough to affect sea levels seriously. Estimates of amounts and rates vary. Calculations suggest that sea levels rose about 130 meters (425 ft) from the melting of 21,000-year-old, northern hemisphere ice of the last Ice Age. The melting of all today's Antarctic ice might raise the sea level by about 80 meters (260 ft).

The East Antarctic ice sheet lies on bedrock that is mostly at sea level or above, whereas much of West Antarctica's ice is based on bedrock well below sea level, and in places far below it. The decay of the ice in East Antarctica might result in a 60-meter (200-ft) sea level rise, but at a very slow rate, perhaps over 10,000 years, as the ice would remain mostly grounded. Glaciologists worry, however, that West Antarctica's ice could melt at a catastrophic speed due to flotation from its bedrock below sea level. A rise in sea level of approximately 15 meters (50 ft) might occur in a few tens to a hundred years. The melting of floating ice, such as the immense Ross Ice Shelf, would not affect sea levels, as the ice is already in flotational equilibrium.

When its ice does melt, Antarctica will be a much smaller continent than it is today. East Antarctica will be the main landmass, but joined to a peninsula of today's Ellsworth Mountains. Today's Antarctic Peninsula will be a separate island mass, and the Marie Byrd Land region will be a sea dotted with volcanic islands, separated from the "Antarctic Peninsula" by a deep marine trough, even after glacial rebound. Many mountain ranges now under ice, such as the Gamburtsev Subglacial Mountains, would rise above the plains of this continent. AF

▼ AN AWFUL LOT OF ICE
Ninety percent of the whole world's ice can be found in the Antarctic. Scientists estimate that, if all the ice in Antarctica were to melt, sea levels could rise by as much as 80 meters (260 ft).

The Antarctic ozone hole

Ozone, a form of oxygen, is a comparatively sparse constituent of the earth's atmosphere—there are only about three molecules of it in every ten million molecules of air. However, ozone is extremely important to the atmosphere and, for the past 600 million years or so, has been the atmospheric sentinel that has protected life on the planet from the biological damage caused by the sun's ultraviolet radiation. Many experimental studies of plants, animals, and humans have shown that there are harmful effects from excessive exposure to ultraviolet-B radiation. This is why the maintenance of the ozone layer is of vital concern, particularly in view of recent evidence that synthetic chemicals have caused serious loss of ozone from the atmosphere.

Damaging the atmosphere

During the early 1970s, atmospheric scientists engaged in theoretical research first became aware that possible damage to the ozone layer might be occurring as a result of human activities. However, it was 1985 before the British Antarctic Survey (BAS) discovered the Antarctic ozone hole and solid evidence was found that the ozone layer might be under threat. In 1987, the United States carried out a research campaign, where NASA aircraft, fitted with sophisticated scientific equipment, flew through the stratosphere from South America into the ozone hole. These experiments produced irrefutable evidence that the ozone hole was mainly caused by the chemical destruction of ozone by atmospheric chlorine. Because the quantity of chlorine required to destroy this amount of ozone could not originate from natural sources, another culprit had to be identified, and scientists then reached the inescapable conclusion that pollution from synthetic chemicals was causing ozone depletion. The prime offenders were the so-called halocarbon compounds, such as the chlorofluorocarbons (CFCs) that contain chlorine atoms.

CFCs were nontoxic and nonflammable and could easily be converted from a liquid to a gas and vice versa. They had many useful industrial and commercial applications—for example, as foam-blowing and cleaning agents, as the cooling fluid in refrigerators, or as aerosol-spray propellants. Because CFCs were designed to be relatively chemically nonreactive with most substances in the environment, this also meant that any of these substances would last a very long time in the atmosphere if they escaped or were released.

In the atmosphere, natural atmospheric circulation transports these chemicals high into the stratosphere,

▲ TELLING INSTRUMENT
A scientist measures how much ozone is overhead at the Arrival Heights Laboratory in Antarctica, using a Dobson spectrophotometer. The Antarctic ozone hole was first discovered from Dobson spectrophotometer data.

▶ NASA RECORDS LARGEST OZONE HOLE TO DATE
In early September 2000 the Antarctic ozone hole reached a record size of about 29 million square kilometers (11 million sq. miles)—more than three times larger than the land area of the United States. The ozone hole (dark blue) can be seen over the southern tip of Chile, and causes ultraviolet radiation increases.

Making ozone

The sun's ultraviolet radiation is more intense in the upper atmosphere, where it is not absorbed or reflected by atmospheric gases and particles to the extent that occurs lower down. This means that it can separate the atoms in oxygen molecules more often than it can lower down in the atmosphere, and the result is more free oxygen atoms. However, the joining of free oxygen atoms with oxygen molecules to produce ozone occurs more often lower down, where the atmosphere is denser, and these atoms and molecules collide more often. Because there are more free oxygen atoms higher up, yet more ozone-producing collisions between oxygen atoms and molecules lower down, the creation of ozone is at its greatest at the in-between altitudes—and it is this that forms an "ozone layer." The altitude range of the atmosphere in which the ozone layer resides is between about 10 and 50 kilometers (6 and 31 miles), and is known as the "stratosphere."

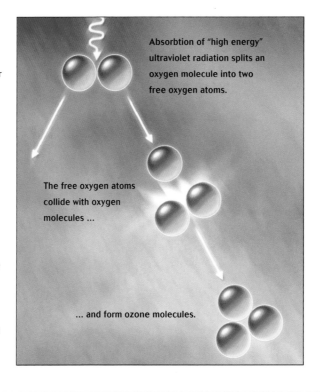

Absorbtion of "high energy" ultraviolet radiation splits an oxygen molecule into two free oxygen atoms.

The free oxygen atoms collide with oxygen molecules ...

... and form ozone molecules.

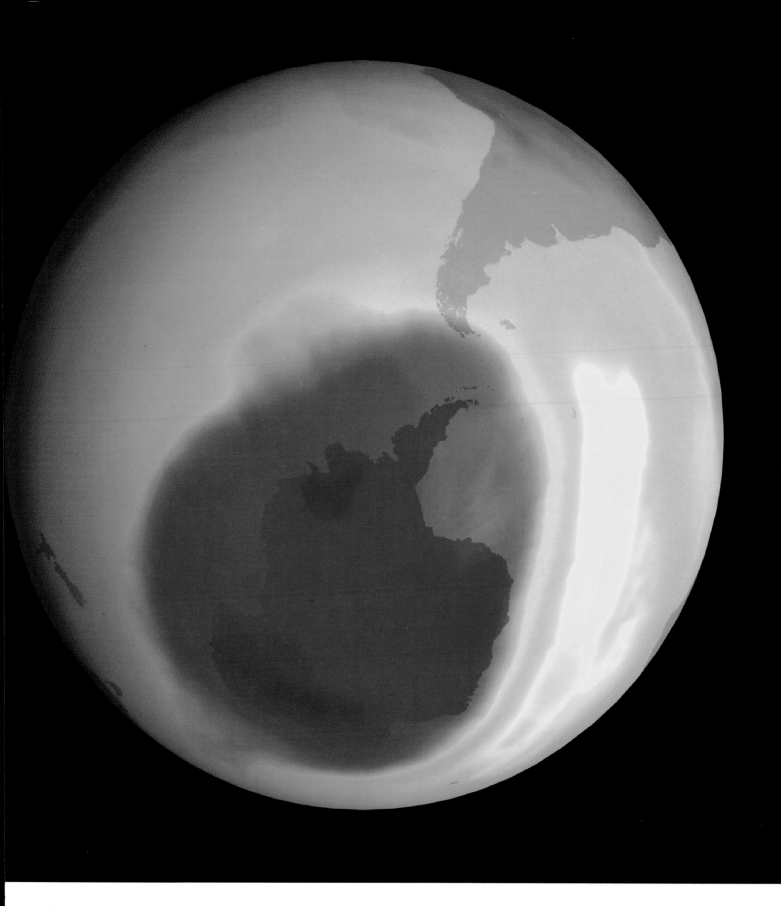

where the ultraviolet radiation is comparatively strong and can strip chlorine atoms off the CFC molecules. Thus, chlorine atoms from synthetic chemicals are liberated into the stratosphere, and are able to attack ozone molecules immediately, and convert them to oxygen by pulling off single oxygen atoms. Through further chemical reactions in the atmosphere, these chlorine atoms are effectively recycled to destroy more ozone molecules—one free chlorine atom in the stratosphere can destroy tens of thousands of ozone molecules before nitrogen compounds remove it from its destructive cycle. This catalytic process is the dominant, fundamental mechanism of ozone destruction. Through it, large numbers of ozone molecules are converted into oxygen, and during the conversion their ability to absorb ultraviolet radiation is lost. Ozone depletion is also caused, to a lesser extent, by pollution from chemical components other than chlorine, such as bromine. As a result, ultraviolet radiation at the earth's surface increases to a level that has the potential to be detrimental to life.

▲ UP, UP, AND AWAY
Researchers from the
University of Wyoming prepare
to launch a helium-filled
balloon that will carry a payload
of instruments to measure
ozone and properties of cloud
particles involved in ozone
destruction. The instruments
are lifted about 35 kilometers
(22 miles) into the air, at which
time the balloon's volume has
expanded about 200 times.

Patterns of distribution

Ozone depletion in the stratosphere
depends on latitude and season. For
example, at latitudes 30°S to 60°S,
total ozone reductions year-round
were about 5 percent between the
late 1970s and the mid-1990s.

The Antarctic ozone hole forms
in early spring over the Antarctic
Continent when the polar dawn
occurs in August and sunlight splits
chlorine molecules into the ozone-
destroying chlorine atoms that
drastically deplete ozone in the stratosphere. In the years leading to and
including 2000, the ozone hole has
suffered an ozone loss of up to
70 percent, and has sometimes
covered 29 million square kilometers
(11,300,000 sq. miles), more than three times the area of
the United States. The ozone hole slowly reaches its
greatest depth and area in late September to early
October, and then slowly recovers as fresh ozone from
lower latitudes gradually enters the ozone hole region.
On occasions, the ozone hole passes over Chile in South
America, where people are alerted through the media to
the concern of increased ultraviolet radiation. Typically,
the ozone hole finally breaks up after mid-November,
when the wind system that contains it (called the polar
vortex) weakens and breaks down. It is at this time that
ozone-depleted air can move to lower latitudes and reduce
ozone levels over Australia, New Zealand, and other
southern hemisphere countries. The effect of this ozone
decrease appears to persist into the following year.

In the northern hemisphere,
ozone losses over the Arctic during
spring have reached up to about
30 percent, and are considerably
less than the losses that occur over
Antarctica. Because the ozone-
destructive chemistry observed over
the polar regions is facilitated by
lower temperatures, the Arctic
stratosphere suffers less ozone loss
because it is generally warmer
(approximately 10°C) than the
Antarctic stratosphere, particularly
in March, when sufficient sunlight
is available to cause large ozone
depletion. It is, however, difficult to
predict what may lie ahead, because
the future climate and temperature
of the Arctic stratosphere cannot be
forecast with confidence.

International problem solving

The evidence that synthetic chemicals are responsible for the ozone
depletion observed since the mid-
1970s is now overwhelming, and an
international treaty was proposed to
progressively eliminate all known
ozone-depleting substances. To this
end, the Vienna Convention for the
Protection of the Ozone Layer was
established in 1985 as a general
treaty to facilitate international
cooperation. Twenty nations signed
an agreement to take "appropriate
measures … to protect human
health and the environment against
adverse effects resulting or likely to
result from human activities which
modify or are likely to modify the
Ozone Layer." The measures, however, were not specified. A protocol
to this treaty, the Montreal Protocol
on Substances that Deplete the
Ozone Layer, was established in
1987 to identify and implement
specific procedures to regulate the
use and production of ozone-depleting substances worldwide. Parties to
the Convention and Protocol now number 155, of
which more than 100 are developing countries. It is
the first time that nations have agreed, in principle, to
cooperate to solve a global environmental problem.

Oct 1981

Oct 1986

Oct 1991

Oct 1997

Dobson units

500
450
400
350
300
250
200
150
100

Decreasing ozone

Progression of the size and depth of the Antarctic ozone hole from
1981 to 1997 (measured by the TOMS instrument on the NASA
Nimbus 7 satellite). The ozone hole, as defined by ozone losses
greater than about 30 percent (outlined here in yellow), generally
reaches its greatest depth in October.

Evidence is now mounting that the so-called green-house effect may be reducing the recovery of global ozone by cooling the stratosphere where the ozone layer resides, and making the chemical reactions that cause ozone depletion more efficient. The greenhouse effect may also result in more water entering the stratosphere over the tropics, further exacerbating the chemical processes that destroy ozone.

These effects, coupled with the longevity of many of the ozone-depleting substances in the earth's atmosphere and the burden of future emission of chemicals (even though measures are now in hand to reduce such

emissions), mean that the length of time required for stratospheric ozone to recover to historical levels (pre-1980) will depend on many factors. Current computer model estimates place such a recovery around the middle of the twenty-first century, but many discoveries were made during the 1990s that significantly altered the experts' understanding and forecasting of ozone depletion, and more of these may be yet to come.

Whatever the future outcomes of research, it is vital that people cease to release CFCs and other ozone-destroying chemicals into the stratosphere so that the ozone layer has a chance to repair itself.   PL

▲ **EXTRA LOAD**
Daily measurements of Antarctic ozone up to 35 kilometers (22 miles) are made using a miniature chemical processing package called an ozone sonde. These can be suspended from conventional weather balloons, as shown here by members of the University of Wyoming research team.

Lights in the sky

Aurora, named for the Roman goddess of dawn, is a dramatic natural phenomenon that swirls or pulsates in the night sky, creating curtains and patches of colored light, predominantly in mixtures of green, red, and violet. An active display is an awe-inspiring sight.

More analytically, aurora is the light emitted by atoms, molecules, and ions that have been excited by charged particles, principally electrons, traveling along magnetic field lines into the earth's upper atmosphere. Light is emitted when excited atmospheric constituents release energy while returning to a lower energy state. This is similar to the process that generates light in a neon tube. For neon, the dominant wavelength emitted corresponds to a red color.

What causes aurora?

The solar corona—the outer region of the sun—is hot enough to continually emit solar wind plasma (a gas consisting of charged particles composed mainly of protons and electrons), which streams into space at speeds of around 300 kilometers (187 miles) per second. The earth's magnetic field also confines plasma, which travels along the field lines and is bounced between the hemispheres by the increasing field strength encountered in polar regions.

Electrons in the solar wind or trapped by the earth's magnetic field do not typically have sufficient energy to generate aurora. Processes associated with the interaction of solar wind and earth's magnetic field accelerate these

▼ **CELESTIAL CURTAINS**
Reflected in the newly frozen bay, sweeping arcs wave majestically as the expansive phase of an auroral substorm begins. The individual bands brighten and develop billowing folds, often twisting into spectacular spiral formations as waves of light surge along their entire length.

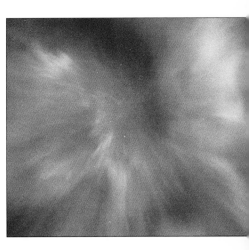

electrons. Solar plasma also carries a magnetic field. When the solar wind's magnetic field opposes the earth's magnetic field, energy is transferred by a process called magnetic reconnection. The extraction of energy from reconnecting magnetic fields is readily demonstrated by gradually bringing the opposite poles of two bar magnets together. Earth's magnetic field becomes linked to the solar wind by magnetic reconnection on the day-side and disconnected on the night-side, accelerating electrons on both occasions. Electrons may also be accelerated by an interaction that resembles friction, as solar wind plasma that has not been magnetically reconnected to the earth races past the flanks of its magnetic field. This manner of energy transfer may be likened to wind passing over water and generating waves.

Global occurrences

Aurora, in the form of an oval around the magnetic pole in both hemispheres, results from these interactions. Most commonly, aurora is located at magnetic latitudes of about 67° on the night-side and about 75° on the day-side. When considering the chances of observing aurora at a particular location, it is important to account for the separation of magnetic and geographic poles.

Aurora is known as aurora australis, or the southern lights, in the southern hemisphere, and as aurora borealis, or the northern lights, in the northern hemisphere. Auroral activity in the two hemispheres is strongly linked in space and time, as electrons can travel freely in either direction along magnetic field lines. Brightenings and movements in the two hemispheres coincide within seconds in magnetically linked regions. This was confirmed in the 1970s by measurements made on specially equipped aircraft following flight paths linked by the earth's magnetic field out of Christchurch, New Zealand, and Fairbanks, Alaska.

Sunspots (cooler regions of intense magnetic field) are signs of enhanced solar wind activity. They have a cycle of about 11 years, which most recently peaked in 2000. Aurora is more common at lower latitudes at peak sunspot times and during the subsequent two years; it slightly intensifies in spring and autumn.

Observing aurora

From the ground, it is possible to observe about 700 kilometers (435 miles) of the approximately 8,000-kilometer (4,970-mile) wide auroral oval. In quiet times this appears as a shimmering curtain extending in an east–west direction.

More active displays, known as auroral substorms, occur when stored energy is rapidly released. In an explosive onset, rays and arcs may twist, move, and break into small segments as enhanced electric currents flowing near auroral heights modify the field lines along which charged particles are traveling. Pulsating patches of auroral glow may persist during the gradual recovery phase, and eventually quiet arcs re-form. A substorm may last an hour, and may recur up to three or four times each night.

The aurora produces many colors, each corresponding to specific energy level transitions of excited atoms, molecules, or ions. Three, however, are most significant. A green line emitted by an oxygen atom is dominant when a sharp lower border can be discerned, usually at about 100 kilometers (60 miles). Red is generated by another atomic oxygen transition, and is prominent at altitudes above 250 kilometers (155 miles). A violet band emitted by a nitro-gen molecular ion is about five times less intense than the green line, but can be observed on the leading edge of active aurora because of the delay of about a second in the average emission time of auroral green light. GB

▲ CRIMSON CORONA
The expansive phase of an auroral substorm develops in minutes to the break-up stage. Individual folds in previ-ously quiescent auroral arcs can split into spirals that splay outward from a single point.

Above left

▶ VANISHING POINT
The coronal point is where overhead auroral rays seem to meet. Along the line of the earth's local magnetic field in this direction, individual rays delineate neighboring field lines that appear to converge like skyscraper spires viewed from immediately below.

▶ A GIFTED DOCTOR
Edward Wilson, a close confidant of Robert Falcon Scott, was a medical doctor, a naturalist, and an artist. During Scott's expedition of 1901–04, Wilson sketched the *Discovery* beneath a stunning aurora australis.

Part II

POLAR REGIONS

For humans the polar regions are the most inhospitable on earth, yet indigenous populations have flourished north of the Arctic Circle for thousands of years, and at the beginning of the twenty-first century the Antarctic region houses more than 40 permanent stations where people live and work all year. From the desolate, scattered islands south of the Antarctic Circle to the Antarctic Peninsula, to the Antarctic Continent, and to the South Pole itself, explorers have left their mark. Scientists have followed, in the hope of unlocking the secrets of Antarctica, and tourists can now visit this most challenging of environments.

The Antarctic Peninsula

The north

Previous pages

◄ **HISTORIC AMBIENCE**
A visit to Petermann Island
recalls Jean-Baptiste Charcot's
second French Antarctic
Expedition of 1908–10. Visitors
can view the remains of
Charcot's mooring points on
the rocks in the cove, visit his
cairn, or try to replicate the
expedition photograph looking down at Port Circumcision.

Hope Bay

This notch in the tip of the Antarctic Peninsula,
3 kilometers wide by 5 kilometers long (1¾ miles by
3 miles), was discovered in January 1902 by the Swedish
expedition led by Otto Nordenskjöld. He named it
Hope Bay, after three of his party inadvertently wintered
there during 1903; they were dropped off not long
before their ship, *Antarctic*, sank and the sound beyond
the bay bears the name of their doomed vessel. Their
makeshift stone hut (where hope was about all they
had) stands near the dock—it was largely rebuilt over
the summer season of 1966–67.

In December 1952, Argentina's Esperanza (Spanish
for "hope") station was founded on Hope Bay's rocky
shore, and has operated continuously ever since.
Antarctica's first Catholic chapel was consecrated there
in 1976. Two years later, a substantial expansion program
brought whole families to Esperanza, and a school was
built under the auspices of Tierra del Fuego's Education
Ministry. The population ranges from about 55 in
winter to 90 in summer, of which about a third are
children or spouses.

In a move to strengthen Argentina's territorial claim
to the Antarctic Peninsula, Silvia Morello de Palma, wife
of the base commander, came to Esperanza
when she was seven months pregnant. She
gave birth to the first "Antarctican," Emilio
Marcos de Palma, on 7 January 1978. Since
then, other babies have been born at Esperanza,
and the presence of families gives this base
the atmosphere of a tiny town. It has about
1.5 kilometers (1 mile) of gravel roads and
43 buildings—including two scientific laboratories and an infirmary staffed by a doctor
and a paramedic. LRA 36 (Radio Nacional
Arcangel San Gabriel), the base's radio station,
began transmission on 20 October 1979.

Esperanza is supplied by ship twice a year,
in December and January, and is also served
year-round by Twin Otter ski planes, which
land on a glacier airstrip 1.5 kilometers (1 mile)
from the base. Hope Bay often experiences
northeasterly winds that may exceed 200 kilometers per hour (125 mph), so landings are
not always possible.

Uruguay's Lieutenant Juan Ruperto
Elichiribehety station is also about
1.5 kilometers from Esperanza. This summer
base, named after the Uruguayan captain
who tried to rescue Shackleton's men from
Elephant Island, houses up to 12 people.
The building was originally Trinity House,
built in 1952 by the British to replace
an Operation Tabarin hut that burned
down in 1949. The British operation
ceased in 1964, and the base was
transferred to Uruguay in 1997.

66°S

◄ **OLD AND UNSTABLE**
Navigating around an aging iceberg
like this one can be dangerous. Over
time, water action creates deep
indentations, and new ledges form
every time a section breaks
off, upsetting the balance of
the iceberg and often resulting
in unpredictable adjustments
in orientation.

72°W

68°S

72°W

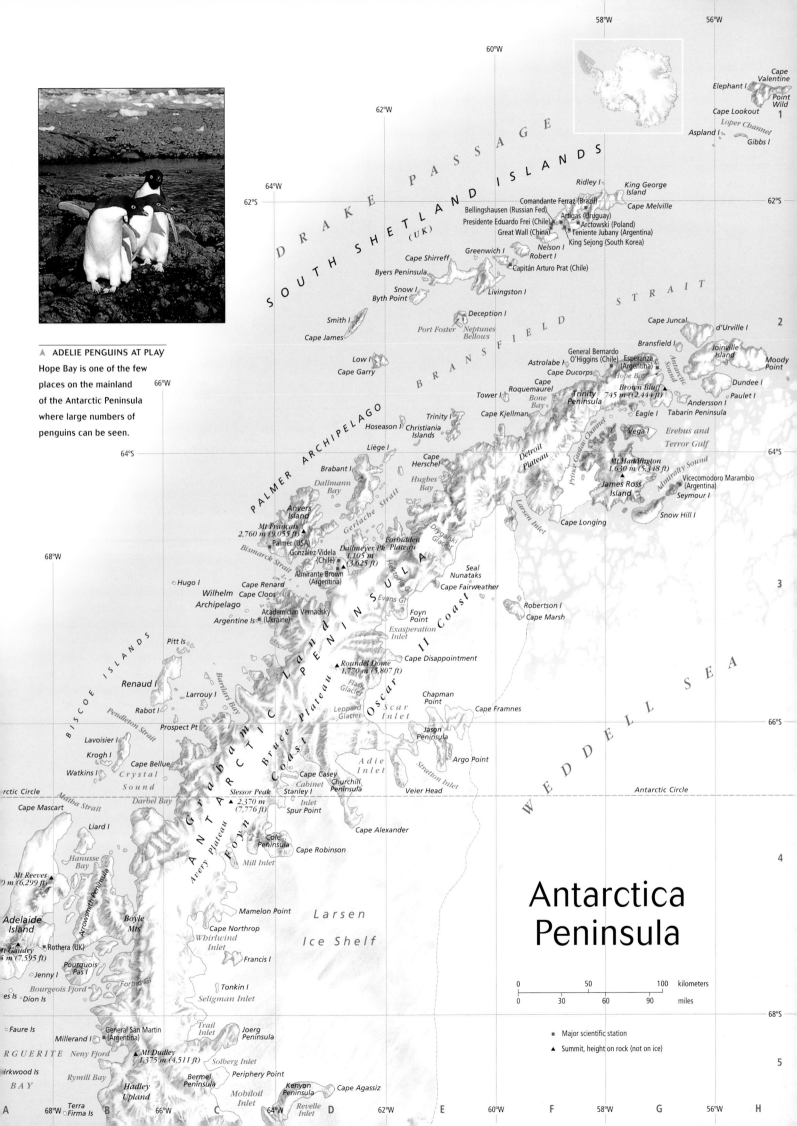

ADELIE PENGUINS AT PLAY
Hope Bay is one of the few places on the mainland of the Antarctic Peninsula where large numbers of penguins can be seen.

Antarctica Peninsula

DRAKE PASSAGE

SOUTH SHETLAND ISLANDS (UK)

BRANSFIELD STRAIT

PALMER ARCHIPELAGO

WEDDELL SEA

Cape Valentine
Elephant I
Point Wild
Cape Lookout
Loper Channel
Aspland I
Gibbs I

Ridley I
King George Island
Cape Melville
Comandante Ferraz (Brazil)
Bellingshausen (Russian Fed)
Artigas (Uruguay)
Arctowski (Poland)
Presidente Eduardo Frei (Chile)
Great Wall (China)
Teniente Jubany (Argentina)
King Sejong (South Korea)
Nelson I
Robert I

Cape Shirreff
Greenwich I
Byers Peninsula
Capitán Arturo Prat (Chile)
Snow I
Livingston I
Byth Point
Deception I

Smith I
Port Foster
Neptunes Bellows
Cape James

Low I
Cape Garry

Cape Juncal
d'Urville I
Bransfield I
Joinville Island
Moody Point
Dundee I
Paulet I
Andersson I
Tabarin Peninsula

General Bernardo O'Higgins (Chile)
Astrolabe I
Esperanza (Argentina)
Cape Ducorps
Hope Bay
Cape Roquemaurel
Brown Bluff 745 m (2,444 ft)
Tower I
Bone Bay
Trinity Peninsula
Cape Kjellman
Eagle I
Vega I
Erebus and Terror Gulf

Trinity I
Hoseason I
Christiania Islands
Liège I

Cape Herschel
Hughes Bay
Detroit Plateau
Prince Gustav Channel
Mt Haddington 1,630 m (5,348 ft)
James Ross Island
Admiralty Sound
Antarctic Sound

Brabant I
Dallmann Bay
Anvers Island
Mt Francais 2,760 m (9,055 ft)
Palmer (USA)
Gonzáles Videla (Chile)
Almirante Brown (Argentina)
Forbidden Plateau
Drygalski Glacier
Larsen Inlet
Cape Longing

Vicecomodoro Marambio (Argentina)
Seymour I
Snow Hill I

Bismarck Strait
Dallmeyer Pk 1,105 m (3,625 ft)
Gerlache Strait

Hugo I
Wilhelm Archipelago
Cape Renard
Cape Cloos
Academician Vernadsky (Ukraine)
Argentine Is

Seal Nunataks
Cape Fairweather
Evans Gl
Foyn Point
Robertson I
Cape Marsh

Pitt Is
Exasperation Inlet
Cape Disappointment

Renaud I
Larrouy I
Rabot I
Roundel Dome 1,770 m (5,807 ft)
Flask Glacier
Chapman Point
Cape Framnes

Prospect Pt
Leppard Glacier
Scar Inlet
Jason Peninsula
Argo Point

Lavoisier I
Krogh I
Watkins I
Cape Bellue
Crystal Sound
Pendleton Strait

Adie Inlet
Cape Casey
Cabinet Inlet
Churchill Peninsula
Stanley I
Veier Head

Arctic Circle
Maiba Strait
Darbel Bay
Slessor Peak
2,370 m (7,776 ft)
Spur Point
Cape Alexander
Antarctic Circle

Cape Mascart
Liard I
Cole Peninsula
Cape Robinson
Mill Inlet

Mt Reeves m (6,299 ft)
Hanusse Bay
Mamelon Point
Larsen Ice Shelf

Adelaide Island
Boyle Mts
Cape Northrop
Whirlwind Inlet
Francis I

Mt Gaudry m (7,595 ft)
Rothera (UK)
Pourquois Pas I
Jenny I
Forbes Gl
Tonkin I
Seligman Inlet

Bourgeois Fjord
Dion Is
Faure Is
Milleran I
General San Martin (Argentina)
Trail Inlet
Joerg Peninsula

MARGUERITE BAY
Neny Fjord
Mt Dudley 1,375 m (4,511 ft)
Solberg Inlet
Kirkwood Is
Rymill Bay
Hadley Upland
Bermel Peninsula
Periphery Point
Kenyon Peninsula
Cape Agassiz
Terra Firma Is
Mobiloil Inlet
Revelle Inlet

ANTARCTIC PENINSULA
GRAHAM LAND
PALMER LAND
Detroit Plateau
Oscar II Coast
Bruce Plateau
Foyn Coast
Avery Plateau
Arrowsmith Peninsula

BISCOE ISLANDS

0   50   100 kilometers
0 30 60 90 miles

■ Major scientific station
▲ Summit, height on rock (not on ice)

▲ CHILLING OUT
Out of the water, Weddell seals, like this dozing pup, appear to do little more than rest and sleep. They have been observed on the beach below Brown Bluff, and are common in most locations along the northeastern end of the Antarctic Peninsula.

Brown Bluff

Situated in the Antarctic Sound, some 15 kilometers (9 miles) southeast of Hope Bay, Brown Bluff is where the eastern edge of the Tabarin Peninsula drops almost sheer to the water from an ice-capped summit. From a rocky beach, a steep scree slope rises to a towering, rust-colored cliff of volcanic rock. This could be central Australia or the United States Badlands, were it not for the penguins and Weddell seals along the shore and the ice cap high above.

Adélie penguins number tens of thousands on Brown Bluff, and several hundred Gentoo penguins also live there. The slope is covered with loose rubble and rock slips are common. Scientists believe that landslides and falling boulders may have obliterated some penguin rookeries.

Paulet Island

James Clark Ross, aboard *Erebus* in search of the South Magnetic Pole, named Paulet Island after Lord George Paulet of the British Royal Navy. A circular volcanic cone, the island rises from the sea 3 kilometers wide and 353 meters high (1¾ miles and 1,158 ft).

This is as far as many Antarctic travelers will be able to venture into the Weddell Sea, because the clockwise current ensures that even this northern region has its share of spectacularly large icebergs. Huge tabular bergs encountered on the way through the Antarctic Sound give some idea of the difficulties faced by the early explorers, most notably the expeditions of Nordenskjöld and Shackleton.

One of Antarctica's largest colonies of Adélie penguins occupies Paulet Island—a fact only too apparent from the smell downwind, up to 1 kilometer (⅔ mile) offshore. At least 100,000 breeding pairs crowd the beach and nearby steep slopes for the short summer—a fine example of avian tenement living. Snowy sheathbills, skuas, and Giant petrels live off the rookery, and opportunistic Leopard seals patrol the shores in the certainty of easy prey.

When *Antarctic* sank, 40 kilometers (25 miles) away in 1903, Captain Carl Larsen and his crew came to Paulet Island. The small hut, 20 by 7 meters (65 by 23 ft), that the 20 marooned men constructed from the regularly shaped volcanic rocks became their winter quarters. The grave of Ole Wennersgaard, who died on 7 June 1903, is still clearly visible along the shore about 300 meters (984 ft) from this ruined building. In summer, the nesting penguins make it difficult to approach the hut or walk past it to an attractive lake.

Seymour and Snow Hill islands

Visiting Seymour Island and Snow Hill Island is possible only when benign sea ice conditions in the Weddell Sea allow access to the southern end of Erebus and Terror Gulf. When James Clark Ross first saw the northeastern part of Seymour Island in January 1843, he named it Cape Seymour after a British rear admiral. Later, when Larsen defined the whole island, 16 kilometers long by 8 kilometers wide (10 by 5 miles), Ross's nomenclature was adopted for it.

Brown Bluff is one of the most notable features of the Tabarin Peninsula. The peninsula takes its name from Britain's wartime Operation Tabarin (itself called after a Paris nightclub), but the sheer bluff is named for its characteristic reddish-brown volcanic rock. 745 meters (2,440 ft) high, the bluff is the remnant of an ancient volcano subsequently shaped by glaciation.

Astrolabe Island

French explorer Dumont d'Urville sighted this island in the Bransfield Strait during his 1837–40 Antarctic voyage, and named it after his expedition flagship, *Astrolabe*. The rocky island is 5 kilometers (3 miles) long, with cliffs plunging more than 100 meters (330 ft) directly to the sea. Chinstrap penguins, fur seals, and Weddell seals breed on the low ground, and skuas and fulmars nest along the craggy heights. Shore landings are possible here, and the surrounding ice can provide a spectacular Zodiac cruise.

Mikkelsen Harbor, Trinity Island

Mikkelsen Harbor, on the south side of Trinity Island, was named by the 1901–04 Swedish South Polar Expedition led by Otto Nordenskjöld, but the source of the name has been lost. The term harbor is deceptive, as the landing site is in fact a small island in a bay that has no harbor at all. Rocky shoals lead to a rocky beach, with glaciers forming a magnificent backdrop on a fine day.

There are a few unoccupied huts on the island, and a radio mast at its low summit. The harbor's chief residents are Gentoo penguins on its slopes, and slumbering Weddell and elephant seals on the beach.

The cross marking the grave of Ole Wennersgaard, who died during the winter of 1903 while shipwrecked on Paulet Island, rises above a milling throng of Adélie penguins. At least 100,000 pairs nest on the island in summer.

The rocky outcrops of Seymour Island contain fossil remains of penguins—giant birds compared with the size of today's species. Nordenskjöld found the first of these in December 1902, and paleontologists continue to make exciting discoveries here. Argentina operates Vicecomodoro Marambio station on the northern side of the island, which includes a small airstrip accessible by C-130 Hercules aircraft from Rio Gallegos. Some 40,000 Adélie penguins nest on Penguin Point at the southeastern corner of the island.

While Seymour Island is largely ice-free and muddy, Snow Hill Island is almost completely covered in snow and supports no wildlife. This island, lying southwest across Admiralty Sound, is 32 kilometers long by 10 kilometers wide (20 by 6 miles); James Clark Ross, uncertain whether it was part of the mainland, named it Snow Hill. In 1902 Otto Nordenskjöld based his Swedish Antarctic Expedition there. Now, the Argentinians care for their hut, and a few artefacts from that expedition, including a primus stove and coffee grinder, are kept at the Antarctic Office on the dock in Ushuaia. Aviator Lincoln Ellsworth used Snow Hill Island in 1935 as the starting point for one of his attempts to fly across the Antarctic Continent.

In this small hut on Paulet Island, 20 shipwrecked sailors (and a cat) lived for nine months over a long polar winter. Fortunately for the stranded men, building the hut was quite easy: regular freezing and thawing of the island's basalt caused the rock to break into even-sized flat stones—ready-made building blocks.

The Gerlache Strait

Curtiss Bay

On Argentinian charts, Curtiss Bay is rather harshly called Bahia Inutil ("Useless Bay"). Its British name was coined in 1960 for Glenn Curtiss (1878–1930), the first American builder of seaplanes, hence Seaplane Point within the bay. In the small glacier-lined cove behind Seaplane Point, ice cliffs 3 kilometers (1¾ miles) long are very active, and large blocks of ice frequently break off and crash into the smooth, reflective water. The bay is little visited, but to the north of the cove there is a small rocky promontory covered in snow where landings are possible. Curtiss Bay is often crowded with large icebergs; whales visit regularly, and Leopard and other seals rest on the floes.

Cierva Cove

Crammed with spectacular icebergs, Cierva Cove lies at the top of Hughes Bay on the southern side of Cape Herschel. The cove is on the opposite side of 91-meter (300-ft) Mount Penaud from Curtiss Bay; its name is similarly associated with aviation—Spaniard Juan de la Cierva invented the autogiro, the precursor of the helicopter, and first flew it in 1923. Cierva Cove is the site of Argentina's summer station Primavera.

Hydrurga Rocks

In 1960 the British named these low-lying, bare rocks, very near the much larger Two Hummock Island of the Palmer Archipelago, after the Leopard seal: *Hydrurga leptonyx*. Leopard seals are present offshore; they prey on the Chinstrap penguins that establish small rookeries wherever bare rock is accessible, as it is here. Antarctic shags also nest on Hydrurga Rocks.

Alcock Island

This island, less than 1 kilometer (½ mile) across, in the middle of Hughes Bay commemorates another aviator: Sir John Alcock was one of the team that completed the first nonstop transatlantic flight from Newfoundland to Ireland in 1919. Whalers called it Penguin Island, but the name was changed to avoid confusion with an island of the same name in the South Shetland Islands.

Alcock Island's coastline is mostly towering cliffs that Chinstrap penguins seem able to scale with ease. When access is not blocked by icebergs, a narrow cleft leads to a challenging landing beach. A scramble across ice and rocks will take an adventurous climber to the 97-meter (318-ft) cliff top and a panoramic view of the glacier-lined Davis Coast. When landing is not possible, cruising

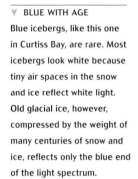

▼ BLUE WITH AGE
Blue icebergs, like this one in Curtiss Bay, are rare. Most icebergs look white because tiny air spaces in the snow and ice reflect white light. Old glacial ice, however, compressed by the weight of many centuries of snow and ice, reflects only the blue end of the light spectrum.

around the icicle-clad cliffs and through the icebergs that invariably gather around the island has its own rewards.

Portal Point

On the western coast of Graham Land on the Reclus Peninsula, Portal Point, the entrance to Charlotte Bay, received its name from the British after they built a tiny refuge hut there in 1956 at the start of a route they established up to the plateau. A survey group wintered there in 1957, but now only the foundations of their hut remain; the building itself was taken to Stanley in the Falkland Islands in 1997, and was reconstructed in the Falkland Islands Museum the following year.

Mountains, glaciers, and icebergs make Charlotte Bay one of the most attractive inlets along the Antarctic Peninsula. The bay indents for about 20 kilometers (12 miles) into the coast, and was named for the fiancée of George Lecointe, second-in-command and hydrographer of the 1898 Belgian Antarctic Expedition. Because there is little resident wildlife in the area, Portal Point is sometimes used by expeditioners as a camp site.

Wilhelmina Bay

South of the Reclus Peninsula, and bounded on the south by the Arctowski Peninsula, Wilhelmina Bay is frequented by whales and littered with islands— Brooklyn, Pelseneer, Emma, Nansen, and Enterprise— that provide some shelter in almost any weather. The only exception is when katabatic winds howl down from the Forbidden Plateau, more than 2,100 meters (7,000 ft) above the bay, and whip the water into foam. Belgian Baron de Gerlache de Gomery named the bay in 1898 for 18-year-old Queen Wilhelmina of the Netherlands, whose reign lasted until 1948.

▲ **EFFICIENT HUNTER**
The Leopard seal's gaping jaws give it a threatening demeanor, but although this seal preys on larger animals, its teeth are also designed for sieving krill, which makes up about half of its diet.

▼ **HARSH NATURAL SYSTEM**
Penguin colonies afford easy pickings for Leopard seals and Orcas. Birds that escape with injuries may recover—or, like this doomed Chinstrap, they may become food for predators and scavengers such as giant petrels and skuas.

Spigot Peak and Orne Harbor

On the northwestern side of the Arctowski Peninsula, the towering black spire of Spigot Peak marks the southern entrance of Orne Harbor. The 286-meter (938-ft) peak takes its name from its resemblance to a spigot or cask plug. Norwegian whalers are believed to have named Orne Harbor and Orne Islands, between Gerlache Strait and Cuverville Island, before 1913. Orne Harbor, discovered by de Gerlache in 1898, is a mere indentation on the west coast of Graham Land. There are two parts to the harbor, and although there are no landing sites except near Spigot Peak, and not much wildlife, it offers some excellent views of glaciers from cruising Zodiacs. Antarctic shags nesting here can be observed quite closely from Zodiac craft; a small landing area around the point gives access to an energetic climb to the saddle for views over Errera Channel and beyond.

◄ BENEATH THE ERRERA ICE
In temperate waters, diving time is governed by air and depth, but in the near-freezing waters of Antarctica, the condition of a diver's hands usually limits diving time to no more than 20 minutes—after that, pain and numbness set in.

Errera Channel

Narrow Errera Channel lies between Rongé Island and the nearby coast of the Arctowski Peninsula—two largely matching shorelines. The channel, roughly L-shaped and about 9 kilometers (5½ miles) long, contains Cuverville and Danco islands, and is one of the Antarctic Peninsula's most frequented sites by visitors longing to see pods of Humpback whales. Errera Channel is an awesome environment, with ice in the water and mountains towering more than 900 meters (3,000 ft) on either side of the channel, which is only 1,000 meters (3,280 ft) wide at its narrowest part in the south. The channel takes its name from Professor Léo Errera, a member of the *Belgica* Commission.

65°S

Rongé Island

Rongé Island was named for Madame de Rongé, a sponsor of the Belgian Antarctic Expedition of 1897–99. George's Point lies at its northern tip, and is home to a small Gentoo penguin population and several little Chinstrap penguin rookeries.

The dark bulk of Spigot Peak marks the southern
end of Orne Harbor. During the brief summer months,
ts cliffs and crags are home to several bird species,
including a colony of shags that nest close to the
waterline. Crabeater seals are common around the
point, and toward the end of the season fur seals
rest along the shore.

Gerlache Strait

62°W
61°W

63°W
64°W
64°W

Low
Island
Cape Garry

1

kilometers
miles
0   25
0   10

■ Major scientific station
▲ Summit, height on rock (not on ice)

Cape Wollaston
Tower Hill
1,125 m (3,691 ft)
Milburn Bay

Hoseason
Island
Stopford Peak
495 m (1,624 ft)
Angot Point

Spert I
Trinity
Island

Christiania
Islands   Intercurrence I

Mikkelsen Harbor

Cape Andreas   64°S
Curtiss Bay
Seaplane Pt

Liège Island   Neyt Point

Orléans Strait

Cape Roux   Cape Cockburn
Pavlov Peak
852 m (2,795 ft)   Small I

Cape Herschel

Davis Coast

Pasteur
Peninsula
Bouquet
Bay
Macleod Point

Two Hummock
Island
Hydrurga Rocks

Cierva Cove

Guyou Bay
Lister Gl
Davis I
Spallanzani
Point

Claude Point

Bart Bay
Laennec Gl   Hill Bay

HUGHES
BAY   Alcock I

Brialmont
Cove

Lanusse
Bay   Brabant
Island   Mitchell Point

Spring Point

Tournachon Peak
858 m (2,818 ft)

P A L M E R   A R C H I P E L A G O

Melchior
Islands

Mackenzie Gl
Lecointe
Island
Valdivia
Point

Freud Passage
Eta I
Omega I
Hippocrate
Glacier
Hunt I

Bluff
Island
Salvesen
Cove

Eckener Point

Mt Zeppelin
1,265 m (4,150 ft)

Danco Coast

D a l l m a n n
Gand I
Pinel Point

B a y
Solvay
Mountains   Buls Bay

Recess Cove
Wellman Glacier

Portal Pt
Nobile Glacier

Perrier Bay

Hulot
Peninsula
Mt Bulcke
1,032 m (3,386 ft)

Gerlache Strait

Bancroft
Bay
Charlotte Bay

Reclus Peninsula

Hamburg Bay

Strath Point
Enterprise I

Gourdon Pen

Delaite I   Nansen
Island

Brooklyn I

Danco Coast

Thompson Pen
Cape Anna

Emma I

2

Fournier Bay

Orne Harbor
Spigot Peak
286 m (938 ft)   Wilhelmina
Bay

Pelseneer I

Leopardo
Glacier
Plateau

Renard Glacier

Anvers Island

Mt Francais
2,760 m (9,055 ft)
Lion I
Cuverville I   Arctowski
Peninsula

William Glacier

Mt Tennant
688 m (2,257 ft)
Rongé
Island   Errera
Channel

Ketley Point   Danco
Piccard
Cove

Forbidden

Leonardo Glacier

Copper Peak
1,125 m (3,690 ft)   Beneden Head

Mt Moberly
1,533 m (5,029 ft)
Börgen
Bay
Nemo Peak
866 m (2,841 ft)

Lausedat
Heights

Mt Walker
2,359 m (7,740 ft)

Hooper Glacier

Waterboat Pt
Andvord Bay

Arago Glacier

Danco Coast

Wylie Bay

Arthur
Harbor

Lemaire I

Deville Glacier

Palmer (USA)
Mt William
1,600 m (5,249 ft)
Neumayer Channel   González Videla
(Chile)
Neko
Harbor

Moser Glacier

GRAHAM

Biscoe Bay
Port
Lockroy
Bryde
Island

Access Point

Almirante Brown
(Argentina)
Dallmeyer Peak
1,105 m (3,625 ft)

Cape Lancaster   Doumer I
Wiencke
Island
Fridtjof I

Rudolph Glacier

Py Point

Bismarck
Strait   Truant I
Cape
Willems

Kershaw
Peaks

Grubb Glacier

Bigourdan Glacier

Knight I   Cape Errera   Bob I

Nordenskjöld Coast

ANTARCTIC

Wauwermans
Islands   Butler
Passage

Nimrod Passage
Puzzle Is   Flandres
Bay

annebrog
Islands   Cape Renard
False Cape Renard   Cape
Fairweather   65°S

Booth I   Guyou Is

Danco Coast

Evans Glacier

Pléneau I
el Is   Cape Cloos

Hovgaard I
termann I   Mt Scott
882 m (2,894 ft)
Danco Coast

LARSEN

ICE SHELF   3

Foyn Point

cian
sky   Mt Shackleton
1,465 m (4,806 ft)

ne)   Yalour
Is

Oscar II Coast

Exasperation
Inlet

Cape
Tuxen
Wiggins Glacier

erthelot Is   Bussey Glacier

Cape Disappointment

Collins Bay   Frooz Glacier

A   64°W   B   63°W   C   62°W   D   61°W   E

◄ FOR THE BIRDS
The humpbacked shape of
Cuverville Island marks the
northern end of the Errera
Channel. Gentoo penguins,
shags, and sheathbills nest
on the available low land,
while petrels, gulls, terns,
and skuas claim spots higher
up on the island's steep,
lichen-covered slopes.

Cuverville Island

De Gerlache named Cuverville Island after a vice admiral in the French navy who helped to provision his expedition. It stands like a stopper in the northern end of the Errera Channel. Five thousand pairs of Gentoo penguins breed on the long, shingly beach, and in coves at the northern end of the island. Cuverville has a single landing site; it is not easy to disembark anywhere else because the terrain rises so steeply to the 252-meter (827-ft) rocky dome that makes up most of the island.

The most visited penguin rookery, at the eastern end of the beach, overlooks a picturesque but tiny rocky cove. The western rookery, protected until recently, is perhaps even more spectacular, as the mountains of Rongé Island loom close across the narrow channel. Up the slope, behind the scattered rookeries, territorial skuas

guard their nests. Antarctic shags also raise young on the island's cliffs, and Pintados and Snow petrels nest at the highest points. Crabeater and Leopard seals often bask on ice trapped on shallow rocks around the island.

In late summer, when most of the snow has melted, patches of vegetation are visible. Mosses and lichens are joined by Antarctica's two flowering plants: Antarctic hair grass (*Deschampsia antarctica*), which looks like tufts of familiar grass, and the pearlwort (*Colobanthus quitensis*), which resembles a moss but has tiny, white-green flowers. Like all Antarctic vegetation, these plants are extremely slow-growing, and may take years to recover from the crushing damage of a single footstep.

Between 1992 and 1995, researchers from the Scott Polar Research Institute camped on Cuverville Island and at other locations in the region to monitor the

effects of tourists on penguins and other wildlife. They found that visitors observing the code of conduct set out by the International Association of Antarctic Tour Operators (IATO) had little effect on penguin rookeries, whereas other birds—particularly Antarctic terns, giant petrels, and Kelp gulls—were often disturbed by the presence of humans.

Danco Island

This small island, about 1.6 kilometers (1 mile) long and 180 meters (590 ft) high, lies in the Errera Channel, south of Cuverville Island. First charted by de Gerlache's *Belgica* expedition in 1898, it was later named by Britain for Lieutenant Emile Danco, the Belgian geologist who died on board *Belgica* while she was trapped in the ice of the Bellingshausen Sea.

Danco Island is sometimes used as an alternative landing point to Cuverville Island. During the breeding season, nesting Gentoo penguins surround the hut at the northern end of the island. Built in 1956 by a British Antarctic Survey team, and occupied until 1959, it is still occasionally used by researchers from various nations, but is closed to tourists.

Andvord Bay and Neko Harbor

Heading south from Errera Channel, the next major indentation in the Danco Coast stretch of the Antarctic Peninsula is Andvord Bay. This quite deep expanse of water, surrounded by mountains and glaciers spilling down to the shoreline, is about 5 kilometers (3 miles) wide and some 14 kilometers (9 miles) long, and was named by de Gerlache to honor Rolf Andvord, the Belgian consul in Norway. Andvord Bay cuts so deeply into the Peninsula that the ice shelf at Exasperation Inlet, off the Weddell Sea, lies only about 40 kilometers (25 miles) away, across the mountains that soar to 1,600 meters (5,300 ft) at the head of the bay.

The northern shore of the bay provides some of the most awe-inspiring and photogenic scenery in the whole Peninsula. Distances are deceptive, and it is hard to grasp the sheer scale of this astonishing landscape. The roar of an avalanche or the crash of a calving iceberg may be audible well after the event—proof of the distance the sound has traveled, although the scene may appear closer.

Tucked within the eastern side of Andvord Bay is the sheltered cove of Neko Harbor, a place of great beauty, which nevertheless carries a name associated with slaughter: *Neko* was a whaling factory ship that operated in this region between 1911 and 1924, and frequently moored in this protected

bay. During those years the pristine landscape of today must have been a scene of death and decay, besmirched by stinking smoke and lapped by bloodstained water.

An unoccupied Argentinian hut, painted red, stands by the water near the end of a small point; every year, Snowy sheathbills nest in the hut's foundations and stand guard on its roof. The landing site is near the hut. Behind, a massive glacier, deeply cracked and tilted, looks as though at any moment it might drop ice the size of a suburban house into the bay. Weddell seals frequent the pebbly beach below the hut, and a Gentoo rookery on the steep slope behind the point may total 1,000 birds at peak season. The paths that the birds have carved through the snow from the rookeries on the rocky outcrops above look like a complex freeway system or a demonic bobsled track. Grubby penguins heading down to the beach to fish meet clean, well fed, plump birds just returned from the sea. The shallows of Neko Harbor are popular with Gentoos for bathing and preening, and whales—mainly Minkes and Humpbacks—sometimes enter Andvord Bay.

The glacier that fills the bay to the north of the landing site is very active, and huge waves can wash up from large iceberg calvings. Just beyond the glacier is a black cliff where Snow petrels make their nests, and across from Andvord Bay a panorama of peaks and glaciers forms a perfect backdrop for the antics of the penguins on the beach.

▼ FLYING SOUTH
Snowy sheathbills are the only birds without webbed feet to breed in Antarctica—to get there, they must fly nonstop across Drake Passage. They nest on the edge of penguin colonies, where the rewards are plenty for scavengers—dropped food, unfortunate penguin chicks, even guano.

◄ QUIET RECESSES
Andvord Bay is partially protected from weather coming from the west by Anvers Island. The long, narrow indentation of the bay broadens at its end into a foot-like shape, rather like the shape of Italy. Fantastic icebergs often rest in the glassy waters of the eastern corner.

▶ PARADISE SUNSET
Paradise Harbor has a well
deserved reputation as one
of the most beautiful places
on the Peninsula. Several
mainland glaciers flow into
its sheltered waters. To the
north, at the harbor entrance,
Lemaire Island protects
Waterboat Point from strong
weather, and to the south
Bryde Island shields Almirante
Brown station.

▲ GRISLY REMINDERS
A Gentoo penguin perches
on a whale vertebra, one of
thousands left strewn along
the beaches of the Peninsula—
and a reminder of days past,
when quiet bays and inlets
were used as gathering points
for whaling operations.

Geologists contend that the permanent ice caps that link the three islands of the Antarctic Peninsula are not just bridges, but integral landforms that create a single landmass extending from the Antarctic Continent. So, apart from its scenic appeal on a sunny day, Neko Harbor has special significance. To step ashore there is to set foot on the continent of Antarctica, something that very few people ever do.

Waterboat Point

Coming from the north, the entry point into Paradise Harbor is Waterboat Point, now the site of the Chilean base of González Videla, operated by the Chilean Air Force over the summer. Whether it is a point at all is a matter of contention—some call it an island because the spit joining it to the mainland is covered at high tide. Whichever, it is the end of the Peninsula between Andvord Bay and Paradise Harbor, and was first surveyed by de Gerlache in 1898. The British Imperial Expedition of 1921–22 named Waterboat Point for the abandoned waterboat that housed two of them when the expedition collapsed. The group's leader, John Cope, who had been a surgeon on Shackleton's *Endurance,* initially planned to bring several aircraft down and fly over the South Pole. This brought Hubert Wilkins into the venture; he was later to be the first person to fly a powered aircraft in Antarctica. The other two expedition members were young Englishmen: Thomas Bagshawe, a geologist, and Maxim Lester, a surveyor. Although they were unable to raise the money to buy any aircraft at all, the four hitched a lift with a whaling vessel to Waterboat Point, only to find that they could not cross the mountains to the Weddell Sea as they had intended.

In March 1921, Cope and Wilkins gave up and left Antarctica, but Bagshawe and Lester stayed on to do some of the scientific work they had planned. They lived under the boat in makeshift quarters extended with food boxes until January 1922, and recorded extensive scientific data while living largely on penguins and minced seal meat. The few remnants of their boathouse remain near the station as a Historic Site. A short distance away stands a hut built at Coal Point in 1950 to commemorate the visit of Chilean President Gabriel González Videla, the first head of state ever to visit Antarctica.

Gentoo penguins and Snowy sheathbills crowd around the base buildings, turning the area into a quagmire of mud and guano. From the small landing dock, a concrete pathway leads up to the base, which is comfortably but not lavishly furnished; a small viewing turret is reached by stairs inside the main building. While the bases are interesting, the bay is the true glory of this location—in still weather, its surface reflects peaks and glaciers, and the water is often crowded with icebergs, seals on floes, and even the occasional whale.

Paradise Harbor

Tough whalers were the first to call this wide bay Paradise Harbor; it is also commonly—but incorrectly— referred to as Paradise Bay. Situated behind Bryde and Lemaire islands, the bay is indeed spectacular. The name has been in use since at least 1920, and proves an irresistible lure to expedition ships following the western side of the peninsula.

The national station in the middle of Paradise Harbor is Almirante Brown, an Argentinian base that once operated throughout the year, but did not open at all over the summer of 2000–01. This base and Chile's González Videla at Waterboat Point together receive more tourists than anywhere else in Antarctica, partly because of the reputation of Paradise Harbor, but also

because both places offer the chance to set foot on the Antarctic Continent.

The Chileans, identifying themselves as the Paradise Harbor Port Authority, may radio approaching ships and give permission to enter Bahia Paraiso, even if the intended destination is the Argentinian base across the bay. This is probably to establish a record of occupation and management, in case the question of which nation can claim the Antarctic Peninsula ever arises again.

Occupied or not, Almirante Brown is the more entertaining of the two sites. Directly behind the base, a steep 50-meter (165-ft) snow-covered slope is easily climbed for a fine view of Paradise Harbor; sliding back down the snow takes a matter of seconds. The base was built in 1951 and named after a founder of the

Argentinian navy; the burnt foundations near the existing buildings are a reminder of a fire in April 1984—a bizarre moment in Antarctic history. It is reported that the station doctor started the conflagration after hearing that he was required to stay for another winter because no relief doctor could be found. He and his companions then had to live in the burnt-out ruins of the base until a ship could rescue them.

Almirante Brown appears to be surrounded by glaciers with no obvious way out. Beyond the station lies a red hut, and past that is a navigation marker on a gravel point leading up to a glacier, where Weddell seals sometimes pull up on the beach. The rocky headland that towers above the base is nesting territory for Antarctic shags, Snowy sheathbills, and Kelp gulls.

Anvers Island and environs

Anvers Island

It can be difficult at times to distinguish the mainland of the Antarctic Peninsula from the offshore islands of the Palmer Archipelago. In this world of black basalt and white glaciers, only charts may reveal the difference. This is particularly true of Anvers Island, one of the largest islands north of the Antarctic Circle on the western side of the Peninsula, the island measures about 60 kilometers (37 miles) from north to south, and is separated from the Danco Coast by the Gerlache Strait. De Gerlache named the island in 1898 after a province in Belgium. It receives few visitors except for those who have pre-arranged a visit to Palmer station.

Palmer Station

The United States Antarctic Program consists of several bases and ships. Palmer station, and the icebreaker *Nathaniel B. Palmer*, take their names from the young New England skipper of *Hero*, who was one of the earliest explorers to reach this part of Antarctica.

Palmer station is situated on southwest Anvers Island on the shores of Arthur Harbor. The harbor contains the remains of one of Antarctica's few ecological disasters. On 28 January 1989 the Argentinian naval vessel *Bahia Paraiso*, which had been active in the Falklands Islands conflict a few years earlier, ran onto a reef, tearing a hole in its side through which a vast quantity of diesel oil leaked into the sea. No one was injured, but the coastline was covered in oil and terrible damage was inflicted on the local wildlife. Today, most species have recovered, and all that is visible of the inverted ship is the glint off the smooth, exposed hull on the western side of the channel.

Britain named Arthur Harbor in 1955 after Oswald Arthur, governor of the Falkland Islands. "Station N," a small British base there, was occupied from February 1955 until 1958 and again from 1969 to 1973, when it was used for air operations until the runway deterio- rated. The runway has been used intermittently since;

a British Antarctic Survey (BAS) Twin Otter landed there in 2001. The British building was lent to nearby Palmer station in 1963, and was used as a laboratory. However, fire destroyed it during repairs in 1971, and the United States removed the remains in 1991.

Palmer station began in 1964 as a single hut, and was extended the following year. The current station, which has linking walkways above the snow level, was built in 1968 and has recently been expanded and renovated. It is quite small, with just two main buildings and a population that ranges from 10 in winter to a maximum of 43 in summer. Many visitors to Antarctica are United States citizens, and Palmer is an extremely popular destination—so popular, in fact, that the station has had to initiate a roster system and applications to visit are lodged months in advance. Tourists are shown a well run operation doing useful research and a fascinating aquarium of Antarctic marine life. Palmer also has the best souvenir shop in Antarctica.

Since 1990, Palmer has been the base for a long-term ecological research (LTER) program that is looking at Antarctic marine life to see how changing sea ice cover affects the ecosystem. The study area covers a region of 180,000 square kilometers (69,500 sq. miles) that includes land and sea, and a wide range of ice cover. With its focus on monitoring the surrounding ocean, it is hardly surprising that the station has excellent marine tanks. They contain various types of krill, starfish, sponges, and sea spiders. From the deck near the aquarium, several seal species, including Crabeaters and Leopards, are usually visible; fur seals join them later in the season. Most visitors to Palmer are taken across the bay to Torgersen Island, where there is a colony of Adélie penguins. The station staff is monitoring the effects of tourism, so visitors are asked to stay to the left (north) of the green flags delineating the control group.

Neumayer Channel

An indentation on the southeastern side of Anvers Island matches the coast of nearby Wiencke and Doumer islands. In between flows the Neumayer Channel, about 25 kilometers (15½ miles) long and in places only 2.5 kilometers (1½ miles) wide, with the snowy summit of Mount Français, the highest peak on Anvers Island, towering 2,760 meters (9,055 ft) above it. The largest vessel to sail through the channel was Holland America Line's *Rotterdam* in January 2000. However, the more than 2,000 passengers and crew on board did not land anywhere, far less visit tiny, crowded Port Lockroy.

▼ **RESEARCH BASE**
Palmer station is one of the few bases conducting long-term research close to the Antarctic Peninsula itself, rather than on King George Island in the South Shetlands. One current study is investiga- ting how Antarctic marine life is being affected by changes in sea ice cover.

➤ **SHEER FROM THE SEA**
The stark contrast between sharp, dark rocks and flowing glaciers on the peaks of the Graham Land coast creates some of the most spectacular scenery in the world. Toward the end of summer, avalanches occur daily on the steep slopes.

Port Lockroy

Although Port Lockroy is the name for the whole harbor within the southwestern coast of Wiencke Island, it is now universally used for the old British station on Goudier Island, one of a string of rocks off Jougla Point. During World War II, Britain was concerned about Germany and Axis-friendly Argentina controlling both sides of the strategically important Drake Passage. In 1943, the Royal Navy mounted Operation Tabarin to establish several small stations in Antarctica to provide meteorological and reconnaissance information. Station A at Port Lockroy and Station B on Deception Island were Britain's first permanent bases in Antarctica.

Port Lockroy was established on 16 February 1944 and operated until January 1962. It is the oldest British

Goudier Island is little more than a barren rock with some patches of snow and a few buildings. However, it has found favor with many Gentoo penguins; they nest everywhere—down by the boatshed, up at the flagpole, around the building foundations, and even on paths and steps. The birds are quite innovative in their selection of nest-building material, incorporating old nails and bits of wood from the station into their design. After being told to stay at least 10 meters (35 ft) from wildlife in keeping with IATO guidelines, visitors are often surprised at Lockroy when penguins stand on their feet and they see eggs close to the path. Over the past few summer seasons, a research project has been examining the effect of human contact on penguin breeding. Preliminary results suggest that there is little or no effect.

▼ DISPUTED TERRITORY
The British survey hut at Port Lockroy was built in the middle of a Gentoo penguin colony, and the birds use the footpaths as their own. When staff arrive each spring, they must gently discourage a few penguin pairs from nesting on the wire-mesh stairs leading to the front door.

building in the Antarctic Peninsula area, and is listed under the Antarctic Treaty as a historic site and monument. The original hut was named Bransfield House after the ship that brought the expeditioners to Antarctica—the ship was named in honor of Edward Bransfield, the first British surveyor of the Peninsula region. This hut remains the core of the station that was extended in 1952–53. Port Lockroy operated with a staff of from four to nine on a normal tour of duty of two and a half years. In 1996, work began to restore the base to the way it was when it closed in 1962. The United Kingdom Antarctic Heritage Trust now operates it during the summer months as a museum and visitor center. When Port Lockroy opened in 1944, a post office was established. Today, thousands of letters and postcards are received and franked by the station's two staff; mail makes its way to its destinations via the Falkland Islands and the United Kingdom.

▲ SIMPLE KITCHEN
Since 1962 Port Lockroy has been maintained by staff of the United Kingdom Antarctic Heritage Trust. It is run as a cross between a museum and a souvenir shop, complete with post office.

A similar program at the United States Palmer station also has early results indicating that visitors who observe the guidelines do not significantly disturb penguins.

Across from the station, the landing beach at the tip of Jougla Point is littered with whalebones; some have been assembled with anatomical inaccuracy into a rough whale skeleton. There are many more Gentoos there, and a colony of Antarctic shags inhabits the low, rocky ridge at the end of the point. Mud, snow, and rocks make it difficult to approach the shags, which are much more wary of humans than penguins are.

Most of the names in these parts were chosen by the 1903–05 French Antarctic Expedition led by Jean-Baptiste Charcot. Mount Français was named after his ship; Goudier was his chief engineer. Edouard Lockroy

was vice-president of the French Chamber of Deputies, and helped Charcot to obtain government funding for his expedition. Nearby Dorian Bay was named after another member of the chamber of deputies.

Damoy Point, marking the northern end of Port Lockroy, offers a quieter landing option on Wiencke Island. From the beach, there is a gentle slope to some scattered Gentoo rookeries. The British built the large hut behind the beach in 1975 as an air facility. When heavy sea ice prevented ships reaching Rothera station, supplies were landed on Wiencke Island and then flown down by ski plane operating from nearby Doumer Island. Though they can be covered in deep snow, and are more open to the wind than Port Lockroy, Damoy Point and nearby Dorian Bay are both very beautiful.

▼ COOL CUSTOMERS
Visitors to Port Lockroy are constantly surprised at how little notice Gentoo penguins take of the passing parade, and wonder if the birds are as relaxed as they appear. Studies so far suggest that human disturbance has little impact on the number of chicks fledged.

Lemaire and beyond

▼ WELL-FED CRABEATER
A Crabeater seal stretched out on an ice floe in sheltered Peninsula waters. The rusty red staining around the seal's mouth is from the pinkish krill that makes up the bulk of the animal's diet.

▼ CHANGEABLE CHANNEL
The ice that can make the short passage through the Lemaire Channel so memorable can also move rapidly under the influence of wind and currents, and within hours block the way with icebergs. This ship is approaching the narrow northern end.

Lemaire Channel

Vessels proceeding south from Palmer station invariably sail eastwards along the Bismarck Strait, named in 1874 by Dallman's German expedition for Prince Otto von Bismarck. There they join those sailing due south from Port Lockroy and those coming from points further north along the Gerlache Strait, to proceed down the Butler Passage toward Cape Renard on the Peninsula coast. The tall, black twin peaks of Cape Renard mark the northern entrance to Lemaire Channel—one of the highlights of a voyage along the Antarctic Peninsula.

Perversely named in 1898 by de Gerlache for Charles Lemaire, a Belgian explorer of the Congo, this narrow passage between Booth Island and the Peninsula is a mere cleft between towering cliffs. That de Gerlache was prepared to risk sailing through it seems an impressive act of nineteenth-century seamanship and courage. Sometimes it is impassable—currents and wind can fill the channel with sea ice and larger icebergs within hours—but when the ice permits, the hour's voyage each way is an unforgettable experience. The ice often supports basking Crabeater seals, and Orca, Minke, and Humpback whales navigate the passage.

Although Lemaire Channel is defined as the whole passage between Cape Cloos in the south and False Cape Renard in the north, its most spectacular stretch is the 7 kilometers (4 miles) where the sheer rock and ice walls of Booth Island tower above it. Midway along this section, the channel opens up on the eastern side into Deloncie Bay, which cuts back into the Peninsula to the Hotine Glacier. Immediately south of there, the channel narrows dramatically to less than 800 meters (2,600 ft) wide, with Mount Cloos on the mainland and Wandel Peak on Booth Island both looming 300 meters (985 ft) overhead. On a clear, still day, the glaciers and peaks that line the route are reflected perfectly in the deep, dark water. On a foggy day, visibility closes in, and there is an eerie sensation of sailing off the end of the earth.

Pléneau Island

At its southern end, Lemaire Channel opens into the Penola Strait, named by John Rymill after his 130-tonne (127-ton) schooner and his Australian sheep station, which were both called *Penola*. The island immediately to the right of the channel's entrance is Pléneau Island, a snow-capped rocky islet just over 1 kilometer (⅔ mile) long, which was once thought to be a peninsula of Hovgaard Island to the south. Charcot named it in 1904 for Paul Pléneau, company director, supporter, and friend, who had sailed with him as the expedition photographer. The island is home to numerous Gentoo penguins, Weddell seals may be found alone or in pairs along the shore, and elephant seals form communal wallows slightly inland. Its highest point commands a stunning view across to Booth Island and the almost indiscernible entrance to Lemaire Channel.

The most spectacular feature of Pléneau Island is not on the land at all but a field of grounded icebergs, west of Lemaire Channel between Pléneau and Booth Island.

▼ **TOWERING PEAKS**
The spectacular northern entrance of the Lemaire Channel is marked by Cape Renard (in the middle of the photograph). Beyond the soaring twin peaks, the channel narrows dramatically to become "False Cape Renard. "

Because the Pléneau bergs are grounded, they are more stable than floating ones, and it is possible to navigate Zodiac craft through them. They form a fairyland of arches and pinnacles and range in color from pellucid turquoise, where shallow water laps at their feet, to brilliant blue in the cracks in the ice. The area is prone to strong winds that can sweep off Booth Island at any time and churn the smooth waters of the channel between the islands into a lather of foam.

Pléneau Island, less than 167 kilometers (104 miles) north of the Antarctic Circle, is often the turning point for Antarctic cruises. There are some good reasons for continuing south, but Hovgaard Island is not one of them. It is large and rises to 369 meters (1,210 ft), but has neither the icebergs and seals of Pléneau to the north, nor the Adélie penguins and history of Petermann Island to the south.

Petermann Island

German geographer August Petermann was a strong supporter of his country's polar exploration in the mid-nineteenth century. His name lives on in the red, dry, dusty Petermann Range in Western Australia, and the contrasting Petermann Ranges of East Antarctica's Dronning Maud Land. Snowy, cold Petermann Island, 1.6 kilometers (1 mile) long and 133 meters (436 ft) high, discovered and named by the German Dallman expedition of 1873–74, lies close to the Antarctic Circle.

However, most of Petermann Island's historical links are French, not German. The landing site on the southeastern side of the island is named Port Circumcision because the small cove was discovered by Charcot on 1 January 1909, the annual feast day of the Circumcision (when, tradition holds, Christ was circumcised). Charcot securely moored his ship, *Pourquoi Pas?* ("Why Not?"), in the cove, ran iron hawsers across the mouth of the little harbor to keep out icebergs, and settled in for the winter of 1909.

Charcot named a rocky hill in the south of the snow-covered island Megalestris Hill, after the Latin term for a skua. He built a stone cairn to commemorate his expedition—it was restored by a British Antarctic Survey (BAS) team in 1958, and is now an Antarctic Historic Monument. Charcot's original plaque has been taken to France, and a replica will be installed.

The hut by the cove is an abandoned Argentinian *refugio*; built in 1955, it now provides shelter to scores of Gentoo penguins and their chicks, and to Snowy sheathbills seeking scraps. For much of the summer, it is impossible to walk past the penguins to sign the visitors' book inside the hut. The memorial cross close by is for three British scientists who died near there on 14 August 1982. Across Penola Strait, on the mainland, stands the U-shaped massif of Mount Scott, 892 meters (2,926 ft) high, and behind it the sheer-walled peak of Mount Shackleton, 1,465 meters (4,806 ft). Charcot named both for the British Heroic Age explorers.

▲ PRIVATE WONDERLAND
A maze of islands, rocks, and grounded icebergs stretches south beyond Petermann Island. Accessible only to comparatively small, shallow-drafted yachts, the area has several sheltered anchorages with outstanding views. Many private expeditions and commercial yachts venture south of the Lemaire Channel to explore this magical place.

◄ GROUNDED ICEBERGS
Large icebergs, often driven in from the west by wind and currents, run aground in the shallow waters north and west of Pléneau Island. They remain there for years as they slowly melt, creating an impressive spectacle of form and color.

While Gentoo penguins have claimed the low rocks by the Argentinian hut, Adélie penguins dominate the rocky summit up the slope to the north. Due east from the Adélie hordes, in a tiny rocky bay formed by a basalt dike, Antarctic shags claim almost every crag at peak season. Around the island's convoluted southern coast, large groups of Crabeater seals rest on floes or streak through the water, and the occasional Leopard seal may be out cruising for wayward penguins.

Yalour Islands

The rocks and islets scattered over 2.4 kilometers (1½ miles) to the south of Wilhelm Archipelago hardly merit the title of islands, but they are a good place to observe the antics of some 8,000 pairs of Adélie penguins and their chicks who live there each summer.

Charcot named the Yalour Islands after Lieutenant Jorge Yalour, an officer on the Argentinian corvette *Uruguay*, which rescued the Nordenskjöld expedition in November 1903. Charcot's first Antarctic venture, in 1903, intended to go to the aid of the Swedes, and diverted to this side of the Antarctic Peninsula only after learning the expedition members had been evacuated.

Argentine Islands

Charcot named the Argentine Islands to thank the Argentinian government for their assistance with his *Français* expedition of 1903–05. Even in the middle of summer, reaching them through the ice is tricky. The main reason to go to there is to visit the Ukrainian station of Academician Vernadsky, one of the most comfortable and welcoming bases in Antarctica. Until 1996 this was the British Faraday station, named after Michael Faraday, discoverer of electromagnetism; then it was sold to the Ukraine for one pound. The Ukrainian government was keen to maintain an Antarctic presence following the break-up of the Soviet Union, when all the Soviet stations became Russian.

The first British building on the Argentine Islands was constructed on Winter Island by the British Graham Land Expedition of 1935–36, but it was destroyed in 1946, perhaps by a tidal wave. Undeterred, in 1947 Britain erected another hut in the same location, and named it Wordie House after Sir James Wordie, who was on Shackleton's *Endurance* expedition (1914–16), and served on the Advisory Committee for Operation Tabarin (1943–45). In 1954, the station was transferred to Galindez Island, and it is this base that has now become Academician Vernadsky, named for Vladimir Vernadsky, first President of the Ukrainian Academy of Sciences. BAS teams have restored Wordie House to the way it would have been in the 1950s. Vernadsky's personnel now maintain it, and visiting ships bring passengers there by Zodiac down a picturesque narrow channel or by the shorter and more spectacular Cornice Channel. Faraday was occupied continuously by the British for 49 years and 31 days, just four days more

▼ ENDURING CONTEST
In view of the intense rivalry between Robert Scott and Ernest Shackleton, Scott may have turned in his icy grave when the soaring 1,465-meter (4,806-ft) peak (pictured on the right) was named Mount Shackleton, while the smaller 892-meter (2,926-ft) peak (on the left) was named Mount Scott in his honor.

than Signy in the South Orkney Islands, Britain's second longest continuously occupied base.

At Vernadsky, the bar is the most popular place for expeditioners and visitors alike—the Ukrainians have maintained its distinctly British atmosphere, right down to the billiard table and dartboard. Hanging above the bar is an amazing photograph of a man and a Minke whale in eye-to-eye contact. Although some question the photograph's authenticity, base members swear that it is real. The infamous collection of brassieres, including some of improbable dimensions, continues to grow as some visitors elect to contribute. When ships call, the staff open "the southernmost gift shop in the world," stocked with souvenir cloth badges, postcards, and attractive carved plaques.

Vernadsky continues Britain's long-term research projects. The weather records there reveal that the local mean annual temperature has increased by 2.5°C (4.5°F) since 1947. Visitors prepared to undertake the rather precarious climb into the ceiling of the station can see the equipment used for measuring the ozone. Over the summer of 2000–01, a slight decrease in the size of the hole in the ozone layer was detected.

Antarctic Circle

The Antarctic Circle is really only a theoretical line on a map, or a reading on a global positioning system (GPS), unless seen at the solstice, when the sun stays above, or below, the horizon. The polar circles move because the earth wobbles slightly on its axis. The Antarctic Circle is currently at 66°33′39′′S, and is heading further south. Ships voyaging to the Circle normally pass through Crystal Sound, named by Britain in 1960 because most nearby places were called after researchers studying ice crystals. It is a good region for ice—even in late summer, when north of the Lemaire Channel is comparatively warm, the Antarctic Circle may be distinctly

colder; late in the season some pancake ice persists there, when there is not even frazil or grease ice to the north.

Between Petermann Island and the Circle the only potential landing site is the Fish Islands, at the foot of towering Prospect Point: a small group of rocky islets that are home to nesting Antarctic cormorants and Adélie penguins. The main attraction here is the ice, including impressive tabular icebergs that often drift north from the ice shelves of the Bellingshausen Sea.

Voyaging through or around the ice south of the Circle requires a lot of effort for relatively little reward. Detaille Island station, at 66°52′S, exemplifies some of the problems. Established by Britain in February 1956, it was abandoned in March 1959 when it proved impossible to reach regularly for relief and resupply.

Considerably further south, at 67°34′S, is Britain's Rothera station on Adelaide Island, the principal BAS station in Antarctica. It was opened in October 1975 to replace the Adelaide station (1961–77), when that station's ice runway deteriorated. In 1992 a wharf and a gravel runway were built at Rothera. Twin Otter aircraft fly from the latter to wherever BAS teams are working. The station has accommodation for 120 people, but that will be increased to more than 200 residents when a new accommodation block is completed in 2002. From Rothera, excursions are possible to the Dion Islands at the northern end of Marguerite Bay, where the Peninsula's sole colony of Emperor penguins gathers in winter. Isolated adults and chicks are occasionally observed here during summer.

The only Antarctic Peninsula station south of Rothera operating year-round is Argentina's General San Martin on Barry Island, which was established in 1951. Tourist vessels rarely venture so far south. DM

▲ CRYSTAL CLEAR
Only toward the end of summer is Crystal Sound and the approach to the Antarctic Circle so ice-free. As winter approaches, ice crystals are quick to form. The sound was named by the British for the ice crystal research that was carried out in the area.

▲ WINTER EMPIRE
Emperor penguins come "ashore" onto fast ice to reproduce in March or April; the males remain there with the eggs over winter, fasting for up to three months.

The Ross Sea and East Antarctica

The Ross Sea region

110°W
14
120°W
13
130°W
12
140°W
Newman I
11
150°W
10

▲ SLEEPING VOLCANO
A team of New Zealanders in a Hagglunds ATV (all-terrain vehicle) travel across the vast, flat Ross Ice Shelf toward Mount Discovery. This dormant volcano was named by Captain Scott after his ship—its youngest vents are 1.8 million years old.

Previous pages
◄ APPROACHING THE END Construction planned from 2001 will lead to the demise of the dome at Amundsen–Scott base. This distinctive landmark that has symbolized the South Pole since 1970 is due to be dismantled and shipped out in 2005.

The Ross Sea lies at the southernmost limit of the Pacific Ocean sector of Antarctica. When James Clark Ross sailed into the expanse of the sea that now bears his name, he hoped that it would allow him sailing access to the South Magnetic Pole. However, the impressive line of mountains that he saw to the west soon quelled this ambition. The western shore of the Ross Sea is delineated by the Victoria Land coast up to Cape Adare, and its eastern boundary is Cape Colbeck, at the northwest tip of Edward VII Land. Its total area is almost 1 million square kilometers (386,000 sq. miles). The Ross Sea is quite shallow—between 300 and 900 meters (1,000 and 3,000 ft)—but to the north it plunges sharply to depths of more than 4,000 meters (13,115 ft). At its southernmost edge is the Ross Ice Shelf, one of the most spectacular features of Antarctica.

At the northwest corner of the Ross Ice Shelf is Ross Island, which is the most historically significant location in Antarctica. It was from here that Scott, then Shackleton, then Scott again led parties out on the first great attempts to walk to the South Geographic Pole—and it is here that Mawson, Edgeworth David, and Mackay returned after walking to the South Magnetic Pole. Today, it is the site of McMurdo Station, the largest Antarctic base. All-volcanic, the island is roughly the shape of a four-pointed star about 70 kilometers (43 miles) wide.

The sea ice around Ross Island does not break up until late January and re-forms in early April, and about six ships visit over the short summer season. Across its southern side the island is attached to the ice shelf from near the tip of Hut Point Peninsula to Cape Crozier.

Glaciers spill onto the ice shelf and onto the western shore. It is also linked by ice shelf to Victoria Land, the waterway in between known as McMurdo Sound, about 148 kilometers long and 74 kilometers across at its widest point (92 miles and 46 miles).

Mount Erebus looms over the island. Apsley Cherry-Garrard wrote after his first view of the mountain on Scott's last expedition: "I have seen Fuji, the most dainty and graceful of all mountains; and also Kinchenjunga; only Michael Angelo among men could have conceived such grandeur. But give me Erebus for a friend. Whoever made Erebus knew all the charm of horizontal lines, and the lines of Erebus are for the most part nearer the horizontal than the vertical. And so he is the most restful mountain in the world, and I was glad when I knew that our hut would lie at his feet. And always there floated from his crater the lazy banner of his cloud of steam." The mountain was seared into the world's consciousness on 28 November 1979, when an Air New Zealand sightseeing flight crashed into the mountain, killing all 257 people on board. At the time it was the world's fourth worst air disaster. It was a tragedy that had a profound effect upon the nation that sent the flight and is still recalled with horror by the those who were based at Scott and McMurdo when the crash occurred.

◄ PACKED UP AND READY New Zealand researchers set out from Scott Base with packed "Ski-Doos" for a three-day trip over the sea ice in spring. These motor toboggans are a cold but safe and efficient way to travel across sea ice.

Ross Sea and Ross Island

The Ross Ice Shelf

A glance at most world maps suggests that there is a bay in the Southern Ocean that extends almost to the South Pole, but closer inspection reveals a subtle change of shading. This is the Ross Ice Shelf—the feature that made the South Pole such a challenging destination. It is roughly triangular in shape, and covers about 500,000 square kilometers (190,000 sq. miles)—about the same area as France, and twice the size of Great Britain. Most remarkably, virtually all of it is floating.

iceberg ever recorded calved from the eastern Ross Ice Shelf in late March 2000; it was 295 kilometers long and 37 kilometers wide (183 by 23 miles), with an area of about 11,000 square kilometers (4,250 sq. miles). Prosaically named B-15, it broke up and knocked other bits off the ice shelf as it drifted westward towards Ross Island. The first cracks that gave rise to B-15 were observed by satellite as far back as 1990.

The Bay of Whales is an indentation towards the eastern end of the ice shelf, where the ice is low enough

▲ PERPENDICULAR CLIFF
Nowadays we can marvel at the nature and extent of the Ross Ice Shelf. The early explorers did not know what they had found. When Scott arrived in 1902 he wrote: "such a phenomenon was unique, and for sixty years ... many a theory had been built on the slender foundation of fact ... It was an impressive sight, and the very vastness of what lay at our feet seemed to add to our sense of mystery."

From the ocean, the Ross Ice Shelf looks like an endless white cliff towering 10–60 meters (30–200 ft) above the sea. When James Clark Ross discovered it in 1841, he described it as a "perpendicular cliff of ice" that would be as difficult to penetrate as the Cliffs of Dover, and named it the Victoria Barrier. (The name was later changed to honor Ross himself.) Some sixty years later, Norwegian explorer Roald Amundsen was to express some of the wonder of the spectacle: "Slowly it rose up out of the sea in all its imposing majesty. It is difficult … to give any idea of the impression this mighty wall of ice makes on the observer who is confronted with it for the first time."

The shelf is about 800 kilometers (500 miles) wide, and extends back almost 1,000 kilometers (620 miles) towards the Pole. It appears to be static, but in fact it is fed by glaciers flowing off the polar ice cap, and can grow outwards by more than 1 kilometer (⅔ mile) each year. There has been a somewhat morbid suggestion that the gravesite of Scott and his companions may soon reach the sea towards which they were struggling.

Birth of an iceberg

On average the Ross Ice Shelf is about 300 meters (980 ft) thick at the seaward edge, and up to 700 meters (2,300 ft) thick further to the south. Although it is hinged along the coast, powerful forces operate on this relatively thin skin of ice, breaking off huge blocks that become the tabular bergs that throng in the Ross Sea. Apart from a questionable sighting in 1956, the largest to permit landing. It was discovered by Carsten Borchgrevink in 1900 and named by Ernest Shackleton in 1908, who wrote that "it was a veritable playground of these monsters." Encroaching ice drove Shackleton away, so he could not use the bay as his base, but Amundsen calculated that it was about 100 kilometers (60 miles) closer to the pole than Ross Island, and used it to make his successful bid for the South Pole in 1911. Richard Byrd also used it on his Antarctic expeditions of 1928–30 and 1933–35.

In 2000, B-15 took the existing Bay of Whales with it, and the ice where Amundsen and Byrd camped is long gone. But the bay remains because it marks the junction of two ice systems: not far to the south, within the shelf, the ice must divide and flow around the large, ice-covered Roosevelt Island. With B-15 gone, the edge of the ice shelf currently lies further south than before, and on 11 January 2001 the Russian icebreaker *Kapitan Khlebnikov* set a new southernmost record for shipping— 78°37' south in the reconfigured Bay of Whales.

▼ **VAST AND FEATURELESS**
Two figures near Hut Point Peninsula on Ross Island give some idea of the vastness of the Ross Ice Shelf. About the size of France, it is a flat-topped body of snow-covered glacial ice, mostly floating, except along coastlines and over other shallow areas.

Hut Point Peninsula

Scott Base

Now a modern scientific station, Scott Base was built by Sir Edmund Hillary's Commonwealth Transantarctic Expedition of 1955–58. One of the original buildings erected during that expedition still stands, and houses memorabilia from the base's subsequent history. The base opened on 20 January 1957, and has been in constant use ever since. It has been considerably enlarged over the years, notably during an extensive rebuilding in 1976. Its eight interlinked pale-green buildings can accommodate up to 86 people in summer and about 10 each winter.

The base is situated on the solid rock of Pram Point, so named by Robert Scott because he needed a pram (a small, flat-bottomed boat) for the trip from his base at Hut Point. Since Hillary established it in 1957 (and was the first base commander), New Zealand has used the base for meteorological observations and for research into the aurora, the ionosphere, marine biology, tides, and the environment. The base is also fundamental to New Zealand's claim to a part of Antarctica: it is the only New Zealand base on the Antarctic Continent. Scott Base, and the whole of Ross Island, fall within New Zealand's Ross Dependency, claimed by Britain in 1923 but placed in the care of New Zealand. New Zealand has responsibility for the island's historic huts, and tourists can enter them only in the company of a supervisor from the New Zealand Heritage Trust.

Scott Base operates in close cooperation with the United States' McMurdo Station, to which it is linked by a gravel road 5 kilometers (3 miles) long. Over the summer, New Zealand aircraft make some 15 flights from

Christchurch airport, and there are more flights from the United States, carrying personnel and supplies for both Scott Base and McMurdo Station to McMurdo's airfields. However, the residents of McMurdo Station are not permitted to use the small rope-tow ski lift at Scott Base.

Hut Point

Hut Point is the site of the "Discovery" hut—a prefabricated wooden hut erected by Robert Scott's British National Antarctic Expedition of 1901–04. Meticulously kept records show that the building, which occupies an area of 11.3 square meters (37 sq. ft), cost a modest 360 pounds 13 shillings and 5 pence sterling. The hut oddly lacks the atmosphere that might be expected from so historic a structure, possibly because it was used by various expeditions over many years, or perhaps because it is dwarfed by the huge modern base—McMurdo Station—that has sprung up on its doorstep.

Incongruously, the original hut was designed after the classic Australian outback homestead (with verandas!), but its construction methods were modified to suit its purpose. The walls and floor were made of two layers of tongue-and-groove Douglas fir packed with felt for insulation, and there was a heating stove for the officers' room and a cooking stove for the men's room. Even so, it was very cold, and Scott decided that the expeditioners would spend the winter aboard the *Discovery*, immobilized in the winter ice nearby. On other occasions the hut provided accommodation for the Shackleton expedition of 1908 and Scott's 1911 expedition, but it was most used by the marooned members of Shackleton's Ross Sea shore party in 1915, who had to survive on seal meat and the meager supplies Scott had abandoned more than a decade before.

After 1915 the hut was unvisited until 1947, but when McMurdo Station was established it became a target for pilferers. Restoration began in 1964, and when the snow and ice were dug out many artefacts were found. The present hut is as it was when Shackleton's party left in 1916, and quite unlike the original layout.

▲ **GETTING AROUND**
Picking up a load of building material for renovations at a field hut, a helicopter leaves Scott Base. Helicopters are the ideal transport for short- to medium-range operations in Antarctica.

◄ **EIGHT GREEN BUILDINGS**
Originally a cluster of huts, Scott Base has evolved over a number of years into a larger, modern station. In this aerial photo taken in early summer the sea ice has not yet "broken out" and still pushes against the shoreline, buckling into sea ice pressure ridges—rather like a line of frozen surf.

⌃ EXPLORING A CREVASSE

On Hut Point Peninsula, a New Zealander lowers himself into a
large crevasse through a hole where the ice roof has collapsed.
Crevasses are an ever-present danger in Antarctica. Snow will
often collect on each lip, building outwards until the gap is
bridged and the crevasse becomes invisible, and potentially very
dangerous. Some snow bridges are thick enough to support a
person or vehicle—and some are not!

⊳ BLOOD DONOR

A New Zealand biologist with
a captured Adélie penguin at
Cape Bird, near the northern
tip of Ross Island. Blood
samples from Adélie penguins
are being used to study the
role of hormones in controlling
the length of feeding trips
at sea.

⌃ HIGH POINT

Castle Rock on Hut Point Peninsula is a
popular destination for off-duty staff from
nearby McMurdo Station and Scott Base.
It is one of the oldest of a line of small
volcanic craters that extend southwards
from the lower slopes of Mount Erebus,
and is composed of volcanic breccia that
formed from eruptions below the ice.

McMurdo Station

Deep within the Ross Sea, McMurdo Station is by far the largest of the Antarctic research bases; at the height of summer there may be as many as 1,200 personnel there. The name of the station comes from adjoining McMurdo Sound, named by James Clark Ross after Lieutenant Archibald McMurdo, an officer on the *Terror*. The site has a strong historic link: at Hut Point, near the station's dock area, stands the Discovery hut from which Scott, Wilson, and Shackleton set out in 1902 on the first serious attempt to reach the South Geographic Pole.

Overlooking the sound and the station is Observation Hill, 230 meters (750 ft) above sea level, and crowned with a cross commemorating Scott and the men who died with him on their way back from the Pole. The base of the hill was the site of Antarctica's only nuclear reactor, which was designed as a portable nuclear power plant, installed in December 1961, and commissioned in March 1962. The reactor operated until 1972, when it was shipped back to the United States, after which time more than 10,000 cubic meters (130,000 cubic yards) of soil contaminated by radioactive waste was removed and freighted out.

The present McMurdo Station first took shape in December 1955 as Naval Air Facility, McMurdo Sound, one of six research stations built by the United States government in preparation for the International Geophysical Year (July 1957 to December 1958). McMurdo was, and still is, the coastal base that supplies the United States South Pole station by air. The National Science Foundation of the United States funded research work through its National Academy of Sciences, and the United States Department of Defense separately provided operational support. Recently there has been a shift away from military support to contractor services, since scientific research is now the primary expression of American presence in Antarctica. A recent government report states, somewhat curtly, that "no armaments are currently stored or in use at McMurdo Station."

Weapons or no weapons, the first sight of McMurdo Station, sprawling in the summer mud on the bare volcanic rock of Hut Point Peninsula, is likely to be a shock for anyone expecting Antarctica to be a pristine white wilderness. This is no tranquil haven from the world, but a cluttered, bustling settlement reminiscent of an Alaskan frontier town. It is littered with above-ground cables carrying telephone connections and electrical power, and heat-taped, insulated pipes for water and sewerage. Pedestrians must look both ways before crossing the road, and there is a speed limit of 30 kilometers per hour (20 mph) for motor vehicles. The station's 85 utilitarian buildings comprise dormitories, administration blocks, workshops, warehouses, and laboratories, as well as more unexpected edifices such as the Chapel of the Snows, a fire station with eight fire trucks, a coffee shop, a gymnasium, and a hydroponic greenhouse. There are also three airfields and a helipad, and the impressive inventory of vehicles includes 177 trucks, 43 personnel and cargo carriers, 65 front-end loaders and forklifts, 27 tractors, and 90 motor toboggans.

> **CHOOSING A SITE**
> Scott's Discovery hut was built in 1901. Though prefabricated, Scott recorded that "[the] erection was no light task, as all the main and veranda supports [had] to be sunk three or four feet into the ground. We soon found a convenient site close to the ship on a small bare plateau of volcanic rubble, but an inch or two below the surface the soil was frozen hard."

Fluctuating population

McMurdo Station has a marked annual population cycle. In August (the southern hemisphere's spring), several Winfly (winter fly-in) flights from New Zealand raise the station's population from its winter maximum of about 200 people, when researchers arrive to begin pre-summer science projects, and support staff flock in to prepare for the summer. In October the population of the station rises quickly, and during summer there are arrivals and departures several times a week. In late December the "Christmas notch" coincides with the melting of the sea ice, and scientists (particularly biologists) who need the ice as a working platform give way to those (such as geologists) who do not. With the approach of the Antarctic winter in late February, shortening days and plummeting temperatures make field research impractical, and the population of the station falls accordingly.

Personnel and supplies are brought in on cargo vessels escorted by icebreakers, or on regular flights

➤ THE ICE RECEDES

As spring progresses, sea access is possible at McMurdo Station. Here, the United States container ship *Green Wave* unloads its cargo onto a floating ice wharf, which was prepared by artificially thickening the sea ice the previous spring and early summer. Though cracks may eventually lessen its strength and usefulness, for most of the short summer season this unusual wharf works very well. In the foreground is Scott's Discovery Hut.

▼ THE END OF WINTER

In the spring, the area of bare sea ice just beyond Scott's hut (with a "streetlight" at its center) is slowly transformed into an ice wharf. Regular flooding and refreezing thickens the sea ice into an artificial flat-topped iceberg, strong and thick enough for normal wharf operations, even when the thinner sea ice around it breaks up.

► RESEARCH STATION
Since 1956, the United States
has had a full time research
operation at McMurdo Station
and from that first year it
has been the largest base
on the continent. Under the
auspices of the National
Science Foundation, a wide
range of scientific research
is conducted here, with
significant logistical support.

▼ ANTARCTIC WORKHORSE
The LC-130 Hercules is the
polar version of the familiar
C-130 cargo plane. Its unique
and robust ski-equipped landing gear allows operation on
snow or ice surfaces throughout Antarctica—pretty much
anywhere that is flat enough
and large enough for landings.
It also has wheels for landing
on prepared hard surfaces.

from Christchurch, New Zealand. For 44 years—from the time the base was established until 16 February 1999—the United States Navy operated these flights in support of the United States Antarctic Program run by the National Science Foundation. They are now operated by the New York Air National Guard's 109th Airlift Wing, based in Schenectady, New York. The 10 planes are LC-130 Hercules cargo aircraft fitted with skis—the only aircraft of this type still operating anywhere in the world. These planes, introduced to the Antarctic Program in 1960, also have wheels for landing on prepared hard surfaces.

At the beginning of the twenty-first century, McMurdo Station is the major national station on the Antarctic Continent with three operating airfields. Williams Field is a ski-way on the Ross Ice Shelf, 16 kilometers (10 miles) from the station. It operates from early December until late January and has a main runway of 3,050 meters (10,000 ft) and a cross-runway of 2,440 meters (8,000 ft). Pegasus, a permanent, hard-ice airstrip for wheeled aircraft, is much further away, but is also on the Ross Ice Shelf. It was completed in 1992 and is usable from mid-January until the end of summer. And in early summer, from late September until early December, there is a sea-ice airfield suitable for use by heavy, wheeled aircraft, including C-130s, C-141s, C-17s, and C-5s. All of McMurdo's icy airfields are firm enough for landing during winter, but winter weather conditions and the lack of both natural light and artificial lighting prevent year-round services. In late August Pegasus opens briefly for Winfly (the winter fly-

in), but everything else occurs over the Antarctic summer, from October to February. At this time New Zealand's Christchurch airport handles about 4,000 passengers to and from Antarctica, as well as a staggering amount of cargo, including 85 tonnes of fresh produce.

A changing focus

McMurdo Station may not be esthetically appealing, but it is extremely productive. Since the end of the Cold War that followed World War II, the emphasis at the station has shifted dramatically away from mere military presence towards vital scientific research, and facilities at the station have been upgraded to meet this challenge. A major new science laboratory, the Albert P. Crary Science and Engineering Center at McMurdo, was opened in November 1991 and named in honor of the geophysicist and glaciologist of the same name, who was the first person to set foot on both the North and South poles. Now scientists can study Antarctica's unique environment in the field, and can return to analyze their findings in a laboratory as well equipped as one anywhere else in the world.

The Crary laboratory boasts state-of-the-art instrumentation: personal computers, a local area network, and e-mail contact via satellite with the rest of the world. It supports studies of marine and terrestrial biology, biomedicine, geology and geophysics, glaciology and glacial geology, meteorology, aeronomy, and upper atmosphere physics, as well as carrying out obligatory environmental monitoring. The five "pods" that constitute the vast laboratory were built in three phases: phase I consists of a two-story core pod and a biology pod; phase II houses earth sciences and atmospheric sciences; and phase III has a research aquarium. As a result, more scientists are turning their attention to Antarctica to conduct increasingly more sophisticated experiments, and the proportion of the United States Antarctic Program budget allocated to scientific research grants rose from 10 percent in 1984–85 to almost 16 percent in 1996–97.

◄ THE STRENGTH OF SEA ICE
The ice here at McMurdo Sound is only 2 meters (6½ ft) thick,
and yet safely allows the operation of United States Air Force
C-5 Galaxy aircraft. The Galaxy was the world's largest aircraft
until the appearance of the Russian Antonov An-124 and An-225.
A Galaxy is 75.5 meters (247 ft) long, has a wingspan of 67.9
meters (222 ft), and stands as tall as a six-story building.

Daily life at McMurdo

Specialized scientific laboratories apart, many of the station's facilities are now showing signs of serious deterioration. An official review in 1997 concluded that a great deal of work was required immediately, including the replacement of 17 above-ground fuel storage tanks. Nevertheless, the continually changing population of temporary residents enjoys facilities beyond the imagination of the original occupants of the Discovery hut. Besides daily e-mail contact with the outside world and several local cable television channels, including an extremely popular weather channel, the station offers a small souvenir shop, clubs, a video store, a post office, and a hairdresser.

Despite its intermittently large population, McMurdo Station now contributes little to the pollution of Antarctica. This represents considerable progress from the situation before 1990, when rubbish accumulated over the winter was simply carried out to the edge of the sea ice to sink in the spring break-up. Today, there is a comprehensive recycling program, and accumulated rubbish is taken away by ship.

▲ ROCK ON!

Icestock '92 was an open-air music concert at McMurdo Station in the summer of 1992—the name was inspired by the larger and more famous 1969 Woodstock Music Festival in upstate New York. The snow is gone and it might be a pleasantly warm midsummer day by Antarctic standards, but it is still Antarctica—note the clothing.

Cape Crozier

▼ BREAKING A HUDDLE
The Emperor penguin colony at Cape Crozier is the world's southernmost penguin colony and the first Emperor penguin colony ever discovered—in 1902 by members of Scott's first Antarctic expedition. Three members of his second Antarctic Expedition almost died in 1911 visiting the colony in the blackness and bitter cold of the Antarctic winter.

Ross Island is a small outcrop of volcanic rock perched between the immensity of the Ross Ice Shelf and the vastness of the Ross Sea. On its eastern side the island bulges around Mount Terror and rounds into Cape Crozier, where ice, sea, and island meet. This high cape was discovered by the Ross expedition in 1841 and named after Francis Crozier, captain of the *Terror*. Thousands of Adélie penguins nest on its gravel slope, and lower down a colony of Emperor penguins occupies the ice shelf during the winter.

Cape Crozier is the subject of one of the most extraordinary travel books ever written: Apsley Cherry-Garrard's *The Worst Journey in the World*. In the Antarctic midwinter of 1911, Cherry-Garrard, with Bill Wilson and "Birdie" Bowers (both of whom perished later with Scott), visited the cape to collect Emperor penguin eggs. Cherry-Garrard thought that "the Emperor is probably the most primitive bird … the embryo … may prove the missing link between birds and the reptiles …" The timing of "the weirdest bird's-nesting expedition that has ever been" was dictated by Wilson's calculation that the birds would lay in early July. The naturalists barely survived, and the remains of their igloo are still there. "I don't think anybody could have made a better igloo with the hard snow blocks and rocks which were all that we had," wrote Cherry-Garrard.

▼ ICY LANDSCAPE
A frozen meltpool tops the Ross Ice Shelf at Cape Crozier. The seaward edge of the Ross Ice Shelf reaches as far as the cape at the eastern tip of Ross Island, where it breaks into long fingers. The Emperor penguin colony is usually found on the sea ice in one of the "ice canyons" between these fingers.

Cape Evans

> **THIRD CHOICE**

Named after Scott's second-in-command, Cape Evans was not the *Terra Nova* expedition's first choice. Only after Cape Crozier was found to be unsuitable and Hut Point inaccessible was the prefabricated hut erected on the coastal slope of Mount Erebus. It was here that Shackleton's Ross Sea Party was marooned when the *Discovery* was swept away in May 1915.

The hut at Cape Evans, 11 kilometers (7 miles) from Cape Royds, is the destination to which Scott and his polar party never returned. Unlike the Discovery hut, it is an evocative place despite the passing of almost a century; so close do visitors feel to the early inhabitants that they sometimes surreptitiously touch the stove to see if it is still warm. No more than a dozen people at a time may enter; it is a place where the strong spirit of the Antarctic's explorers lingers to be appreciated in relative solitude.

The hut is on a promontory named after Scott's second-in-command, Lieutenant Edward ("Teddy") Evans. Nearby, a ridge 50 meters (165 ft) high leads to the base of Mount Erebus. In its time the hut was the largest and best Antarctic accommodation ever built. Measuring 15.25 meters by 7.6 meters (50 by 25 ft), and with multiple-insulated floor, walls, and ceiling to house 25 men in relative comfort, it was well designed and constructed. Davis, the ship's carpenter, declared that "hut" was a misnomer—it was more like a parish hall. It was assembled in England, but some faults were detected and the hut was re-erected in Lyttelton, New Zealand, and the problems rectified.

After the bodies of Scott's polar party were found, the *Terra Nova* left Cape Evans at 5.20 pm on 19 January 1913. "Teddy" Evans wrote: "We have left at Cape Evans an outfit and stores that would see a dozen resourceful men through one summer and winter at least." Prophetic words: when a storm wrenched the *Aurora* off her two anchors, which are still buried in the gravel by the hut, the stranded men had to rely on those supplies for 20 months. Lacking such basics as fuel, soap, and clothes, three of them perished before rescue eventually arrived. The survivors were evacuated in January 1917.

The hut was next visited in the late 1940s, and restoration of the ice-filled hut began in 1960. Despite the desperate plight of the Ross Sea Party, the only re-construction required to re-create the interior as Scott's party left it was to replicate Oates's bed, with the aid of photographs. Artefacts and links to the Heroic Age parties are everywhere, and there is even a mummified penguin on the chart table. A guest book now sits on the magnificent oak wardroom table that features in Ponting's famous photograph of Scott's last birthday party, on 6 June 1911.

> ▲ **DOMESTIC INTERIOR**
> In 1913 Scott wrote: "A large galley-stove is placed on the right, and behind it is the chief touch of color in the hut in the form of rows of tins of food, spices and utensils." In 1960 a New Zealand party cleared the ice from Scott's hut with picks and shovels. The interior now on show resembles the way it looked during Scott's occupation.

Cape Royds

▲ FEATHERED ELEGANCE
The Adélie penguin presents a "dinner-suited gentleman" image. It is superbly adapted to aquatic life: a short stiff tail and rearward feet together act as a rudder, steering a stubby inflexible body. The "flipper" wings are powerful propellers on either side of the body.

▼ FROZEN IN TIME
Still looking as if the men might walk back into the hut at any minute, Shackleton's hut at Cape Royds—like Scott's hut at Cape Evans—is frozen in both time and in reality. The dry air and cold temperatures have preserved the contents well.

The homely little hut from Shackleton's 1907–09 *Nimrod* expedition is situated up a slight rise from Backdoor Bay, in a hollow largely filled with icy Pony Lake. On the slope behind is the world's southernmost Adélie penguin rookery (about 4,000 breeding pairs). Measuring 7 meters by 6 meters (23 by 19 ft), the hut was prefabricated in England and insulated with cork and felt. Shackleton's table and chairs have not survived, but there is extensive evidence that packing cases were the expedition's basic building blocks for beds and cupboards. In March 1908, it was from here, as part of the *Nimrod* expedition, that Edgeworth David led the first ascent of Mount Erebus, the foothills of which start to rise only 2 kilometers (1¼ miles) from the hut, although the summit is 25 kilometers (15½ miles) away.

It was never intended that Shackleton should base his 15-man expedition at Ross Island—indeed, he had specifically promised a representative of Scott that he would not do so. However, heavy sea ice and other factors made the island the only viable option. Named by Scott after Lieutenant Charles Royds, the meteorologist of the *Discovery* expedition, Cape Royds is the westernmost point of Ross Island. The hut was erected in February 1908, and was subsequently modified when the men discovered how cold the unlined floor was. They improvised an airlock by sealing off the underfloor space with boxes around the outside of the hut, a technique that was at least partially successful.

After Shackleton almost reached the Pole, the hut was abandoned in 1909. As always, the provident Shackleton thought of survival and left "a supply of stores sufficient to last 15 men one year. The vicissitudes of life in the Antarctic are such that such a supply might prove of the greatest value to some future expedition." He was right: the hut was visited by Scott's *Terra Nova* expeditioners of 1910–12 and proved invaluable for Shackleton's own marooned Ross Sea Party of 1915–17.

By 1947, when the next visitors called at Cape Royds, there were holes in the roof and the whole edifice was in disrepair. The garage built to house Antarctica's first car (a Scottish 12 horse-power Arrol-Johnston that was taken back to England after its Antarctic sojourn) had collapsed, and the stables had filled with snow. After several more visits the hut was restored and repaired in the summer of 1960–61. There were still chemicals in the darkroom and supplies in the biology laboratory, and many more artefacts were discovered nearby.

A happy hut

All the expeditioners left here alive and largely successful, unlike the occupants of the tragic hut at Cape Evans, just down the coast; any ghosts at Cape Royds would be more likely to offer a tot of rum than a fright. Because so many expeditioners used the hut, it is difficult to determine the provenance of the clothes and objects that make it feel like a cheery museum exhibit. The seal blubber that lay thick around the stove before restoration was almost certainly left there by the stranded men of the *Aurora*, who had to burn blubber for fuel. The canned goods include Bird's Egg Powder, Curried Rabbit, Stewed Rump Steak, and an inordinate amount of canned cabbage—not perhaps an entirely happy choice of nourishment for men confined in close quarters. There are whole tins of Colman's Mustard, and the instructions for serving the ham loaf carry an ironic undertone: "Slice thin and *serve cold.*"

Scott was of Royal Navy training and believed in segregating his officers from his crew, but Shackleton was from Merchant Marine stock and treated all men as equals. The difference is reflected in graffiti on the walls at Cape Royds, including one on the wall close to the spot where *Aurora Australis*, the first book printed in Antarctica, was produced: "Wild & Joyce, printers, book binders etc. Gentlemen only," it reads. There are the signatures of Dr Mackay, who went to the South Magnetic Pole with Mawson and Edgeworth David, and of expedition artist George Marston, as well as of Shackleton, who wrote his autograph upside down on the casing that formed his bed end. Personal glimpses such as these are notably lacking in Scott's huts.

➤ ICE AND SNOW EVERYWHERE
The Cape Royds area looks over the frozen sea ice of McMurdo Sound toward the Transantarctic Mountains on the Antarctic Continent. Shackleton's hut, which is still in remarkably good condition, is wired to the ground in the foreground. Beyond the hut lies Pony Lake, also solidly frozen.

The Dry Valleys

Like all the world's deserts, Antarctica has oases. However, unlike hot, sandy deserts, where oases are places where water is found, the oases of Antarctica are defined as ice-free areas. These are special places on a continent that is 98 percent ice-covered, and where much of the remainder is exposed coastline or rugged mountain ranges. The most extensive Antarctic oasis areas are the Larseman and Vestfold hills, near Australia's

is the one normally visited by helicopter from tourist ships or from Ross Island. This valley, and the Taylor Glacier at its western end, was named by Robert Scott in 1903 for Australian geologist Griffith Taylor. Scott's report of the discovery reflects the awe that everyone experiences here: "I was so fascinated by all these strange new sights that I strode forward without thought of hunger until Evans asked if it was any

▲ WATERLESS WASTES
The lunar landscape of Bull Pass between McKelvey and Wright valleys is characteristic of the Dry Valleys terrain. The valleys receive about 1 centimeter (less than ½ in) of snowfall each year but evaporation greatly exceeds precipitation. The ground surface is dry, but there is permafrost beneath it.

Davis base; the Bunger Hills, near Casey station (also Australian); and the Dry Valleys, near McMurdo Station (United States) and Scott Base (New Zealand), on the western side of McMurdo Sound. The Dry Valleys are by far the largest of these oases, its ice-free area extending more than 4,000 square kilometers (1,550 sq. miles).

The Dry Valleys exist because this section of the Transantarctic Mountains acts as a dam wall containing the polar ice cap. The polar ice sheets have fallen some 500 meters so now only a few glaciers breach the threshold into the valleys.

Like the other Antarctic oases, the valleys are close to the coast, and are now bare because the glaciers that formed them have retreated and most of the continental ice is trapped behind the rock walls at the head of the valleys. The three main valleys—Taylor, Wright, and Victoria—run parallel to each other on an east–west axis. The Taylor Valley is the closest to the coast, and use carrying our lunch further … from our elevated position we could now get an excellent view of this extraordinary valley, and a wilder or in some respect more beautiful scene it would have been difficult to imagine … we have seen no living thing, not even a moss or a lichen; all that we did find, far inland amongst the morrain heaps, was the skeleton of a Weddell seal."

▲ FIELD WORK FASHIONS
United States scientists camp on a frozen tarn in one of the small dry valleys southeast of the main Dry Valley area of the Transantarctic Mountains. A helicopter brings supplies, and welcome new faces, to the people working here. The design of polar tents has not changed greatly since Scott's time, nearly a century ago.

◄ PRACTICAL NAMES
Airdevronsix Icefall was named in 1957 after the United States Air Development Squadron Six. The icefalls come off Wright Upper Glacier and drop to the floor of Wright Valley, both named after physicist and geologist Sir Charles Wright.

A living museum

The Dry Valleys are aptly named: their microclimate is dominated by dry winds, and so evaporation far exceeds precipitation. But the valleys are far from lifeless. In the 1970s scientists discovered minute bacteria, algae, and fungi living in tiny cracks in the rock, and blue-green algae have been found at the bottom of the few lakes that are not frozen solid. At the upper end of this simple food chain are three species of microscopic nematode worms that live on bacteria and can survive being freeze-dried in this extraordinary environment. The few seals that stray into the valleys do not survive, and mummified seal remains that have been found may be hundreds of years old. In the absence of precipitation, the strong winds that blow in the valleys have sculptured the rocks into fantastic shapes and wafer-thin hollow shells.

The Dry Valleys is an environmentally fragile region, and human visits are carefully controlled; some parts of the Valleys are so sensitive that tourist visits are not permitted, and even scientists need special approval for their studies in certain areas. For this reason few people have laid eyes on Antarctica's longest river, the Onyx, a meltwater stream flowing for 30 kilometers (20 miles) down the Wright Valley from the Wright Lower Glacier to Lake Vanda. In the same valley lies Don Juan Pond, a small lake that is so salty that it never freezes, although it is at most 75 centimeters (2½ feet) deep. When a new mineral, a calcium salt, was discovered here it was dubbed antarcticite. The lake is not named after the

◁ SALT ANTIFREEZE
Don Juan Pond is a pool of water lying in isolation in the south fork of the Wright Valley. It is considered the saltiest body of water on earth—and is so salty, in fact, that it never freezes, even in the depths of the Antarctic winter. There is no outlet from this pond; it is fed by groundwater and, for a few weeks of each year, by snow melt.

archetypal seducer; its name is a conflation of the names of two United States lieutenants—*Donald* Roe and *John* Hickey—who accompanied a 1961 field party.

The Dry Valleys constitute an environmental extreme, a place where plant and animal life can survive against incredible odds. In this cold and arid local climate even small climatic changes can have catastrophic consequences, so scientists continue to monitor weather conditions and conduct research in the Dry Valleys, with the hope of recognizing the first effects of climatic change that may be of worldwide significance.

▼ GLACIERS—LARGE AND SMALL

The Taylor Glacier (left) feeds Lake Bonney, which is nearly cut in two by Bonney Reigel, a terminal moraine on the southern side of the valley. The small, steep glacier (right) is the Rhone Glacier, typical of Dry Valleys' "alpine" glaciers: slow-moving because they are frozen to the glacier bed and cannot slide over it.

▼ LAKE OF SURPRISING CONTRASTS

Lake Vanda is ice covered year round (though in this late summer view there is a narrow moat of ice-free water around the shoreline) but the bottom of the lake is warm—a remarkable 25°C (77°F). Although the surface water is fresh, the water at the bottom of the deepest part of the lake is three times as salty as seawater.

The Ross Sea coast

▼ ITALIAN PRESENCE
The Italian station at Terra
Nova Bay is built on a small
granite peninsula at the
southern end of Gerlache
Inlet. One of the world's
largest Emperor penguin
colonies is found on the
sea ice at Cape Washington,
40 kilometers (25 miles) to
the east of the station.

Several important Antarctic sites lie along the western shore of the Ross Sea. Franklin Island, 12 kilometers (7½ miles) long, was named by Ross in honor of Sir John Franklin, governor of Tasmania. Franklin was later to seek (and perish in the attempt) the Northwest Passage with *Erebus* and *Terror*, the very vessels that Ross had used in Antarctica.

Terra Nova Bay is the site of an Italian summer station that was built in 1986–87 and can house 70 people. Scott discovered the bay and named it after his relief ship, which he later used as the main ship for his last expedition. Scott's six-man Northern party also named Inexpressible Island, just off the coast, when they were forced to winter in a cave there from March until

September 1912. They had disembarked for six weeks of summer exploration, but ice conditions prevented their ship from returning, and at the end of winter they had to cross 350 kilometers (220 miles) of sea ice to their expedition headquarters at Cape Evans. Mount Melbourne is a volcanic cone on the northern side of Terra Nova Bay. One of the few volcanoes on the Antarctic Continent itself, it is 2,730 meters (8,950 ft) high. Ross named it for Lord Melbourne, then British Prime Minister.

Coulman Island, offshore from the Victory Mountains, is part of an extinct volcano, 30 kilometers by 13 kilometers (19 by 8 miles) in extent and rising to 2,000 meters (6,560 ft). Today it is best known for its penguins: Adélies along the shore and Emperors on the winter fast ice. Ross named the island after his father-in-law, Thomas Coulman. Cape Hallett and the Hallett Peninsula take their name from the purser on Ross's *Erebus*. An International Geophysical Year joint New Zealand/ United States base was built on Cape Hallett in 1957, but was closed in 1973 and largely dismantled, relinquishing the site to the thousands of Adélie penguins who had abandoned the area when the base was built.

A scattering of nine small islands and numerous volcanic rocks, the Possession Islands lie about 8 kilometers (5 miles) off the coast of Victoria Land. On 12 January 1841 there was too much ice to permit a landing on the mainland, so Ross raised the British flag on the northernmost and largest of the islands and drank a toast to the Queen. Ross noted the presence on the islands of a large population of Adélie penguins that resented the invaders. There are now more than 300,000 birds there, and they still object to human incursions on their territory.

Cape Adare lies at the tip of a peninsula extending northwards from Victoria Land at the top of the Ross Sea. Ross named it in 1841 after his friend Viscount Adare, and the adjoining land was later called the Adare Peninsula. All around are ice and snow, but the cape itself is a barren basalt ridge with a small triangular patch of gravel at sea level on the western side.

This is penguin country; in the peak breeding season it swarms with more than half a million adult Adélies, plus their chicks. It is a malodorous rookery with a history. It was here that the men of the whaler *Antarctic* disembarked in 1895, and squabbled ever afterwards about who was first ashore in what they believed was the first landing on the Southern Continent. Carsten Borchgrevink, one of the landing party, later raised money for the *Southern Cross* expedition to winter here in 1898—the first expedition to winter on the Southern Continent. Their living quarters, designed and prefabricated in Norway of spruce, have been restored by the New Zealand Antarctic Heritage Trust. Borchgrevink's was not a happy party, and the hut, measuring 6.5 meters by 5.5 meters (19 by 18 ft), has little of the evocative atmosphere of the Ross Island huts.

Ridley Beach, named after Borchgrevink's mother, lies in front of a ridge upon which the Antarctic Continent's first burial site stands. When *Southern Cross* biologist Nicolai Hansen was dying in October 1899, he asked to be buried on top of the 350-meter (1,150-ft) ridge. His wish was granted, with a lot of hard work and the use of explosives to create the rocky grave. The brass plaque and iron cross that his fellow expeditioners lugged to the summit and erected are still there.

◄ PORT IN A STORM
From 1957 to 1964, Cape Hallett was the site of a year-round New Zealand/United States base that continued to operate as a summer station until 1973. After closure, many of the shelters were cleared away. Today, the main sign that humans have ever been here is this dome that was left standing to offer sanctuary in an emergency.

▲ MARVELOUS MOUNTAINS
Cape Adare looks across Robertson Bay to the majestic Admiralty Range tumbling to the sea in a series of ridges and glaciers on the north coast of Victoria Land. In 1841 James Clark Ross named the range after his collective superiors and Mount Minto within it for the Earl of Minto, First Lord of the Admiralty.

▶ UPLIFTING GRANDEUR
Even when exhausted, Scott's men could still admire lofty Mount Melbourne. Victor Campbell wrote on 11 January 1912 "The snow and mist cleared away ... giving us a magnificent view up a large glacier, the main body of which seemed to flow past the west slope of Mt Melbourne ... but all hands very tired."

East Antarctica

> **WELCOMING COMMITTEE**
A meandering line of Emperor penguins strays from the main huddle to inspect an observer on the fast ice breeding platform of Auster Rookery, near Mawson station. By late summer, most chicks have safely fledged and are ready to head out to fend for themselves in the cold waters of the mighty Southern Ocean.

At the South Pole there is no east or west—only north. However, geographers divide the world into eastern and western hemispheres, and Antarctica matches these divisions. East Antarctica—also known as Greater Antarctica because most of Antarctica's landmass is here—lies below Africa, Asia, and Australia. It is characterized by an ancient and stable Precambrian rock shield covered by an ice sheet about 2 kilometers (1¼ miles) thick, and has a roughly semicircular shape that quite closely follows the Antarctic Circle. West (or Lesser) Antarctica lies south of the Americas between the International Dateline and the Greenwich Meridian, and includes the Antarctic Peninsula, which is marked by the deep indentations of the Weddell and Ross seas and the extended arm of the Antarctic Peninsula. The Transantarctic Mountains form a natural barrier between East and West Antarctica.

There is a significant difference between the underlying landforms of East and West Antarctica although they are not immediately apparent because of the deep covering of ice across the continent. However, if the ice were to disappear, East Antarctica would be a continent similar in size and nature to Australia, whereas West Antarctica would be a much smaller land mass tapering away to a series of small islands and one large one with a long narrow spine along what is now the Antarctic Peninsula. In East Antarctica, large mountain ranges would stand revealed and there would also be a few large basins that would become virtually inland seas.

East Antarctica contains about 90 percent of Antarctica's ice. It has very little exposed land, virtually all of it along the coast, and it is generally occupied by wildlife and scientific bases in uneasy conjunction. Not many people have ever seen this remote region, and there are only a few weeks each year when even icebreakers can approach. Until quite recently, exploration was largely along the shore, and the place names reflect the nationalities of early Antarctic expeditions. From west to east, the 11 major "lands" on this side of Antarctica are Dronning Maud Land, Enderby Land, Kemp Land, Mac.Robertson Land, Princess Elizabeth Land, Wilhelm II Land, Queen Mary Land, Wilkes Land, Terre Adélie, George V Land, Oates Land, and Victoria Land.

▲ **BARED BY THE SUN**
Towards the end of summer, the maximum amount of underlying rock is exposed at the end of the snow melt. This is the 6.5-kilometer (4-mile) long Browning Peninsula at the southern end of the Windmill Islands, a biologically rich region close to Australia's Casey station.

▼ **ICEBOUND PERIMETER**
Much of the coast of East Antarctica is unapproachable from the sea. The extent of the Windmill Islands, about 27 kilometers (17 miles) long and 10 kilometers (6 miles) wide is immediately apparent from the air. In this dangerous environment, aviation mishaps occur often, but surprisingly few have been fatal.

▲ OBLITERATION OF A STATION

When the United States Wilkes station was built in just 17 days in 1957, it was located in a protective hollow. By the end of the first year it was being overwhelmed by snow accumulation and was largely abandoned in 1965. Unlike this wooden building that is still visible, many of the structures were Jamesway huts of fabric over metal that are now buried or collapsed.

Commonwealth Bay

On 8 January 1912 Douglas Mawson's Australasian Antarctic Expedition reached a bay in the Dumont d'Urville Sea. They called it Commonwealth Bay, and the cape to the east they named Cape Denison, after a Sydney sponsor of the expedition. Time was running out, Mawson was anxious to find a base, and this landfall seemed perfect. Mawson wrote: "The sun shone gloriously in a blue sky as we stepped ashore … The rocky area at Cape Denison … was found to be about one and a third miles in length and half a mile in extreme width. Behind it rose the inland ice, ascending in a regular slope and apparently free of crevasses." The expedition began to unload stores ashore. "The day had been perfect, vibrant with summer and life," wrote Mawson, "but towards evening a chill breeze off the land sprang up …" The "breeze" scarcely relented over the two years that they were there, for Cape Denison lies directly in the path of the katabatic winds that stream down from the polar ice cap; even today these winds can make landings impossible. But those who do land can walk through the historic beginnings of Australia's extensive Antarctic claim.

The main attraction is Mawson's Hut—actually two huts, one was mainly for accommodation and the other a workshop. In Cape Denison's savage climate these structures deteriorated badly before there was any thought of preserving them, but restoration began in the early 1980s and continues to this day. To see how tough the conditions have been, one need only look at the much better condition of the magnetograph house nearby. It has survived well because the topography protects it from the southerly winds. It and the adjoining magnetic absolute hut were built to hold the delicate equipment necessary to measure the changing Magnetic fields so close to the magnetic South Pole. The transit hut was used as an essential windblock while taking star sightings. In memory of two members of Mawson's party B.E.S. Ninnis and Xavier Mertz, who died on the ice, in November 1913, "on the highest point of Azimuth Hill, overlooking the sea, a Memorial Cross was raised to our two lost comrades."

▼ PRESERVING HISTORY
After Mawson's party left Cape Denison in 1913, the huts were visited in 1931 and again in the 1950s. By the 1980s, the roof of the main hut was largely supported by the snow inside. After several restoration projects, in 2001 it was announced that the buildings had been preserved to a state that would endure for at least another decade.

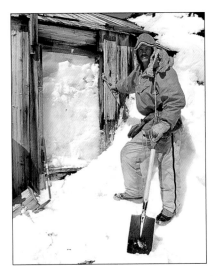

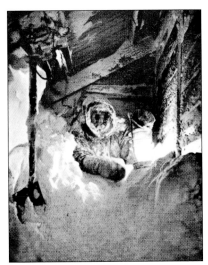

◄ SNOW—AN EVER PRESENT COMPANION
In "The Home of the Blizzard," Mawson captioned this photograph taken by Australian Frank Hurley: "An incident in March soon after the completion of the hut: Alfred Hodgman, the night watchman, returning from his rounds outside. Pushes his way into the veranda through the rapidly accumulating drift snow."

Dumont d'Urville

Terre Adélie, named by the great French Antarctic explorer Jules-Sébastien-César Dumont d'Urville after his wife, is a small slice of French claim between the Australian claims. It lies between Pourquoi Pas Point (136°11′E) and Point Alden (142°02′E). The French have been active there since 1950, and the present Dumont d'Urville station was built in 1956 to replace an earlier one that burned down in 1952 at Port-Martin, 100 km (60 miles) to the east. The current station is on Petrel Island, at the southeastern end of the Géologie Archipelago and 2 kilometers (about 1¼ miles) from the continental coast. Some 49 buildings are now scattered over an island less than 1.5 kilometers (1 mile) long.

The station, which is almost universally known as "DuDu," accommodates about 30 in winter and 120 in summer. It was first used as a winter base in 1958. Its beautiful setting is marred by a gravel runway that was created (amidst considerable controversy) by dynamiting an island and some islets that held Adélie penguin rookeries. In January 1984, when the runway was complete but unused, a huge storm-driven iceberg damaged both runway and hangar. The French government yielded to pressure and decided not to repair them, so the runway remains unused.

Rather ironically, Dumont d'Urville is an important center for the study of the rich local wildlife including seals, petrels—and penguins. Between early December and early March there are four resupply voyages by the French icebreaker *L'Astrolabe*, based in Hobart. Five kilometers (3 miles) away at Cape Prud'homme there is an array of oversnow vehicles jointly owned by France and Italy to service the inland Dome C/Concordia project.

FRENCH PRESENCE
The present Dumont d'Urville station was built by the French in 1956 to replace an earlier one at Port-Martin, 100 kilometers (60 miles) to the east, destroyed by fire in 1952. The buildings and a gravel runway that has never been used occupy tiny Petrel Island at the southeastern end of the Géologie Archipelago.

◄ OBSTACLE COURSE
On a typical day, the surface of Commonwealth Bay is shaped by wind-driven waves. Mackellar Islets, a group of about 30 rocks and small islands, occupy the middle of the bay. Several other uncharted shallowly submerged rocks add to the navigational hazards of these waters.

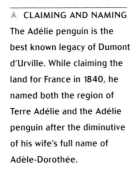

CLAIMING AND NAMING
The Adélie penguin is the best known legacy of Dumont d'Urville. While claiming the land for France in 1840, he named both the region of Terre Adélie and the Adélie penguin after the diminutive of his wife's full name of Adèle-Dorothée.

DECEPTIVE COASTLINE
Approaching d'Urville's landfall, Mawson wrote: "on the port side ... rose a massive barrier of ice. The problem to be solved was, whether it was the seaward face on an ice-covered continent, the ice-capping of a low island, or only a flat-topped iceberg of immense proportions."

Casey to Zhongshan

Casey

Australia's Casey station is situated just outside the Antarctic Circle among low, rocky islands and peninsulas near the edge of the Antarctic ice cap. Law Dome, a circular ice cap 200 kilometers (125 miles) in diameter and 1,395 meters (4,575 ft) high, lies 110 kilometers (70 miles) inland. It was named after Phillip Law, director of the Australian Antarctic Division from 1949 to 1966, who explored much of Australia's Antarctic claim.

The first structure in the area was Wilkes station, erected on Clark Peninsula by the United States in 1957 and operated as a joint base until it was handed over to Australia in 1963. By then it was falling apart and largely buried under snow, so in 1964 the Australian Government approved a new base, to be built a short distance away on rocky Bailey Peninsula. It was named for Lord Casey, Australian Governor-General at the time and a strong supporter of Australia's Antarctic program. Completed in February 1969, it was built on stilts so that snow could blow under its 13 buildings, which were linked by an unheated tunnel, but it vibrated in high winds and panels blew off in blizzards. In 1979 work began on a new Casey station, not far to the west and also on Bailey Peninsula. The new station was opened in December 1988. Its 16 multi-colored buildings are affectionately referred to as Legoland; they house 17 to 20 people over winter and up to 70 in summer. With steel frames on concrete foundations and an external skin of steel-clad polystyrene panels, they are a far cry from the plywood of Wilkes, some of the roofs of which are still visible above the snow.

Mirny

Russia started regular Antarctic research in 1956, and opened Mirny station in February of that year. Russia's current economic problems have forced her to seek international cooperative research to reduce costs, and the average winter population in Mirny's 30 buildings is

60, rising to a maximum of 169 in summer. One supply ship visits each summer when the ice has broken out—otherwise supplies must be transported 40 kilometers (25 miles) across the ice, since fast ice has re-formed by early April. In winter the ice near the Haswell Islands in front of Mirny is home to an important colony of Emperor penguins, so the accumulation of 40 years of abandoned vehicles and debris at the station is cause for concern, but there is no money to clean it up.

Davis

An Australian party under Phillip Law landed in the Vestfold Hills in March 1954 and decided that establishing an Australian base there would pre-empt possible Russian plans to move in. Davis Station, on the edge of the Hills, was built in January 1957 for the International Geophysical Year and named after John King Davis, ship's captain for Shackleton and Mawson. It is both the most southerly and the most temperate of the Australian bases on the Antarctic Continent: expeditioners call it the Riviera of the South. The hills do provide some shelter, but Davis has experienced wind gusts of 206 kilometers (130 miles) per hour. The station has 29 buildings supporting 22 in winter and up to 70 in summer—the largest population of scientists of any Australian continental base. There are three ship's visits, one in mid-October when goods must be transported over the sea ice from the ship to the shore.

The Vestfold Hills were discovered in January 1935 by Norwegian whaling captain Klarius Mikkelsen. He

▲ WASTE DISPOSAL
This sign at Casey station may be a humorous way-marker but it also points to a problem for all field trips: removal of waste. Before Australian stations had full sewerage systems, toilet waste was incinerated at the end of each day.

▼ COSTLY LITTER
This aircraft wreck at Mirny represents an ongoing problem for the weakened Russian economy—finding the considerable funds required to remove accumulated refuse and pollutants from Antarctica. The present station, built in 1971, replaced an earlier one now largely buried under ice.

◄ THE COMFORTS OF CASEY
The current Casey station is the third Australian base to have been built at Vincennes Bay. Scientific programs here include marine biology, atmospheric and space physics, botany, and human adaptation to the cold. Even in winter, the base's hydroponics building can deliver herbs and fresh vegetables.

Snowflakes fall gently on a station entry sign with its implicit promise of comfort. Except during the fiercest blizzards, travel in familiar Antarctic territory today with the help of satellite navigational aids and modern radar is relatively easy, even in inclement weather.

▼ BENEATH THE CHAOS GLACIER IN RANVIK BAY
Diving in polar waters presents specific technical problems: while salt content keeps the sea liquid below 0°C, moisture in air and lungs will freeze at this temperature. Polar divers use several regulators and take other precautions to minimize the hazards.

landed here with his wife, Karoline, who thus became the first woman to set foot on the Antarctic Continent. The largest ice-free area on the Antarctic coast, it is a beautiful place of lakes and fiords; the southernmost nesting area of giant petrels is on nearby Hawker Island.

Zhongshan

China acceded to the Antarctic Treaty on 9 May 1983 and was accepted as a consultative party on 7 October 1985. The Polar Research Institute of China started in Antarctica when it sent two scientists to Australia's Casey station in January 1980. Following the establishment of Great Wall station on King George Island in 1985, Zhongshan Station was completed on the Larsemann Hills of Prydz Bay in February 1989. It can accommodate 15 to 20 people in winter and up to 60 during the summer.

► PRECARIOUS FOOTHOLD
With so little of the Antarctic continent free of permanent ice, modern research stations, such as Zhongshan, perch on rocky outcrops fringing the coastline. Regular iceberg calving from the nearby Dalk Glacier often chokes seaward approaches to the station, making resupply difficult, and at times generating huge waves—one of which in 1988 caused a hasty relocation of the base to the higher ground it now occupies.

Mawson to Belgrano

Mawson

Mawson was named after Australia's most distinguished explorer of the Heroic Age, Sir Douglas Mawson. On New Year's Day 1930, during the BANZARE Expedition, Mawson took off in a seaplane and saw a coast he called Mac.Robertson Land, creating a permanent punctuation mystery; it is named for Sir MacPherson Robertson of Melbourne, a supporter of the expedition. Throughout his life Mawson pressed for a continuing Australian presence in Antarctica, and in the 1950s Phillip Law used photographs from the United States' Operation Highjump to decide to build a year-round Australian station on the firm rock of Horseshoe Harbor. Law records that Mawson was not convinced, but Law prevailed and the two were later photographed examining a photograph of the new station. The Australian flag was raised at Mawson station on 13 February 1954 and 10 expeditioners over-wintered that first year, making Mawson the longest-operating station south of the Antarctic Circle.

In 1956 Antarctica's first aircraft hangar was built on the shore near Mawson. Today the station uses a blue-ice airstrip 5 kilometers (3 miles) away, and light aircraft and helicopters operate from October to February. The 35 buildings house 20 people in winter and 60 in summer. Many of the original buildings are still standing, though now put to different uses—even the husky kennels remain, though unused. The main living area, consisting of bedrooms, a cinema, a bar, a mess, and recreation areas, is unceremoniously known as the Red Shed.

SNOW SAFETY
There are currently about 10 Honda quad all-terrain motorcycles at Mawson and they are useful for short trips around the station. However, the issue of wearing crash helmets became contentious in 1990 when a directive came through from Hobart ordering their use at all times.

SAFE HARBOR
With mooring cables holding firm fore and aft, the re-supply vessel *Icebird* sits snugly in Horseshoe Harbor in front of Mawson station. The pristine snow-covered outcrop of West Arm is highlighted by the dark storm clouds beyond.

TREADING LIGHTLY
Hagglunds are used for field-work. With a load capacity in excess of 2 tonnes, they can tow a trailer of equal weight, yet the footprint of their tracks is less than half the weight of a person's. Easy to drive, power is transmitted to each of the four tracks, so they can go virtually anywhere and tackle most gradients.

Syowa

The Japanese station Syowa, on the northern end of East Ongul Island, opened in January 1957 after the Japanese Antarctic Research Expedition of 1956. The Norwegian cartographers who mapped the area from aerial photographs taken in 1936–37 thought that the two parts of this landform were one island, and called it *ongul*, Norwegian for fishhook, but when the Japanese Antarctic Research Expedition arrived they found that the "island" was split by a strait. Syowa has grown from the original three buildings to 47 buildings today. The impressive main building is three stories high, topped by a domed skylight, and the station houses some 20 expeditioners in winter and up to 60 in summer. Each year it is supplied by an icebreaker that anchors nearby, and there is a sea-ice airstrip that is also quite close. Heavy sea ice is a constant problem that has forced some breaks in operation.

▼ PIONEERING LIVE TELEVISION

With live cameras a feature of many Antarctic station websites today, it is hard to imagine that the first live television transmission was from brightly-colored Syowa station in 1979. Syowa lies on Lützow-Holm Bay within an oasis where some of the coast and islands are ice-free and at odds with the pack ice that often fills the bay.

▲ SCENIC SERENITY ON A GRAND SCALE

Cradled by the arms of a natural amphitheater, a vessel lies securely anchored in Horseshoe Harbor off Mawson station in Mac.Robertson Land. Lightly crevassed slopes lead gently onto the plateau where the peaks of Mount Henderson (left) and the Northern Masson Range (right) pierce the blue continental ice.

▲ ON THE EDGE
SANAE IV is spectacularly located on top of Vesleskarvet Cliff, which is part of the 110 kilometer (70 mile) Ahlmann Ridge in Dronning Maud Land. The ridge runs south from the Fimbul Ice Shelf where earlier South African stations have been located. This area was first seen from the air by the Germans in 1938–39.

▲ HEAVY HAULAGE
The task of constructing SANAE IV, Antarctica's newest station, was considerable as all the materials had to be hauled from the coast by tractors and assembled above the ice and on top of a ridge. During summer, accumulated waste from the station is carried back to the coast and shipped to South Africa.

Novolazarevskaya

Russia's "Novo" station is located at Schirmacher Oasis, 75 kilometers (47 miles) inland on Dronning Maud Land and 5 kilometers (3 miles) from India's Maitri station. Schirmacher is a region of ice-free hills about 17 kilometers (10½ miles) long and dotted by meltwater ponds. The station, built on rock 102 meters (335 ft) above sea level, opened in January 1961. The nine buildings house 30 over winter and up to 70 in summer. A resupply ship visits the coast each year, and supplies are either flown from the ship by helicopter or transported by tractor trailer, depending on the fast-ice conditions.

Maitri Station

The first Indian Antarctic Expedition took place in 1981, and India was admitted to the Antarctic Treaty in August 1983. The same year saw the construction on one of the ice shelves on Princess Astrid Coast of Dakshin Gangotri, India's first permanent Antarctic station. When that closed in 1989, Maitri was commissioned, and was opened on 9 March 1989. A modern station of innovative design, it is built on adjustable telescopic stilts. Winter staff have their own rooms in the main complex, and there is accommodation for 65 when the summer huts are in use. Two doctors are in residence in winter and four in summer, and there is a permanent team of Indian Army engineers to service the equipment.

Like nearby "Novo," Maitri sits on ice-free rock. India's National Center for Antarctic & Ocean Research, which opened in 1997, is in Goa—a world

away climatically. But India has strong links with Antarctica: Maitri's geological interests include the study of Gondwanaland, when India and Antarctica belonged to the same landmass, and low-temperature engineering research in Antarctica is directly relevant to conditions in the Himalayas.

SANAE IV

South Africa was an original signatory of the Antarctic Treaty, and has been conducting research in Antarctica since the International Geophysical Year of 1957–58. It has two sub-Antarctic stations, on Marion and Gough islands, and completed the Antarctic Continent's newest year-round station, SANAE IV, in the summer of 1997–98. The station is 69 meters (226 ft) above sea level on the Vesleskarvet nunatak in Dronning Maud Land. Vesleskarvet was mapped by the Norwegians—the name means "the little barren mountain," but the South Africans simply call it "Vesles."

There have been year-round South African stations on the Fimbul Coastal Ice Shelf of western Dronning Maud Land since 1960. SANAE IV is the first built on rock, not ice, and is located 170 kilometers (106 miles) from the coast, due south of SANAE III. Construction began in the summer of 1993–94 using prefabricated panels made in South Africa, shipped to the Antarctic coast, and carried overland to Vesleskarvet by tractor trains. The station opened in January 1997, and that year saw the first over-winterers. It is designed for 20 residents in winter and up to 80 in summer. It is resupplied twice each summer, and has a helipad and airstrip for flights to and from the coast and Germany's Neumayer station.

SANAE IV's smooth surfaces and rounded corners were designed after wind-tunnel testing. There are three main two-story buildings 14 meters by 44 meters (46 by 145 ft) in area, raised on stilts 4 meters (13 ft) above the rock and linked by passageways at the lower level. The exterior is dark at the base to retain heat and discourage snow, while the roof is fluorescent orange for visibility—it matches the color of the station's vehicles.

Neumayer

Neumayer was a nineteenth century German geophysicist who actively promoted his country's Antarctic exploration; German, Swedish, Belgian, and British expeditions have named Antarctic features after him. Germany established the first Georg von Neumayer Station in Antarctica in 1981 in the northeastern Weddell Sea on the Ekström Ice Shelf, a completely flat ice shelf 200 meters (655 ft) thick.

By 1990 snow build-up and ice movement necessitated a move to a new station 10 kilometers (6 miles) away, and the present station opened in March 1992. It is 40 meters (130 ft) above sea level, and is snow-covered: all that can be seen from a distance are the two towers at the station entrance, standing about 2 meters (6½ ft) above the snow. The station consists of two enormous parallel steel tubes around 90 meters (300 ft) long, divided into living quarters, mess, hospital, workshops, and so on. Utilities are carried in a transverse tube, and there is a tunnel to the garage. In winter a staff of nine, headed by the base doctor, passes nine months in total isolation, except for radio contact.

The station is 8 kilometers (5 miles) inland from its deepwater anchorage at the edge of the ice. Each summer, two ship's visits bring supplies and the summer population, which may reach 50, and the nearby snow runway receives one intercontinental flight and about a dozen intra-Antarctica flights. Neumayer's principal research field at present is the role of the polar regions in world climate.

Halley V

Britain was among the first signatories of the Antarctic Treaty, and it has a long history of Antarctic exploration and research. The first Halley station opened in January 1956 for the International Geophysical Year. It was erected on stilts over the Brunt Ice Shelf, but had to be rebuilt in 1966, 1972, 1982, and 1989, as each successive station sank beneath the snow. The station is named after Edmond Halley, the British astronomer who first mapped the declination of the earth's magnetic field, and also charted the course of Halley's Comet. In 1985 the British Antarctic station at Halley alerted the world to the springtime depletion of ozone in the Antarctic stratosphere, which led to an international agreement to ban chlorofluorocarbons (CFCs).

The present station is Halley V, about 12 kilometers (7½ miles) from the coast and 37 meters (120 ft) above sea level. It is Britain's most isolated Antarctic station. Three of its buildings stand on telescopic stilts so that they can be raised above the rising snow, and a garage, which weighs more than 50 tonnes, and an accommodation building are mounted on skis so they can be towed to a new location every year. The average winter population is 15, rising to a maximum of 65 in summer. Supplies are delivered twice a year, first by ship to the edge of the ice, then across the ice shelf by surface transport, and Twin Otter flights operate in summer from the nearby snow runway. During Halley's 105 days of total darkness each winter the station staff enjoy magnificent auroral

displays, and from May to February they are joined by a nearby Emperor penguin colony which has some 15,000 breeding pairs.

Belgrano I

Founded in January 1955, Argentina's Belgrano Base was briefly the southernmost base in the world until the South Pole station was built in 1956. The first Belgrano was built on the Filchner Ice Shelf, but was buried in snow by 1978 and abandoned—a wise decision, as that bit of the ice shelf subsequently broke away and headed out to sea.

Belgrano II opened in February 1979. It is located on Bertrab nunatak, one of the few exposed areas of rock at the southwestern corner of the Weddell Sea near the Filchner Ice Shelf. Built on solid rock 50 meters (163 ft) above sea level and 120 kilometers (75 miles) inland, it has 12 buildings with a year-round population of 21. The region is noted for violent storms, and the base experiences four months of darkness each winter; even in summer the only wildlife are some gulls, petrels, and skuas. The station is serviced by an airstrip built on the surface of a glacier.

▲ ACCOMMODATING SNOW
Indicative of the extremity of the Antarctic climate, the current Neumeyer station has been deliberately dug into the ice shelf—the station there previously disappeared under accumulated snow. Now the only above-ground structures are aerials, a wind generator, a balloon launching facility, and an atmospheric research laboratory that can be jacked up as snow accumulates.

➤ LIMITED SHELF LIFE
Halley V stands on a steel platform supported by stout telescopic legs. The station is mechanically raised nearly 1 meter (3 ft) a year to maintain a near constant level above the surface snow of the Brunt Ice Shelf, which has already crushed and buried its four predecessors. Halley V is also being carried slowly out to sea by the inexorable motion of the floating ice shelf upon which it rests.

Inland stations

Amundsen–Scott, South Pole

When Sir Edmund Hillary and his party of New Zealanders arrived at the South Pole aboard their second-hand Ferguson tractors on 4 January 1958, they were the first people to arrive overland at the pole since Robert Falcon Scott's team in 1912. However, Hillary was met by the American residents of the Amundsen-Scott station, who had flown to the bottom of the world. There has been a continuously occupied United States station at the geographic South Pole since November 1956, and there are now 33 flights there each year.

The most distinctive feature of what is commonly referred to as South Pole station is The Dome, an aluminium geodesic dome 50 meters (164 ft) wide and 16 meters (53 ft) high, enclosing three two-story buildings and a mass of equipment. Built in 1975, it is unheated and the floor is packed snow, but it does provide shelter from the wind. In winter the station houses 28, but in summer, when accommodation outside the dome can be used, 130 people is the average. No flights to the South Pole are possible between mid-February and late October, so the residents are as isolated as astronauts, but they can communicate by e-mail and telephone via satellite.

The station is 2,835 meters (9,295 ft) above sea level, high enough to cause headaches in many newcomers, and its recorded temperature range is −13.6°C to −82.8°C (7½°F to −117°F). The mean annual temperature is −49°C (−56°F), and the cold is exacerbated by an average wind speed of 5.5 meters per second (18 fps). (The highest wind speed recorded here was 24 meters per second—79 fps.) Beneath the ground snow lies 2,850 meters (9,345 ft) of polar ice cap. Each year the

▼ **WINTERING ALONE**
During the long winter at Amundsen-Scott base temperatures are too low for aircraft to land. Most coastal bases on the Antarctic continent are equally isolated. Telephone and email links ensure remote contacts with the outside world, but once the last flight or ship departs, no new faces will be seen until spring.

ice cap moves 10 meters (33 ft) towards the sea, so the South Pole marker is repositioned annually. The ceremonial pole that features in so many photographs is a reflective globe encircled by the twelve flags of the original Antarctic signatories.

The Dome has outlived its usefulness and is to be replaced by an outside structure—two U-shaped buildings on stilts, connected by enclosed walkways, and with space for 110 winter residents. The South Pole Station Modernization Project is expected to be completed in 2005.

Vostok

The lowest temperature recorded on earth was measured at Russia's Vostok: –89.2°C (–128½°F) on 21 July 1983. This record is unlikely to be broken except at Vostok itself, because the station is situated high on the polar ice cap, 3,488 meters (more than 11,500 ft) above sea level, and 1,399 kilometers (870 miles) from the moderating effects of the ocean. The highest temperature ever recorded there was –22°C (–7½°F). In June 1982 a nightmare scenario unfolded when a fire destroyed the powerhouse at Vostok, and 20 wintering expeditioners had to survive in a small building with the help of an improvised oil heater until they could be evacuated by aircraft the following November.

Vostok began operations on 16 December 1957, and ice core drilling started in 1959. Except in the winter of 1994, the station has operated continuously since then, but its activities have been scaled down in recent years because of the Russian economic crisis. There are five buildings, and the winter population is 13, rising to 25 in summer. The station is supplied by tractor trains from the coast, which take two months to travel from Mirny, 1,410 kilometers (876 miles) away.

Below the station, 4,000 meters (more than 13,000 ft) beneath the ice, lies Lake Vostok, Antarctica's largest "lake."

Geothermal warming has produced a vast freshwater lake measuring 60 kilometers by 280 kilometers (37 by 174 miles) and hundreds of meters deep. Closed off by ice for many thousands of years, these lakes may harbor life forms long extinct elsewhere and provide valuable information about early climatic conditions. But if Lake Vostok is penetrated using contemporary technology this precious resource will be compromised, so drilling has been halted just short of the lake—at ice that was on the surface about 400,000 years ago—while scientists try to devise a way of investigating the lake without contaminating it. Since 1964 Russia has operated a cooperative deep drilling program with France, and since 1991 United States ski-equipped Hercules LC130s have flown winter teams from Mirny to the compacted ice airstrip near Vostok as part of a joint program for ice-core and lake investigations.

Concordia

In March 1993, France and Italy signed an agreement to construct and operate a new Antarctic wintering station at Dome C, to be called Concordia. Supported by the European Union and by 10 European countries, its main goal is core drilling through deep ice. The station's first stage, which opened in December 1997, consists of three buildings, linked by enclosed walkways, that can house up to 45 expeditioners in summer; from 2003 it will operate around the year, with 15 winterers. These are the first of up to 12 proposed buildings. Nearby is a leveled snow airstrip 3,000 meters (9,800 ft) long.

Concordia is a remote inland base inside the polar vortex; the ozone hole can be detected there in the southern spring. It is 3,220 meters (10,560 ft) above sea level. Its closest neighbor is the Russian station Vostok, 560 kilometers (350 miles) away. It is 950 kilometers (590 miles) from the coast, about 1,100 kilometers (685 miles) from Dumont d'Urville, and 1,200 kilometers (750 miles) from Terra Nova Bay. It is serviced by tractor trains from Dumont d'Urville and by ski planes from Terra Nova Bay or Dumont d'Urville.

▲ EYES ON THE SKIES
The sun circles lazily above the horizon casting long shadows from the weather watching instuments at the South Pole. The measurements recorded by these instruments include air temperature, wind speed, incoming ultraviolet radiation, total ozone concentration. The results are fed into global models and used for meteorological forecasting.

The Sub-Antarctic Islands

The sub-Antarctic islands

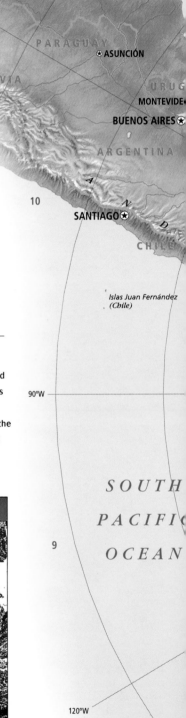

Rio de Janeiro

BRAZIL

60°W

PARAGUAY

⊛ASUNCIÓN

BOLIVIA

URUG

MONTEVIDE◦

BUENOS AIRES ★

ARGENTINA

10

SANTIAGO★

CHIL

Islas Juan Fernández
(Chile)

90°W

SOUTH

PACIFI

9

OCEAN

120°W

8

Previous pages

◄ **ISLAND JEWELS**

Though isolated and bleak, the sub-Antarctic islands are austerely beautiful, and are home to an impressive array of hardy wildlife that returns time and again to these rocky outcrops in a vast ocean.

▲ **REGAL BIRDS**

Largest of penguins except the Emperor, King penguins live only on sub-Antarctic islands. The bird's Latin name, *Aptenodytes patagonicus*, is misleading because it is not found in Patagonia.

The vast Southern Ocean and its neighboring oceans are far from empty. Islands are scattered across these seas—but there is a lot of water in between. Most of these islands were discovered by chance and few are habitable. Many early "discoveries" were mistakes, and over centuries new islands have been added to—and others deleted from—nautical charts. For someone standing on the icy deck of a ship and peering though driving sleet or rain, a distant iceberg seen vaguely through the fog may look a lot like land.

Even defining what constitutes a sub-Antarctic island is something of a challenge. Islands that are truly sub-Antarctic lie south of the Polar Front. Cool temperate islands of the southern oceans, including the Falkland Islands, Macquarie Island, and Iles Kerguelen and Crozet are covered here—all are important sites for Antarctic wildlife, and are frequently visited, significant locations in the human history of Antarctica. The comparatively warm temperate Auckland and Campbell islands, southeast of New Zealand, are also called sub-Antarctic in some contexts, —they are often visited on the way to the Ross Sea. Vegetation on sub-Antarctic islands ranges from mosses to small trees; life is gentler than on the Continent and the islands support a great variety of wildlife.

▼ **CALVING ICEBERGS**

The glacial snouts of South Georgia frequently calve and litter the surrounding waters with icebergs and ice fragments. Glacial retreat over the past 120 years has exposed expanses of bare rock that plants have yet to colonize.

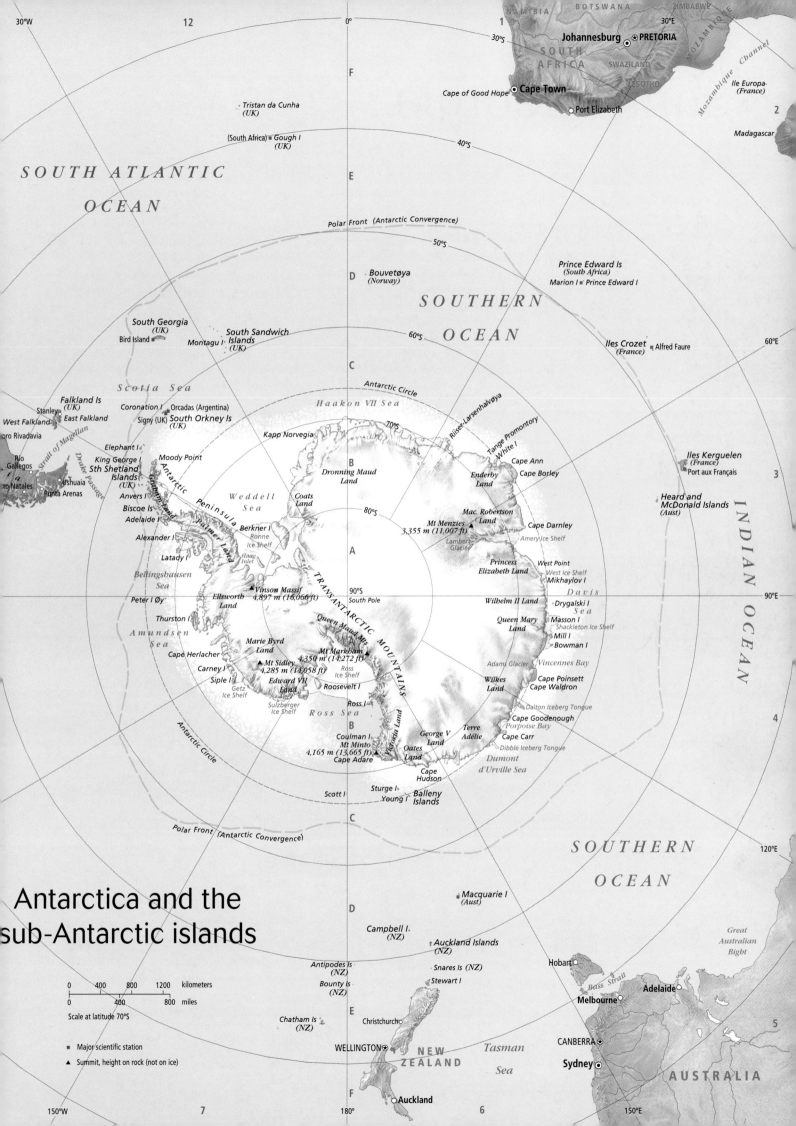

Antarctica and the
sub-Antarctic islands

SOUTH ATLANTIC OCEAN

SOUTHERN OCEAN

INDIAN OCEAN

Tasman Sea

Johannesburg ⊙ PRETORIA

SOUTH AFRICA
SWAZILAND
LESOTHO
NAMIBIA   BOTSWANA   ZIMBABWE
MOZAMBIQUE

Cape of Good Hope
Cape Town
Port Elizabeth

Mozambique Channel
Ile Europa (France)

Madagascar

Tristan da Cunha (UK)

(South Africa) ▪ Gough I (UK)

Bouvetøya (Norway)

Prince Edward Is (South Africa)
Marion I ▪ Prince Edward I

Iles Crozet (France) ▪ Alfred Faure

Polar Front (Antarctic Convergence)

South Georgia (UK)
Bird Island
Montagu I
South Sandwich Islands (UK)

Iles Kerguelen (France)
Port aux Français

Antarctic Circle

Haakon VII Sea

Scotia Sea

Coronation I   Orcadas (Argentina)
Signy (UK)   South Orkney Is (UK)

Kapp Norvegia

Dronning Maud Land

Riiser-Larsenhalvøya

Tange Promontory
White I
Cape Ann

Heard and McDonald Islands (Aust)

Falkland Is (UK)
West Falkland   East Falkland
Stanley
oro Rivadavia
Río Gallegos
to Natales
Ushuaia
Punta Arenas

Elephant I
King George I
Sth Shetland Islands (UK)
Anvers I
Biscoe Is
Adelaide I
Alexander I
Latady I

Strait of Magellan
Drake Passage

Antarctic Peninsula
Graham Land
Palmer Land

Moody Point

Weddell Sea

Coats Land

Enderby Land
Cape Borley

Mac. Robertson Land
Cape Darnley
Amery Ice Shelf

Berkner I
Ronne Ice Shelf
Haag Inlet

Mt Menzies
3,355 m (11,007 ft)
Lambert Glacier

Princess Elizabeth Land

West Point
West Ice Shelf
Mikhaylov I

Bellingshausen Sea

Vinson Massif
4,897 m (16,066 ft)
Ellsworth Land

TRANSANTARCTIC MOUNTAINS

Queen Maud Mts

South Pole

Wilhelm II Land

Davis Sea
Drygalski I

Peter I Øy

Thurston I

Marie Byrd Land

Mt Sidley
4,285 m (14,058 ft)

Mt Markham
4,350 m (14,272 ft)

Queen Mary Land

Masson I
Shackleton Ice Shelf
Mill I
Bowman I

Amundsen Sea

Cape Herlacher
Carney I
Siple I

Edward VII Land
Getz Ice Shelf
Sulzberger Ice Shelf

Roosevelt I

Ross Sea

Ross I

Ross Ice Shelf

Adams Glacier
Vincennes Bay

Wilkes Land

Cape Poinsett
Cape Waldron

Dalton Iceberg Tongue

Cape Goodenough
Porpoise Bay

Dibble Iceberg Tongue

Coulman I
Mt Minto
4,165 m (13,665 ft)
Cape Adare

George V Land
Terre Adélie
Oates Land

Cape Carr

Victoria Land

Cape Hudson

Dumont d'Urville Sea

Scott I
Sturge I
Young I
Balleny Islands

Macquarie I (Aust)

Campbell I (NZ)

Auckland Islands (NZ)

Antipodes Is (NZ)
Bounty Is (NZ)
Snares Is (NZ)
Stewart I

Hobart

Adelaide
Melbourne

CANBERRA
Sydney

Great Australian Bight
Bass Strait

AUSTRALIA

Chatham Is (NZ)

Christchurch

WELLINGTON ⊙

NEW ZEALAND

Auckland

Scale

0   400   800   1200   kilometers
0   400   800   miles
Scale at latitude 70°S

▪ Major scientific station
▲ Summit, height on rock (not on ice)

30°W   12   0°   1   30°E
30°S
40°S
F
E
50°S
D
60°S   60°E
C
70°S
B
80°S
A
90°S   90°E
B
C
120°E
D
E
F
150°W   7   180°   6   150°E   5

South Georgia

Isolated island

South Georgia has been likened to a piece of the Alps dropped in the middle of the ocean—a fair description, as the island is a summit of the Scotia Ridge, a line of mountains 4,350 kilometers (2,700 miles) long, that links the ranges of the Antarctic Peninsula with the Andes through a submarine hairpin loop to the east. The South Sandwich and South Orkney islands are other exposed peaks along the ridge.

South Georgia has no near neighbors. Antarctica, South America, and Africa are about 1,500 kilometers (930 miles), 2,100 kilometers (1,300 miles), and 4,800 kilometers (2,980 miles) away, respectively, and the closest substantial landmass is the Falkland Islands, 1,400 kilometers (870 miles) away.

After the Falklands War, South Georgia ceased to be a Falklands' dependency, and in 1985 became the British Dependent Territory of South Georgia and the South Sandwich Islands. Despite the new title's equal billing, South Georgia's land area—3,500 square kilometers (1,350 sq. miles)—is ten times larger than the 11 islands and several islets at the edge of the South Sandwich Trench that make up the South Sandwich Islands. These islands are a fog-enshrouded arc of very active volcanic, mostly ice-covered isles; they are rarely visited except by millions of breeding Chinstrap, Gentoo, Macaroni, and Adélie penguins.

South Georgia is uplifted rather than volcanic; its main landmass, about 170 kilometers (105 miles) long and 30 kilometers (18½ miles) wide in the middle, is

surrounded by numerous offshore rocks and islets. Mount Paget, 2,934 meters (9,626 ft) high, is the tallest of a spine of mountains running northwest to southeast. Although Antarctic pack ice does not extend this far north, the bays around the coast freeze in winter and glaciers and ice fields cover some two-thirds of the island. The south is higher and more rugged than the north, and in many places the mountains fall sheer to the sea. Because the mountains act as a barrier against the prevailing westerly winds, the northwestern side is much more sheltered than the southeastern side.

Just four people live permanently on South Georgia. The Harbor Master is responsible for island administration, and also serves also as Customs and Immigration Officer, Postmaster, and Fisheries Liaison Officer. He

lives at the base constructed in 1949–50 by the Falkland Islands Dependency Survey (FIDS—now the British Antarctic Survey, or BAS) at King Edward Point overlooking Grytviken within Cumberland Bay. In 1962–63 the base was extended and a residential building, Shackleton House, was built near Hope Point. The last British troops left South Georgia early in 2001; during the previous year the base was improved for the return of the shifting population of scientists.

No wood for a toothpick

The first person to see South Georgia is believed to have been London merchant Antoine de la Roché, who was blown far off course while rounding Cape Horn in 1675. It was next sighted in 1756 by the crew of the Spanish vessel *Léon*, but Captain James Cook was the first to step ashore there, disembarking from *Resolution* on 17 January 1775. With "a discharge of small arms" that reportedly alarmed the penguins of Possession Bay, he claimed and named the land for King George III. Cook hoped that this was the tip of the southern continent that he had traveled so far to find. He recorded his dashed hopes by calling the island's southernmost point, Cape Disappointment (the first of three capes in Antarctica so named), noting: "the disappointment I now met with did not affect me much, for to judge by the bulk of the sample it would not be worth the discovery."

Even before this, South Georgia had not impressed Cook: "I did not think it worth my while to go and examine these places where it did not seem probable that any one would ever be benefited by the discovery … The inner parts of the country were not less savage and horrible: the wild rocks raised their lofty summits till they were lost in the clouds and the valleys lay buried in everlasting snow. Not a tree or shrub was to be seen, no not even big enough to make a toothpick."

Cook did, however, report on the numerous seals, and in 1788 the first sealers arrived from Britain, followed in 1791 by others from the United States. Scanty records indicate that Edmund Fanning, of *Aspasia* from New York, may have had the most lucrative harvest; in 1800 he took 57,000 seal skins from the island. Seal numbers dropped rapidly thereafter, and South Georgia's sealing industry correspondingly declined.

In 1819 the Russian Antarctic circumnavigating expedition led by Thaddeus von Bellingshausen came to South Georgia and charted the southern coast that Cook had ignored. The German expedition for the International Polar Year arrived on *Moltke* in 1892 and left 11 scientists to winter over. Their base at Royal Bay, which is now in ruins, included an observatory.

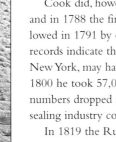

▲ SEAL SLAUGHTER
Between 1786 and 1825, some 1.2 million fur seals were slaughtered on South Georgia, bringing them to the verge of extinction. Building up again from small remnant colonies on Bird and Willis islands, the population may now be approaching 2 million.

◄ LINKING CONTINENTS
There are few parts of South Georgia where mountains do not tumble straight to the sea. It was Heinrich Klutschak (1848–90), an Austrian aboard *Flying Fish*, who first suggested that South Georgia may be part of a land link between Antarctica and South America.

numbers of all types except Minke rapidly declined. In the summer of 1925–26—South Georgia's peak whaling year—23 ships caught and killed 7,825 whales, including 5,709 Fin and 1,855 Blue whales, to produce more than 400,000 barrels of oil. In the following year the catch included 3,689 Blue whales—a significant proportion of the number of Blue whales alive today.

The largest animal ever measured, 33.5 meters (110 ft) from nose to tail and longer than any dinosaur so far discovered, was a Blue whale brought to Grytviken in 1912. Although land-based whaling became increasingly regulated, the 1925 invention of the whaling factory ship accelerated the slaughter. Factory ships operated where they wished, without supervision, and by 1930 more than 40,000 whales were being killed each year. Whale populations declined, and the last of the Norwegian whalers left South Georgia in 1962. Japanese companies took over Grytviken and Leith but found them uneconomic and closed them in 1965.

The most notable visitor to South Georgia during the whaling days was Sir Ernest Shackleton. He arrived there in November 1914 on *Endurance*, intent on making the first trans-Antarctic crossing. After *Endurance* was crushed in the ice, he returned to South Georgia on the tiny 23-foot *James Caird* in May 1916 and, with Frank Worsley and Tom Crean, completed the first island crossing. Shackleton returned to South Georgia with many of the same expeditioners aboard *Quest* in 1922, but on 5 January suffered a fatal heart attack while the ship was at anchor in King Edward Cove. At the request of his widow, Lady Emily Shackleton, he was buried in the Grytviken whalers' cemetery on 5 March. A month later, *Quest* returned from Antarctica and the men built a cairn surmounted by a cross at Hope Point in memory of this exceptional leader. The brass plate is simply inscribed: "Sir Ernest Shackleton, explorer. Died here 5 January, 1922. Erected by his comrades."

▲ ABANDONED STATIONS
The once-thriving whaling stations of South Georgia are falling apart; only Grytviken is considered safe for visitors, and even here they must not venture inside the station ruins. The manager's house has been restored as a museum and post office.

▲ PAST AND PRESENT
Grytviken today is a peaceful, pleasant historic site—a far cry from Frank Hurley's account of it in whaling days: "Slimy waters lapped loathsome foreshores, polluted by offal and refuse—the accumulation of years. Viewed through the reeking atmosphere with the nose firmly gripped, even the magnificent inland scenery seemed to grow tainted and lose its splendor."

Whaling days

In 1894 a Norwegian whaling expedition visited South Georgia. Its commander, Carl Anton Larsen, saw great potential for whaling and founded the Compaña Argentina de Pesca out of Buenos Aires (with a permit from the British consul), setting up operations at Grytviken in 1904. This station continued to operate until 1965, and was joined by six other bases: Stromness (1907–61); Husvik (1907–61); Godthul (1908–29); Leith (1909–65); Ocean Harbor (1909–20); and Prince Olaf Harbor (1912–32). Some of these bases first served as sheltered mooring sites for floating factory ships, but later developed as land factories.

Whale oil was used worldwide for tanning, lighting, and lubrication, and demand greatly increased once it was discovered that it could be turned into margarine and soap. Bones and meat were processed into fertilizer. Different whale species were hunted over the years and

▶ SHADED BY MOUNTAINS
The southern end of South Georgia is cleft by Drygalski Fiord. Less than 1.6 kilometers (1 mile) wide and 11 kilometers (7 miles) long, this beautiful waterway is shaded to the east by the Salvesen Range, which rises to Mount Carse (2,332 meters/7,650 ft), and to the west by Mount Sabatier.

▼ MIXED QUARTERS
Black-browed albatrosses,
identifiable by the fiercely
angled dark patch around
their eyes, have a circumpolar
distribution and breed on
many sub-Antarctic islands.
They are the most abundant
of all albatrosses, and often
nest very near, or even form
mixed colonies with, other
seabirds, including shags and
Rockhopper penguins.

A wealth of wildlife

South Georgia lies 300 kilometers (186 miles) south of the Polar Front, where currents from the Weddell and Ross seas meet and swirl together to create a rich soup of nutrients. This attracts hordes of creatures from the top of the Antarctic food chain—penguins, albatrosses, fur seals, and whales.

Many Chinstrap penguins inhabit the South Sandwich Islands, but there are relatively few (perhaps 2,000 pairs) on South Georgia. King penguins—the second largest and most colorful of all penguins—breed prolifically on South Georgia. There are around 30 colonies accommodating over 200,000 pairs, with the biggest groups at Royal Bay (9,000 pairs), St Andrews Bay (39,000 pairs), and the Bay of Isles, most notably at Salisbury Plains (39,000 pairs). Many more Macaroni penguins nest in the area (some 2.5 million pairs) but they are mainly found in large communities on inaccessible steep seaside slopes, such as on Cooper Island. About 120,000 pairs of Gentoo penguin raise their chicks on South Georgia, mainly in small colonies in sheltered bays.

South Georgia is also home to many flying birds. The largest of them—the Wandering albatross—has a resident breeding population of some 4,300 pairs. A decline in their numbers has prompted a management plan at Albatross Island in the Bay of Isles to ensure that visitors do not disturb the birds. Other albatross species, including the Black-browed albatross, also nest on South Georgia. Light-mantled sooty albatrosses are found near Grytviken and Gold Harbor, while Gray-headed albatrosses frequent the northern end of the island, especially at Elseshul. Petrels, skuas, gulls, terns, and shags can also be seen, as well as Speckled teal, and the South Georgia pintail. The latter, more obviously carnivorous than other species of duck, has been observed eating seal carcasses. The South Georgia pipit, a drab brown bird resembling a sparrow and the world's most southerly songbird, is special. It lives only on South Georgia, but rats introduced from ships have reduced its breeding range from all over the mainland to some of the offshore islands and isolated parts of the rugged south coast.

By the early twentieth century, seals had been hunted for their pelts to near extinction, but numbers have now risen again to some 1.8 million. More than 95 percent of the world's population of Antarctic fur seals is found on South Georgia and the little islands around it, and their numbers are continuing to increase by an average of 10 percent each year. In the weeks before Christmas, territorial adult males each jealously guard a stretch of beach. The males have departed by mid-January, but the remaining females rigorously enforce their private space for their pups. Smaller numbers of Leopard and Weddell seals are also found around South Georgia, and

there are more than 300,000 elephant seals, although the huge dominant adult males leave the beaches early in summer.

It is thought that the population of fur seals has recovered very rapidly because the whales with which the seals competed for food have also been slaughtered to near extinction. Whales breed much more slowly than seals, so whale populations have been slower to recover. Humpbacks, Minkes, and Right whales are sometimes seen in South Georgian coastal waters.

War zone

In 1982 South Georgia was the flashpoint for the Falklands War. An Argentinian company had been given an option to collect scrap metal from the island's abandoned whaling stations, but when an Argentinian naval ship arrived at Leith Harbor on 18 March 1982 and

▶ RESIDENT YEAR-ROUND
Unlike most southern penguin
colonies, which seem to fill
and empty of birds almost by
clockwork, South Georgia
King penguin colonies like this
are never completely devoid
of penguins. Kings have a
breeding cycle more than one
year long, and both adults and
half-grown chicks can be
found in the colonies over
winter when feeding is poor.

hoisted the Argentine flag, Britain registered an official protest. On 31 March, 22 British marines landed at King Edward Point. On 3 April (a day after the Falkland Islands were invaded), the *Bahia Paraiso* sailed into Cumberland Bay and called on the island's administrator to surrender. In the ensuing three-hour battle, 200 Argentinian troops landed and captured the station and many of its staff, but only after two Argentinian helicopters had been shot down and several Argentinian soldiers killed. Subsequently six Royal Navy vessels sailed to South Georgia and recaptured King Edward Point on 25 April and Leith the following day. The British captives were held by the Argentinians for 12 days and then released in Uruguay. The 185 Argentinian troops captured by the British were also released in Uruguay. On 14 June, Argentinian forces on the Falkland Islands surrendered to Britain.

◀ **EXPERT SWIMMERS**
Some of the population of 300,000 Southern elephant seals on South Georgia have been fitted with transmitters. This has shown them to be exceptional swimmers—one traveled to the southern end of the Antarctic Peninsula, a journey of 2,600 kilometers (1,600 miles). They can also stay under water for long periods—often diving deep for up to two hours, and surfacing for only a few minutes.

South Orkney Islands

▼ SERIOUS SIGNAGE

Damp and desolate Laurie Island is a very long way from any of the cultural centers shown on the signpost in front of Orcadas Base. The 11 buildings of the base house scientific personnel who continue meteorological observations begun by William Bruce in 1903.

⊳ GLACIAL BEACH

Glaciers form a grand backdrop to a narrow beach at Shingle Cove, part of Iceberg Bay on the south side of Coronation Island. George Powell landed on Coronation Island on 7 December 1821 and claimed the group of islands for Britain.

Naming the islands

On 6 December 1821 American sealer Nathaniel Palmer and British sealer George Powell, sailing together in *James Monroe* and *Dove* respectively, sighted these islands and charted them as Powell's Group. When another British sealer, James Weddell, arrived aboard *Jane* in February 1822, he chose the current name because the islands were at the same latitude in the south that Britain's Orkney Islands occupy in the north. Weddell's name won out in popular usage. However, Powell was not completely overlooked; one of the four main islands bears his name.

The entire group is about 600 kilometers (370 miles) northeast of the Antarctic Peninsula and covers a land area of 622 kilometers (386 miles), almost 90 percent of it glaciated. Even in late summer, the islands can be surrounded by icebergs and sea ice, making landings difficult. Coronation Island, by far the largest, rises to the highest point: Mount Nivea, 1,265 meters (4,150 ft). Signy Island

⊲ PENGUIN HAVEN

Coronation Island is a windswept rock in the middle of the Southern Ocean, but a few bare patches make it a haven for Adélie penguins nesting above the beach in Shingle Cove.

lies to the south; Powell Island and Laurie Island are to the east. A towering glacier backs the landing beach at Shingle Cove in Iceberg Bay on Coronation Island; once ashore, there is an easy walk up a rocky ridge to scattered Adélie penguin rookeries.

During the nineteenth century sealers annihilated virtually every seal; whalers arrived in the twentieth century. An unsuccessful whaling station lasted from 1920 to 1926—the owner, Peter Sørlle, named Signy Island after his young daughter.

Territorial claims

In 1903 the *Scotia* brought William Bruce and the Scottish National Expedition to the South Orkneys. From 1 April that year, a meteorological station operated on Laurie Island—Ormond House, above the beach in the lee of a hill, is now in ruins. Britain turned down Bruce's offer to use the base as an ongoing meteorological station, but Argentina subsequently accepted his proposal. What is now known as Orcadas station began operations on 22 February 1904 and is the longest continually-operated Antarctic station, accommodating 14 people over winter and up to 45 in summer.

In March 1947 Britain set up Base H, a year-round meteorological station, at the site of Sørlle's whaling station at Factory Cove, Signy Island. Now known simply as Signy, the base is closed to unofficial visitors.

Britain's and Argentina's rival claims to the South Orkneys are held in abeyance by the Antarctic Treaty.

Falkland Islands

A tiny slice of Britain

The Falkland Islands lie north of the Polar Front and are not, therefore, sub-Antarctic Islands, but they are a popular stop en route to South Georgia. In 1982 this archipelago hit the headlines when British troops fought to retain ownership against invading Argentinian forces. Both nations invoke history to support their claims, but the Falkland Islanders are staunchly British. Union Jacks fly in many front yards, yet the Falkland Islands still feature nightly as Las Islas Malvinas on Argentinian weather reports—which record rain or snow for more than half the year.

Besides the large islands of West Falkland and East Falkland, there are at least 200 much smaller islands, adding up to a total landmass of 12,170 square kilometers (4,699 sq. miles). The total population of the islands is about 2,500, some 1,500 people living in Stanley, the only town. The Falklands War has left a legacy of mined beaches and fields behind barbed wire and warning signs near Stanley—finds of live ammunition can be reported to the Ordinance Office in the main street.

▼ **BLACK-BROWED ALBATROSS ON SAUNDERS**
Several species of albatross are occasional visitors to the Falkland Islands, but only one, the Black-browed albatross, calls the islands home. This medium-large albatross breeds in 12 locations on eight of the islands, nesting on remote cliffs and utilizing the updraughts for landing.

Rich in wildlife

West Point Island, on the western side of the group and owned by the Napier family for generations, is a convenient first landing site. Across the island, in a cleft high above the sea, Rockhopper penguins and Black-browed albatross nest in noisy profusion.

New Island, also on west side of the group, is a nature reserve owned by expert naturalists Tony Soper and Ian Strange. The resident wildlife includes fur seals, Gentoo, Magellanic, and Rockhopper penguins, Black-browed albatrosses, prions, and cormorants.

Carcass Island, at the northwest corner of the group, is named for nothing more sinister than one of the Royal Naval vessels that first colonized the Falklands. Today's owners welcome visitors; the island's wide range of bird life includes Magellanic and Gentoo penguins.

Saunders Island's Port Egmont was the first British settlement in the Falklands (established 1766); some ruins from that era remain near the present owner's farmhouse. However, most visitors disembark in the north, where there are Gentoo penguins on one shore of the spit and Rockhopper and Magellanic penguins on the other. Rockhoppers surf straight onto the rocks, even though there is a smooth beach close by. A colony of King penguins completes Saunders Island's array of flightless birds. On the grassy hill, where sheep graze among Magellanic penguin burrows, are Rockhopper rookeries and nesting pairs of Black-browed albatrosses.

▲ **AN ARCH OF BONE**
Christchurch Cathedral in Stanley was built in 1872. The distinctive arch in front of the cathedral is fashioned from the lower jawbones of two Sperm whales. It was brought from South Georgia to celebrate the centenary of the Falkland Islands in 1933.

South Shetland Islands

Gateway to Antarctica

On the voyage from South America to the Antarctic Peninsula, the looming black volcanic crags and extensive glaciers of the more southerly South Shetlands Islands, such as Elephant Island so often shrouded in fog, provide a dramatic introduction to Antarctica. Many ships heading for the southern continent maneuver among the icebergs to make their first landfall on this part of the South Shetland archipelago, it may also be the last glimpse of Antarctica for vessels as they re-enter the Drake Passage on the homeward journey. Annexed by British Captain William Smith in 1819 as New South Britain, the string of islands was renamed the South Shetlands in the following year.

▲ A FORBIDDING SHORE
This view of Elephant Island from d'Urville's *Atlas* shows the western swell that prevented *Astrolabe* from closer approach in 1838. Like so many that come here only to be driven away by wind, waves, and fog, the French ships merely skirted the island.

Elephant Island

Captain George Powell probably named this island in 1821 for the number of "sea elephants" he found there. Only 12 by 20 kilometers (7½ by 12½ miles), it lies at the northeastern end of the South Shetlands archipelago and is not on most itineraries. Even when conditions look good, ice in the bay and treacherous currents make landing tricky; it becomes well nigh impossible when seals and penguins crowd the beach. A more convenient location for wildlife-watching, Cape Lookout in the south, is frequented by many fur and elephant seals, and has a large population of penguins—Chinstraps, some Gentoos, and the occasional Macaroni.

Ernest Shackleton brought his men to Elephant Island after they escaped from the ice of the Weddell Sea in 1916, ensuring that this remote island would always be a place of pilgrimage. Frank Wild and 21 others from the group spent 105 winter days in this inhospitable spot until they were finally rescued. Shackleton's expeditioners made their first landfall at Cape Valentine in the northeastern corner. Lack of high ground forced them to move 14.5 kilometers (9 miles) west along the north coast to a small promontory, where a hurricane shredded their remaining tents the night they arrived.

The new residence was dubbed Cape Wild—a name that Australian photographer Frank Hurley regarded as "at once an apt description and a tribute to a great-hearted comrade." Hurley further described the campsite as "a precarious foothold on an exposed ledge of barren rock, in the world's wildest ocean … Our refuge was like the scrimped courtyard of a prison—a narrow strip of beach 200 paces long by 30 yards wide … Behind us, the island peaks rose 3,000 feet into the air, and down their riven valleys, across their creeping glaciers, the wind devils raced and shrieked, lashed us with hail, and smothered us with snowdrift … Inhospitable, desolate and hemmed in with glaciers, our refuge was as uninviting as it well could be."

The narrow 250-meter (820-ft) gravel spit, rising to a tiny rocky headland of 6 meters (20 ft), has since been downgraded to Point Wild. The view of the point from approaching ships is often obscured by Gnomon Island, which featured prominently in several of Frank Hurley's photographs.

◄ A ROSY VIEW
Approaching Elephant Island, Ernest Shackleton wrote: "Rose-pink in the growing light, the lofty peak of Clarence Island told us of the coming glory of the sun." A humpback whale enjoys the view close to where Shackleton "would have given all the tea in China for a lump of ice to melt into water."

➤ CLOSE INSPECTION
Adélie penguins gather at the water's edge to inspect the depths before diving in. Adélies are at home standing on snow like this—most of the species nest on Antarctica, along the Peninsula and elsewhere around the Continent. Only a small part of their population is found as far north as the sub-Antarctic islands.

Penguin Island

Rounding the northern end of King George Island, the first accessible landing sites are Turret Point on King George Island and Penguin Island, just 1 kilometer (½ mile) away. Edward Bransfield named the latter in 1820 for the penguins he found there. Currently about 8,000 breeding pairs of Chinstrap penguins and about half that number of Adélies hide away at the island's southern end. At the beginning of summer, the whole of this tiny island, about 1.6 kilometers (1 mile) across, is usually still under snow. This snow cover and nesting Southern giant petrels can make it difficult to move on to the island from the landing site. Later in the season, when delicate mosses, grasses, and lichens are exposed, there may also be fur seals to contend with on the beach. The island's summit is Deacon Peak, a 169-meter (554-ft) volcanic cone filled with ice or water.

King George Island

Largest of the South Shetland group and named in 1820 for the British monarch at the time, King George Island is roughly 69 kilometers long by 26 kilometers wide (43 by 16 miles), and 90 percent covered with ice. The glaciers of the island have retreated significantly during the past 15 years.

Across the narrow passage from Penguin Island, Turret Point is distinguished by a cluster of tall rock stacks at either end of a gravel beach. The point marks the eastern end of King George Bay and was named by the *Discovery II* expedition in 1937. When the katabatic winds off nearby glaciers allow a landing, elephant and Weddell seals, Adélie and Chinstrap penguins, Kelp gulls, Antarctic terns, and Southern giant petrels may be observed. Nesting Antarctic shags crowd one of the offshore stacks.

Admiralty Bay is the largest harbor in the South Shetland islands, extending over 122 square kilometers (47 sq. miles) and reaching depths of over 500 meters

(1,640 ft). Named by Captain George Powell in 1822, this expansive anchorage was used by whalers in the early twentieth century—icebreakers are never required to gain entry to it. Since 1996 the whole of Admiralty Bay has been an Antarctic Specially Managed Area.

Base G, the first permanent base on King George Island, was built by the British in 1947 at Keller Peninsula. Later the same summer, Argentina set up a hut 25 meters (80 ft) away and used it periodically. Base G closed in 1961 and was removed in 1996; only its foundations remain. Meanwhile, Brazil built Comandante Ferraz station nearby in 1984, with living quarters for 13 in winter and 25 in summer. There are also Ecuadorian, Peruvian, and United States summer stations here; the American station, quaintly known as Copacabana, lies within a designated Site of Special Scientific Interest, and is thus out of bounds. Polish Arctowski station, named after Henryk Arctowski, the geologist on board *Belgica*, has operated at Point Thomas since 1977. Upgraded in 1998, it can accommodate up to 14 expeditioners in winter and 20 in summer.

Lichens and grasses make Point Thomas an attractive spot; wildlife there includes fur, elephant, Weddell, Crabeater, and Leopard seals, Adélie, Gentoo, and Chinstrap penguins, and many flying birds.

▲ SCAVENGING FOR FOOD
Kelp gulls loiter by penguin and seal colonies, hoping to snatch any booty that might come their way. Giant petrels and skuas dominate southern scavengers, but lucky Kelp gulls may score their scraps, meals intended for penguin chicks, or even the remains of unlucky eggs or birds.

Above left

▲ PENGUIN POLICEMEN
Chinstrap penguins are called "police penguins" in Russian—a reference to the strap-like feather pattern that reaches under the chin like a Russian policeman's cap. They are probably the noisiest of the smaller penguins, and are known for aggressively defending territory from all comers, including people.

▲ FAR FROM HOME
While Russians are used to wintry conditions, this signpost at Bellingshausen station plaintively reveals that home is a long way away: Murmansk is 16,500 kilometers (1,025 miles) in one direction, and Vostok station is 4,375 kilometers (2,718 miles) to the south; the balmy warmth of Havana is 11,500 kilometers (7,140 miles) away.

James Weddell, sealing captain of the *Jane*, named Maxwell Bay in 1822 after Lieutenant Francis Maxwell, who had earlier served with him. Lying at the southern end of King George Island, the bay is protected on its southern side by Nelson Island. Unlike anywhere else in Antarctica, the area is crowded with buildings, muddy roads, and an airport. So many bases are located here that it could well be called Antarctic Treaty Bay; in the past few summers it has even been the site of an international marathon for tourists.

Russia's Bellingshausen station (1967) and Chile's Presidente Eduardo Frei station (1969) are close neighbors. Air traffic from Chile's Teniente Rodolfo Marsh Martin station, behind the hill, passes low overhead, and one of the station's buildings sometimes operates as Antarctica's only hotel. China's Great Wall station (1985) is located in nearby Hydrographers Cove, and Uruguay's Artigas station (1985) stands to the northeast, on the far side of Suffield Point. The newest base is South Korea's King Sejong station (1988) in Marian Cove. Argentina's Teniente Jubany station was built at Potter Cove, near Maxwell Bay, in 1948.

Aitcho Islands
The thrusting, rocky spires of the tiny Aitcho Islands dramatically confirm the volcanic origins of the South Shetland Islands. The group lies at the northern end of English Strait, between Robert and Greenwich islands. Pronounced simply: aitch(h)-o, the group was named in 1935 by the *Discovery II* expedition in honor of the Admiralty's Hydrographic Office (better known as the H.O.)

Going ashore involves pushing through fringing kelp, but the resident wildlife is worth the effort. Large rookeries of Chinstrap penguins surround the most frequented landing site at the eastern end of the main island, while Gentoos nest among the bones at Whalebone Beach. On the island's western side, elephant seals wallow near another landing beach but, after the snow has melted, footprints can all too easily damage the large beds of cushion moss, and the presence of humans may disturb nesting giant petrels.

Yankee Harbor, Greenwich Island
Yankee Harbor is a very sheltered bay on the southern side of Greenwich Island. The bay extends well back, its mouth protected by a crescent-shaped, narrow gravel spit about 1 kilometer (½ mile) long. American sealers visited here from 1820, and a rusting trypot, used for boiling blubber down to oil, marks the start of the spit. Here territorial skuas and feisty fur and other seals challenge visitors. Large populations of Gentoo penguins and attendant Snowy sheathbills occupy the terraces above the beach.

▶ A STUNNING PROSPECT
From the red building on the shore, which is part of Camara station, to the rocky point on the right, Half Moon Island provides a grandstand view of Livingston Island (seen on the far right).

> **A FRAGILE SHORE**
> The relatively temperate climate of the South Shetland Islands supports a greater number of plants than anywhere else so near to the Antarctic mainland. But the plant life is delicate—moss beds such as this can bear the scars from a single footstep for many decades.

Half Moon Island

Named for its crescent shape, this island appears on South American charts as Isla Media Luna. It is 2 kilometers (just over 1 mile) long, and lies on the northern side of the eastern tip of Livingston Island, whose glaciers loom in the background when visibility is good.

The landing site is near the eastern tip of the island on a beach sheltered by crags. There, a small wooden boat wreck lies near a colony of nesting Chinstrap penguins. The ridge behind the beach can be crossed when the penguin colony is not too densely occupied. Argentina's Teniente Camára station (established in 1953 but not regularly occupied in recent years) lies to the west. The hill behind the station provides a sweeping view of the island, but has now been claimed during the nesting season by aggressive skuas. The beach in front of the station sometimes has Gentoo penguins on it, and fur seals arrive towards the end of summer.

> ▲ **AN ISLAND WALK**
> Unlike other Antarctic islands, it is easy to walk across Aitcho Island. Skirting moss beds and molting elephant seals, it is possible to reach one of the island summits, where nesting giant petrels can be seen.

▶ ROCKS AND FOSSILS
Looking eastward from the main landing area at Hannah Point into Walker Bay, the rocky buttress in the background lies beyond the far end of the island's walking trail. The large rocks below the glacier on the right hold a collection of fossils and rock samples.

Livingston Island

Livingston Island is about 61 kilometers (38 miles) long, between 3 and 32 kilometers (2 and 20 miles) wide, and has been known since 1820. Spain built the Juan Carlos I summer station on the south side of the island in 1988.

Hannah Point, on the eastern side of Walker Bay, is named after a sealing vessel that was wrecked there on Christmas Day 1820. Landings are possible if wind and waves allow, but each shipload of passengers has the potential to do more damage to wildlife than at other, less crowded locations. The top of the rise, where a thin, red band of pure jasper runs through the rocky ledge, commands a grand panorama of the northern part of Livingston Island.

The main landing point is near a rock, 15 meters (49 ft) high, right on the shoreline and surrounded by rookeries. A few nesting Macaroni penguins—the only crested penguins that visitors to the Antarctic Peninsula are likely to see—sometimes reside among the multitudes of Gentoo and Chinstraps. Macaroni penguins can be difficult to find or to approach closely because they are generally surrounded by large rookeries of aggressive Chinstraps.

An alternative landing site at Walker Bay is well to the west on a long, open beach, backed by a meltwater pond where the land behind rises to a towering ridge and glaciers. Scientists regularly camp here, and have set out bones, rocks, and minerals on two flat-topped boulders. Later in the summer, when the snow and ice have melted, the rocks in the area reveal extensive growths of mosses and lichens. Antarctica's only flowering plants also eke out an existence in the rocky soil—the grass-like *Deschampsia antarctica* and the tiny *Colobanthus quitensis*—a type of pearlwort so small that its minute flowers can only be seen at very close range. When the penguins start to leave the rookeries after midsummer, it is possible to walk from one landing site to the other, though care must be taken not to alarm the giant petrels that nest nearby or to disturb the still-crowded rookeries toward the headland. On a rise just above water level, at the western end of the rookeries, a dark brown streak usually marks the presence of a large wallow of molting elephant seals—a scene of perpetual squabbling and out-and-out blood-drawing fights.

Deception Island

Deception Island combines the novelty of sailing into an active volcano with the possibility of swimming, or at least wallowing, in thermally heated water. With the possible exception of Maxwell Bay on King George Island, Deception Island has little in common with the rest of Antarctica, and is dominated by derelict buildings and obsolete whaling works.

The island, which lies some 15 kilometers (9 miles) south of Livingston Island, is nearly 13 kilometers (8 miles) in diameter and rises to a height of 602 meters (1,975 ft) at Mount Pond. Deception Island's distinctive horseshoe shape stands out on charts. Its name, which has been in use since 1821, probably came about because the mouth of the harbor is easy to miss.

Port Foster comprises the whole flooded caldera within the island. It is named after Henry Foster of the *Chanticleer*, who conducted some scientific experiments there in 1829. Entering the caldera through a gap in the wall known as Neptunes Bellows, so named because of the violent wind that sometimes blows across the mouth of the entrance, is invariably exciting. Ships must negotiate a gap of less than 400 meters (1,310 ft) wide, hugging the northern side (where Cape petrels often gather) to avoid a submerged rock less than 2 meters ($6\frac{1}{2}$ ft) below the surface almost in the middle of the Bellows. Jean-Baptiste Charcot named Ravn Rock in 1910 after the whale-catcher *Ravn* that was based in Port Foster at the time he surveyed the island. The wreck of the *Southern Hunter*, a British whaler that ran aground avoiding a vessel of the Argentinian navy in 1957, lies on the southern side of the entrance.

Once through the Bellows, ships generally turn hard to starboard to stop in Whalers Bay. Above the black volcanic sand beach, which is often shrouded in fumes and vapor from natural vents in the sand that is frequently the repository of starfish, another missing chunk of caldera wall forms Neptunes Window. On a day with reasonable visibility the Antarctic Peninsula, 90 kilometers (56 miles) away, can be seen from this vantage point near the remains of a small dry dock by the water's edge. Some say that the young American sealing captain Nathaniel B. Palmer, who was the first to enter Port Forster, may also have been the first person to glimpse the Antarctic Continent from Neptunes Window in 1820.

From 1906 the Norwegian-Chilean whaling factory ship *Gobernador Bories* used the bay as a base. Britain

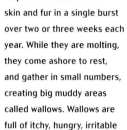

▲ IRRITABLE SEALS
Elephant seals shed their skin and fur in a single burst over two or three weeks each year. While they are molting, they come ashore to rest, and gather in small numbers, creating big muddy areas called wallows. Wallows are full of itchy, hungry, irritable elephant seals, and are best observed from a distance.

claimed the island in 1908 as part of the Falklands Island Dependency, and when the Norwegian Hektor whaling company wanted to set up a whaling station in 1911, Britain gave it a 21-year lease. This stipulated that Hektor must dispose of the 3,000 or so whale carcasses that littered the beach; the almost complete absence of whalebones around the bay testifies that the company honored its contract. When the whaling station closed in 1931 after falling prices and new technology made it redundant, the equipment used to extract oil from the bones was abandoned. This unlovely collection of

➤ WARMED BY VOLCANIC ACTION
The *Akademik Ioffe* sails across the rim of the caldera into Port Foster. Several sea stacks, like the one partially obscuring the ship, can be found around Deception Island. Volcanic warmth ensures that the island has less snow cover than its neighbors.

◄ THE WESTWARD LIMIT
The remains of a BAS Otter fuselage slowly deteriorate beside the hangar on Deception Island. The aircraft marks the westward limit for visitors to Whalers Bay—the area beyond is protected to enable scientists to study plant reestablishment after eruptions and mudslides.

▲ A SURPRISING SWIM
The black sand of Pendulum Cove frames a wading pool, heated to a comfortable "swimming" temperature by underground volcanic activity. The nearby ruins of Chile's Pedro Aguirre Cerda station, which was destroyed by a volcanic eruption in 1967, reveal that the area can even become too hot for comfort.

rusting metal looks like a nightmare version of Frankenstein's laboratory. During the whaling days, some 45 men died in these parts, and were buried in a cemetery behind the station. Although the cemetery was engulfed by a mudslide during the volcanic eruptions of the 1960s, an empty coffin that had been ejected by the slide poignantly marked the spot until 2000. The coffin has now disappeared, presumably demolished by winter storms.

Some of the Whalers Bay storage tanks contained fuel, and were destroyed by Britain during World War II to prevent German vessels using Deception Island as a refueling depot. From February 1944, the British occupied some of the Norwegian buildings as Station B, which functioned continuously until 5 December 1967, when a volcanic eruption forced a rapid evacuation.

The first building occupied by the British expeditioners of 1944 burned down in September 1946. The whaling station dormitory was then renamed Biscoe House, after the Royal Navy officer who explored Graham Land in 1832, and the building was used until 1969. The slab to the east of Biscoe House once held a prefabricated accommodation block that was erected in 1966 and named Priestley House after Sir Raymond Priestley, the geologist on Scott's last expedition, who later directed Britain's Antarctic operations.

Britain's occupation of Deception Island came to a violent natural end in the late 1960s. Volcanic rumblings over the years—including one in 1923 that stripped the paint from ships in Port Foster—indicated that Deception Island was far from quiescent. Then, in December 1967, two big eruptions wiped out Chile's Presidente Pedro Aguirre Cerda station, and caused the British and the staff of Argentina's Decepción station to evacuate, picking up the Chileans as they went. The British returned the next year, but only two and a half months later, on 23 February 1969, there was a further eruption accompanied by mudslides that left the station looking much as it does today. The British surrendered the wrecked base to the elements; the land around the abandoned Chilean station is now a Site of Special

Scientific Interest, and Decepción station operates irregularly in summer. Deception Island's only other base—Spain's Gabriel de Castilla station—is also only used in summer.

The last building to the west at Whalers Bay is the old British Antarctic Survey (BAS) aircraft hangar erected in March 1962; the wings and tailplane of a BAS de Havilland Otter that was destroyed in the 1960s eruptions is stored there. Beyond this landmark, another Site of Special Scientific Interest protects the mosses and lichens that are re-establishing near the lake. Between lake and hangar, the runway that was hacked out for Antarctica's first powered flight, on 16 November 1928 by Australian aviation pioneer Sir Hubert Wilkins, is still just discernible. He was a brave man—the runway included two bends, a couple of gullies, and the certainty of a ducking in Port Foster if the take-off failed.

Deep inside Port Foster lies Pendulum Cove, named for Foster's magnetic and pendulum experiments in 1829. The old Chilean station there is out of bounds, and Pendulum Cove's sole attraction is a trough of warm water near the shore, created by subterranean thermal activity when the tide is right. This is rarely deep enough to fully immerse a human body, and the surrounding water retains its extreme polar chill.

Antarctic creatures find the water of Whalers Bay uncomfortably warm, and are reluctant to venture inside. In late summer a few fur seals, or even the occasional Weddell or Leopard seal, may haul out on the beach, and there are always one or two Chinstrap or Gentoo penguins about. A full experience of Deception Island's wildlife, however, depends on the sea being calm enough for a landing at Rancho Point, known always by its unofficial name: Baily Head. Even on a calm day, Zodiacs have to surf onto the beach, and if there is any swell, the waves turn into dumpers that can swamp even these stable craft.

Nevertheless, Baily Head is worth any effort to get there. Fur seals sometimes crowd the long, black sand beach, and multitudes of Chinstrap penguins make it their domain. The small valley near the high, rocky headland extends inland and becomes a two-way penguin superhighway during the summer months. The traffic—a horde of diminutive figures in formal black-tie evening gear—moves with great agility to and from a huge bowl-like amphitheater completely covered with nesting Chinstraps. If the weather is particularly good, it is possible to walk from there across the snow-covered ridge and down into Whalers Bay.

▶ BAILY HEAD
The broad melt water streambed that runs through the Baily Head Chinstrap rookery provides a useful highway for birds and human visitors alike. The 168-meter (550-ft) headland provides welcome shelter for the penguins that crowd the valley's floor and slopes.

Other sub-Antarctic islands

Peter I Øy (Peter I Island)

Peter I Øy, the first land ever seen inside the Antarctic Circle, was discovered by Thaddeus Bellingshausen in January 1821, and named after the Russian Czar, Peter the Great. The island is an extinct volcano, 18 kilometers long by 8 kilometers wide (11 by 5 miles), with Lars Christensen Peak rising 1,755 meters (5,760 ft) from its center. Its shores are predominantly ice cliffs, making visits extremely difficult. Ola Olstad and his *Norvegia* expedition achieved the first landing in 1929 and successfully claimed the island for Norway.

Scott Island

Scott Island, named for Robert Falcon Scott, is a mere dot in the ocean just inside the Antarctic Circle at the entrance to the Ross Sea, about 400 meters long, less than 200 meters wide and about 50 meters high (1,312 by 656 by 164 ft). It was discovered in 1902 by Lieutenant Colbeck aboard Scott's relief ship, *Morning*. In 1999–2000, the island had its moment in the sun—literally and figuratively—when it was the first land to see the dawn of the new millennium.

▲ RISING FROM THE SEA
Haggits Pillar is Scott Island's sole neighbor. Scott Island was discovered on Christmas Day 1902 by the crew of *Morning,* who thought that it "might have been a large discolored berg ... but it was soon observed to be a typical Antarctic Island."

Balleny Islands

Five islands and several islets form the Balleny group, straddling the Antarctic Circle south of New Zealand, and scattered across 160 kilometers (100 miles). Sturge Island, the southernmost, is the largest of them, at 32 by 6.5 kilometers (20 by 4 miles), and also the highest, with Brown Peak rising to 1,705 meters (5,595 ft). Buckle Island lies midway along the group and measures 21 by 5 kilometers (13 by 3 miles). Borradaile Island, to the north, is much smaller—3.2 by 1.6 kilometers (2 by 1 miles)—but is still larger than nearby Row Island, which about 1.6 kilometers (1 mile) in diameter. At the northern end, Young Island measures 30.5 by 6.5 kilometers (19 by 4 miles). Tiny Sabrina Island is reserved for nesting birds—landings cannot be made there without special permission—but the shores of all the Balleny Islands are so sheer that few people have ever set foot on any of them. "One sight in bad weather of that sinister coast is enough to make a landsman dream for weeks of shipwrecks, perils, and death"—so wrote Louis Bernacchi after sailing past the islands on the *Southern Cross* expedition with Carsten Borchgrevink.

◄ VOLCANIC ISLANDS
Volcanic in origin and almost completely covered in ice, the Balleny Islands are cloaked in cloud for much of the year. Over the winter months pack ice surrounds these islands, which are still seismically active—a major earthquake was recorded there in recent years.

▼ REMOTE BUT WORTH A VISIT

In 1999, a two-year-old elephant seal hauled out on this sliver of beach on Peter I Øy. A tag identified the seal as one captured on Macquarie Island, over 4,000 kilometers (2,485 miles) to the northwest.

◄ ALBATROSS HAVENS
Auckland Island (foreground and left distance) and Adams Island (right) are the largest of New Zealand's sub-Antarctic islands. Adams Island houses the world's largest breeding population of Wandering albatrosses, and several colonies of White-capped albatrosses breed on its steep slopes and cliffs.

In 1849, Charles Enderby, of the famous British whaling family, set up an unsuccessful settlement at Port Ross at the northern end of Auckland Island, but nothing of it remains today. Many early visitors were shipwrecked sailors, and wreck sites dot the coast. Both Campbell Island and the Auckland Islands were farmed, and even now native wildlife on many of the islands suffers as a result of introduced species.

Campbell Island and the Auckland Islands

New Zealand has five sub-Antarctic island groups, but those most often visited by ships traveling to the Ross Sea are Campbell Island and the Auckland Islands. These are basaltic volcanic islands, covered in peat, that stand on the submarine Campbell Plateau to the north of the Polar Front. They are cold, wet, and windy, but are ice-free (although shaped in the past by glaciers), and can support trees. Indeed, they have a wealth of vegetation that exists nowhere else, including many endemic vascular plants and the world's southernmost forests.

Campbell Island is the most southerly of the group, lying about 700 kilometers (435 miles) south of Bluff at the bottom of New Zealand's South Island. It has an area of 114 square kilometers (44 sq. miles)—considerably less than main Auckland Island's 510 square kilometers (197 sq. miles).

The Auckland Islands were first sighted on 18 August 1806 by Abraham Bristow, captain of the whaler *Ocean,* who rather obsequiously named them after Lord Auckland, "my friend through my father." Later scientific visitors included the Charles Wilkes and Dumont d'Urville expeditions, only days apart, and Joseph Dalton Hooker, the celebrated naturalist, who came here on the expedition led by James Clark Ross. Campbell Island was discovered on 4 January 1810 when Captain Frederick Hasselburg of *Perseverance* (the person who first reported Australia's Macquarie Island) named it after the owner of his sealing company. The seal populations of both island groups were decimated immediately after their discovery.

The Auckland Islands are home to the world's largest breeding populations of Wandering albatrosses and Shy albatrosses. More than 90 percent of the rare Hooker's sea lion population also breeds there. Campbell Island has the world's largest breeding population of Royal albatrosses; from the meteorological station at Perseverance Harbor, a boardwalk track leads up to the Col-Lyall Saddle, where these birds nest in abundance.

All the New Zealand islands are National Nature Reserves; landing permits are required, and visitors must be accompanied by a representative of the Department of Conservation. Visitors to Auckland Island can inspect the old settlement site at Erebus Cove, Port Ross. But, most prefer nearby Enderby Island for the variety of rare wildlife there, and for the chance to walk through the mysterious rata forest with its bright red flowers—the last opportunity to stand under a tree before Antarctica. Stella Hut, at Sandy Bay, is a small hut once used as a supply depot and nearby are several modern huts that scientists use during the summer. The beach nearby is occupied by a colony of sometime aggressive Hooker's sea lions, and vigilant observers may see the most elusive penguin of all, the extremely timid Yellow-eyed penguin. Royal albatrosses nest on the slopes behind the rata forest.

▼ LIFE-SAVERS
Depots like Enderby Island's Stella Hut provided supplies for shipwrecked sailors. The eight survivors of *Derry Castle*, wrecked off the north tip of the island in March 1887, built tussock huts around the depot, and were rescued.

► SLEEPING BEAUTY
A White-capped albatross snoozes on its nest on Auckland Island. The lavish plant growth around it shows that some vegetation thrives in this challenging environment. The Aucklands are renowned for their large flowering plants: here, the Pink Antarctic daisy (*Pleurophyllum speciosum*) flowers just above the albatross.

MATES FOR LIFE

The Royal albatross breeds only around New Zealand. It is one of the longest-lived of all birds, sometimes surviving for as many as 60 years. It spends its first three or four years at sea, riding the winds and resting only on the water, and then chooses a lifetime partner on the island where it was born.

A DIVING DIVA

The Hooker's sea lion is one of the rarest pinnipeds (mammals with feet that look like fins) in the world. It has a highly localized distribution—more than 90 percent of the population breeds on the Auckland Islands—and is vulnerable to any kind of disturbance. These animals are superb divers, plunging to depths of over 450 meters (1,475 ft) in quest of their prey on or near the sea floor.

ONE OF A KIND

Campbell Island shags belong to a large family of pelican-like birds, many of which are confined to isolated islands of the sub-Antarctic. These shags are fine swimmers, hunting in large groups in sheltered bays or far out to sea, and covering great distances under water in search of food.

PRETTY AS A PICTURE

The flower heads of the sub-Antarctic daisy (*Pleurophyllum speciosum*) grow to the height of a small child, and its corrugated leaves, designed to act as solar panels to soak up the sun's rays, are the size of large serving platters. Such "megaherbs"—large herbaceous plants—are chara-cteristic of the New Zealand sub-Antarctic region.

SOUTHERN FORESTS

Clad in mosses and lichens, the southern rata forests of the Auckland Islands are unique. Rare species such as Hooker's sea lions and Yellow-eyed penguins can be seen from their depths, and overhead Royal albatrosses soar toward their breeding areas on the nearby upland moors.

> NEW FURS FOR OLD

> NEW FURS FOR OLD

Elephant seals haul out onto Macquarie Island's Sandy Beach for their yearly molt. By huddling and rubbing against each other, the seals shed their old coats. In the background, tourists take in the sights under the guidance of a National Park Ranger.

▲ BREEDING SPACE

Less than two percent of Antarctica consists of exposed rock, and these limited areas are the sites of fierce competition for breeding grounds. Penguins defend their breeding territories aggressively, against both their own species and others, as this battling Gentoo (left) and Rockhopper pair demonstrate.

Macquarie Island

Long, thin Macquarie Island—34 kilometers by up to 5.5 kilometers (21 miles by $3\frac{1}{2}$ miles)—lies about 1,500 kilometers (930 miles) south of Tasmania, just north of the Polar Front and about halfway to the Antarctic Continent. Its 128 square kilometers ($79\frac{1}{2}$ sq. miles) form the most remote State Reserve of Australia's island state of Tasmania. The ocean around the island became an Australian Marine Park in 1999.

Like most other sub-Antarctic islands, Macquarie Island has no trees. It has a mean annual temperature of around 5°C (41°F), and some form of precipitation on 300 out of 365 days. Its World Heritage listing in 1997 was largely geologically based, because it is the only island that consists solely of rocks from deep within the earth's mantle and oceanic crust. It emerged from the sea some 600,000 years ago as the crest of a tall ridge, and continues to rise at the rate of 0.8 millimeters ($\frac{1}{300}$ in) each year. Unlike other sub-Antarctic islands, Macquarie Island is a plateau shaped by waves rather than ice.

The first report of Macquarie Island came from Frederick Hasselburg, captain of the sealer *Perseverance* out of Sydney, who arrived here on 11 July 1810. Hasselburg noted the wrecked vessel "of ancient design" of an earlier visitor, but the identity of this voyager has never been discovered. Hasselburg named the island after Lachlan Macquarie, then Governor of New South Wales. Subsequently, every seal was taken for skins and oil, and boiling down penguins for oil continued well into the twentieth century. The destruction of the island's native wildlife continues from the rats, mice, cats, and rabbits that the sealers left behind.

The seals returned—currently the island has a large population of elephant seals and three types of fur seals: New Zealand, Antarctic, and Sub-Antarctic. Penguin numbers have also recovered—there are now 850,000

breeding pairs of Royal penguins, 5,000 pairs of Gentoo penguins (the only ones found on the Pacific side of Antarctica), 100,000 pairs of King penguins, and large colonies of Rockhopper penguins. Other birds include several species of albatrosses and some 800 pairs of the Macquarie Island cormorant, an endemic subspecies of the King cormorant.

Bellingshausen called at Macquarie Island in 1820, Wilkes in 1840, Scott with *Discovery* in 1901, and Shackleton in *Nimrod* on his return from Antarctica in 1909. Douglas Mawson arrived in *Aurora* in 1911 and established a radio relay station on Wireless Hill at the northern end of the island, which he used to communicate between his base at Commonwealth Bay and Australia. Australia's Macquarie Island station was built on a spit below Wireless Hill on 25 March 1948, and has since operated continuously. Affectionately known as Macca, the station houses about 20 people during winter and 40 in summer.

◄ SPLENDID ISOLATION

A narrow isthmus links Macquarie Island to North Head, where Australia's National Antarctic Research Expedition base was built in 1948. Heavy seas sometimes breach the isthmus, leaving the small base even more isolated than usual.

Heard and McDonald Islands

Heard Island lies 4,100 kilometers (2,500 miles) south-west of Perth, Australia. The tiny, ice-free McDonald Islands are 40 kilometers (25 miles) to the west, and the even smaller Shag Islands are 10 kilometers (16 miles) north of it. Like the much larger Iles Kerguelen, which lie 440 kilometers (270 miles) to the north, these islands stand on the undersea Kerguelen Plateau. They were annexed by Britain in 1908, transferred to Australia in 1947, and remain external territories of Australia.

Heard Island is roughly circular, and has an area of 368 square kilometers (142 sq. miles), 80 percent of this area ice-covered; much of the shoreline consists of ice cliffs. The island's major feature is the cone of Big Ben—Australia's only active volcano—topped by Mawson Peak, at 2,745 meters (9,005 ft) considerably higher than any other mountain in Australia.

Even in Antarctic terms, human contact with Heard Island is very recent. The first person to see the island is believed to have been British sealer Peter Kemp, aboard *Magnet* on 27 November 1833. Twenty years later, on 25 November 1853, American sealer John Heard of *Oriental*, saw the island and subsequently had it named after him. Only a few weeks later, in January 1854, William McDonald, British captain of *Samarang*, discovered the islands that now bear his name—the largest of them covers less than 1 square kilometer (⅓ sq. mile). The first people to land on Heard Island were the crew of the American sealer *Corinthian*, captained by Erasmus Darwin Rogers, in 1855. Sealing reached a peak in 1858, and by 1880 too few seals remained for viable hunting.

Challenger came and collected scientific samples in 1874, as did Drygalski in *Gauss* in 1902. Mawson stopped for a week on the BANZARE voyages in 1929; the remains of his hut survive at Atlas Cove. Australia established a base on this site in 1947, but it closed in 1955 and visits since then have been sporadic. Since 1997, Heard Island has been a World Heritage listed Wilderness Reserve. An Australian team cleaning up the old base in 2000–01 found that glaciers have retreated and vegetation has dramatically increased.

These islands house the rare Heard shag and a subspecies of Sheathbill, as well as more than 2 million Macaroni penguins and a rapidly increasing population of King penguins—now more than 100,000 breeding pairs. These are also the only sub-Antarctic islands without any feral species. Because landings are difficult and the weather terrible, the Australian government's limit of 400 visitors each year has never been reached.

▼ AN EPIC LANDSCAPE
The topography of Heard Island ranges from ice-capped volcanic peaks to lush sub-Antarctic vegetation. Glaciers cover 80 percent of the island, and the rest consists of rocky cliffs, sand and gravel beaches, peninsulas, lagoons, and glacial tongues. Big Ben, the island's dominant volcanic peak, towers to 2,745 meters (9,005 ft), belching steam and clothing its upper slopes in looming low cloud.

▽ ENDANGERED HABITAT

The Iles Kerguelen are the largest areas of dry land in the southern Indian Ocean, and more than 30 seabird species flock to the group's rocky shores. Introduced species—particularly rabbits—have altered the landscape of the island and have made life even more difficult for native fauna, especially burrowing petrels, whose nesting habitat has been greatly reduced.

Iles Kerguelen (Kerguelen Islands)

In 1771, Captain Yves-Joseph de Kerguelen-Trémarec sailed south with instructions to establish trade with the natives of the great southern continent. The ambitious Frenchman found neither. Instead, on 12 February 1772, he caught sight of a foggy, volcanic archipelago, miscalculated its position, put some of his crew ashore briefly to claim it for France, and sailed back to Mauritius with a fanciful tale of a land perfect for settlement that he had named Southern France. In 1773 Kerguelen was sent back, with three ships and 700 men, to colonize his discovery. Again, he did not step ashore, but those who did soon decided that the land was worthless, and the fleet went on to Madagascar. When he returned to France with this more accurate report of the island's potential, Kerguelen was court-martialed and imprisoned in Saumur for nearly four years.

When James Cook came to Kerguelen in *Resolution* and *Discovery* on Christmas Day 1776, he claimed it for Britain, and wrote: "I could have very properly called the island Desolation Island … but in order not to deprive M. de Kerguelen of the glory of having discovered it, I have called it Kerguelen Land."

The islands are now part of the French Southern and Antarctic Lands (Terres Australes et Antarctiques Françaises). They consist of the main island, measuring 120 by 110 kilometers (75 by 68 miles), and some 300 small islets and reefs. Deep fiords thrust into the coast of the main island; the interior is heavily glaciated, and rises to Mount Ross, 1,850 meters (6,070 ft) above sea level. The islands lie just to the north of the Polar Front, and the weather there is typically wet and windy, although the sea remains ice-free.

Between 1791 and 1817 hunters wiped out the fur seal population, and then switched their attention to elephant seals and whales. An attempt to farm sheep failed, but feral sheep now roam the island; at various times rabbits, rats, cats (to control the rabbits), reindeer, cattle, and even salmon and trout have been introduced. During World War II, German vessels visited the island's fiords, until Australia laid mines at some anchorages. France built Port-aux-Français, on the east coast, in the summer of 1949–50. Today it is a large, well appointed base, with accommodation for 80 people in winter and 120 in summer; its facilities include a cinema, a sports center, and a chapel. It operates alongside a CNES (Center National d'Etudes Spatiales) satellite tracking station.

The island's most famous plant species is the Kerguelen cabbage (*Pringlea antiscorbutica*), which—as its scientific name suggests—contains the vitamin C needed in the human diet to ward off scurvy. It belongs to the same family as the common cabbage, and was discovered by James Cook and reported by Joseph Dalton Hooker. It has adapted to wind-rather than insect-pollination, apparently because of the lack of winged insects on the islands.

Iles Crozet

The Iles Crozet are the top of an underwater volcanic plateau; their highest point is Pic Marion-Dufresne, 1,090 meters (3,755 ft) high, on Ile de l'Est. Like the Iles Kerguelen, these unglaciated islands belong to the Terres Australes et Antarctiques Françaises. The chain is made up of an eastern group comprising

◄ SCRAMBLING ASHORE
The Iles Kerguelen have occasional sandy landing sites where penguins can emerge from the sea to breed. But the landscape is precipitous, and rough, rocky ground can pose a problem for penguins like these Kings, who cannot easily negotiate steep ground.

Ile de la Possession, the largest of the whole group at 150 square kilometers (58 sq. miles) and Ile de l'Est, and a western group 120 kilometers (75 miles) away consisting of the Ile aux Cochons, measuring 67 square kilometers (26 sq. miles), and the much smaller Iles des Pingouins and Iles des Apôtres, together with other tiny rocks and shoals.

French Captain Nicolas Marion-Dufresne, aboard *Mascarin*, discovered the Iles Crozet in 1772 and named them for his second-in-command, Jules Marie Crozet, who took command when Dufresne was killed by New Zealand Maori. For many years the only human visitors to the islands were occasional sealers and shipwrecked mariners. Whaling stations and scientific bases were built on several of the islands, but now only Alfred-Faure station, above Port Alfred on Ile de la Possession, is occupied year-round. The islands, declared a national park by France in 1938, have the largest colonies of King penguins in the world, with perhaps a third of the world's population of this species breeding there.

Bouvetøya (Bouvet Island)

Bouvetøya is the tip of an inactive volcano that pierces the South Atlantic; it is 780 meters (2,559 ft) tall at Olavtoppen, its highest peak. About 5 by 8 kilometers (3 by 5 miles) and 49 square kilometers (19 sq. miles) in area, it is regarded as the world's most isolated landmass because, except for diminutive Larsøya Island nearby, the closest land is the Antarctic Continent, some 1,600 kilometers (995 miles) to the south. The ice sheet covering Bouvetøya ensures that nothing grows there. Wilhelm II Plateau, an ice-filled volcanic crater, dominates the center of the island, and the sheer sea cliffs of rock and ice make any landing difficult.

This cloud-capped island was first sighted by Jean-Baptiste-Charles Bouvet de Lozier on New Year's Day 1739; hoping that he had seen the tip of a southern continent, he named the north-western point Cape Circoncision after the holy day on which it was discovered. But after battling fog and ice for 12 fruitless days, he gave up hope of landing, or even circumnavigating the point to see if it really was a cape joined to a larger landmass. He made a mistake with the island's coordinates and placed it too far east, so that James Cook subsequently failed to find it in 1772 and 1775. American sealer Benjamin Morrell—a known liar—claimed to be the first to land there in 1822. The island's position was accurately charted by Captain Lindsay, a whaler with the Enderby Company, in 1808, and George Norris, another Enderby captain, made a documented landing in 1825, claimed the island for Britain, and renamed it Liverpool Island.

Landings are so difficult that Bouvetøya has been largely left alone, and remains uninhabited. However, due to its interest in Antarctic whaling, Norway annexed the island in 1928; Britain relinquished its claim, and Norway declared the island a nature reserve in 1971. After several attempts to erect a weather station, the first automatic station was built there in 1977.   DM

▲ A GREAT LIAR
The enigmatic Benjamin Morell was an American sealer who traveled widely in the sub-Antarctic islands and lived for some time on the Iles Kerguelen. He claimed to have been the first person to land on Bouvetøya, in 1822—but he later made many other outrageous claims, and one contemporary called him "the biggest liar in the Southern Ocean."

The Arctic

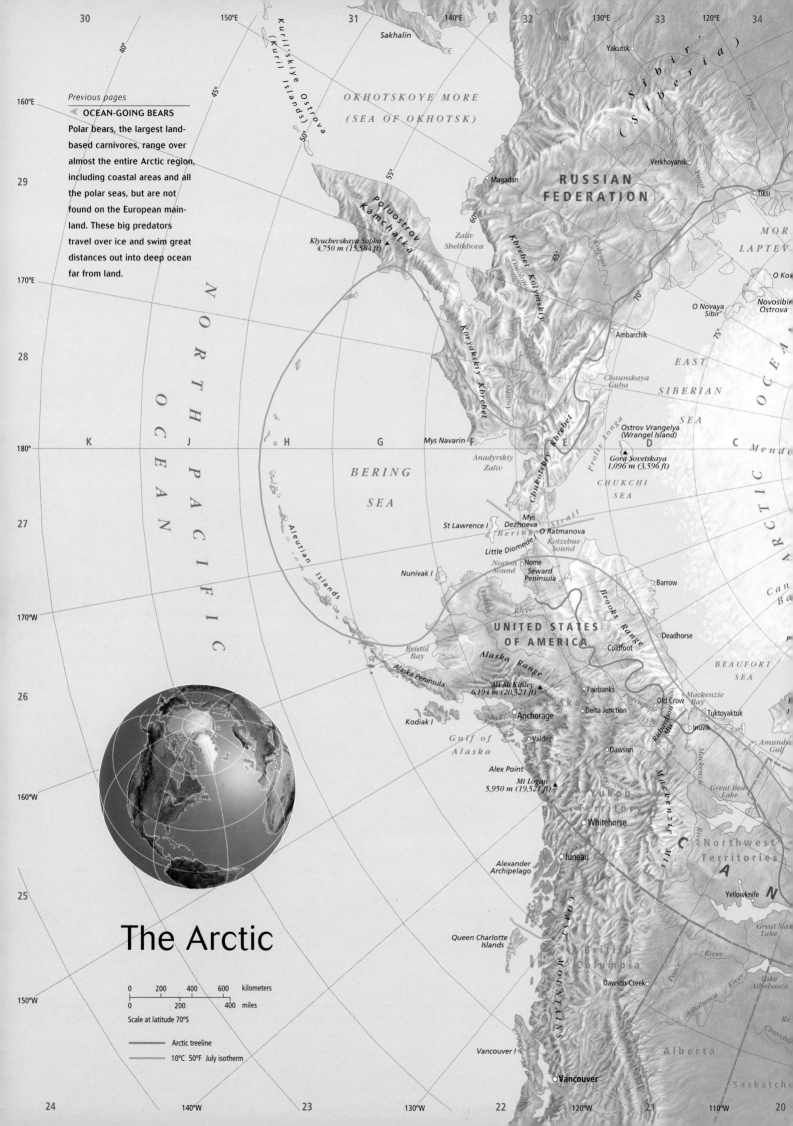

Sakhalin

Yakutsk

Kuril'skiye Ostrova (Kuril Islands)

OKHOTSKOYE MORE
(SEA OF OKHOTSK)

40°

45°

50°

55°

60°

Verkhoyansk

RUSSIAN
FEDERATION

Si bi r (S i b e r i a)

Magadan

Tiksi

Previous pages

◄ OCEAN-GOING BEARS
Polar bears, the largest land-
based carnivores, range over
almost the entire Arctic region,
including coastal areas and all
the polar seas, but are not
found on the European mainland. These big predators
travel over ice and swim great
distances out into deep ocean
far from land.

*Zaliv
Shelikhova*

*Poluostrov
Kamchatka*

Khrebet Kolymskiy

Kolyma

65°

MOR
LAPTEV

O Kot

O Novaya
Sibir'

Novosibir
Ostrova

Klyuchevskaya Sopka ▲
4,750 m (15,584 ft)

Koryakskiy Khrebet

Anadyr

Ambarchik

70°

EAST

*Chaunskaya
Guba*

75°

N
O
R
T
H

P
A
C
I
F
I
C

O
C
E
A
N

K   J   H   G   F

Mys Navarin

*Anadyrskiy
Zaliv*

BERING

SEA

E

Chukotskiy Khrebet

Ostrov Vrangelya
(Wrangel Island)

Gora Sovetskaya
1,096 m (3,596 ft) ▲

Prolw Longa

SIBERIAN

SEA

D   C

Mende

CHUKCHI
SEA

A
R
C
T
I
C

Aleutian Islands

St Lawrence I

Mys
Dezhneva
Bering Strait
O Ratmanova

Little Diomede I
O Ratmanova

*Rotzebue
Sound*

Can
Ba

Nunivak I

*Norton
Sound*

Nome
Seward
Peninsula

Barrow

Brooks Range

Deadhorse

BEAUFORT
SEA

UNITED STATES
OF AMERICA

Coldfoot

Yukon

River

Bristol
Bay

Old Crow

*Mackenzie
Bay*

Tuktoyaktuk

Alaska Peninsula

Alaska Range

Fairbanks

Porcupine

Inuvik

Amundse
Gulf

Mt McKinley ▲
6,194 m (20,321 ft)

Delta Junction

Alaska

Richardson Mts

Kodiak I

Anchorage

Valdez

*Gulf of
Alaska*

Mackenzie

Great Bear
Lake

Alex Point

Mt Logan ▲
5,950 m (19,521 ft)

Dawson

Yukon

Territory

Mackenzie Mts

River

Whitehorse

BRITISH
COLUMBIA

*Northwest
Territories*

N

Juneau

*Alexander
Archipelago*

Yellowknife

Great Sla
Lake

The Arctic

Queen Charlotte
Islands

COAST MOUNTAINS

*British
Columbia*

River

| 0 | 200 | 400 | 600 | kilometers |
| 0 | | 200 | | 400 miles |

Scale at latitude 70°S

——— Arctic treeline

——— 10°C 50°F July isotherm

Vancouver I

Dawson Creek

Peace

River

Athabasca

River

Lake
Athabasca

Vancouver

Alberta

Churchi

Saskatchew

150°W

160°W

170°W

180°

170°E

160°E

29

28

27

26

25

24

36 90°E 1 80°E 2 70°E 3 60°E 4 50°E 5 Nizhniy 40°E 6 30°E
                                                   Novgorod

RUSSIAN
FEDERATION   ⊛ **MOSKVA**   UKRAINE
                                                              ⊛ **KYYIV**   7

Dudinka   BELARUS
                          ○ Vorkuta   ⊛ **MINSK**
Gydanskiy   ○ Arkhangel'sk
Poluostrov
Poluostrov   *Beloye More*   Sankt-Peterburg   ⊛ **WARSZAWA** 20°E
Yamal   *(White Sea)*   ● Sankt-Peterburg   LATVIA LITHUANIA
                    Karskoye   *Ladozhskoye* ESTONIA
                    More   Matochin Shar   Kolskiy   *Ozero*   POLAND   8
Poluostrov Taymyr   *Poluostrov*   HELSINKI ⊛   *Baltic Sea*
Vil'kitskogo   *Novaya Zemlya*   Severomorsk ○ Murmansk FINLAND   Åland   Gotland
                                            Nordkapp ○ Honningsvåg   *Gulf of Bothnia*   ⊛ **BERLIN**
Severnaya   BARENTS   Hammerfest   **STOCKHOLM** ⊛   GERMANY
Zemlya   SEA   ○ Troms   ⊛ **KØBENHAVN** 10°E
Zemlya   Zemlya Frantsa-Iosifa   SWEDEN   ○ Bodø   DENMARK
Vil'cheka   *(Franz Josef Land)*   Bjørnøya   *Vänern*
Ostrov Greem Bell   Zemlya Georga   *(Bear Island)*   ○ Bodø   **OSLO** ⊛   *Skagerrak*
                    Zemlya Aleksandry   NORWAY   NORTH
                Ostrov   ○ Bergen   SEA   9
Nansen   Dzheksona   NORWEGIAN
Basin   Nordaustlandet   9
                            Newtontoppen   SEA   SEA
                        1,717 m (5,633 ft) ○ Longyearbyen   SEA
Fram   North   Spitsbergen Ny Ålesund   Svalbard
Basin   90° Pole A   B   Kongsfjorden **(Nor)** C   NORWEGIAN D   E   F   G   Shetland Is H 0°
        A   GREENLAND   ● **LONDON**
Lomonosov   *Wandel*   SEA   SEA   Orkney Is   Edinburgh
Karov Basin   *Sea*   **Jan Mayen**   UNITED
Ridge   **(Nor)**   Nord-Jan   KINGDOM 10
                        ○ Nord   Beerenberg ● Sør-Jan   Faeroe Is   Belfast
Lincoln   2,277 m (7,470 ft)   *(Den)*   Outer   ⊛ **DUBLIN**
Sea   Cape Columbia   Hebrides
                                                                                                      IRELAND
Ellesmere   Kong Frederik VIII Land   ICELAND
Ellef   Island   11
Ringnes I   Kong Christian X Land   Gunnbjørn Fjeld   ● **REYKJAVÍK**
Axel   3,700 m (12,139 ft)
Heiberg I   **Greenland**   ▲
QUEEN ELIZABETH   **(Kalaallit Nunaat)**
ISLANDS   Pituffik   **(Den)**   *Denmark Strait*
RY   Qimusseriarsuaq   Kong Christian IX Land
NDS   Bathurst   NORTH 20°W
Island   Cornwallis I Devon Island   *Baffin*
Qausuittuq   *Bay*   ATLANTIC
(Resolute)   Bylot I   Qeqertarsuaq
Somerset   Nanisivik   *(Disko)* ○ Ilulissat
Island Arctic Bay Borden   *Qeqertarsuup*
Prince of   Pen   *Tunua*   Kong Frederik VI Kyst   OCEAN
Wales I Brodeur   ○ Ammassalik   12
Pen   *Baffin Island*   Kong Frederik VI Kyst
Boothia   *Davis*
Pen   *Strait*
King   Uummannarsuaq
William I   ● **NUUK**   *(Cape Farewell)*
Gjoa Haven   Melville   Prince   Narsarsuaq
Nunavut   Peninsula   Charles I   30°W
R   *Cumberland Sound*
                Southampton   *Foxe*
                Island   Iqaluit   *Frobisher Bay*
                            Foxe   **LABRADOR**
A   *Pen*   *Hudson Strait*   **SEA**
                Coats I   ○ Cape Chidley   13
                    Mansel I
Tronns Point   *Ungava*
Churchill   *Bay*   **Newfoundland**
        HUDSON   14
        BAY   *Québec*
nitoba   *James Bay*   Ontario   *Island of*
                                            Newfoundland
19 90°W 18 80°W 17 70°W 16 60°W 15 50°W 14 40°W

Europe

The European Arctic region extends across several countries and many islands. Its southern reaches along the Arctic Circle are the most accessible of all polar regions. Tarred roads in Finland, Sweden, and Norway enable motorists to drive for days north of the Arctic Circle. Iceland lies below the Arctic Circle, with only the tip of the tiny island of Grimsey, off the Icelandic north coast, extending into the Arctic Circle.

▲ TARGET SPECIES
Walrus ivory is second in quality only to elephant ivory, and this, together with their thick hides and rich oil, made walruses targets for European hunters as long ago as the Middle Ages. By a century ago, they had been hunted almost to extinction.

North Cape (Nordkapp)

Norway's North Cape, at 71°11′N, is at the top of continental Europe, and is the closest that any non-expedition traveler can get to the North Pole. It is further from Oslo to North Cape by the winding coastal road than it is from Rome to Oslo. This far north, the only other points of continental human habitation are Barrow in Alaska and a few settlements in the Russian Federation—and none have road access. North of the Circle is Lapland, not itself a national entity, but a cultural region that takes no account of the political borders between Sweden, Norway, and Finland. Finally, at the top of Norway, a short ferry trip takes adventurous travelers to Honningsvåg, the northernmost village in the world. North Cape is 40 kilometers (25 miles) further north, where the road ends on a barren, uninhabited headland 300 meters (100 ft) above the steel-gray Arctic Ocean.

Svalbard

Svalbard ("cold coast"), the archipelago north of Norway, is dominated by Spitsbergen, the main island of the group. Although the islands were apparently known from the twelfth century, the first confirmed visitor was Dutch navigator Willem Barents, in 1596. Whalers soon followed him and, more recently, coal miners. After World War I, the Treaty of Spitsbergen gave sovereignty to Norway, which agreed to use the islands only for non-military purposes.

Svalbard is administered by the Sysselmann (a representative of the Oslo government), but in practice it is mining companies that control most of the islands. The main settlement is Longyearbyen, with a population of 1,200 people. The scientific community of Ny Alesund on Konsfjorden (King's Bay) is on the western side of Svalbard. On the metal frame near what is now the camping ground, the airships *Norge* (in which Roald Amundsen flew over the North Pole in May 1926), and the doomed *Italia* (under the command of Umberto Nobile) were once tethered. The moss-covered remains of the hut where Nansen and Johansen spent the winter of 1895–96 was rediscovered on Jackson Island in 1990.

Sixty percent of Spitsbergen's dramatic landscape is glaciated; Newtontoppen and Perriertoppen, its highest mountains, are both 1,717 meters (5,633 ft) tall. Wildlife includes Polar bears, European reindeer, and Arctic foxes; walruses were almost wiped out by early hunters, but are still sometimes seen.

An organized tour or cruise is the only practical way to see Svalbard. There are no rural roads and no scheduled boat services between islands, and helicopters may land only in settlements.

Bear Island (Bjørnøya)

Bear Island, 180 square kilometers (70 sq. miles) in area, lies about halfway between Spitsbergen and North Cape and is administered as part of Svalbard. Its southern part terminates in cliffs 400 meters (1,300 ft) high, and even the flatter land in the north drops 30 meters (100 ft) to the shore. Two air masses meet there, producing frequent fogs, but the annual precipitation is only about 400 millimeters (16 in).

◀ COLORFUL HONNINGSVÅG
Honningsvåg is 40 kilometers (25 miles) southeast of North Cape, along a winding, scenic road. The community once thrived on seasonal fishing, but its proximity to North Cape has meant an increase in tourist interest.

Bear Island, too, was discovered by Willem Barents, and was then occupied by whalers, sealers, and miners. Scientists and hunters have lived on the island at various times, and a few buildings in the north are now historic sites. Currently, it is inhabited only by large numbers of many bird species.

Jan Mayen

North of Iceland, east of Greenland, and level with North Cape, the island of Jan Mayen is 55 kilometers (34 miles) long and measures only 16 kilometers (10 miles) across at its widest point, giving it a total area of 380 square kilometers (146 sq. miles). A narrow, high land bridge separates Nord-Jan (the northern end) from the much-smaller Sør-Jan (the southern end). Norway's only active volcano, Beerenberg, rises to 2,277 meters (7,470 ft) on Nord-Jan. Last erupting in 1985, its summit is cloud-free for a only few days each year.

Whalers reached Jan Mayen by the early seventeenth century—the ruins of five whaling stations remain. The island took its name from a Dutch whaling captain, Jan Jacobs May van Schellinkhout, who was involved in mapping the island in 1614. Norway assumed sovereignty in 1930, and the island is administered from the city of Bodø. The Cold War airstrip is still used and Jan Mayen is important to Norwegian fishing, but today the only residents are a small crew manning the meteorological station of Gammle Metten.

⬆ ISLAND OF MINERALS
Spitsbergen receives little precipitation, but what does fall may come as snow, even in summer. The light snow cover made early mineral exploration comparatively easy. Coal mining was once all-important, and several other minerals, including gold and copper, are present.

Greenland (Kalaallit Nunaat)

Greenland is the largest island in the world, covering 2,175,600 square kilometers (840,000 sq. miles). Because of distortion by the Mercator projection used on many maps, it looks out of all proportion to its real size—appearing larger than the continent of Australia, which is actually four times its area.

The landscape is dominated by a towering polar ice cap between 1,500 meters (4,920 ft) and 3,000 meters (9,840 ft) thick. Sixty thousand Greenlanders live around its narrow rim, which is about 20 kilometers (12 miles) wide and deeply indented by fiords. Locals call their country Kalaallit Nunaat ("Land of the People")—a curious name, perhaps, for a land so sparsely populated. There are few roads, so air and water provide the main

⬆ LOW GROWERS
Only six to seven percent of the land of Svalbard has any significant vegetation. There are 165 known species of flowering plants on the islands, and none are more than 30 centimeters (12 in) tall. Cold, windy conditions keep all plant life close to the ground for protection from the elements.

▲ ICY KULUSUK

The weather and sea ice along the east coast of Greenland are much more severe than along the west coast, and the tiny village of Kulusuk, on the eastern coast, spends much of the year encased in ice. The local people still make their living by hunting, and they also make money by running dog-sledging tours when tourists fly into town.

means of transport. The west coast is warmed by an Atlantic current, but an Arctic current keeps the east coast largely ice-bound.

Eric the Red, banished from Iceland, was the first recorded visitor to Greenland in 982 AD. He decided to live there, and rather inaccurately named it Greenland to encourage others to follow. The ruins of his farm remain at Brattahlid, near Narsarsuaq in the southwest. Greenland's climate was milder then—there was even forest—and Vikings arrived to establish a settlement that endured for 500 years. Little physical evidence of Greenland's earlier indigenous culture has survived.

Greenland is a self-governing part of Denmark, with increasing autonomy. In 1985 it became the first member to withdraw from the EEC, and today it looks more to a circumpolar Inuit community than to Europe for its future. Fishing, the dominant industry, produces more than 80 percent of its export income. Greenland's capital and largest town was originally named Godthåb ("Good Hope"), but residents prefer to call it Nuuk, meaning "promontory"—an apt description, as the town spills across a point. Many of its 13,000 inhabitants live in huge apartment complexes.

The south of Greenland lies below the Arctic Circle, and Nuuk is below the latitude of Iceland. Disko Bay, sheltered by Disko Island, has a very active glacier that spawns numerous icebergs. The quaint little town of Ilulissat is tucked deep within this bay, and is the birthplace of the renowned Greenland explorer, Knud Rasmussen, who traced possible routes of Arctic migration from Greenland to Alaska. His family home is now a museum.

Less than 30 kilometers (18 miles) of water separates Greenland from Canada's Ellesmere Island. The United States Thule Airbase at Pituffik was an important post during the Cold War, and continues to operate to this day. The main eastern town is Tasiilaq, on Ammassalik Island, with a population of 3,000 people. Greenland's eastern side is absolute wilderness; it is the world's largest national park, 1,400 kilometers (870 miles) from one end to the other, and covering more than 972,000 square kilometers (375,289 sq. miles).

It was long believed that the northernmost tip of Greenland was the land closest to the North Pole, but some slivers of dirt have now been found further north of the island itself.

Russia

Russia's vast Arctic region spans six time zones—almost half of the coast surrounding the Arctic Circle belongs to Russia. Six seas lap its northern shores—the White, Barents, Kara, Laptev, East Siberian and Chukchi seas. Cape Chelyuskin is the northernmost point of the Eurasian landmass.

Because much of the interior is inaccessible, except in winter, when the summer swamps that lie over the permafrost freeze, the Russian Arctic effectively begins at Murmansk and ends in the Bering Strait. Both north and east were front lines during the Cold War, and a former series of bases along this northern coast are now military junkyards. The large tracts of glorious Arctic wilderness are best seen from a ship passing through the Northeast Passage.

Murmansk

Murmansk, on the Kola Peninsula, is the largest city north of the Arctic Circle; its population of about 400,000 outnumbers all the people living elsewhere in the Arctic. It is a metropolis of drab apartment blocks and starkly utilitarian public buildings, but it is outdone in unattractiveness by nearby Severomorsk, home port of the large Russian Northern Fleet. Severomorsk harbors nuclear submarines that are sinking into the mud, rusting naval and fishing vessels, and every sort of icebreaker (nuclear, conventional, twin-hulled, and Teflon-coated)—nearly all in terminal decline.

New Land (Novaya Zemlya)

The dividing line between the Barents and Kara seas is formed by the distinctive islands of Novaya Zemlya. At first glance, the two larger islands look like a single, long, narrow finger of land, but closer inspection reveals the narrow dividing strait of Matochkin Shar. Only very detailed maps are able to show the smaller islands of the archipelago.

Russians first visited these islands a millennium ago, but it was not until 1877 that a small population of Samoyed people came to live on the southern island, the slightly smaller of the two, and boasting a gentler landscape and a milder climate.

In summer, the rolling hills of the archipelago are carpeted by wildflowers, diverting attention from a sinister presence. From 1955, Novaya Zemlya was the Soviet Union's main nuclear testing ground, and it is still used for testing non-nuclear weapons. Some of the 130 nuclear tests there included large atmospheric detonations in the early 1960s, which reportedly shattered windows more than 1,000 kilometers (620 miles) away. The last blast was in 1990, and several radioactive zones of exclusion have been established. The dumping of nuclear waste is a more insidious threat—numerous sites off the Siberian coast are prominently marked on Russian shipping charts.

Ice Harbor, on the northern shore of Novaya Zemlya is where Willem Barents built a hut in 1596. It was not rediscovered for more than 300 years, but it is now a place of pilgrimage for intrepid Arctic enthusiasts.

Franz Josef Land (Zemlya Frantsa-Iosifa)

The archipelago of Franz Josef Land is the northernmost land in the Russian Arctic sector, and the most northernpart of Eurasia. Sixty islands and numerous islets make up a total landmass of about 16,000 square kilometers (6,180 sq. miles). The four largest islands are (in decreasing order of size) Prince George Island, Wilczek Island, Graham Bell Island, and Alexander Island. The group extends further north than Svalbard, and is always surrounded by ice and often inaccessible.

⊼ RADIOACTIVE SLOPES
The islands of Novaya Zemyla are largely mountainous and covered in glaciers, although in some southern areas plant life flourishes over the very short summer season. There are high levels of radioactive waste in the area, both from early Cold War testing and from more recent dumping of radioactive waste.

➢ ARCTIC ICEBREAKER
The *Kapitan Dranitsyn* is one of Russia's conventionally powered icebreakers. She carries two helicopters, which are used to seek out the best route when the ice is thick. In these times of economic difficulty for the former Soviet Union, the *Kapitan Dranitsyn* is mostly used to carry tourists to hard-to-reach polar regions.

A VERY QUIET BAY
One of several abandoned Russian stations in Franz Josef Land, this bay was named Tichaya ("Quiet Bay") because of the lack of ice movement in the area. Tichaya Station was begun in the early 1930s, and not completely abandoned until 1959. In its heyday in the mid-1930s, the station housed more than two dozen people, and several babies were born in this bleak research station.

Only rarely does the temperature rise above 0°C (32°F), and 85 percent of the land is glaciated. The archipelago's highest point is 670 meters (2,200 ft), and its longest river is less than 20 kilometers (12 miles) in length.

Named for their emperor by the Austro-Hungarian expedition that discovered the islands in 1873, Franz Josef Land is now visited only by explorers, adventurous hunters, and scientists. Russia claimed the archipelago in 1926, despite protests from Norway and other nations. During the Cold War, five year-round bases were supported there by icebreakers and aircraft—even the *Graf Zeppelin* airship visited in 1931. The islands were not opened up again until 1990, and a single station, Krenkel, survives on Cheysa (Hayes) Island. An Austrian research station operates on Ziegler Island.

Taymyr Peninsula (Puluostrov Taymyr)

The northernmost continental land on earth concludes at an army camp at the tip of Cape Chelyuskin. In this remotest of locations, Russian conscripts mount guard from wooden watchtowers looking toward the North Pole, in order to guard against the extremely unlikely possibility of an attack from the north. The Taymyr Peninsula and the islands of Severnaya Zemlya (North Land), across the Vilkitskiy Strait, separate the Kara Sea from the Laptev Sea. Severnaya Zemlya is a continuation of the mainland, but it is generally so icebound that few tourists are ever able to see it—even from the decks of powerful icebreakers in late summer.

New Siberian Islands (Novosibirskiye Ostrova)

The treeless shores of the several New Siberian Islands are littered with driftwood that has floated down the Lena River and the much smaller Yana River, both of which flow from the taiga forests further south in Siberia. Tall cliffs and strangely eroded volcanic spires are features of these rugged islands, where many tusks of long-extinct mammoths have been found. These islands mark the western end of the East Siberian Sea.

Wrangel Island (Ostrov Vrangelya)

Wrangel Island is cut by the 180° meridian and skirted by the international dateline to the east. It was named by the captain of an American whaling ship for Baron Ferdinand von Wrangel, a Russian explorer who had led an Arctic expedition in 1821–23. The island marks the eastern extent of the East Siberian Sea and the western limit of the Chukchi Sea. It is relatively ice-free, and was claimed by Canadians and Americans until Russian sovereignty was assured when the Soviet Union annexed the island and landed Chukchi and Russian families there in 1926. Their village on the south coast is still occupied today.

Wrangel Island, separated from the Siberian mainland by Long Strait, is about 125 kilometers (80 miles) wide, and about 7,252 square kilometers (2,800 sq. miles) in area. Three mountain ranges shelter wide valleys, the highest point being Mount Sovetskaya, at 1,096 meters (3,600 ft). Unglaciated tundra dotted with lakes supports about 400 plant species. The summer bird population is large; mammals include Polar bears, lemmings, European reindeer, and Arctic foxes.

Chukotski Peninsula (Chukotskiy Khrebet)

The remote and barren coast and interior tundra of Chukotski Peninsula is home to the Chukchi people, an ethnically homogeneous Paleo-Siberian group. The inhabitants who live inland are nomadic reindeer herders, while the coastal villagers rely largely on fishing, hunting, and subsistence whaling. Intermittent and irregular government support from faraway Moscow has left these people with barely enough money to operate their fishing boats.

The Chukotski Peninsula terminates at Cape Dezhev, which is less than 90 kilometers (55 miles) from Alaska's Seward Peninsula. Almost exactly midway between, in Bering Strait, are two tiny islands—steep-sided mountains rising to about 350 meters (1,150 ft)—separated by about 4 kilometers (2½ miles) of water. Big Diomede Island (Ostrov Ratmanova) is in Russia; Little Diomede Island is part of Alaska. They are separated not only by a national border and the international dateline, but also by the enmity generated by the Cold War years. Surveillance and military staff on both of the islands have been greatly cut back since the collapse of the Soviet Union.

Vitus Bering glimpsed these islands on St Diomede's Day (16 August) in 1728. Big Diomede Island has an area of 10 square kilometers (almost 4 sq. miles) and supports a weather station. About 200 residents occupy Little Diomede Island's 2 square kilometers (¾ sq. mile).

> WHAT'S IN A NAME?

The name Chukchi means "rich in reindeer," and was originally used to distinguish the inland reindeer herders from the fishing-oriented coastal peoples of far eastern Russia, but it is now used to denote most of the area's indigenous people. Chukchis living in coastal communities like this one on the Chukotski Peninsula have always been economically dependent on the sea.

▼ REINDEER ROPING

The Chukchi people in the inland of the Chukotski Peninsula manage their reindeer using methods that are centuries old. Lacking pens to herd the animals into, they capture their reindeer for inspection by throwing a rope over their horns.

North America

The Bering Strait averages a depth of between 30 and 50 meters (100–165 ft). During the Ice Age, 25,000 to 10,000 years ago, when the sea level was lower than it is today, this strait was a land bridge from Asia into North America. Many animal and plant species crossed it, and for many millennia it formed a path for the first human settlers to trickle into North America.

Alaska

Even when the Bering Strait eventually filled with water, it proved little barrier to trade or raid. The link between Russia and Alaska was so strong that it was not until 1867 that Alaska was sold to the United States for $US7.2 million—a purchase that was widely ridiculed in America as "Seward's Folly," after the Secretary of State, William H. Seward. The Territory of Alaska was established in 1912, and Alaska became the United States' forty-ninth state in 1959.

Most of Alaska is wilderness, and lies south of the Arctic Circle. The only road linking Arctic Alaska with the bulk of North America is the Dalton Highway, which runs north of Fairbanks. Until a few years ago, this was "the haul road" for building, and then servicing, the oilfields of Prudhoe Bay and the Alaskan Pipeline that runs almost 1,285 kilometers (800 miles) from the northern oilfields to the southern port of Valdez. The pipeline is a marvel of modern engineering, and the accompanying road, opened all the way for public access in 1994, leads through the treeless beauty of the Brooks Range. For private vehicles the road ends at a locked

▼ UNTOUCHED MOUNTAINS
The Brooks Range in Alaska runs east–west, north of the Arctic Circle and across the top of the state. This vast area is a maze of woven stream beds and stark, gray mountains. The region remains largely pristine, and supports caribou, Dall sheep, and moose, as well as wolves and brown bears.

gate into the oilfields, just before the shores of the Beaufort Sea and the Arctic Ocean, but there are conducted tours of the oilfields for those who wish. The buildings at Deadhorse, Prudhoe Bay, and the small, drab village of Coldfoot, midway along the highway, are uninspiring, but the land on either side of the highway is magnificent. The Brooks Range is in the 33,993 square kilometers (13,125 sq. miles) of the Arctic National Park, a wilderness reserve inhabited by moose, wolves, and grizzly bears.

Barrow, the most northern town of the United States, is an Inupiaq community of about 4,500 residents. It is accessible only by air, and most travelers who go there do so to make the extra 16-kilometer (10-mile) journey to the tip of Cape Barrow, at 71°23′N. During summer there are 82 days when the sun does not drop below the horizon.

Canada
Canada, the world's second-largest country, has vast tracts of land inside the Arctic Circle, including the Yukon Territory, the Northwest Territories, and Nunavut. The 60°N parallel forms their southern border and the uppermost limit of Canada's provinces. The DEW (Distant Early Warning) Line was constructed by the United States from Greenland to the Bering Strait between 1952 and 1957, at a cost of $US600 million dollars—a substantial amount at the time. It became largely obsolete with the development of more advanced rocket technology, but some stations continued to operate after the collapse of the Soviet Union in 1990. The most accessible station is the one that stands just outside Churchill, in Manitoba.

Yukon
Most of the mountains and some of northern Canada's most interesting towns are within the Yukon, including Whitehorse on the Alaska Highway and Dawson City to the north. The Alaska

▶ INUIT HUNTER
This Canadian Inuit is one of many hunters supporting his small village. Harvesting the local wildlife generates a significant proportion of the village income, and hunters still search out seals, whales, caribou, foxes, wolves, and Polar bears.

Highway was built in just eight months in 1942 as a strategic move after Japan bombed Pearl Harbor and occupied Alaska's Aleutian Islands. It runs from Dawson Creek in British Columbia to Delta Junction, just short of Fairbanks, Alaska. There is now a very ambitious plan to match the highway with a railroad that would continue past Fairbanks to Nome, across the Bering Strait to Russia, and onward to the Russian rail network.

Very little of the Arctic Yukon is readily accessible, except for 60 kilometers (37 miles) of the Dempster Highway that passes through the Yukon, north of the Circle. This unpaved highway opened in 1979 and runs 456 miles (734 kilometers) from the Klondike Highway, south of Dawson City, through to Inuvik in the Northwest Territories.

The northern end of the Yukon is marked by the Continental Divide in the Richardson Mountains, and its most northern settlement is Old Crow, an Indian settlement on the Porcupine River, with a population of about 250 and no road access at all.

▼ DOWNTOWN NOME
The coastal town of Nome, with a population of more than 3,500 people, is one of the largest communities in the Arctic. The town has no road access to the outside world, and all supplies for a large part of western Alaska are unloaded from barges in Nome's dock area, and then trucked out on short regional roads during the summer.

▲ A HAVEN FOR BIRDS
Areas of clear water (polynyas) and leads usually develop in the icy waters off Bylot Island, northeast of Baffin Island, Nunavut, providing dependable access to the water for wildlife. The island, a bird sanctuary, has large nesting populations of Thick-billed murres and Black-legged kittiwakes, and the world's largest colony of nesting Greater snow geese.

Northwest Territories

The Northwest Territories extends across the top of three Canadian provinces to the North Pole, covering 1,171,918 square kilometers (452,480 sq. miles). There are only about 42,000 people living there—17,000 of them in Yellowknife, the only city, and the capital. Yellowknife is a modern city; its name is derived from the yellow copper from which the local Athapaskan band of Indians made their tools. Even today, half the territory's population is indigenous.

The Northwest Territories takes in Banks Island, some of Victoria Island, and the Parry Islands, and Canada claims the sector of the polar ice beyond these to the North Pole. However, almost the only roads are in the south, and there are very few of these. While the south is extensively forested, the Arctic lies beyond the taiga and comprises the lakes and summer swamps characteristic of the tundra.

From the Yukon's Dawson City, the unpaved Dempster Highway (which first opened in 1979) goes to Inuvik in the Northwest Territories on the vast Mackenzie River Delta. Inuvik is an Inuit community of 3,000 people living 100 kilometers (60 miles) short of the Beaufort Sea. In the winter it is possible to drive on the frozen river to the even smaller and more remote coastal community of Tuktoyaktuk.

Nunavut

Canada's most recently established territory covers one-fifth of Canada's landmass—nearly two million square kilometers (772,200 sq. miles). Nunavut was created in 1993 to come into existence on 1 April 1999, and comprises all or part of several administrative regions formerly part of the Northwest Territories. In the native Inuktitut language, Nunavut means "our land." The capital, in Frobisher Bay on Baffin Island, was named Frobisher Bay too, when it was a Hudson Bay Company post, but in 1984, the 523 voters there decided that their town should become Iqaluit, meaning "the place of fish." In 2001 Iqaluit had a population of 4,000, but it is growing rapidly as its infrastructure expands.

Apart from some tiny outposts, there are just 26 communities in Nunavut. There are no roads into the territory and a single 21-kilometer (13-mile) road links the two communities of Arctic Bay and Nanisivik. Travel to and within Nunavut is largely by air, but ships also explore the region's myriad islands and ice-filled waterways. Baffin Island's Admiralty Inlet is one of the world's longest fiords, and Nunavut's waters are home to exotic species such as narwhals and beluga whales.

The main hub of Nunavut's high Arctic is Resolute, known locally as Qausuittuq—a community of about 200 people on Cornwallis Island, located right on the Northwest Passage. Gjoa (pronounced "Joe") Haven, on King William Island, has grown rapidly. Roald Amundsen named the harbor after his ship, describing it as "the finest little harbor in the world." Sheltered from the pack ice outside, the wooden *Gjoa* wintered here in 1903–04 during Amundsen's first successful traverse of the Northwest Passage. A nearby park contains Amundsen's rough magnetic observatory and a cairn he erected to honor the German geographer Georg von Neumayer. Graves in this park are thought to be those of some members of the ill-fated John Franklin expedition. DM

AT HOME ON THE ICE

Polar bears spend large amounts of time on the ice, hunting or just waiting for prey. Their adaptations to an icy lifestyle include hair almost completely covering their feet, long, water-repellent guard hairs, and teeth specialized for their largely carnivorous diet.

FAST FLOWERS

Purple saxifrage is a cushion-type plant well adapted to the short Arctic summer. Advanced flower buds spend winter protected by the foliage, and as soon as the snow melts, the plant flowers in as few as five days. The plant is usually found on barren exposed rock and in damp crevices in cliffs.

HISTORICAL SITE

Dundas Harbor, on Devon Island, Nunavut, is one of the few places where it is possible to see musk oxen, as well as Narwhals, Belugas, Walruses, and Polar bears. The area is also of archeological and historical interest: some of its Thule culture sites date back more than 1,000 years, and there are remains of an abandoned Royal Canadian Mounted Police post.

Part III
POLAR WILDLIFE

The ends of the earth are home to many remarkable species: the polar bears and tusked walruses of the Arctic; the whales and seals of the polar oceans; the seabirds that must land to breed on the rocky shores and ice shelves of the frozen lands; the flightless penguins that must huddle for warmth on the Antarctic rock and ice. Plants find a precarious roothold in hostile soil, and tiny invertebrates and hardy mosses and lichens cling to life in a savage climate. All these life forms have evolved techniques to survive the challenging conditions that prevail in high latitudes.

Antarctic Ecology

Plants and invertebrates

For land plants, life becomes harder the further south they grow. Plants can grow only where there is enough liquid water, and they must contend with freezing conditions, prolonged darkness, and extremes of light. As latitudes rise, plant life becomes less varied and abundant, with fewer species, fewer major plant groups, a lower biomass, a smaller proportion of ground covered by plants, and a shorter average plant stature. These changes resemble those along a moisture gradient from well watered regions into arid deserts—a closer analogy than it might seem, because the Antarctic region is extremely dry. As well, low temperatures and extreme fluctuations in light levels at high latitudes affect chemical processes such as photosynthesis, and plants are forced to take protective measures against these stresses.

The Antarctic regions

In terms of plant and animal life, the Antarctic region can be divided latitudinally into three distinct zones.

The sub-Antarctic islands comprise six island groups near the Polar Front: Marion and Prince Edward islands; Iles Crozet; Iles Kerguelen; Heard and McDonald islands; Macquarie Island; and South Georgia. These islands support abundant but treeless vegetation, verdant in summer but brown when foliage dies in winter. The islands south and southeast from New Zealand (Snares, Auckland, Campbell, Antipodes, and Bounty islands) support similar vegetation, with the addition of upright shrubs and trees; they are in some contexts also called sub-Antarctic. All the major plant groups except cone-bearing trees (gymnosperms) occur in the sub-Antarctic. Microscopic invertebrates and larger invertebrates such as snails, earthworms, flies, beetles, and spiders are plentiful on the sub-Antarctic islands.

Maritime Antarctica consists of the west coast of the Antarctic Peninsula and its offshore islands, together with the islands south of the maximum extent of the sea ice: the South Sandwich, South Orkney, and South Shetland islands, and Bouvetøya (Bouvet Island). In this region, mosses and liverworts (bryophytes) are reasonably abundant and widespread, but only two species of non-woody flowering plants (soft-stemmed angiosperms) grow, and then only sparsely. Microscopic invertebrates and larger invertebrates such as springtails and free-living mites live in soil and amongst rocks and vegetation.

Continental Antarctica is the "Great Circle" of East and West Antarctica—the Antarctic Continent, with the exception of the west coast of the Peninsula. Near the coast in continental Antarctica, mosses, liverworts, lichens, and algae can grow, but are generally sparse. On the high Antarctic ice sheet at the Pole itself, there are probably no plants at all.

Growing near freezing point

Many plants go on metabolizing and growing, albeit slowly, at temperatures well below 0°C (32°F). The moss *Bryum argenteum* photosynthesizes to at least –2.5°C (27.5°F) and the lichen *Umbilicaria aprina* to at least –17°C (1.5°F). Although water freezes at 0°C, dissolved organic and inorganic molecules, ions, and hydrated macro-molecules lower the freezing point of cellular fluids in Antarctic plants.

Most Antarctic invertebrates avoid freezing by supercooling, surviving temperatures as low as –40°C (–40°F). The ability to supercool is achieved by having solutes, such as alcohols and sugars, in body cells. The mite *Maudheimia wilsoni* can withstand temperatures as low as –30°C (–22°F) without freezing. Alternatively, they may be freezing-tolerant, freezing at relatively high subzero temperatures, but having ice nucleators and cryoprotectants so tissues are not damaged. The nematode *Panagrolaimus davidi* is freezing-tolerant to –40°C, and the rotifers *Philodina gregaria* and *Adenita grandis* to even lower temperatures.

In summer, dark-colored surfaces become surprisingly warm. The surface temperature of rocks facing the midday sun can rise to almost 30°C (86°F). The daily heat penetrates rocks to a depth of more than 1 meter (3 ft), and is slowly released at night when the sun is low in the sky or just below the horizon. Plants growing near these rock surfaces benefit from the warmth, but such sites may also dry out rapidly. In a large rocky oasis with little permanent snow, such as the Vestfold Hills, the whole area has a slightly but significantly higher summer temperature than the neighboring ice-covered coastal edge of the Antarctic ice sheet.

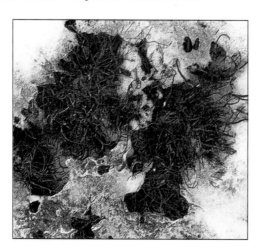

> **LIFE ON THE EDGE**
> In the dry, cold Antarctic environment, lichens (the circular dark patch on the nearest rock) benefit from the warmth absorbed and later released by dark rock surfaces. Snow melting at the rock surface provides much-needed moisture.

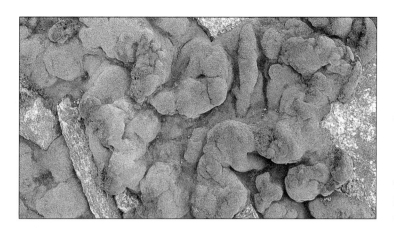

◁ ANTARCTIC CARPET
Lush growths of moss occur
at Antarctic sites where
moisture and temperature
conditions are favorable
and there is some protection
from wind-blown particles.
Individual moss stems grow
close together forming a
carpet which becomes
wrinkled, folded, and broken
by repeated cycles of freezing
and thawing.

▷ A THRIVING COMMUNITY
Megaherbs *Stilbocarpa polaris* (bright green
leaves) and *Pleurophyllum hookeri*, (gray-
green leaves), and the tussock grass *Poa
foliosa* grow along Finch Creek, Macquarie
Island. Royal penguins use the creek bed
as a route from the beach to breeding
areas. Fur seals are also seen there.

Microscopic invertebrates—protozoans, rotifers, tardigrades, nematodes, platyhelminths, and crustaceans — live where moisture is available on land and in lakes, but there are no land-based vertebrates in continental Antarctica: seals, penguins, and petrels come ashore to breed, but they must live and feed at sea.

Marine plants of the Antarctic

For plants, conditions in the ocean at high southern latitudes are much more favorable than on land. In the intertidal zones and shallow waters off sub-Antarctic islands, some of the world's largest marine algae form dense kelp communities of high biomass. Even in the colder shallow water off continental Antarctica, below the depth of the sea ice and where the sea bed is not scoured by icebergs, the biomass, stature, and diversity of marine algae are greater than that of land plants. Marine plants receive less light than land plants because light intensity is much diminished by passing through water, which suggests that low temperatures during the summer growing season and lack of liquid water are much more significant factors than low light intensity in the sparsity of Antarctic land vegetation.

Sub-Antarctic islands

The sub-Antarctic islands vary greatly in altitude and area covered by permanent ice. They range from

Antarctic plants and invertebrates

Current estimates of numbers of species

	SUB-ANTARCTIC ISLANDS	MARITIME ANTARCTICA	CONTINENTAL ANTARCTICA
PLANTS			
Flowering plants	56	2	0
Ferns and relatives	16	0	0
Mosses and relatives	~400	~100	~30
Lichens	~300	~150	~125
INVERTEBRATES			
Millipedes (Myriapoda)	3	0	0
Mites (Acari)	52	24	21
Flies (Diptera)	44	2	0
Springtails (Collembola)	92+	?	24
Crustaceans (Crustacea)	41	10	14
Snails (Mollusca)	3	0	0
Annelid worms (Oligochaeta)	4	0	0
Nematode worms (Nematoda)	22	40	10
Tardigrades (Tardigrada)	40+	17	6
Rotifers (Rotifera)	102	46	41

> **INTERTIDAL GIANTS**
Luxuriant beds of bull kelp (*Durvillaea antarctica*) are attached to rocky sub-Antarctic shores in the intertidal zone. Many invertebrates live amongst the hold-fasts that attach these plants to the shore. On treeless sub-Antarctic islands, these marine algae are the largest plants.

unglaciated Macquarie Island, 433 meters (1,420 ft) high, and often fondly called "the great green sponge," to heavily glaciated Heard Island—2,745 meters (9,005 ft) at its tallest point. Their climate at sea level reflects that of the surrounding Southern Ocean. Average air temperature varies by only a few degrees from summer to winter, and the islands receive frequent precipitation as rain or snow/ice, so that plants have no extended periods without water.

From sea level to approximately 200 meters (650 ft) altitude, the ground is completely covered by herbaceous flowering plants, bryophytes, and some ferns, except where there are rocky bluffs, screes, moraines, fresh lava flows, mobile sand sheets, seal and penguin disturbance, or landslides. The tallest vegetation, up to 2 meters (6½ ft) high, comprises tall tussock grass and large-leaved megaherbs (dicotyledons). In less favorable conditions there is short herb vegetation approximately 0.3 meters (1 ft) high: grasses and sedges (monocotyledons), small dicotyledons, mosses, or ground-hugging cushion plants.

At higher altitudes, lower temperature and increased exposure to wind reduce both the percentage of ground covered by plants and plant stature. Below 100 meters (330 ft) on Macquarie Island, the large herb *Stilbocarpa polaris* may have leaf stalks 1 meter (3 ft) long, each supporting a leaf blade the area of three dinner plates, whereas at 300 meters (985 ft), *Stilbocarpa* plants are rare, and grow slowly to about 20 centimeters (8 in) in diameter only where they are sheltered from abrasive, wind-driven sand, gravel, and ice pellets. On glaciated islands, ice sets a limit to plant distribution, but in recent decades the ice has been retreating and much new land has become available for colonization by plants.

◄ **LARGE AND LUSH**
Near sea level on Macquarie, Auckland, and Campbell islands *Stilbocarpa polaris* grows to about 1 meter (3 ft) tall. By mid-summer, showy yellow flowers are abundant; at the end of the season seeds fall to germinate the following summer.

With increasing altitude and abrasion, the proportion of flowering plants drops while the proportion of bryophytes rises. Cushion-form flowering plants like *Azorella selago* or *A. macquariensis* become most common—their tight mat of small, tough leaves resists wind and wind-driven abrasives, and their extremely long contractile roots, radiating shallowly in the soil, anchor them securely. Under the most extreme conditions, a surface of apparently bare gravel may be held together by a fine but dense, thread-like network of liverworts, their minute, reddish leaves sheltering in the crevices between the small stones.

> FELDMARK COMMUNITIES
Where bare ground exceeds 50 percent, polar (and alpine) vegetation is called feldmark. Vegetated areas may be scattered, or may form stripes or steps-and-stairs terracing that marches spectacularly across hillsides roughly parallel to the contours.

▼ CUSHION PLANTS

Several species of *Colobanthus*, known as pearl-wort, grow as small, compact clumps on sub-Antarctic islands. *Colobanthus quitensis* is one of only two species of flowering plants growing in Maritime Antarctica.

Maritime Antarctica

The Peninsula forms a geographical and botanical transition zone between the far south of South America (Tierra del Fuego to Isla Hornos) and the great circular sweep of East and West Antarctica. Although the Drake Passage is a fearsome barrier, it is the narrowest ocean gap between Antarctica and other lands that plants might cross. The Peninsula is a mountainous, glaciated spine, essentially devoid of plants except near sea level, and most of its offshore islands are of similar physiography. The western fringe of the Peninsula near sea level is a relatively favorable environment for plants; the eastern fringe is colder.

Antarctica's native flowering plants and its only two vascular species—the small grass *Deschampsia antarctica* and the small cushion plant *Colobanthus quitensis* (pearl-wort)—are found on the western side of the Peninsula. By contrast, Tierra del Fuego, just to the north of the Drake Passage, has 417 species of vascular plants, of which 386 are flowering plants. Antarctica's two species of flowering plants are found as far as 68°S; at a similar latitude in the northern hemisphere, Iceland supports some 329 species, and a

◄ FINDING A ROOTHOLD

The tiny tussocks of the grass *Deschampsia antarctica* grow amongst rocks and in soil-filled cracks in bedrock. Right to the southern limit of its distribution, *D. antarctica* can grow densely enough to form closed swards.

further 10 degrees north, 110 species of flowering plants grow in Spitsbergen. Suitability of soil and temperature and adequacy of moisture limit the distribution of Antarctica's grass and pearlwort. During recent climate changes, pearlwort populations have expanded—a trend that is likely to continue with future warmer summer temperatures.

On the eastern side of the Peninsula, and on Deception Island and Ross Island, volcanic activity enables some species of bryophytes to grow that are otherwise known only from further north. Most sites are unsuitable for temperate species, but it appears that there is a constant rain of plant fragments over Antarctica from warmer regions to the north, and some of these plants manage to become established on steam-warmed ground where the temperature is right and moisture is available from condensed steam.

Continental Antarctica

In East and West Antarctica plants are confined to snow-free and ice-free land, including mountains that project through the ice sheet, and to the land around the coastal fringes of the continent and of the Ross, Filchner, Ronne, and Amery ice shelves. In Victoria Land, volcanically heated sites can support bryophytes. Apart from these areas, the only known plant forms in continental Antarctica are algae that grow on snow and ice, feeding on mineral nutrients from the snow and its trapped dust, and coloring the snow gray, green, orange, pink, or red.

Biologically, the most productive areas of Antarctica are the few small ice-free patches of land around the coastal fringe. These areas, called oases, are breeding areas for marine animals, which carry ashore plant nutrients, particularly phosphorus. Humans use the oases for building national bases and for tourism—in the process bringing in pollutants and substantially disrupting the local ecology.

Certain environmental conditions favor plant growth in continental Antarctica. Water is available at sites with adequate snow, a northerly aspect that assists melt, and a microtopography that channels and concentrates or retains melt water. Heat is retained by rock surfaces when the sun is high in the sky and is slowly released as the sun circles low in the sky each summer day. Also necessary are supplementary nutrients, such as nitrogen provided by nitrogen-fixing lichens and cyanobacteria that are usually components of Antarctic vegetation, and phosphorus brought in from the ocean by marine birds.

Vigorous moss-dominated vegetation forming a green carpet is usually found only in scattered small patches or strips, often following drainage lines. Extensive areas of abundant moss and/or lichen—

exciting finds for botanists—are much less common. Two such areas in continental Antarctica are in the Windmill Islands oasis around Casey station, and at Granite Harbor in southern Victoria Land.

By contrast, some areas are notable for a remarkable scarcity of plants. In the Vestfold Hills oasis, 400 square kilometers (155 sq. miles), a coastal strip about 10 kilometers (6 miles) wide has little vegetation, but there is abundant mossy vegetation further inland, towards the edge of the ice sheet. Plant growth in the coastal strip is restricted by salinity inherited from more than 5,000 years ago, when ocean covered parts of this zone.

▲ TINTED SNOW
Unlike most Antarctic plants, snow algae are able to grow in snow itself, gleaning moisture from the snow as it melts. Various pigments in the cells of snow algae trap light energy for photosynthesis. *En masse*, the pigments color the snow—here, a pink hue.

Obtaining water in the driest continent

Although Antarctica's volume of ice is huge, low precipitation and considerable evaporation into the dry atmosphere make it the driest continent. Plants grow where liquid water is available during the summer, but the very low relative humidity dries them out quickly when there is no water.

Most vascular plants (flowering plants, conifers, ferns) try to keep their cells moist by having leaves covered with a waterproof cuticle, and by controlling variable openings in leaf surfaces, called stomates. For vascular plants to produce necessary sugars by photosynthesis, carbon dioxide gas, a raw material, must enter leaves via open stomates. But open stomates means water vapor can escape from moist cells inside to dry air outside, leaving the plant vulnerable to death by desiccation in the absence of a water supply.

Mosses and lichens have no cuticles or stomates and

desiccate quickly without water. But many mosses and lichens in Antarctica survive repeated cycles of wetting and drying by suspending metabolic activity when desiccated and resuming it when rehydrated. They take advantage of any available water and light—for example, when a dusting of snow over their bodies melts in midday warmth. Plants with this ability also exist outside the polar regions, and are called resurrection or poikilohydrous plants.

◄ COLOR FOR PROTECTION
The striking colors of Antarctic lichens—yellow, orange, and black—and of some Antarctic mosses—bronze and black—come from sun-screening pigments produced by the plants to protect their photosynthetic apparatus from damage by light of high intensity and by ultraviolet light.

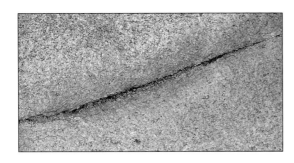

▲ FILLING THE GAP
In the harsh continental
Antarctic environment
endolithic (inside rock) and
sublithic (underneath rock)
habitats provide shelter for
some lichens and algae. Here,
dark gray lichen is living in a
crack in a rock.

▼ COMMUNAL LIVING
Around Adélie penguin nests,
the green alga *Prasiola crispa*
benefits from the nutrients in
guano. It is one of the plants
that survives months of
continuous light in summer
and months of winter darkness in continental Antarctica.

In Antarctica, as in
hot deserts, the ultimate
refuge of plant life is
below translucent rocks
or within the tiny pores
in rocks. Here green algae
and lichens grow in thin
layers, receiving some
light, warmth from the
heated rock, and more
moisture than they would in the open air.

The invertebrates that live amongst Antarctic vegetation feed on dead plant material, soil bacteria, algae, and
fungi. The carbon dioxide exhaled by the invertebrates
and bacteria living amongst the
stems of Antarctic moss micro-
forests increases the rate of
photosynthesis in the moss leaves,
enhancing the production of
sugars during the brief Antarctic
summer growth period.

Lakes are common in ice-free
areas, the best known being those
in the Dry Valleys of southern
Victoria Land and in the Vestfold
Hills of East Antarctica. Each lake
contains a population of microorganisms: green algae, fungi,
protozoa, and cyanobacteria
and other prokaryotes (simple
organisms that can survive in
conditions too extreme for other
forms of life, including volcanic
soils with high concentrations of
heavy metals, volcanically heated
soils and water, and the extreme
cold of Antarctic rock outcrops).
The mix of organisms depends
upon the water's chemical and
physical properties: some lakes
are far more saline than seawater,

others are as pure as distilled water; some are well
mixed, others are stratified, with stable layers of different
temperatures, light levels, oxygen content, and other
chemical properties. Some non-saline well-mixed lakes
have underwater forests of moss more than a meter (3 ft)
tall growing from the lake bed below the reach of the
floating winter ice. The moss, *Bryum pseudotriquetrum*, is
normally terrestrial, where it grows only a few centimeters tall. Sheets of photosynthetic cyanobacteria
commonly form on lake beds, thin layer upon thin layer,
forming mats 10 centimeters (4 in) thick or more; if
lithified (turned to rock), they would be called stroma-
tolites. Dried fragments of these sheets, like pieces of
torn paper, accumulate as lines of flotsam on lake shores.

Coping with light fluctuations

Antarctic plants have to cope with
low light levels, or even no light at all.
They do this by storing enough carbo-
hydrates to meet their very low respira-
tory needs during the lightless winter,
thereby keeping their photosynthetic
machinery intact during prolonged
darkness. Photosynthesis resumes
when the sun rises above the horizon
and the plants rehydrate, even when
buried by snow. Low light merely
prolongs the winter no-growth period,
or causes very slow growth.

Extremely high summer light, with
its ultraviolet component, is a more
substantial problem, apparently exacerbated by the combination of high light
levels and low temperatures. Light
absorption by the photosynthetic pigments (chlorophylls) occurs equally well

at high and low temperatures, and
generates energized, highly reactive
states in these pigment molecules. This
energy is used in the biochemical
reactions that convert carbon dioxide
gas and water into sugars. However,
these biochemical reactions slow down
as temperature falls, and the unused
energy damages the pigment complexes
that absorb it, so that high light at low
temperatures may inhibit or harm
plants' photosynthetic machinery. High
ultraviolet levels in the incoming light,
especially when there is severe ozone
depletion in spring and early summer,
compound the problem. Antarctic
plants, and many elsewhere, respond
by secreting large amounts of colored
sun-screening chemicals that absorb
much of the damaging light.

UNUSUAL ADAPTATIONS
Beneath the frozen surface of an Antarctic lake (right of illustration) is liquid water in which populations of microorganisms live. Some photosynthesize, some ingest other organisms (heterotrophs), and fascinatingly some photosynthesize during summer when light is available but are heterotrophic during the dark winter.

The only plant communities known on the surface of the ice sheet are transient patches of snow algae, found at low altitudes where some summer melt occurs. But the whole ice sheet contains airborne microscopic spores of fungi and other single-celled organisms, such as bacteria, and no doubt tiny reproductive structures and fragments of plants. These organic bodies were probably trapped in snow, in the same way as the dust particles that also occur throughout the ice sheet, which have proved valuable markers of global arid periods in the past. The distribution of organic particles deep in ancient Antarctic ice may prove to be another useful archive of data about earth's past environments. This great ice sheet also covers and seals lakes of liquid water, the largest being Lake Vostok. Fossil or living organisms occurring in its water and sediment await discovery.

The future of Antarctic life forms

High-latitude environments are undergoing change, but the consequences for polar organisms are difficult to predict. Global climate change is bringing warmer temperatures, changed precipitation, and changed ice cover. Depletion of the ozone layer over both the Antarctic and Arctic results in increased levels of ultraviolet light. On sub-Antarctic islands introduced animals are another threat; rabbits damage vegetation and promote soil erosion, cats prey on ground-nesting birds, and invertebrates accidentally introduced in building materials influence vegetation by consuming litter and live plants and compete with indigenous species. Toxic waste and fuel spills from Antarctic bases cause serious habitat destruction for Antarctic plants and animals. And, finally, tourism may become a problem: tourism is now ship-based, but serious environmental damage to plant and invertebrate habitats is inevitable if tourism ever becomes largely land-based. PS and DA

Arctic *versus* Antarctic

Arctic landmasses and surrounding oceans are warmed in summer by heat carried north by the Gulf Stream and Kuroshio Current, whereas comparable latitudes in the southern hemisphere are occupied by the cold Southern Ocean and the continental ice sheet. Northern and southern vegetation at similar latitudes are spectacularly different. In northern Scotland, about 55° from the equator, forests of tall trees grow, but sub-Antarctic islands such as Macquarie Island support only soft-stemmed angiosperms no more than 2 meters (6½ ft) high. At about latitude 70°, Canadian Arctic islands have ground-hugging dwarfed trees, but at the equivalent latitude in Antarctica, tiny mosses are the tallest and most complex plant life.

Life in the Southern Ocean

▼ JOURNEY OF DISCOVERY
On Scott's inaugural journey to Antarctica in 1905, the explorers were delighted to discover a rich and unique fauna. These illustrations, from Scott's *The Voyage of the Discovery*, 1905, show some representatives of the invertebrate groups discovered on this expedition.

Identifying the food web

The first explorers to go to Antarctica, more than two centuries ago, were impressed by the abundant wildlife—whales, seals, penguins, and winged seabirds. The rich, ochrous colour on the underside of the sea ice also intrigued them. They recognized that it came from high concentrations of microscopic, single-celled plants called diatoms, which, they rightly concluded, represented the base of a rich food web.

From that time until less than 20 years ago, the Southern Ocean around Antarctica was thought to be one of the richest marine ecosystems on earth, with a simple food web in which diatoms fed shrimp-sized crustaceans called krill, and fish, whales, seals, and birds consumed the krill. We now know that this is a gross oversimplification. While the Southern Ocean has some extremely productive areas, it is no more productive, overall, than more nutrient-poor parts of the world's oceans, and its biology is just as complex as that of warmer waters. This ocean is really several interconnected ecosystems rather than as a single large one.

Biological habitat

The extent of the Southern Ocean is not clearly defined because it takes in the southern parts of the Atlantic, Pacific, and Indian oceans. Depending on how its northern boundary is defined, the Southern Ocean accounts for 10 to 20 percent of the global ocean. It is

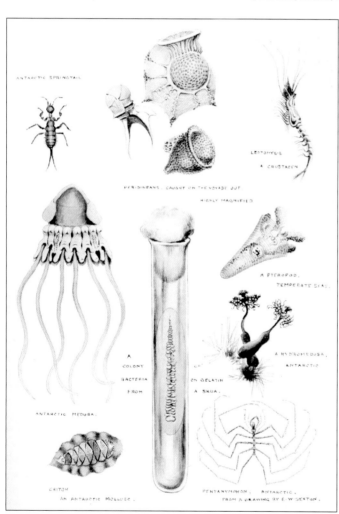

▷ ALBATROSSES IN MOTION
These huge birds travel vast distances, apparently effortlessly gliding on long, narrow wings using the updrafts of the "Roaring Forties," a belt of strong westerly winds that occurs between about 40° and 50° south. Albatrosses are generally large and live for many decades. With their relatives, the petrels, they form one of two main groups of birds in Antarctica; the other group is the penguins.

dominated by the huge eastward flow of the Antarctic Circumpolar Current—the earth's largest current, which provides the main connection between the Atlantic, Pacific, and Indian oceans. Near the Antarctic Continent, the coastal current flows in the opposite direction, toward the west. The shape of Antarctica, with its large indentations, the Ross and Weddell seas, the northward-jutting Antarctic Peninsula, and the varying contours of the sea floor, change the breadth and speed of the circumpolar currents and give rise to localized currents. Oceanic boundaries between these currents and Antarctica's coastal features determine the sea-ice conditions and the level of biological activity. As a consequence, marine life around Antarctica has a very patchy distribution.

◄ HUMPBACK DIVING
Humpbacks are readily identified by their low, broad blow, huge flippers, and also the way they arch their backs and show their flukes when diving. During summer they feed in Antarctic waters, and in early winter they migrate to their breeding grounds in coastal tropical waters. Populations were drastically depleted by whaling, and recovery is slow.

▼ SHARING THE SEAS
Both albatrosses and Dusky
dolphins have a circumpolar
distribution that extends north
of the Antarctic sea-ice zone.
Albatrosses feed mostly on
prey found at the surface
of the sea, whereas Dusky
dolphins eat fish and squid.

▼ SHARING THE SEAS
Both albatrosses and Dusky
dolphins have a circumpolar
distribution that extends north
of the Antarctic sea-ice zone.
Albatrosses feed mostly on
prey found at the surface
of the sea, whereas Dusky
dolphins eat fish and squid.

The Southern Ocean and its Antarctic Circumpolar
Current developed after the break up of the supercontinent Gondwana, some 40 million years ago. Fossil
evidence indicates that the Circumpolar Current was
established about 27 million years ago, and by roughly
22 million years ago the Polar Front (also known as
the Antarctic Convergence) had formed. Sinking
warmer northern water and colder southern water
meet in the circumpolar zone. It is not uncommon
for the surface water temperature to drop by about
2°C (35.5°F) across the Front.
The Polar Front is one of
the earth's major oceanic
boundaries and is a significant
biological barrier.

The Southern Ocean,
south of the Polar Front, can
be divided into three zones
that differ significantly in
their biology, because of their
environmental differences.
The Permanently Open

◄ SMALL BUT NUMEROUS
Minkes are the smallest and most
numerous of the baleen whales,
and their numbers seem to be
increasing. Unlike most baleen
whales, their blow is low; it has a
distinctly fishy smell.

▼ BIOLOGICAL ZONES
The Polar Front is a major oceanic
boundary in the Southern Ocean
between the warmer northern
waters and the cold Antarctic
waters. South of the Polar Front is
an ice-free zone surrounding the
Seasonal Ice Zone, which is ice-
covered for most of the
year. Close to the
coast there are
regions that
are perma-
nently ice-
covered.

Ocean Zone (POOZ) is nutrient-rich but has relatively
low levels of primary production. The main water flow is
the Antarctic Circumpolar Current. The phytoplankton
(single-celled plants) are generally tiny—nanoplankton,
2–20 micrometers in size. They are grazed by several
groups of animals, but generally not by Antarctic krill
(*Euphausia superba*).

The Seasonal Ice Zone (SIZ), south of the POOZ,
is ice-covered in winter but is essentially open water in
the summer months. It is the most productive zone of
the Southern Ocean. Blooms of phytoplankton are
common in the shallow, less saline water produced
by the southward-retreating ice edge in spring and
early summer. Large diatoms and blooms of *Phaeocystis*
algae are the dominant phytoplankton, on which various

Biological zones

0°

30°W   30°E

60°W   60°E

Haakon VII Sea

70°S

*Weddell
Sea*

80°S

*Ronne
Ice Shelf*   *Amery Ice Shelf*

*Bellingshausen
Sea*

90°W   90°S   90°E

*Davis
Sea*

*Amundsen
Sea*   *Shackleton
Ice Shelf*

*Ross
Ice Shelf*

120°W   *Ross Sea*   120°E

*Dumont
D'Urville Sea*

Polar Front

150°W   180°   150°E

☐ Permanently open ocean zone (POOZ)
☐ Seasonal ice zone (SIZ)
☐ Coastal and continental shelf zone (CCSZ)

groups of planktonic animals graze. Krill are abundant in this zone, especially in the more southern parts. Baleen whales, Crabeater seals, penguins, and other birds exploit the massive krill stocks.

The most southerly region, the Coastal and Continental Shelf Zone (CCSZ) is also known as the Permanent Pack Ice Zone—somewhat of a misnomer, as the region is not covered by pack ice at the time of its maximum retreat. The ice remaining is often fast ice. Phytoplankton blooms may be intense, but are generally short-lived. Antarctic krill are uncommon, being replaced by their smaller relative, *Euphausia crystallorophias*. In the absence of krill, birds and mammals are less abundant than in the SIZ.

The temperature at which seawater freezes depends principally on its salinity. Around Antarctica, the freezing point of seawater is close to –2°C (28.4°F), and it is rarely warmer than 4°C (39.2°F). As a result, Antarctic marine organisms experience constantly low temperatures.

For most of the year, the surface of the sea around Antarctica is frozen to a depth of up to several meters, but with an average thickness of less than a meter (3 ft). The freeze begins in March and reaches its maximum extent in September, when the sea ice covers almost 20 million square kilometers (8 million sq. miles)—which is nearly two and a half times the area of Australia. The sea ice is a major habitat for microscopic organisms, including the diatoms seen by the early explorers. They proliferate on its underside, as well as at the boundary between the sea ice itself and the snow that settles on its surface, which depresses this

▲ IN SEARCH OF FOOD
Adélie penguins breed on and around the Antarctic Continent. They will travel 200–300 kilometers (120–190 miles) away from their nesting sites in search of food—krill and other small crustaceans, and also fish. Their breeding success rate is poor during years when the sea ice is extensive.

Generally, there is more biological activity on the underside of sea ice where the snow has blown away and more light can penetrate. The age and texture of the sea ice also influence the species' composition and the abundance of these microbial communities.

As the sea freezes in autumn and winter, cold, highly saline water is excluded from the forming ice, which has a lower salinity than that of the seawater from which it is produced. The cold, saline water sinks to the sea floor and flows north into the northern hemisphere, as Antarctic Bottom Water (ABW). This is one of the main processes that create the vertical circulation and mixing of the global ocean.

During spring and summer, melting sea ice in the SIZ produces a layer of less saline surface water. Phytoplankton bloom in this nutrient-rich, stable, shallow layer, which receives 24 hours of sunshine and produces localized areas of high output, capable of sustaining the biological diversity and richness that so impressed the first explorers of this region.

There is serious concern that global warming may reduce the amount of sea ice around Antarctica, thus diminishing the production of ABW. Mathematical models indicate that a lessening in the vertical mixing of the oceans could lead to stagnation of Southern Ocean bottom water, depleting it of oxygen and therefore making it uninhabitable by most living things. Any decline in the amount of sea ice would also mean less habitat for the organisms that use it, as well as leading to a reduction of the relatively less saline water that promotes the ice-edge blooms.

▲ ALGAE UNDER SEA ICE
For long periods, the sea surface around Antarctica is frozen. In some areas, algae grow underneath the ice in profusion. These plants are an important winter food source for herbivores. When the summer melt occurs, the algae are released for deeper dwellers.

▲ PASTURES OF THE SEA
Phytoplankton are floating, microscopic, single-celled plants that constitute the base of oceanic food webs. This photomicrograph is at 500 times the magnification of living phytoplankton. It shows a diversity of shapes and sizes of phytoplankton, here mostly diatoms. Some diatoms are needle-shaped, others form long, ribbon-like chains.

boundary into the water. Cracks and other features in the ice are refuges for the small animals that graze the microorganisms. In addition, the sea ice is a platform on which some species of seals and penguins breed.

Despite its vast area and obvious biological activity, our understanding of sea-ice ecology is somewhat limited. The amount of light in the water column under the sea ice is effectively too low for photosynthesis by phytoplankton. However, there are environments favoring algal growth on, within, and on the underside of the sea ice. Some extremely high concentrations of algae can develop in these places. On sunny days, oxygen is produced by sea-ice algae faster than it can diffuse away, and is trapped as bubbles under the ice, surrounding the organisms that produced it. As oxygen at such concentrations is toxic, it is likely to inhibit photosynthesis. The algae and its associated microbial community of protozoa (single-celled animals) and bacteria on the underside of the sea ice is effectively the only food available to grazers while the sea is covered with ice, which can be for nine or ten months of the year. The under-ice algae on fast ice (that is, sea ice attached "fast" to the Antarctic continent) supports crustacean grazers—copepods, krill, and amphipods—as well as fish. The algae living on the underside of pack ice (which is sea ice found further offshore than fast ice that drifts with the currents) is grazed by krill. The distribution of sea-ice algae and their associated microorganisms is extremely patchy.

➤ DIATOM DISKS
Diatoms are the most abundant phyto-plankton in the sea around Antarctica. This scanning electron micrograph shows cells of the diatom *Porosira pseu-dodenticulata* (50 micrometers across). When these cells divide, they form short chains that are easily broken up. The glassy walls of silica accumulate as thick deposits on the bottom of the sea.

Protists

At the level of the single cell, the distinction between plant and animal becomes clouded. These days they are grouped together and called protists.

Phytoplankton are microscopic, floating, single-celled plants. The 200 or so species identified from Antarctic waters are startlingly diverse in size, shape, lifestyle, and food value to grazers. The smallest measure about 1 micrometer (1 thousandth of a millimeter), and they vary in shape from nearly spherical to hair-like, up to about 4 millimeters (about ⅛ in). For convenience, phytoplankton are divided into three size categories: picoplankton (0.2–2 micrometers), nanoplankton (2–20 micrometers), and microplankton (20–200 micro-meters). Nanoplanktonic and picoplanktonic organisms are generally the most abundant and diverse.

Cells of one group of phytoplankton, the diatoms, are encased in intricately sculptured, glassy walls; others are covered with finely patterned scales. Some can swim; others drift enclosed in mucus. Scientists are only just beginning to understand the factors that determine which species of phytoplankton grow in a particular region at a particular time of year.

Like all other plants, phytoplankton require light for the process of photosynthesis, in which they convert carbon dioxide and water into sugars and oxygen. Very little light is available in Antarctic waters during the Antarctic winter, when the sun is below, or low on, the horizon. At this time, the sea is also covered with ice and snow, further reducing the amount of light that penetrates it. In summer, when the sun shines for 24 hours and the sea ice is decaying, phytoplankton grows rapidly. Summertime blooms of microplankton are an important food source for krill.

The concentrations of the major nutrients for phyto-plankton growth—nitrogen and phosphorus, and in the case of diatoms, silica—are high in the Southern Ocean, yet the concentrations of phytoplankton are much lower than in most other parts of the world's oceans with similar nutrient levels. The Southern Ocean is one of the oceanic areas described as High Nutrient–Low Chlorophyll (HNLC). This so-called "Antarctic Paradox" can be explained by a combination of physical, chemical, and biological factors. The Southern Ocean is a very windy place, and wind mixes the surface of the ocean. Phytoplankton can be wind-mixed to depths greater than 100 meters (330 ft), where there is not enough light for photosynthesis. Although concentrations of major nutrients are high, the micronutrient iron (essential for phytoplankton growth) is in very low concentration over much of the Southern Ocean, and the lower the light, the more iron phytoplankton needs. Thus iron and light appear to act in concert to limit phytoplankton growth.

As well as these single-celled plants, there are single celled animals (protozoa) that feed on the phyto-plankton, bacteria, and detritus in the water. They, too, are highly diverse in shape, size, and lifestyle. Some filter their food from the water, others glide over surfaces grazing as they go, and others are voracious hunters. Some plants are able to hunt for

▲ ALGAL BLOOMS
Shield-like scales cover this single-celled alga (about 15 micrometers across). It occurs throughout the world's oceans where the water tem-perature is above 2°C (35.5°F). It can be so abundant that the sea appears milky white, or so extensive that blooms are visible from space.

Top

▲ CELL SCULPTURES
The lorica, or house, of this single-celled tintinnid is 40 micrometers long. It is secreted by the animal and decorated with the scales and wall fragments of its prey. Tintinnids attract prey items by beating minute, hair-like structures, called cilia, to create a water current.

◄ THE EDGE OF THE SEA
Sunlight reflects on broken sea ice, just off Cape Royds, deep in the Ross Sea. Tidal movements break up the ice where it meets the land, providing breathing holes for seals and access into and out of the water for penguins.

Influencing the climate

As well as being the basis of the food web on which essentially all the other marine organisms depend, marine microbial components play a role in influencing climate. The Southern Ocean is a major oceanic site for the absorption of atmospheric carbon dioxide. And it is principally the activities of microorganisms which determine whether carbon dioxide is rapidly re-released to the atmosphere, or retained in the ocean for hundreds or even thousands of years. Additionally, some marine microorganisms produce sulfur-containing compounds, which, when they are released into the atmosphere, form aerosol particles that promote the formation of clouds. Clouds play a significant role in climate by reflecting the sun's energy back into space. The Southern Ocean is the source of some 17 percent of the total global production of these sulfur compounds.

▲ BACTERIA AND VIRUSES
The smallest, but most abundant, biological components of the marine environment are the bacteria and viruses. A fluorescent stain makes them visible under the microscope—the bacteria are the larger bright dots. There are ten times more viruses than bacteria in the sea.

food—an ideal adaptation in Antarctica, where light levels are too low for photosynthesis for much of the year. At the other extreme are protozoa, which retain the photosynthetic systems from the phytoplankton they consume, and use their photosynthetic products as their food. Grazing by plants and photosynthesis by animals is called mixotrophy. Protist grazing removes a substantial amount of the bacteria and smaller phytoplankton. Bacteria feed on the dissolved organic material that is produced by all organisms, particularly the phytoplankton. As protozoa are eaten by krill and other grazers, they and the bacteria represent another pathway in the food web between phytoplankton and the grazers. This pathway is called the microbial loop.

Bacteria and viruses

As elsewhere in the world's oceans, bacteria are a major component of the Southern Ocean food web. Their concentration is about a billion per liter (2 pints), and they break down detritus, remobilizing the nutrients. In addition, they play several other roles that can have ecosystem-wide ramifications. Some bacteria live alongside living phytoplankton, where they can influence the nutrient concentrations. By dissolving mucus on diatom surfaces, bacteria can change the stickiness of the phytoplankton, thereby altering their ability to form clumps. As a consequence, bacterial activity can influence the dynamics and fate of algal blooms. Recently, bacteria have been implicated in dissolving silica, the main constituent of the cell walls of diatoms. Bacterial recycling of this nutrient promotes the growth of diatoms, rather than the silica being removed from surface waters when dead diatoms sink to deep water. Bacteria secrete digestive

enzymes to dissolve organic material, which they may then absorb as food. However, these enzymes also break up material, which would have otherwise sunk to the deep ocean. Thus bacteria are crucial to marine biological processes, playing a pivotal role in the grazing food chain, the microbial loop, particle sinking, and carbon fixation and storage.

The smallest biological entities in the sea are viruses. Viruses consist of genetic material (DNA or RNA) enveloped in a protein coat. They are not living, in the sense that they cannot metabolize, or reproduce by themselves. They infect a host cell, hijack its reproductive system, and produce masses of new viruses which, when the host cell bursts, can infect yet more host cells. Their abundance in the Southern Ocean is similar to that elsewhere in the sea, about 10 billion per liter (2 pints). The roles these viruses play are only now being explored. They probably infect all types of marine organisms, and influence the species composition of microbial communities and nutrient cycling, and are involved in the transfer of genetic material between host organisms. Virus infection accounts for about 50 percent of the mortality of marine bacteria, and may inhibit phytoplankton production by up to 80 percent. Viral destruction of bacteria may be one of the major sources of dissolved organic matter in the sea, which is, in turn, consumed by bacteria. Only about a decade or so ago, viruses in the sea were considered unimportant; now the question is—do viruses control life in the oceans?

The importance of microscopic organisms in the sea has become increasingly clear over the past two decades. It is now known that much, often more than half, of the total flux of matter and energy passes through the microbial loop. The contemporary view is of an organic matter continuum from small simple molecules of amino acids (the building blocks of proteins) and sugars through colloids to transparent polymer particles. This so-called oceanic "dark matter" spans a size-range from a few tens of nanometers (1 nanometer is 1 millionth of a millimeter) to about a millimeter ($\frac{1}{32}$ in). Recently, it has been proposed that these organic constituents of seawater interact to form a very dilute gel with a structure that has microscale hotspots, which are sites of heightened microbial activity. Microorganisms, like larger plants and animals, do not live by themselves but in communities with other organisms, each having its own role.

Krill and other grazers

Krill—food for fish, birds, and mammals, and the target species of a commercial fishery—are one of the central elements in the Southern Ocean ecosystem. There is probably a greater mass of Antarctic krill (*Euphausia superba*) than any other single species on earth. Krill has a patchy, offshore, circumpolar distribution. Abundance and distribution vary from year to year, in some cases with profound consequence for its predators.

Krill are a group of about 80 different species of crustaceans, the euphausiids, which resemble shrimps, and live in the open ocean. In the Antarctic context, the term krill refers to Antarctic krill, the largest and the most abundant of the five species of euphausiids living in the Southern Ocean. The dominant euphausiid in Antarctic coastal waters is *Euphausia crystallorophias*, so-called Crystal krill, a much smaller species than *E. superba*.

Considering their abundance and importance, remarkably little is known about krill. Estimating their abundance is fraught with difficulties because of their very patchy distribution and their ability to avoid capture in research-size nets. The latest estimate, based on surveys using sophisticated echo-sounders and a re-examination of historical information on their range, is between 60 million and 155 million tonnes. Some studies indicate that there may have been declines in some areas over the past 20 years or so, but whether these declines are part of a cycle is not known.

In order for a sustainable krill fishery to be properly managed it is essential to know how many krill there are, how old they grow, and the extent to which they are food for other animals. Krill have proved very difficult to keep in captivity, but researchers are trying to work out how long they live, their reproductive strategies, winter survival, and behavioral aspects of their lifestyle. Only in the past decade has it been established that they live for five to ten years and are sexually mature at three years old. The conditions they need for breeding in captivity have only been learned in the last year.

Krill have a complex life cycle, which is not yet completely understood. Spawning is believed to take place near the surface, from late December to March. Females can produce several broods of up to 10,000

▲ **AS COMMON AS KRILL**
The Antarctic krill (*Euphausia superba*) could be the most abundant single species on earth. Krill graze on microorganisms and are themselves food for whales, seals, birds, and fish. They gather food with their front legs, which form a feeding basket, and swim with their rear legs. This species is 50 millimeters (2 in) long.

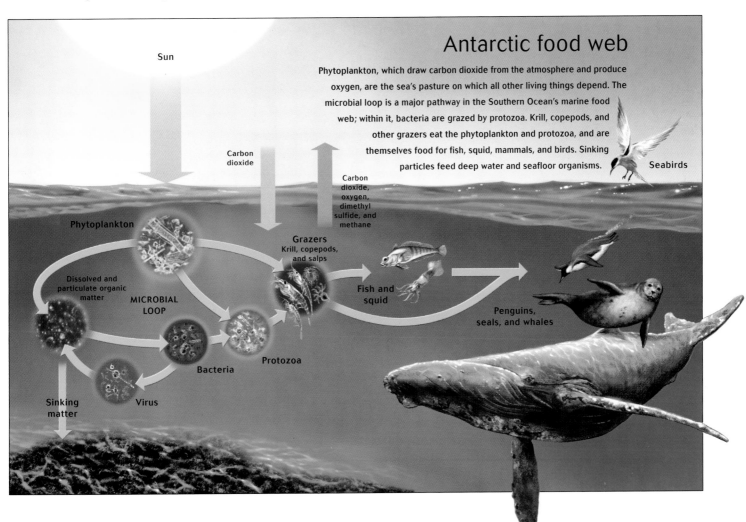

Antarctic food web

Phytoplankton, which draw carbon dioxide from the atmosphere and produce oxygen, are the sea's pasture on which all other living things depend. The microbial loop is a major pathway in the Southern Ocean's marine food web; within it, bacteria are grazed by protozoa. Krill, copepods, and other grazers eat the phytoplankton and protozoa, and are themselves food for fish, squid, mammals, and birds. Sinking particles feed deep water and seafloor organisms.

Sun

Carbon dioxide

Carbon dioxide, oxygen, dimethyl sulfide, and methane

Seabirds

Phytoplankton

Dissolved and particulate organic matter

MICROBIAL LOOP

Grazers
Krill, copepods, and salps

Fish and squid

Protozoa

Bacteria

Virus

Sinking matter

Penguins, seals, and whales

> ► KRILL SWARM
Antarctic krill characteristically form dense swarms with tens of thousands of animals per cubic meter. When these swarms are close to the sea surface, the water appears a brick-red color. Such swarms are highly attractive to predators. Swarms vary in size, and superswarms can cover many hundreds of square kilometers.

eggs at a time, depending on food availability. The eggs, about 0.5 millimeters ($\frac{1}{64}$ in) in diameter, sink to depths of about 1,000 meters (3,300 ft), where they hatch; the embryos go through a series of non-feeding larval stages before rising again to arrive near the surface as larvae needing to feed. This descent and ascent is thought to take 3 to 5 months. In their second summer, they develop into juveniles and can form swarms. At the start of their second winter, juvenile krill are about 25 millimeters (1 in) long, and look like miniature adults.

Krill characteristically form swarms in which the concentrations of animals can be about 30,000 individuals per cubic meter ($1\frac{1}{3}$ cu. yards); an easy target for whales, seals, and penguins—and fishing trawlers. Often swarms consist of all males, all females, or all juveniles. A single swarm of krill encountered near Prydz Bay in the southern Indian Ocean sector of the Southern Ocean was calculated to weigh about 57,000 tonnes and was composed entirely of large, sexually immature males. Such swarms color the sea reddish brown. Occasionally "superswarms" form; one investigated in 1981, off Elephant Island, near the tip of the Antarctic Peninsula, covered 450 square kilometers (180 sq. miles) and was estimated to contain more than 2 million tonnes of krill. This superswarm contained both sexes; however, there was some segregation by sex and age.

Adult Antarctic krill are about 60 millimeters ($2\frac{1}{2}$ in) long. All crustaceans, including krill, have a shell (exoskeleton). They grow by molting the old shell, expanding the new one as it is forming, and then growing into it. This is done repeatedly during the life of the animal until it reaches maturity—so, the larger the crustacean, the older it is. Antarctic krill, however, are different; it is not possible to tell their age from their size. For much of the year their principal food—phytoplankton—is scarce because of the wintertime absence of sunlight for photosynthesis. In laboratory studies, krill have coped with more than 200 days of starvation. Rather than rely on accumulated fat reserves, they use up muscle protein.

Fishing for krill

About 38 percent of the dry weight of krill is protein. In the 1960s, a krill fishery was proposed as a readily available source of protein for human consumption. Commercial harvesting in the mid-twentieth century had hugely reduced the population of baleen whales, and it was considered that there was a krill "surplus"—that is, the krill that the whales would have consumed. For most of the 1980s, the annual krill catch was more than 400,000 tonnes, peaking at more than 500,000 tonnes but falling rapidly to less than 100,000 tonnes after the break up of the Soviet Union, when former Soviet fisheries began to operate on a commercial basis. Catches of low-value products in distant waters declined. Krill are now harvested for human food, aquaculture feed, and sport-fishing bait. Once caught, they must be processed immediately to avoid decomposition and spoilage. The exoskeleton has a high concentration of fluoride, which must be removed to prevent it contaminating the flesh. The future for krill fishing is uncertain. The present costs of harvesting and, particularly, processing are high for a low-value product. Advances in processing technology and marketing of a greater range of high value-added products, such as pharmaceuticals, would be needed for the industry to expand substantially.

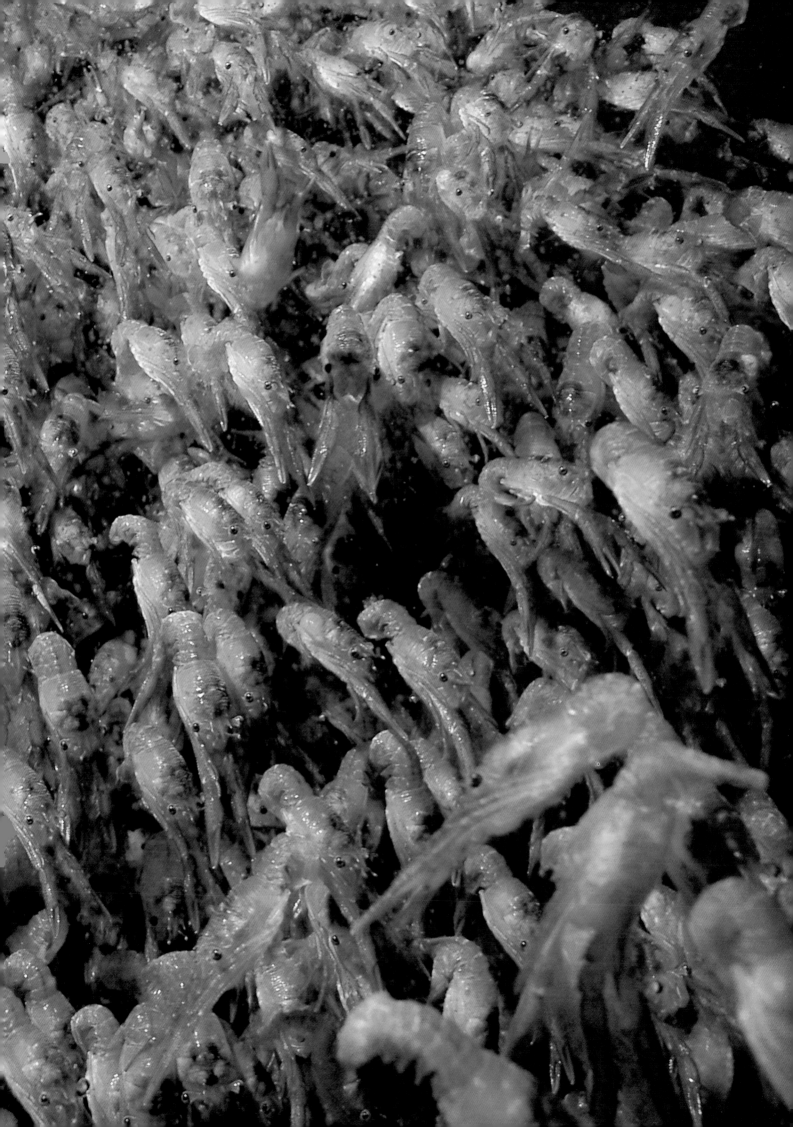

They continue to molt about once a month, but instead of growing, as they do when food is abundant, they shrink. Also, at the end of summer, mature krill lose their sexual characteristics, which they regain in spring for spawning in summer. So a five-year-old-mature krill, when it has lost its sexual characteristics and has shrunk, cannot be distinguished from a sexually immature two-year-old.

Because of the ecological importance, resource potential, and biomass of krill in the Southern Ocean, other zooplankton tend receive much less attention. These include other euphausiids, copepods, salps, amphipods, chaetognaths (arrow worms), and larval fish. Of these, copepods and salps can reach high abundances and play key roles. Non-krill zooplankton have been reported to account for 75 percent of the total zooplankton biomass in the Scotia Sea, north of the Antarctic Peninsula, and copepods have been found to constitute 60–87 percent of the total summer zooplankton biomass in Admiralty Bay on King George Island. Recent studies show that the contribution of krill to the zooplankton biomass in the Southern Ocean has been overestimated. It has become apparent that there is dramatic variability in the distribution and abundance of zooplankton geographically, as well as seasonally and interannually.

Copepods, like krill, are also crustaceans. More than 70 species are recognized in Antarctic waters. Generally microscopic, only a few species are larger than 1 millimeter ($\frac{1}{32}$ in). Some species graze on protists, others are carnivores, yet others are omnivorous. We have a poor understanding of the proportion of the protistan biomass that is grazed by krill and how much by copepods. Estimates based on the abundance of copepods and their energy requirements indicate that they may consume more than eight times as many protists as krill. Unlike krill, with only few exceptions, copepods are not the principal food of the top predators. Instead, they are grazed by fish, including the so-called Antarctic silverfish or herring (*Pleurogramma* species), which, in turn, are eaten by seals, penguins, and other seabirds.

Salps are planktonic tunicates (sea squirts). They feed by pumping water through a fine mucus net filter, and are able take in a wide size range of particles. They are delicate, gelatinous organisms with a life history alternating between solitary organisms and colonies that reproduce asexually. The production of colonies coincides with the springtime phytoplankton bloom. As a solitary salp can produce hundreds of colonies, high concentrations of these organisms can develop rapidly in spring and summer in response to food availability. Their high, localized abundance and high filtration rates enable them to consume 10–100 percent of the protists in these

▲ LIFE'S CYCLE
An Adélie penguin feeds its chick. Foraging by penguins and other seabirds can have a major impact on prey species near their breeding colonies. They also have a substantial influence on the terrestrial environment of their breeding areas. Nest-building activities modify the landscape, and the excrement, egg shells, feathers, and carcasses of dead birds are major sources of nutrients for plant growth.

areas. Thus salps are able to reduce concentrations of food for other planktonic grazers. These observations, plus finding that areas dominated by krill are usually essentially devoid of salps, and that areas in which salps are abundant are low in krill, have led to a reappraisal of the ecological role of salps in the Southern Ocean.

Krill reproduction and the survival of larvae are apparently also affected by blooms of salps. Investigations conducted in the south of the Atlantic Ocean, as well as in the Indian Ocean sector of the Southern Ocean, point to an interaction between sea-ice extent and the dominance of either krill or salps. Patterns of oceanic circulation and sea-ice conditions have a profound influence on the biology. It appears that in years when the southern boundary of the Antarctic Circumpolar Current tends to be offshore there is a greater than average development of sea ice. In these conditions, krill populations are dominant and salps are further offshore. However, when the southern boundary of the Antarctic Circumpolar Current is close to the Antarctic coast and sea ice is less extensive, salps, rather than krill, are more abundant. At least in the region of the Antarctic Peninsula, there is evidence, spanning the past 50 years, of fewer and fewer winters having extensive sea ice. This decrease in sea ice correlates with the increase in air temperatures over this time. As decreased krill abundance is related to a decrease in sea-ice extent, there is concern about the food availability for krill predators.

Although salps, unlike krill, are not a major constituent in the diet of vertebrate predators, they play another important role in the Southern Ocean. As their feces sink quickly, they contribute to the transport of carbon from surface waters to the deep ocean. This vertical flux of carbon in the world's oceans is one of the major pathways in the global carbon cycle—atmospheric carbon dioxide, taken up by phytoplankton, is transported to the deep ocean, where it is only slowly recycled back to carbon dioxide by the activity of deep-sea organisms.

Antarctic fish

More than 20,000 species of fish occur worldwide, but only about 120 of them are found south of the Polar Front. This Front represents a major oceanic barrier to the movement of fish species, especially those that live in coastal and shallow habitats. Also, the Antarctic continental shelf is deeper than the shelf around other continents, and has no shallow connection to these continental shelves. This isolation of Antarctic fish has played a major role in their evolution and in the composition of Antarctic fish communities. For example, many of the fish that live on the Antarctic continental shelf have adaptations found in deepwater fish from other parts of the world. Unlike other parts of the world's oceans, Antarctic fish are uncommon in surface waters.

Fish that live in Antarctic coastal bottom waters are by far the most diverse and abundant, accounting for more than 60 percent of species and 90 percent of the abundance of the Antarctic fish fauna. In marked contrast to other parts of the global ocean, the Southern Ocean contains extremely few true schooling pelagic (open-sea) fish—the few pelagic fish that do occur have

▲ BOTTOM-DWELLING ANTARCTIC COD
Most Antarctic fish live close to the ocean bottom. They are slow-moving (unless threatened), grow slowly, have long lives, become sexually mature late in life, and have low reproductive rates.

▲ DINNERTIME FOR SEALS
A Weddell seal devours its prey, an Antarctic cod. These seals live further south than other seals. They tend to stay close to their breeding sites, feeding on a wide range of prey items that vary from large to small fish, squid, and tiny crustaceans, such as amphipods and euphausiids.

evolved from bottom-dwelling groups to a partial or full-time life away from the bottom. The sea-ice zone is the habitat of adults of many of these, but they spend their first years of life in deep water. The Antarctic herring or silverfish (*Pleurogramma* species) is pelagic and often, especially as juveniles, associated with sea ice. It feeds on copepods, euphausiids, and juvenile fish, and is, in turn, eaten by marine mammals and birds. Lacking a swim bladder, it achieves neutral buoyancy with fat deposits. *Pleurogramma* reaches a length of approximately 25 centimeters (10 in) and is an abundant and important organism in the Antarctic marine food web. Myctophids (lanternfish), 2–30 centimeters ($\frac{3}{4}$–12 in) long, are the other important pelagic group of fish in Antarctic waters. They are called lanternfish because of the groups of luminous organs (photophores) on their head and body. They are opportunistic feeders, consuming copepods, euphausiids and other grazing species, as well as fish eggs and juvenile fish. During the day, lanternfish

can be found at depths of about 200 meters (650 ft) but migrate to be near the surface at night. They are unusual among Antarctic fish in that they have a swim bladder.

Unlike most Antarctic invertebrates, body fluids of Antarctic fish are less saline than seawater—their freezing point depression is about –1°C (30.2°F). They have all evolved antifreeze compounds called glycoproteins, which depress the freezing point of the water in their body fluids. These antifreezes work by binding to the surface of forming ice crystals, thus preventing their growth.

Most Antarctic fish are less than 30 centimeters (12 in) long, but a few species grow to more than 1.5 meters (5 ft) long and weigh more than 50 kilograms (110 lb). Antarctic fish generally grow slowly, take three to eight years to become sexually mature, have long life spans, and low metabolic rates. They produce only a few large, yolky eggs, so their reproduction rate is generally low. These characteristics put Antarctic fish at serious risk of overfishing. Marbled rock cod (*Notothenia*

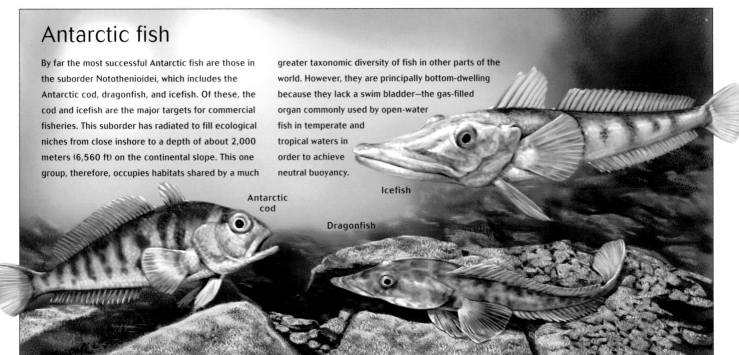

Antarctic fish

By far the most successful Antarctic fish are those in the suborder Notothenioidei, which includes the Antarctic cod, dragonfish, and icefish. Of these, the cod and icefish are the major targets for commercial fisheries. This suborder has radiated to fill ecological niches from close inshore to a depth of about 2,000 meters (6,560 ft) on the continental slope. This one group, therefore, occupies habitats shared by a much greater taxonomic diversity of fish in other parts of the world. However, they are principally bottom-dwelling because they lack a swim bladder—the gas-filled organ commonly used by open-water fish in temperate and tropical waters in order to achieve neutral buoyancy.

Antarctic cod

Icefish

Dragonfish

rossii), Mackerel icefish (*Champsocephalus gunnari*), and Patagonian toothfish (*Dissostichus eleginoides*) have all suffered the predictable pattern of exploitation—initially high catches that rapidly decline to an uneconomic point, whereupon fishing effort is transferred to a new target species, and the pattern is repeated.

Antarctic squid

Squid are almost certainly one of the most important components of the Southern Ocean ecosystems, yet they remain one of the big unknowns. Because of their ability to avoid capture in nets, very little is known of their abundance, longevity, and lifestyles. Most species that are netted are not the squid taken by toothed whales. Also squid caught in nets are much smaller than those found in dietary studies of whales. There is a strong relationship between a squid's body and its beak size; some of the best information on squid species composition and size comes from beaks recovered from the stomachs of seals and toothed whales. Up to 18,000 squid beaks have been recovered from inside a single Sperm whale.

Some 70 species of squid occur south of the Polar Front; about half of these are endemic to the area. Apparently, many species produce egg masses in water deeper than 1,000 meters (3,300 ft). The young squid grow rapidly, reach sexual maturity in a year, spawn, and die. Some species, however, may live for several years. Squid form a major part of the diet of birds, particularly albatrosses, seals, and toothed whales (except Orcas). It is very difficult to tell how much squid is consumed by predators, but crude estimates put it at more than 40 million tonnes per year, and suggest that the total squid biomass of the Southern Ocean is about 100 million tonnes. Squid eat voraciously, some species consuming up to 30 percent of their body weight a day; they were once thought to be major consumers of krill, with consumption estimates ranging as high as 100 million tonnes per year. However, recent investigations indicate that myctophids and deepwater fish are their major prey. As these fish feed mainly on copepods, it is emerging that squid may represent an important food web link between copepods and toothed whales, seals, and birds.

Although squid are fished around New Zealand and the Falkland Islands, there is no commercial squid fishery at present in Antarctic waters, despite years of speculation about its commercial potential. Squid populations can fluctuate dramatically because of their short life-cycle, and the fact that they spawn only once and die soon after. They are especially susceptible to over-fishing. Any commercial fishery would have to be particularly carefully managed to ensure it was sustainable and to prevent any adverse impact on predators of squid.

▲ SQUID FISHING
Around the Falkland Islands (pictured), and south of New Zealand, there are squid fisheries. Squid are usually taken at night by jigging—a fishing method whereby the squid are attracted to the boat by lights and snagged by barbs set in the water. The automatic jigging machines and lights are set along the side of the boat.

Benthic communities

Just as the abundant wildlife impressed the explorers of Antarctica, so did the organisms that live on the sea floor, brought up in grabs, dredges, and nets. The coastal and continental shelf areas of Antarctica have rich benthic (bottom-dwelling) communities, generally dominated by animals, which feed on particle of matter raining down from the surface waters.

The Antarctic continental shelf is generally both deeper and narrower than the continental shelves of other continents. The huge amount of debris deposited on the seabed around Antarctica is only inorganic (rock) material transported by glacial activity, rather than the inorganic and biologically-derived organic deposits surrounding the other continents, which are carried there by rivers and wind action. Ice exerts a major influence on Antarctic benthic communities in several ways. These include shading the light available for benthic algae, scouring by sea ice and icebergs, and the formation of ice on the shallow seabed. The water temperature on the continental shelf is generally low and stable. Light for photosynthesis and the fallout of material from surface waters is a sharp seasonal pulse. Unlike coastal waters in other parts of the world, scouring by sea ice ensures that the intertidal community around most of Antarctica is sparse.

One of the best-studied benthic environments at high latitude in Antarctica is McMurdo Sound. Generally the area, regularly scoured by ice down to a depth of about 15 meters (50 ft), is almost devoid of algae, except for those capable of fast growth in the short summer. Any animals there are mobile foragers, such as sea urchins, starfish, various worms, crustaceans, and fish. However, below this zone is a rich and diverse flora and fauna. In shallow coastal environments, benthic micro-algae make a substantial seasonal contribution to primary production. Some 700 species of seaweeds have been reported from Antarctic waters, of which about 35 percent are endemic. Seaweed and bottom-dwelling animal community composition varies with latitude and water depth. Seaweeds are sparse below about 40 meters (130 ft) and virtually absent below 100 meters (330 ft), where there is scarcely enough light for them to photosynthesize. Many of the seaweeds found south of the 0°C (32°F) surface isotherm do not occur north of this boundary, and vice versa. The zone below that scoured by ice is colonized by a diverse array of filter feeders, including anemones, soft corals, molluscs (shellfish), ascidians (sea squirts), and tube worms. Here too are mobile scavengers, including starfish, sea urchins, pycnogonids (sea spiders), and fish. Below about 30 meters (100 ft) depth, the dominant animals are sponges covering up to 55 percent of the bottom in McMurdo Sound. These sponges vary widely in size and shape and grow only very slowly. The largest are shaped like a volcano, and are up to 2 meters (6½ ft) tall and 1.5 meters (5 ft) across. This sponge community provides homes for many mobile species, and anemones, tube worms, bryozoans, molluscs, and ascidians also occupy this zone.

Some sponges' skeletons are composed of silica fibers called spicules that provide support for the sponge and

> **LONG AND HUNGRY**
The large nemertean or proboscis worm (*Parborlasia corrugatus*), pictured on the sea bed below the diver, can grow to several meters in length. It is a carnivore that feeds by enveloping its prey in its proboscis. Although it has a soft body and would appear to be an easy target for other predators, it protects itself by exuding acidic mucus.

deter predators. Some Antarctic sponges contain algae that provide them with nutrients. More recently, it has been discovered that the glassy spicules on the sponges also function as optical fibres, transmitting light to the algae that adhere to the spicules deep within the sponge.

Areas around Antarctica where the sea floor is sandy or muddy contain burrowing animals. The densities of these animals are among the highest found anywhere on earth.

As is the case in temperate and tropical waters, predator–prey interactions are thought principally to determine the composition of Antarctic benthic communities. However, in Antarctic waters, icebergs scour the continental shelf, which has a depth of about 500 meters (1,640 ft). Statistically, this scouring would be expected to occur at least once every 230 years, though depth and proximity to active iceberg calving areas would mean much more frequent scouring of some parts of the continental shelf. Areas scoured by icebergs look like ploughed fields with parallel grooves, some bordered by raised embankments. The passage of the iceberg strips all organisms from the bottom. There is a sequence of recolonization of iceberg scours with mobile animals, including fish, sea urchins, and mollusks. They are followed by animals attached to the bottom, including tube worms and sea squirts. Habitat destruction by iceberg scouring and subsequent recolonization may be a principal cause of the great differences in community composition between habitats on the Antarctic continental shelf.

In addition to the seaweeds and large bottom-dwelling animals, the Antarctic sea floor is covered with microscopic algae, protozoa, and other animals and bacteria. The protozoa and bacteria also live in benthic sediments. The benthic microalgae are generally subdivided into two groups: those that live in the top few layers of sediment, and those called ephiphytes that live attached to other living things, such as seaweeds, or on the surface of rocky seafloors. These algae can be extremely abundant, especially in some habitats, such as beds of sponge spicules, where the surface area of the sediment is increased dramatically by the long, needle-like spicules to which the algae are attached. In addition, the spicules

discourage grazing animals. The benthic microalgae are thought to make a significant contribution to primary production in shallow coastal areas and play an important role in providing nutrients to the benthic animals, as well as contributing to the organisms living in the overlying water column.

Benthic bacterial concentrations are usually many times higher than the concentrations in the overlying water column, but similar to those found in sediments in other parts of the world. Bacterial activity in sediments is usually confined to the top 5 centimeters (2 in). The rates of bacterial processes in Antarctic marine sediments in breaking down organic material and recycling it back to carbon dioxide are thought to be similar to that in the deep oceans of the world. To better understand the role of the Southern Ocean in the global carbon cycle, much more research is required to measure the rates of benthic microbial activities. HM

▲ **PRICKLY RELATIVES**
The common Antarctic sea-urchin (left) eats algae that it scrapes off rocks as it crawls. The large, many-armed star-fish is a carnivorous predator; the bulge in its center suggests it is has just eaten.

Above left

▲ **A SPIKY FIND**
D.G. Lillie did biological work on Scott's 1910–12 Antarctic expedition. He noted: "almost every trawl brought up quantities of siliceous sponges covered with glassy spicules."

MANY LEGS

Echinoderms are unusual in the animal kingdom for being radially symmetrical—that is, they are symmetrical around their center, in much the same way as a flowerhead. The primitive starfish have five arms; however, they can have up to 40, like this species, which is found along the Antarctic coast.

SWAYING ANEMONE

A *Phyctenactis* anemone, surrounded by kelp, sways in the currents off the Antarctic Peninsula. Although anemones look like plants—in the way they cling to the substrate— they are actually animals. They are filter-feeders, which trap detritus from the water as the current passes over them.

SPIDER OF THE SEA

Sea spiders from Antarctica are not spiders at all but crustaceans, which span as much as 20 centimeters (8 inches) across. Like other Antarctic bottom-dwelling invertebrates, they grow slowly and become larger than their warm-water relatives. Their growth rate is thought to be controlled more by food availability than the low sea temperature.

FIERCE PREDATORS

The seabed communities of Antarctica are surprisingly diverse and colorful. Many species of starfish live in coastal areas, and some, such as this *Odontaster validus*, are found in large feeding aggregations. Although they do not move quickly, they are voracious predators and will prey on other starfish much larger than themselves.

Seals and
Sea Lions

Seals

ORDER: Pinnipedia

Previous pages

◀ **SENSITIVE SEAL**

The long and sensitive whiskers of this bull New Zealand fur seal help it detect movements made by prey. Whiskers also seem to assist seals to find their way in dark or murky waters.

Seals are a significant and highly visible part of the wildlife of Antarctica. They are classified into three families: Phocidae, the true seals, or hair seals; Otariidae, the eared seals; and Odobenidae, the walrus. Of the 19 true seal species, five are found in Antarctica. The Crabeater, Weddell, Ross, and Leopard seals each have their own genus, while the fifth, the Southern elephant seal, shares its genus with the Northern elephant seal (*Mirounga angustirostris*). The eared seals include sea lions and fur seals. There are five sea lion genera, and each has a single species. Named for the long mane of fur covering the beefy necks of the males, sea lions have a broad distribution, generally preferring temperate coasts but occasionally occupying equatorial niches or sub-polar zones. The nine fur seal species, distinguishable from sea lions by a thick waterproof pelt under long guard hairs, form two groups: the Northern fur seal has its own genus, while the eight southern fur seals form a second genus, *Arctocephalus*. Eared seals are less polar in their habits than many true seals, but most southern fur seals live in, or occasionally reach, sub-Antarctic or Antarctic regions. The Walrus, with its great tusks, is unique among seals. An exclusively Arctic animal, it is most closely related to the eared seals.

Seal physiognomy

Seals are well adapted for ocean life. The true seals have a thick layer of blubber for insulation in water, whereas the fur seals rely on their dense fur for insulation. The hind limbs of all seals are long, thin, and flattened, rather like paddles. Their faces are short, with reduced (in fur seals) or non-existent (in true seals) external ears. Their bodies are smooth and sinuous, and they swim by using their flippers and undulating their bodies. Most have a thick, heavy neck, but the vertebrae are only loosely interlocked,

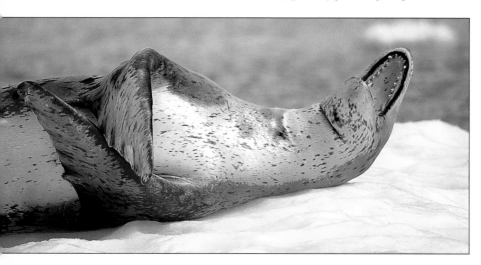

◀ **LARGE MOUTH, LARGE PREY**

All seals are carnivores, and the size of the mouth of each species is related to the size of its preferred prey. Adult Leopard seals usually feed on a range of prey that includes larger fish, penguins, and seals.

Seal family tree

Little is certain about the origins of seals. Many scientists place them in their own taxonomic order, Pinnipedia, while other authorities consider them a suborder of Carnivora. Some scientists believe that the closest living relatives to all seals are the weasels, badgers, skunks, and otters, while both modern and fossil cranial material suggests that all seals are descended from a bear-like ancestor.

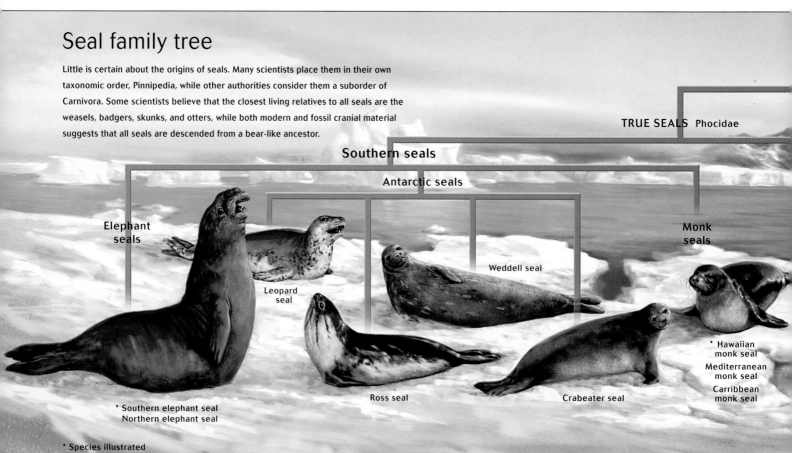

Southern seals

Antarctic seals

TRUE SEALS Phocidae

Elephant seals

Monk seals

Leopard seal

Weddell seal

Ross seal

Crabeater seal

* Hawaiian monk seal
Mediterranean monk seal
Carribbean monk seal

* Southern elephant seal
Northern elephant seal

* Species illustrated

making them extremely flexible for a mammal. Agile and very strong, seals can withstand the impact of waves and currents and maneuver around ice and rocky shores.

Seals have slitted nostrils that close under water—another adaptation to marine life. When the nostrils are in the relaxed position they are closed, and water pressure increases the force keeping them closed. This means that when resting in water a seal does not need to concentrate on keeping its nostrils closed. In order to breathe, the seal must actively contract a pair of muscles. While at sea, seals can hold their breath for extended periods—elephant seals, for example, have been recorded holding their breath for two hours. When a seal does open its mouth under water, the tongue and soft palate close off the back of the throat, which means that the seal can take a mouthful of food without swallowing seawater. In addition, the larynx presses against the epiglottis to stop food and water from entering the lungs.

Eared seals have powerful muscles in their shoulders to support their strong front flippers, which they use as paddles in water and as feet on land. They can also bring their hind flippers forward and use them like legs to run across land. True seals, however, have lost this facility and when swimming, they use their hind limbs to undulate

▲ STILL TIED TO THE LAND

As is true of all true and eared seals, this fur seal has a flexible, streamlined body with reduced peripheral appendages, which makes it well adapted to life in the water. But seals, unlike whales, are still connected to land. While a few seal species appear to mate in the water, most mate on land, and all come ashore (or onto ice) to give birth.

PINNIPEDS Pinnipedia

WALRUS Odobenidae

EARED SEALS Otariidae

Northern seals

Sea lions

Northern sea lion   Australian sea lion   * Hooker's sea lion   South American sea lion   California sea lion

Hooded seal   Gray seal   **White-coated seals**

Northern fur seal   Southern fur seals

* Common seal
Caspian seal
Ribbon seal
Harp seal
Ringed seal
Spotted seal
Baikal seal

Bearded seal

Walrus

* Antarctic fur seal
South American fur seal
New Zealand fur seal
Galapagos fur seal
Juan Fernandez fur seal
Australian fur seal
Guadalupe fur seal
Subantarctic fur seal

◄ **FAT BUT WARM**
Elephant seals use blubber as
their primary insulation. It may
look awkward on land, but it
does not interfere with swimming.
These seals often cover 70–80
kilometers (44–50 miles) per day
when traveling to feeding locations.

their hindquarters, where their main muscles are. This gives them a highly efficient swimming motion, but their front flippers do not have the strength or reach to be effective on land, and their rear flippers cannot bend forward in the way that the flippers of eared seals can. True seals do not run, but undulate across land, humping their body up with their flippers and throwing it forward with surprising speed.

The eyes have it

Seals' eyes have several adaptations for seeing under water, including a flattened cornea and pupils that can open very wide to admit as much light as possible. In land-dwelling animals the curved surface of the eye refracts light to project an image on to the retina in the back of the eyeball and the lens is used only for focusing, but in water the lens directs the image to the back of the eyeball. All seals have large, well developed eyes, which gather light more effectively than small ones. Elephant seals and Ross seals, both deep divers, have very large eyes with large lenses and can see well in deep water. Seals have retinas like those of terrestrial carnivores but lack cones, the sensory cells necessary to detect color. Their eyes consist entirely of rods—sensory cells that function well in low light. Elephant seal eyes have a modified photopigment like that of deep sea fish, perhaps to discern the bioluminescence of deep sea squid, their prey.

Out for the count

Antarctica poses unique problems in estimating population sizes. If you count those on land, what about seals in the water? How do you count seals that are hauled out on sea ice far away? The solution is to survey long narrow strips of sea from a ship or an airplane and count the seals hauled out on ice. Seals prefer to haul out at certain times of day. The time is noted and the proportion of seals presumed to be on the ice is factored into the estimate. The ice type and cover are also taken into account, as seal densities can be related to the area of pack ice available for them to haul out onto. Final estimates are likely to be low, as the proportion of seals actually on the ice will never be 100 percent; the best available estimates place the Crabeater seal population at about 14 million, but change the assumptions and the same counts suggest that the number may easily be as high as 40 million.

Eating for energy

Seals are active hunters, and as a group prey on everything from plankton to higher vertebrates. However, each species occupies its own ecological niche, and therefore has a tendency to exploit a particular type of prey. For example, Crabeater seals eat primarily krill, whereas Ross seals inhabiting the same regions will take mostly squid and fish.

Seal milk is a very high-energy drink, with a fat component of 30 to 60 percent and a protein content of at least 5 to 15 percent. (In comparison, cow's milk contains 2 to 4 percent fat and 1 to 3 percent protein.)

Some species—for example, some of the southern fur seals, many southern sea lions, and elephant seals—ingest rocks and sand. Rocks the size of a small orange have been recovered from the stomachs of seals, and one Southern elephant seal had 35 kilograms (80 lb) of rocks in his gut. These rocks, which are known as gastroliths, are found in such quantities that they are probably ingested intentionally, rather than by accident, and may provide stabilization while diving, or ballast to minimize the energy required for deep dives, or they may be for breaking up food or grinding up parasitic worms. Breeding and molting seals are more likely to have gastroliths, possibly because they need to fill their stomachs while they cannot feed at sea.

Mating and gestating

A female becomes pregnant immediately after mating, but the embryo stops growing after a few days and does not implant in the uterus for several months. It then develops in the normal way, and at the normal speed. Delayed implantation means that the pup is born almost a year after mating, although the period of fetal growth is much less than that, generally eight to nine months. Thus, birth and the next mating occur around the same time, so that the pup is born when it has the best chance of survival and the female has the most protected and shortest possible breeding and pupping period.

Females occasionally adopt orphaned pups if their own die. As most seals that lose pups are young mothers, this behavior may help them by providing much needed experience. Some pups find a foster mother after a normal period of nursing by their blood mother, and so are suckled for much longer than normal and gain many advantages from the resultant extra size and energy stores. Most foster mothers adopt a single pup, usually only when they lose their own pup when it is very young. In a few seals, however, the mothering instinct is so strong that they will nurse so many orphans they cannot produce enough milk to feed them properly.

> ICE FISHING

A Weddell seal with a large Antarctic toothfish (*Dissotichus mawsoni*) which it caught in 300-meter (980-ft) deep water under this hole in the fast ice near Cape Armitage, Ross Island. Fish are the main food of Weddell seals, but they also catch squid, bottom-living prawns, and octopus.

▲ BACHELOR RIGHTS

A young bachelor Southern elephant seal rests on top of a weaner. Sometimes seals of different ages and sexes come ashore to rest and molt at the same time. A peaceful scene like this one may erupt into a short-lived aggressive outburst at any time. When caught in the fray, a smaller individual often becomes the victim of a larger seal's temper tantrum.

Eared seals

◄ WATERPROOF JACKET
Waterproof and nearly windproof, a fur seal's fur is made up of two layers—an outer layer of coarse guard hairs and a velvety underfur, some 2–3 centimeters (about 1 in) thick.

▼ AWAITING LUNCH

Two New Zealand fur seal pups peer out from their hideaway above the high tide mark. After a mother has suckled her pup for 10 days, she mates—usually with the bull controlling the territory. She then leaves on her first feeding trip, returning a few days later to suckle her pup again.

There are two main groups of eared seals, the fur seals and the sea lions. They are physically very similar, although sea lions have rounder snouts and shorter front flippers, and are usually slightly larger than the fur seals. The feature most often used to distinguish the two groups, however, is the fur seal's thick, almost waterproof pelt, which has given the animal its English name.

This pelt, which is commercially valuable, consists of two types of hairs. Each long, stout guard hair has about 50 shorter, softer, underfur fibers growing from the same hair shaft. This gives the seal roughly 46,000 per square centimeter of skin (300,000 hairs per sq. inch). (In comparison, a human head has about 15,000 hairs per square centimeter/100,000 hairs per sq. inch.) Sebaceous glands in the seal's skin produce enough oil to make the fine hairs into a waterproof barrier that traps air, so that only the tops of the guard hairs get wet. In addition, the strong guard hairs serve to support the fine underfur. In this manner, the fur seal's pelt is custom-made to provide insulation when the animal is diving. Sea lions, on the other hand, lack this underfur, and historically their pelts have little financial value.

Fur seals and sea lions probably diverged into unique groups only about two million years ago—a relatively short time in evolutionary terms—and the two groups still retain many similarities. Their social and reproductive behaviors, for example, are very similar, and natural hybrids have been observed where their populations overlap in the wild.

On average, eared seals live for about 18 years. Males probably live shorter lives than females because they must vigorously compete with other males for breeding rights. It is a stressful process for the males, and many do not survive past early reproductive age.

The males of both the fur seals and the sea lions are always much larger and heavier than the females. The males have particularly solid-looking chests that are much bigger than those of the females and are covered with longer fur to increase their apparent size. The males use these big chests for displaying to rival males in order to intimidate them. The robust chest also offers protection on the rare occasion that the displays degenerate into active fighting between competing males.

> PLAYTIME
Eared seals can be playful, often surfing, tossing and dragging rocks, sticks, or other animals, chasing each other, wrestling, rolling over, waving their flippers, or biting their own tails. The purpose seems to be the formation of social bonds, and possibly the development of motor skills.

Living spaces

The Otariidae occur from tropical, through temperate, to polar environments, but are most commonly found in temperate or sub-polar conditions. Tropical populations of fur seals and sea lions tend to be quite small, whereas sub-polar populations are usually very large. These environments tend to be more productive with nutrient-rich waters that encourage large populations of prey species, which eared seals can successfully exploit on a seasonal basis. Tropical environments are, on average, less productive and subsequently have smaller populations of potential prey, leading to smaller populations of seals.

All seals need to replace their hair, or fur, in a process know as molting. This is particularly important for fur seals, which rely on the waterproofing in their soft underfur for temperature regulation. Fur seals and sea lions molt very gradually so that they can continue to swim throughout this process without compromising the insulative capacity of their fur. Northern fur seals have the slowest known molt in the Otariidae; it can take up to three years for their fur to be completely replaced. More generally, the process takes in the order of a month.

Claiming territories

Eared seals come together in breeding colonies on favored beaches to give birth and mate. The males will generally arrive before the females and spend the first few days fighting to establish territories, and they then maintain rank by vocalizing, posturing, and threatening. These breeding territories are often defined by natural boundaries, such as rocks or crevices, which allow easy identification of interlopers and provide clear-cut zones for posturing. In many species, the males return to the same area for as long as they can continue to win their place—usually only two or three years.

▲ FLUSHING THE SYSTEM
Like all mammals seals need water—yet many do not seem to drink it. Some seals get enough water from eating fishes that contain very little salt. The many species that live on invertebrates consume salty food, but seal kidneys are so effective that they can excrete salt in extremely concentrated urine and even extract fresh water from the seawater. Male eared seals are especially likely to drink seawater during their long breeding fast. Although they can derive some water through the oxidation of stored fat, they must replace water lost through panting, urination, and sweating. Presumably the net gain in fresh water from drinking seawater, while very small, is enough to be worthwhile.

◄ STRONG DIFFERENCES
Hooker's sea lions show a strong sexual dimorphism, beginning from birth. Male pups are born bigger than females, and older males become progressively darker, as well as far larger, than the females. Fewer than 20 per-cent of Hooker's sea lion males will successfully breed during the mating season.

The number of pups that a male fathers depends on the quality of his territory and the length of his tenure. The quality of a territory is reflected in the number of females that it attracts, which in turn depends on the location of the territory and the distribution of females across the space. The distribution of the females is partially influenced by climate and access to water. The length of time a male retains his tenure depends on his size and age, as these factors affect his competitive and fasting abilities. Experience counts: young males start with poor breeding territories and acquire better ones over time, until finally they submit to a younger and fitter contender.

Females influence their breeding success by choosing the territory they use, sometimes moving through the territories of several males before settling. Males of many species try to herd females, usually without much luck because females will return to the territory of their choice. Of all the Otariids, South American sea lion bulls are possibly the most successful herders.

All eared seal bulls must go without food during the breeding period, because to leave would mean a new series of dangerous and exhausting battles to re-establish dominance. Leaving the colony also means that the male runs the risk of missing the females' receptive period. This is one reason why male Otariids are so much larger than females—their size gives them energy reserves that allow larger males to remain on the beach for a long period and to father more offspring than smaller males that must abandon their posts to feed.

The territory maintained by breeding males can contain more than 20 females, so the rewards for winning a territory are considerable. Superfluous males usually hover on the edges trying to engineer sly opportunities to mate, although in some species bachelors form non-breeding herds with the youngsters from other beaches. The females arrive some days after the males and give birth a few days after that. They then stay with their pup for a week or two, nursing their young while the males jostle for position and control of females. Females then come into estrus and mate.

In many eared seal species, the female bellows loudly when a male tries to mount her. If that male is not the local territory holder, her call will notify him that another male is trying to mate with one of his females. Calling helps to ensure that her pup will be fathered by the strongest male available, increasing the pups' likelihood of also being a dominant adult and in turn fathering many pups.

A mated female stays with her pup for a couple of days and then starts to alternate between going to sea for several days to feed and returning to nurse her pup. This alternating pattern of absence and attendance means that the period of pup dependence can last for many months. In some sea lions it takes females more than a year to wean their pup. In polar species, lactation is restricted to the summer months, and can take as few as four months.

Cherchez la femme

Breeding fur seal males typically defend a specific patch of beach—taking a gamble that it will fill with females. Male sea lions are less obviously territorial, but still work hard to isolate a group of breeding females and maintain exclusive access to those females.

➤ WHILE MUM'S AWAY
Hooker's sea lion pups are very active animals, and when their mothers go to sea to forage, they take their pups to vegetated areas behind the beaches. In larger colonies, the pups will gather together in groups and play in the zone behind the beach, scrambling around in streambeds and along rocky outcrops.

Immediately after birth, the mother and her pup bond by nuzzling and vocalizing. When a female goes to sea to feed for the first time after giving birth, the pup stays close to where it suckled and the mother returns to the same area and locates her pup by sound and smell. When she calls, all the nearby pups respond, but she accepts only her own offspring. It is critical that the pup responds, because if it does not call the mother bites it or tosses it away, a behavior that has developed to stop her from feeding greedy or orphaned pups and thereby lessening the energy she has to nurture her own pup.

➤ TAILOR-MADE
All seals have flippers adapted more for swimming than walking—the foot bones are elongated and flattened, and the tissue surrounding the toes is broadly webbed. Here, the seal uses its small claws to scratch itself in the same manner as a dog, making both the length and the flatness of the hind foot clearly visible.

Fur seals

ORDER: Pinnipedia

FAMILY: Otariidae

Of the eight southern fur seals, genus *Arctocephalus*, five either permanently occupy or regularly visit sub-polar or polar regions.

South American fur seal

ARCTOCEPHALUS AUSTRALIS

OTHER NAMES: Southern fur seal, Falkland fur seal

LENGTH: female 1.4 m/4 ½ ft; male 1.9 m/6 ft

WEIGHT: female 50 kg/112 lb; male 160 kg/360 lb

STATUS: up to 750,000, sensitive to El Niño events

IUCN: Lower Risk—least concern   CITES: not listed

South American fur seals range up the coasts of South America from Lima to Rio de Janeiro, throughout Tierra del Fuego, and to the Falkland Islands. They breed on islands and remote coasts, and congregate in small groups with the females well spaced. Pups are born in spring and early summer (mid-October to December), the peak time being November and early December. Mating occurs six to eight days after the female gives birth. Females nurse for at least eight months, and sometimes up to two years—a very long period for a fur seal. Most of the South American fur seal population is non-migratory, with some limited dispersal in winter, but most females remain near the colonies all year.

This species is unusual in that noisy gangs of young males—usually about ten, but sometimes as few as two or as many as forty—regularly raid breeding territories, trying to herd females away or mate with them under the protection of the gang. These raids may occur up to every two hours during the peak of female receptiveness.

SOUTH AMERICAN AND NEW ZEALAND FUR SEALS

- ▦ South American fur seal
- ▦ New Zealand fur seal

New Zealand fur seal

ARCTOCEPHALUS FORSTERI

OTHER NAME: Western Australian fur seal

LENGTH: female to 1.5 m/5 ft; male to 2.5 m/8 ft

WEIGHT: female to 70 kg/160 lb; male to 185 kg/400 lb

STATUS: 85,000–135,000; increasing at most sites

IUCN: not listed   CITES: Appendix II

New Zealand fur seal populations are found in New Zealand and nearby sub-Antarctic islands, in South Australia and Western Australia, and recently at Macquarie Island. These seals breed on islands and remote coasts in small to very large groups, with the females being well spaced. The males have massive necks and thick manes that are wider than the rest of their body, and which are used in display threats when establishing territories: they undulate the neck and utter guttural barks to intimidate all opposing males. New Zealand fur seals are non-migratory, with some dispersal after breeding and a return just before breeding, but most females stay near the colonies all year. They eat squid, octopus, fish, and even seabirds.

Because the species is largely non-migratory, males spend much of the year together. Male hierarchies are based on size, with the biggest males being the most dominant outside the breeding period. During breeding, however, dominance is territorially based.

Females pup in late spring and summer (mid-November to mid-January), with births peaking in the second half of December. Mating occurs eight to nine days after birth, and peaks in late December to early January. Pups are weaned at about twelve months.

▼ **FAST AND ACCURATE**

Otariids are high speed predators in the water, able to modify the depth of their dives to match the depth of their prey. Depth-recorders indicate they usually make V-shaped dives, presumably passing through schools of krill or fish, turning, and passing through again on the way back up to the surface.

ᐱ DISTINCTIVE FEATURES

The eared seals are generally smaller than the true or hair seals, and they are all very lithe and supple. The small external ear that characterizes the group can be seen on this Antarctic fur seal, as can the thick coat that made this species a popular target for sealers.

◄ AGILE ON LAND

Fur seals are extremely flexible, and can bring their hind limbs forward underneath their bodies and rise up on their front flippers. When moving fast, they usually bound forward with both hind limbs together, a motion well suited to moving around in their rocky habitats.

ᐱ CHILD'S PLAY

When not sleeping, these New Zealand fur seal pups spend much of their time in and around tidal pools, preparing for a life at sea from the comfort and safety of a sun-warmed tidal pool. Seal milk is rich in both fat and protein, so these playful, stocky pups grow rapidly.

Antarctic fur seal

ARCTOCEPHALUS GAZELLA

OTHER NAME: Kerguelen fur seal

LENGTH: female 1.3 m/4 ft; male 1.9 m/6 ft

WEIGHT: female 35 kg/80 lb; male 135 kg/300 lb

STATUS: at least 2–4 million

IUCN: Lower Risk—least concern

CITES: Appendix II

This colonial species breeds on islands south of the Polar Front, including the South Shetland Islands, South Georgia, the South Orkney and South Sandwich islands, Bouvetøya, and Heard and McDonald Islands, but is also found north of the Polar Front, on Marion and Prince Edward islands, Iles Kerguelen, and Macquarie Island. Colonies vary considerably in size and spacing.

Both males and females move away from breeding colonies in winter. On land, they sometimes travel inland to lie on top of tussock grass behind the breeding beaches. They can move at 20 kilometers per hour (12 mph) on beach flats, and can outmaneuver humans through the tussock grass.

They make comparatively long trips during the breeding season, in which they eat krill, cephalopods, and fish. While at sea feeding females make more than 400 dives of between 5 and 100 meters (15 and 330 ft) in five to six days. They can travel 60 kilometers (37 miles) in over eight hours before making their first dive, and

▲ NOT ALL ARE HIS

Large male fur seals claim breeding territories in an attempt to father all of their females' pups; however, up to 30 percent of offspring from a bull seal's territory may be fathered by other males.

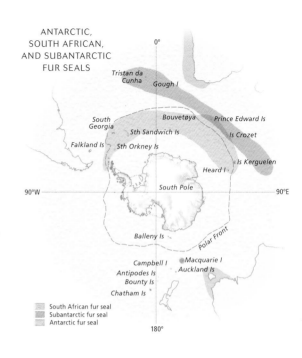

ANTARCTIC, SOUTH AFRICAN, AND SUBANTARCTIC FUR SEALS

South African fur seal
Subantarctic fur seal
Antarctic fur seal

▼ A BLONDE IN THE MIDDLE

While most Antarctic fur seals have medium to dark brown fur, a small number are natural-born blondes. Pale, almost white fur seals are highly visible on dark rock and stand out in the crowd, but they don't seem to suffer any more difficulties catching food than their more camouflaged relatives.

then regularly dive for four minutes at a time, to an average depth of 30 meters (100 ft).

Antarctic fur seals give birth in late spring to early summer (late November to December), peaking in early December. Mating occurs six to seven days after the birth of the pup, and peaks in mid-December. Females nurse for four months or less, and the pups seem to initiate weaning. Pups are extremely active and begin practicing swimming in January, but do not become skilled swimmers until March. There is some overlap in range between the Antarctic fur seal and the Subantarctic fur seal, and hybrids have occurred in the wild.

South African fur seal

ARCTOCEPHALUS PUSILLUS

OTHER NAMES: Cape fur seal, Australian fur seal, Tasmanian
   fur seal, Victorian fur seal

LENGTH: female 1.8 m/6 ft; male to 2.4 m/8 ft

WEIGHT: female to 120 kg/270 lb; male to 360 kg/800 lb

STATUS: to 2 million in southern Africa, 50,000 in Australia

   IUCN: Lower Risk—least concern   CITES: not listed

The South African fur seal ranges over southwestern Africa from Angola to Algoa Bay, and to southeastern Australia and Tasmania. It breeds on islands in medium to large colonies, with females relatively densely packed. As is true of other fur seals, it is polygynous, with female group sizes ranging from five to more than fifty. The species is non-migratory but shows some seasonal movement and some long-distance dispersal in winter. Individuals have been found 1,500 kilometers (950 miles) from the nearest colony. Feeding females generally stay at sea for approximately five days before returning to the colony. Unusually for an *Arctocephalus* species, mature bulls will tolerate the presence of young males more than one year old.

Pupping takes place in late spring and early summer (November and December), peaking in late November and early December. The peak mating period follows five to six days after birthing. South African fur seals generally nurse their offspring for 8 to 10 months, and sometimes longer if no new pup is born.

Subantarctic fur seal

ARCTOCEPHALUS TROPICALIS

OTHER NAME: Amsterdam Island fur seal

LENGTH: female to 1.5 m/5 ft; male to 1.8 m/6 ft

WEIGHT: female to 50 kg/112 lb; male to 150 kg/375 lb

STATUS: slightly more than 310,000

   IUCN: Lower Risk—least concern   CITES: Appendix II

This species breeds on islands and island groups north of the Polar Front, including Tristan da Cunha, Gough and Marion islands, Iles Crozet, Iles Amsterdam, and Macquarie Island. Breeding colonies vary in size and tend to be well spaced. Males move out to sea after breeding, while females and pups remain in the colonies. Their diet ranges from crested penguins and squid at Iles Amsterdam to cephalopods, fish, and krill at Marion Island. Females

➤ BIG, BEAUTIFUL EYES
This fur seal's large eyes are designed for seeking out prey in dim waters. Big eyes collect light more effectively than smaller ones, and are well adapted for taking advantage of every bit of light available in deep or muddy, murky water. Seal's eyes are also a slightly different shape to a land mammal's, to give the seal clear vision under water.

give birth in late spring to summer (late November to January), peaking in mid-December, and the peak mating period occurs 8 to 12 days after the birth. Pups are normally nursed for 9 to 11 months. Non-territorial males are not tolerated in the breeding colonies and form bachelor colonies with immature seals, spending 94 percent of their time completely inactive. Breeding males actively herd females and defend territories defined by natural barriers.

Diving data

All fur seal females go to sea to feed several days at a time when raising a pup. While at sea the females need to be adroit and effective swimmers to pursue krill, fish, and squid. For most female fur seals, the mean duration of a dive is two to four minutes, with a maximum duration of five to seven and a half minutes. The mean depth of most dives is 30–45 meters (98–150 ft), with a maximum depth of 100–200 meters (330–660 ft), and a mean number of 15 to 20 dives per hour.

Most experimental data on fur seals tend to be on females because they are comparatively easy to catch compared with males, and much easier to follow when they haul out to the breeding beaches. Females also gather in very large numbers, making it possible to gather a large amount of data on many individuals relatively easily.

Sea lions

ORDER: Pinnipedia
FAMILY: Otariidae

Hooker's sea lion

PHOCARCTOS HOOKERI

OTHER NAME: New Zealand sea lion

LENGTH: female to 2 m/6½ ft; male to 2.5 m/8 ft

WEIGHT: female 230 kg/500 lb; male 400 kg/880 lb

STATUS: less than 14,000; IUCN: Vulnerable   CITES: not listed

Hooker's sea lions breed on the sub-Antarctic islands of New Zealand, including Stewart, Auckland, Snares, and Campbell, with some births also occurring on the New Zealand mainland. They gather in large groups on remote coasts, with the females remaining in very dense groups. The males of this species are highly polygynous, each having a harem of at least 15 females to oversee, and sexual dimorphism is pronounced. There is no migration, but some dispersal; non-breeding males occur on both Macquarie Island and the South Island of New Zealand.

When foraging, females of this species dive for two to three days without a break, making dives as deep as 460 meters (1,500 ft) and staying down for as long as 12 minutes. Pupping occurs in December or early January and peaks in late December. Mating takes place in early January, normally six to seven days after birth. The young quickly become strong and active, swimming to nearby islands at two months. They may suckle for up to a year, but they begin feeding independently long before they are fully weaned.

▼ FRIENDS OR ENEMIES?
Hooker's sea lions are very sociable animals that spend much of their time ashore in groups as tightly packed as sardines in a tin. Unlike the New Zealand fur seal, these sea lions favor sandy beaches over rocky shorelines. They often move a long way inland and can sometimes be found high on hillsides.

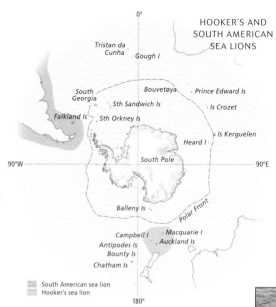

HOOKER'S AND SOUTH AMERICAN SEA LIONS

[Map labels: 0°, Tristan da Cunha, Gough I, Bouvetøya, Prince Edward Is, South Georgia, Sth Sandwich Is, Is Crozet, Falkland Is, Sth Orkney Is, Is Kerguelen, Heard I, 90°W, South Pole, 90°E, Balleny Is, Polar Front, Campbell I, Macquarie I, Antipodes Is, Auckland Is, Bounty Is, Chatham Is, 180°]

□ South American sea lion
□ Hooker's sea lion

South American sea lion

OTARIA FLAVESCENS

OTHER NAME: Southern sea lion

LENGTH: female to 2 m/6 ft; male to 2.5 m/8 ft

WEIGHT: female to 140 kg/310 lb; male to 350 kg/770 lb

STATUS: approximately 265,000, sensitive to El Niño events; IUCN: Lower risk—least concern   CITES: not listed

The South American sea lion breeds on islands and remote coasts of South America from Peru, south through Tierra del Fuego to Uruguay, occasionally further north to the coast of Brazil, and is also found in the Falkland Islands. It forms medium to large colonies, with the females densely packed. South American sea lions consume fish, squid, crustaceans, and, occasionally, young fur seals. They are non-migratory, but disperse up to 1,600 kilometers (1,000 miles) after breeding, with males covering greater distances, and they are known to enter freshwater rivers. During the breeding season males shift from defending their territory to sequestering up to 18 females, and will attempt to take them by force. Successful males may fast and get very little sleep for up to two months. Pupping occurs in summer (mid-December to early February), peaking in mid-January. Mating is at its peak six days after birth, and mothers generally nurse their young for at least six months, or up to two years if no new pup is born.

⋗ DANGER IN NUMBERS

In 1998, due to a bacterial infection that spread through the breeding colonies, 53 percent of Hooker's sea lion pups died, as well as more than 20 percent of other groups such as juveniles and adult males. Widespread death from a single factor—whether hunter, predator, infection, or even weather—is more of a risk for species that live in close-knit colonies.

▽ ROOM FOR ONE MORE

A Hooker' sea lion pup leaves its mother's side when it is as young as two days old, and within ten days congregates with other pups. About six percent of Hooker's sea lion females will nurse more than one pup at a time—her own and a foster pup.

True seals

True seals, sometimes called hair seals to distinguish them from fur seals, belong to the family Phocidae. There are five Antarctic species of true seal: the Crabeater seal, the Leopard seal, the Weddell seal, the Ross seal, and the Southern elephant seal. All Antarctic true seals, together with the northern-hemisphere Monk seals and Northern elephant seal, are grouped together and called Monacine seals. This grouping is based primarily on dental and skeletal similarities.

Very streamlined, true seals are well designed for life in the oceans, and some are truly ice-dwelling. To cope with the extreme conditions in which they live, they have well developed layers of blubber instead of the thick fur that insulates the fur seals. Even the thickest fur would be inadequate to meet the challenges of truly polar conditions—certainly, no true seal has fur that is thick enough to function as satisfactory insulation in these regions.

True seals are more at home in water than they are on the land. They swim using their strongly muscled hind limbs, applying this power in a lateral stroke with their toes expanded. Their entire hind end undulates laterally, and they then straighten out their body in a recovery stroke, with the toes bunched together to

minimize drag from the water. This is an extremely powerful stroke, which makes the true seals very capable swimmers.

True seals, however, are not as competent on land as they are in water because they have lost the ability to bring their rear flippers forward and use them as legs, in the way that eared seals do. True seals lumber along on the ground, using their whole body for movement, and their front flippers as levers. Those that live on ice have longer, stronger claws than those that live on land, and these long claws are extremely effective for clambering up on ice and moving around on ice ledges.

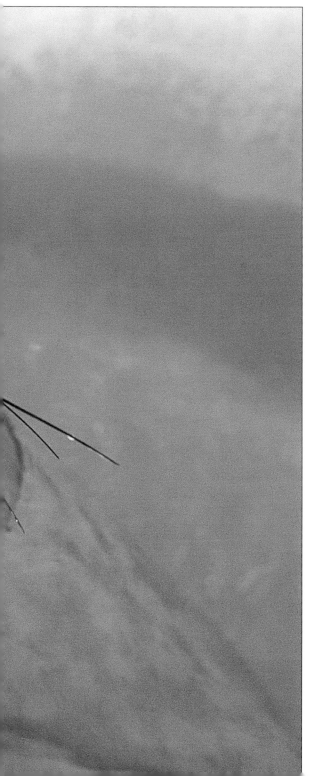

Streamlined design

True seals have wonderfully streamlined bodies that make them extremely efficient in water. They have fewer external parts, which would be difficult to heat in such extreme conditions, than fur seals. For example, instead of external ears (pinnae), they have only small ear openings, called meatuses, which lie flush with the sides of their head. In northern seals the meatus can be up to 1 centimeter wide (less than $\frac{1}{2}$ inch) but in Antarctic seals it is a mere 2 millimeters wide (less than $\frac{1}{8}$ inch) and is concealed and protected by the hair.

In true seals, the genital organs are concealed within their body contours. The testes are situated inside the skin muscles, but outside the main body muscles. This differs slightly from genital streamlining in whales, where the testes are completely internalized within the body muscles. All that can be seen when the sexual organs are not in use is a small hole in the belly of the animal. Antarctic true seals have a single pair of nipples, while fur seals and sea lions have two pairs. However, true seals have the most productive mammary glands of any mammal, producing exceptionally rich milk that is high in both fat and protein.

Mothering behavior

Unlike the Otariidae, true seals have a very short period of pup dependence, which can be as short as four days or as long as several months depending on the species. The main task of the female during this time is the efficient production and transfer of milk from mother to pup. However, their milk is highly nourishing, and this accelerates the nurturing process: true seals produce milk that ranges from 40 to 60 percent fat and 5 to 14 percent protein, with the fat levels usually increasing towards the end of the nursing period. Not surprisingly, suckling pups can gain as much as 25 percent of their birth weight each

▲ RICH MILK
Elephant seal mothers care for their young for about 20 to 25 days, with larger females nursing for a few days longer than smaller ones. During this time the pup grows from about 40 kg (90 lb) to about 120 kg (265 lb). Pups are abruptly weaned when their mothers leave them to go to sea.

◄ A WARRIOR'S LIFE
A large male Weddell seal rests on the surface in its breathing hole. While mothers rear their pups on the sea ice, fierce battles rage beneath the ice. Mature males spend their time underwater, fiercely defending the territories surrounding their breathing holes. The breeding season starts about September and is over by December. The next two months are spent resting and recovering both from rearing pups and from injuries received in underwater fights.

day. Most true seal mothers fast while nursing, alternately resting and feeding the pup several times a day. This is an unusual pattern for mammals—because of the amount of energy required to feed young mammals, most mammalian mothers need to at least double their food intake when nursing. Due to this fasting, female seals lose a great deal of weight while nursing. They wean their young abruptly, simply by going to sea, leaving the pup to fend for itself.

A mother's hunting skills are of great importance to her pups. The survival of the pups depends on the foraging success of the mother over the whole preceding year. For a true seal pup, the chances of surviving after weaning depend on a combination of weaning mass (the amount of weight they have managed to gain during the nursing period), and the availability of prey while the pup is learning to catch its own food. Each year, a new cohort (the complete group of youngsters born to a population in that year) takes to the water at approximately the same time, when they must survive—or not—without any maternal guidance.

Surprisingly, in many Antarctic seal populations the survival rates of various cohorts seem to be related to El Niño/ Southern Ocean Oscillation (ENSO) events. Cohorts of Weddell seals and Crabeater seals, for example, have shown fluctuations over four to six years that match the occurrences of ENSO events. Leopard seal numbers found on the sub-Antarctic islands also correlate with unusual pack ice movement associated with ENSO events.

Analyzing seal diets
Working out what a seal eats is not as easy as one might imagine. Marine species, unlike land animals, are very difficult to observe when they are feeding. One way of finding out a seal's diet is to examine the contents of the stomach, intestines, or feces, and identify the hard remains left after digestion. Bones, squid beaks, crustacean exoskeletons, and shells, any hard items, can be used to identify what an animal has swallowed.

To count fish species, scientists most often count sagittal otoliths, which are small bones from the head of fish. From these they can also identify the species, as well as estimate the size of the fish. This technique can provide estimates of both the numbers of fish consumed and the approximate weight of each type of prey. However, this method has shortcomings. It provides information only about what an individual animal has eaten recently, and it is likely that the animal's diet will change when it goes out to sea. In addition, different hard items break down at different speeds, so the contents of an animal's stomach at any given time might be misleading. It is also highly likely that this type of information will hold only for a limited area or season, and may even be heavily influenced by individual animals and their personal preferences, since different seals are known to prefer different food types when a range of choices is available.

▲ A HAZARDOUS LIFE
The scavenged remains of a seal, possibly injured by an Orca or Leopard seal, while learning to swim. Upon coming ashore to recover, too tired to defend itself, the seal may have been pecked to death by giant petrels and other scavenging birds. Eventually, an array of scavengers will eat everything but the bone.

Diving adaptations
Seals have 15 pairs of ribs, rather than the usual mammalian number of 13 pairs. This allows extra space for their lungs, which are slightly larger than in other mammals. A human diver breathes deeply before diving and takes a large breath before actually going down. This is a disadvantage, because the air in the lungs makes the diver more buoyant so that he or she needs more energy to descend. More seriously, full lungs, or a SCUBA tank, can lead to a dangerous pressure problem, called the bends, when surfacing after a dive: nitrogen dissolved into the blood under pressure can be rapidly released from solution into the bloodstream and can cause tissue damage. When a seal dives, however, the pressure of other organs collapses its diaphragm against its lungs, forcing any remaining air out of its lungs. This protects the seal from any risk of the bends.

Seals do hyperventilate before diving, but they store the oxygen in their blood and muscles and then expel the air. Seals have more blood than similar-sized terrestrial mammals. For example, seals have proportionally about 1.6 times as much blood as a human. Seal blood also carries a higher proportion of hemoglobin, which carries the oxygen. The combined advantages of these two features means that seals can carry about 3.5 times more oxygen per unit of body weight than humans can. Furthermore, about 25 percent of a true seal's weight is blubber, which is effectively inert tissue that does not require a large amount of energy to maintain. If the blubber component is discounted, seals are effectively able to store approximately 5.3 times as much oxygen as humans on a per weight basis. Seals can also store much more oxygen in the high amounts of myoglobin in their muscle tissue. The muscle of a Weddell seal, which carries approximately 10 times as much oxygen as human muscle, is so red that it looks almost black.

When a seal dives for long periods, its heart rate drops from 60 to 70 beats per minute to as low as 15 beats per minute. The brain and heart receive fully oxygenated blood for as long as possible, while blood flow to the viscera, skeletal muscle, skin, and flippers falls by 90 percent.

▲ MOTHER AND CHILD
A female Weddell seal remains
very close to her pup for
the first two weeks after birth,
then she occasionally goes to
sea and leaves the pup alone.
The pup may be given its first
swimming lessons when less
than ten days old, and will be
fed for a total of six or seven
weeks before having to go to
sea alone.

➤ STORING OXYGEN
Weddell seals have very large
spleens, where they store red
blood cells for later release
into general circulation during
deep dives. This technique for
maintaining access to oxygen
during longer dives is used by
other true seals, with spleen
size being related to diving
ability, but it is not used by the
eared seals.

▲ COOLING OFF

Elephant seals are well insulated against the cold water, but conversely may overheat in the hot Antarctic sun. They have a special internal mechanism to combat this, but will also employ other measures to keep cool, such as covering themselves in damp sand, which cools by evaporation and after it is dry serves as a sunshade.

In extreme cases, these tissues will run out of oxygen and acquire energy by converting glucose to lactic acid, a process known as glycolysis. When a Weddell seal dives, lactic acid does not start to build up for 25 minutes, which indicates that oxygen is still available in these tissues for at least that long. If a Weddell seal spends 45 minutes under water, it spends approximately 20 minutes drawing on energy derived from glycolysis, and must rest on the surface for at least 105 minutes to recover. Recuperation involves an increased heart rate to return oxygen to the body, and the removal of lactic acid from body tissues.

Maintaining body temperature in Antarctica

Under normal conditions, the Antarctic environment is much cooler than the body temperature of a seal, which is around 38°C (100°F), and it must use a range of mechanisms to keep its vital organs, particularly heart and brain, at this temperature. Seals have developed some clever adaptations to minimize heat loss and conserve energy, including the ability to constrict their blood vessels and cut off the supply of warm blood to skin lying against the ice. This skin becomes very cold, close to

freezing point, but the seal's layer of blubber effectively insulates the critical internal organs, which remain heated by warm blood, without the constant energy drain of attempting to maintain 38°C in the exposed skin.

Paradoxically, Antarctica receives a great deal of solar energy in high summer, and seals coming ashore on warm Antarctic days would overheat very rapidly if it were not for a special mechanism. In mammals, blood normally flows outward from the heart through the arteries to arterioles and the capillary bed, and is returned through venules to the veins and back to the lungs, where its oxygen is replenished. But seals have an alternative blood flow system; they can dilate special blood vessels that are very close to the surface of the skin and that bypass the capillary bed. These bypasses, called arterio-venous anastomoses, allow the animals to direct warm blood to their skin surface very rapidly. This enables them to disperse excess heat into the environment and quickly return the cooled blood to their body core, where it can draw off more heat and return to the surface to repeat the cooling process. This type of system is essential in animals that are as well insulated as seals.

Baby fur

Most true seal pups have no insulating blubber, but they do have very fine, thick baby fur, called lanugo, that is very long in comparison with that of the adult seals. At birth the baby seal has virtually all the hair follicles that it will have as an adult, so these are much closer together than they will be when the animal has grown to full size. This combination of features means that the pup has a more effective insulating coat of fur than have its parents, and it very seldom gets cold. If a pup does begin to get cold, it can produce additional heat by increasing its metabolic rate.

▶ **RESTING BETWEEN DIVES**
If Weddell seals keep their dives short, they can dive all day, needing only a few minutes breathing air at the surface between dives. After a long dive (25 to 80 minutes) the Weddell seal needs more time to recover, sometimes well over an hour.

Wonderful whiskers

Whiskers, known scientifically as vibrissae, are very important to seals. They are prominent in most species, visibly clustered on the upper lip and above the eyes, with a single pair on top of the nose. There are many nerve endings at the base of whiskers, and they are extremely sensitive to motion. In marine mammals, vibrissae can pick up minute changes in the flow of water—even the minor disturbance caused by a fish's tail while it is swimming may be detected through a seal's whiskers. The vibrissae are pushed forward when a seal is chasing prey, possibly making them more effective at detecting fine differences in motion. This may enable the seal to predict the exact location of a fish by changes in motion rather than by sight, which is often of limited use especially at greater depths. Experimental trimming of the vibrissae of wild seals has shown that seals without their whiskers find it much harder to catch fish in the dark than do seals with their whiskers intact.

Crabeater seal

LOBODON CARCINOPHAGUS

LENGTH: 2.35 m/7 $\frac{1}{2}$ ft

WEIGHT: 220 kg/500 lb

STATUS: estimates range from 10 to 40 million, but probably about
15 million; IUCN: Lower Risk—least concern   CITES: not listed

> ➤ **ENVIRONMENTALLY SUITED**
> Crabeater seals, unlike their nearest relatives, the
> Weddell and Leopard seals, avoid land, preferring to
> come out of the water onto ice. The musculature of
> the Crabeater seal is designed for swimming not
> walking—movement on ice rather than on land is
> much easier for them because they can slide along,
> without having to lift their body at all.

The Crabeater seal is a truly Antarctic seal species. They populate the outer fringes of the pack ice, ranging among the icebergs and smaller floes, and recent evidence suggests they may like the deep ice close to the Continent. Crabeaters are exploratory: lost individuals sometimes wander north to New Zealand, Australia, and even as far as the River Plate in Argentina. Even within their main habitat they are often found as solitary individuals, sometimes as pairs or triads, and seldom in small groups. They do not truly migrate, but they do show some seasonal movement with the annual expansion and retreat of the ice.

Crabeaters have a huge area to call home. In summer, the Antarctic pack ice covers an area of about 4 million square kilometers (1½ million sq. miles). In winter, this increases to more than 22 million square kilometers (8½ million sq. miles). They have been observed at population densities ranging from one to seven animals per square kilometer (3 to 18 per sq. mile) but under the right conditions on good firm ice, up to a thousand Crabeaters will gather in a small area.

Despite their common name, Crabeater seals do not live on crabs: they are specialist feeders that live mainly on krill. They also eat a small proportion of fish and squid, and other invertebrates. Their teeth are highly adapted for their diet, each tooth basically the same shape—extremely convoluted with a few twisted gaps. The complex cusps are designed to strain krill from the water, and it is thought that a seal catches many krill in one mouthful, closes its teeth, then pushes its tongue up towards the roof of its mouth, as baleen whales do. Seawater is forced out through the teeth, then the krill are swallowed.

Up to 78 percent of all Crabeater seals bear the scars of attacks by Leopard seals or Orcas. Leopard seals target smaller and younger Crabeaters. Pups under six months old probably do not survive most Leopard seal attacks, and while older, stronger Crabeaters may survive, the Leopard often inflicts permanent damage.

Like all true seals, Crabeaters cannot bring their hind limbs forward to assist with locomotion on land. Nevertheless, they can move at 25 kilometers per hour (15½ mph), and are probably the fastest-moving seals on ice. They use a sinuous, galloping movement, with alternate backward thrusts of their front flippers, and side-to-side movements of their hindquarters.

Crabeater seals do not breed in colonies as Otariids do; instead, they breed on the pack ice in isolated pairs. The breeding system is one of serial monogamy, in which the males search out and find a female, remain with her until she comes into estrus, mate with her, and then move on in search of another female. Generally, breeding Crabeater females are dispersed so that there is at least 1 kilometer (⅔ mile) between mating pairs. Females give birth in spring (late September through to early November), peaking in October. They fast for the four-week period while they are nursing, and lose about 50 percent of their body weight. Pups, on the other hand, gain almost 100 kilograms (220 lb) while nursing. Nursing females are very aggressive towards males. They nurse their pups for approximately 28 days, then come into estrus and mate in late October and November, when they stop lactating.

During the lead-up to mating, males fight for the opportunity to attach themselves to a female and her pup. The successful male protects both the female and her pup from predators, as well as from other males. It is not known if the male forces the pup to wean by driving it away when the female comes into heat, or the female weans the pup herself before she comes into estrus.

At any rate, at the point that the pup is weaned the male becomes aggressive, separating the mother from her pup and then mating. Some studies suggest that the male and female may stay together for a few days after mating and then take to the sea together.

CRABEATER SEAL

0°
Tristan da Cunha
Gough I
South Georgia
Bouvetøya
Prince Edward Is
Sth Sandwich Is
Is Crozet
Falkland Is
Sth Orkney Is
Heard I
Is Kerguelen
90°W
South Pole
90°E
Balleny Is
Polar Front
Campbell I
Macquarie I
Antipodes Is
Auckland Is
Bounty Is
Chatham Is
180°
Distribution
Vagrants

◄ NOTHING TO HIDE

The fur of Crabeater seals is very stiff, and because it is not used as a primary source of insulation it lacks the underfur layer that fur seals have. A Crabeater's coat is mostly medium brown when new, and fades over the year to a light yellow or even white-blonde by summer. Older Crabeaters often have very scarred hides—healed wounds from attacks by Leopard seals and possibly Orcas.

Leopard seal

HYDRURGA LEPTONYX

OTHER NAME: Sea leopard
LENGTH: female 3 m/10 ft; male 2.8 m/9 ft
WEIGHT: female 370 kg/83 lb; male 325 kg/73 lb
STATUS: 220,000 to 440,000
   IUCN: Lower Risk—least concern   CITES: not listed

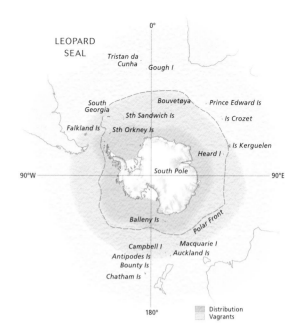

▼ **WHERE ARE YOU GOING?**
Leopard seals are closely related to Crabeater and Weddell seals, but have a sleek reptilian grace that their relatives lack. They do not form large breeding colonies, and their seasonal movements are not well known—although some individuals appear to disperse northward during winter, and young animals seem more likely to move about than adults.

The Leopard seal is the largest of the five true seals in Antarctica after the Southern elephant seal. It is a true hunter, possessed of powerful jaws and a long, sinuous neck that it uses to great effect, retreating backwards then striking like a snake. This powerful animal can leap onto ice 2 meters (7 ft) above water level—an exit speed from water of about 6 meters (20 ft) a second. Leopard seals would rather be on ice than on land, but will haul out of the ocean onto land if ice is not available.

The Leopard seal's main habitat is the the Antarctic pack ice, and its icebergs and smaller ice floes. The animal is circumpolar, ranging right through the southern oceans, but it has often been observed on sub-Antarctic islands, and in Australia, Chile, Tristan da Cunha, southern Africa, and New Zealand. Individuals have drifted as far north as Raratonga, in the Cook Islands. But despite this broad range, Leopard seals are not truly migratory, though they do disperse northwards from time to time—probably for reasons associated with the availability of food and

the location of pack ice. Generally speaking, mature Leopard seals prefer to remain in the Antarctic pack ice, while youngsters can be observed throughout the sub-Antarctic zone from June to October each year. Adults ranging from three to nine years old utilize the Antarctic coast in summer.

Leopard seals are known for their hunting prowess, and this is probably the only seal that sometimes preys on warm-blooded animals. They take Crabeater seal pups

> **FAST AND FEROCIOUS**
This Leopard seal displays teeth well adapted to seizing and tearing flesh, but also for straining krill—an important part of its diet. Leopard seals are fast, fierce hunters that take a wide variety of prey, including penguins, fish, squid, and even other seals—usually Crabeater pups. They are strong, graceful swimmers and can leap onto ice floes 2 meters (7 ft) above the water.

and also penguins. They usually capture adult penguins in the water, rarely hunting on land, where they are relatively cumbersome. There is one extraordinary record of four Leopard seals consuming 15,000 Adélie penguins in 15 weeks; one animal was found with 16 adult penguins in its stomach.

Despite their reputation, most Leopard seals do not regularly hunt warm-blooded animals. The diet of Leopard seals consists of 50 percent krill, 20 percent penguins, 15 percent other seals (mostly Crabeater pups), nine percent fish, and six percent squid. Juvenile Leopard seals generally do not have sophisticated hunting skills and probably rely on krill for food. However, they are not as expert at catching them as the highly successful specialist Crabeater seal.

The Leopard seal's teeth provide a clue to its diet. In shape they are similar to the Crabeater's, with extremely convoluted, lobed cusps that they use as krill strainers. However, unlike the Crabeater's teeth, the Leopard seal's teeth have very sharp edges and points, which the seal uses for grasping seals, penguins, and fish.

Leopard seals are solitary animals, and the only groups that they form are mother–pup pairs and temporary mating pairs. They breed in the pack ice, with the females widely separated from each other. It is not known whether Leopard seals are serially monogamous, like Crabeater seals. The females, which are slightly larger than the males, pup in spring and early summer (September to December), with births peaking in November. They nurse their pups for about a month, and then mate when they stop lactating, generally in late December.

▲ RICH PICKINGS
During the later part of the summer, many Leopard seals hunt by waiting offshore for young penguins and seals to enter the water. Some individual Leopard seals seem to return to the same location each winter, while other individuals just drift around.

Weddell seal

LEPTONYCHOTES WEDDELLI

LENGTH: female 3.3 m/11 ft; male 3 m/10 ft
WEIGHT: 400–500 kg/900–1,125 lb
STATUS: approximately 800,000
    IUCN: Lower Risk—least concern   CITES: not listed

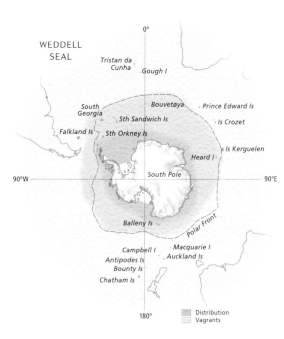

WEDDELL SEAL

Distribution
Vagrants

▶ HOME BODIES
While Ross and Leopard seal populations tend to move with the flow of the ice—presumably to stay in their favorite zone within the pack ice—Weddell seals seem to prefer to stay put on inshore fast ice.

▼ A MOTHER'S FAST
A Weddell seal pup weighs 25–30 kilograms (55–66 lb) at birth and when weaned, at six or seven weeks, is about 100 kilograms (220 lb). While the mother nurses, she does not feed and drops from about 450 kilograms (990 lb), to about 300 kilograms (660 lb).

This species is the most southerly breeding of all mammals. They tend to inhabit areas of coastal fast ice (ice that is connected to land) rather than the moving pack ice. Although Weddell seals favor the coastal fast ice areas of the Antarctic Continent and the nearby islands, Weddell seals are occasionally found in South America, Australia, and New Zealand.

Particularly during winter, these seals breathe and enter the water through holes in the ice. They need to make or maintain these holes themselves, cutting the ice with their incisor teeth and enlarging and maintaining the openings with their canine teeth. They will not begin to cut a hole in ice that is more than 10 centimeters (4 inches) thick, but once they have made a hole they will maintain it until the ice is up to 2 meters (6½ ft) thick.

Outside the breeding season, Weddell seals are not particularly gregarious, but neither do they actively avoid each other, and in good conditions they will comfortably occupy ice at densities of 15 to 35 animals per square kilometer (38 to 90 per sq. mile). However, they do maintain some distance from each other.

However, during the breeding season, the females gather in small, loosely knit groups of as many as 100 individuals to give birth in the very stable fast ice around the Southern Continent. They share the breathing holes that occur along cracks in the ice, while breeding males wait in the water beneath these entry points. Males defend an aquatic territory against other males, and by defending an entry hole, a single male can monopolize a large number of females. It is possible that male Weddell seals are a little smaller than the females in order to give them an advantage when maneuvering to defend their underwater territory.

During the pupping season, the breathing holes become the scene of fierce displays of aggression. The breeding males guard their holes against each other and drive younger animals away from the area. The sub-adults remain in the background until after the breeding season; at this time the tension eases somewhat, and the males permit the young to come back into the area. The groups gathered at the breathing holes do not really disperse after breeding, but they do eventually spread out over a larger area as more cracks appear in the ice with warmer weather, creating new breathing holes.

The peak pupping period is from September through to November, but pupping may take place at any time from late August onward. The lactation period lasts for approximately 50 days, much longer than for most other phocids. This means that lactating Weddell seals must feed during the nursing period, unlike the other species of Antarctic true seals. Males that have won a territory wait for the females to wean their pups, and mate with them in November and December, when they enter the water.

Weddell seals are believed to consume a diet of fish, cephalopods, krill, and other invertebrates. They locate and catch this prey using a range of diving behaviors.

Some dives can be quite long (20 to 73 minutes), and comparatively shallow. These dives are generally aerobic (using oxygen stored in the blood and muscle), but a small proportion are longer than the 25-minute aerobic limit. Other dives are deeper and may last for 5 to 25 minutes, and be as deep as 400 meters (1,300 ft). Weddell seals often feed at a depth of 200–400 meters (650–1,300 ft) to catch Antarctic cod—prey that is well worth the effort, as these fish may reach as much as 30 kilograms (66 lb) in weight. When diving, the seals swim at about 35 meters (110 ft) per minute. Thus, a normal dive covers the surface equivalent of about

◀ DENTAL PROBLEMS
Weddell seals live only to about 22 years. A contributing factor may be their practice of maintaining holes and ramps in ice by sawing and grinding, thus causing damage to their teeth. Dental problems, such as broken, worn and abscessed teeth, eventually lead to an early death.

3–6 kilometers (2–4 miles), although Weddell seals also have been known to travel up to 12 kilometers (7 miles) from their access hole in the ice.

Diving patterns change over the year, probably according to the availability and distribution of the seals' prey. In summer, Weddell seals hunt fish and squid throughout the day, diving to a mid-depth of about 150–300 meters (500–1,000 ft). In spring and autumn, however, when the main target is krill, diving depths fluctuate considerably in the course of a day, from between 100 meters (300 ft) and 500 meters (1,600 ft). Weddell seals have good underwater eyesight and will make much deeper and riskier dives in strong daylight. It is quite likely that in the winter dark they dive for only short periods and stay well within areas with which they are familiar.

Seal song

Most seals vocalize from time to time, either above or below the water. Underwater, their utterances vary considerably, some species being almost completely silent, others extremely vocal. The noises made by seals include barks, whinnies, clicks, trills, moans, hums, chirps, belches, growls, squeals, and roars, as well as many other sounds not so easily described. Some sounds are similar to those that whales use for echolocation. However, suggestions that seals echo-locate are still extremely speculative. Rather, many of the reported instances of underwater vocalization by seals appear to be related to establishing and maintaining breeding territories.

Weddell seals are particularly noisy under water and produce a constant stream of varied sounds such as whistles, buzzes, tweets, trills, chirps, and growls. Male Weddell seals produce more complex vocalizations than do females and these sounds can be detected as much as 30 kilometers (20 miles) away.

Furthermore, the vocalizations can even penetrate the fast ice and so might well play a part in territorial displays.

Seals also chirp, whine, bark, hum, growl, squeal, and roar when they are above the water. Not surprisingly, communal species are much more vocal on land than solitary species, and seals that live in colonies spend much more time vocalizing than those that live widely spaced apart. When individual animals from noisy communal species are isolated from other individuals, they fall almost completely silent. However, if they are returned to a group situation they become extremely noisy again, which suggests that their vocalizations are a form of direct communication.

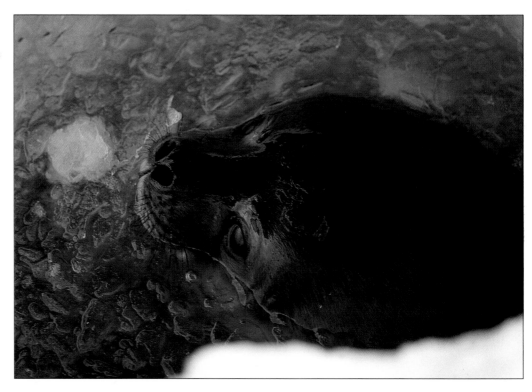

Ross seal

OMMATOPHOCA ROSSI

OTHER NAMES: Singing seal, Big-eyed seal

LENGTH: to 2.4 m/8 ft

WEIGHT: 220 kg/500 lb

STATUS: thought to be approximately 220,000

IUCN: Lower Risk—least concern   CITES: not listed

Ross seals live deep in the Antarctic pack ice, and are patchily distributed right around the southern polar regions. They challenge the Weddell seal for the title of southernmost mammal. The Ross seal is a true pack-ice species and only rarely hauls out on land. Scientific knowledge of this species is limited because its preferred habitat of dense pack ice is difficult for humans to easily reach in ships or from land.

At least part of the Ross seal population appears to undertake some seasonal movement, presumably in response to movements of pack ice and fluctuations in food availability. It is thought that individuals may maintain separate underwater territories, and possibly communicate with each other from within those territories, across great distances. Ross seals can reach considerable depths in pursuit of their favored prey, which consists of cephalopods, fish, krill, and bottom-dwelling invertebrates.

The most distinctive physical feature of the Ross seal—a feature that is unique among seals—is the dark, longitudinal streaks on its throat and the sides of its head. The Ross seal is not as heavy as most of the other true seals in Antarctica, but it does have a thick neck and throat that shows its striped coloration very clearly. It has very long back flippers—the longest of any seal in proportion to its body size. Its body hair and vibrissae (whiskers) are very short. It has a small mouth with very highly developed muscles of the tongue and pharynx, together with a short hard palate at the top of the mouth and a very long soft palate behind that.

When they are approached, Ross seals open their mouths very wide and fill their soft palate with air. They then use their inflated soft palate and their pharynx as resonating chambers, pushing the back of their tongue against their soft palate and producing a series of remarkable trills and thumps, without needing to expel any air at all.

Ross seals breed in the pack ice, where the females are either solitary or accompanied only by their pup or a very small group of females that remain widely separated. Not much is known about their mating behavior, but it is known that the females are slightly larger than the males, and that the pups are born in late spring and early summer (November and December), and that they learn to swim almost immediately. The peak mating period occurs in December, about a month after the females have pupped, and it is believed that mating marks the point at which the pup is weaned.

▲ **JUST WANDERING ABOUT**

Ross seals don't appear to migrate, but they live so deep in the ice that it is hard to be certain. Populations do, however, move over large areas, and surveys often show very patchy distributions within the ice. Studies suggest that compared with Crabeater and Leopard seals, numbers of Ross seals may be declining, if only locally.

ROSS SEAL

0°

Tristan da Cunha
Gough I

Bouvetøya   Prince Edward Is

South Georgia   Sth Sandwich Is   Is Crozet

Falkland Is   Sth Orkney Is

Is Kerguelen

Heard I

90°W   South Pole   90°E

Balleny Is   Polar Front

Campbell I   Macquarie I
Antipodes Is   Auckland Is
Bounty Is

Chatham Is

Distribution
Vagrants   180°

Southern elephant seal

MIROUNGA LEONINA

OTHER NAME: Southern sea elephant

LENGTH: female 2–3 m/6$^{1}/_{2}$–10 ft; male to 5 m/ 15 ft

WEIGHT: female 400 kg/900 lb; male to 3,700 kg/8,000 lb

STATUS: 750,000 in 1985; some populations are declining

IUCN: Lower Risk—least concern   CITES: Appendix II

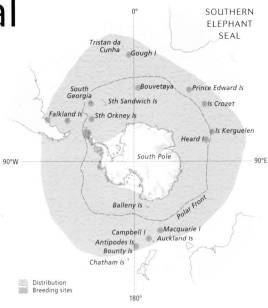

SOUTHERN ELEPHANT SEAL

Distribution
Breeding sites

▲ VOCAL REMINDER
A female elephant seal loudly claims her spot on the beach.

▼ UP CLOSE AND PERSONAL
No matter how large the beach, elephant seals usually search out others and lie so they are touching—they then spend their time grumbling and scuffling over positions.

The Southern elephant seal is the largest species of seal. Both sexes are solid-looking animals with a thick layer of blubber, but the bulls are very bulky, have exceptionally thick necks and chests, and are over nine times larger than the females. Despite their bulk, however, both sexes are surprisingly quick and nimble, even when they are out of the water. Around their necks, mature bulls usually show a great deal of heavy scarring from fighting, and cows may have some small scars on the head and neck incurred during mating.

The male elephant seal has an inflatable nose. This developed nose is a sign of adulthood, and it extends fully only when the bulls are completely mature, at approximately eight years of age. At this time, the tip of the nose can hang down below the animal's mouth, and its nostrils point straight down. The nose has a massively enlarged nasal cavity, which a breeding male can fill with air. The whole nose can then be made erect by increasing the blood pressure to the area. At this point, a large, raised cushion along the ridge of the nose is formed. The enlarged cavity may act as a resonating chamber for the male's roaring, which he uses to intimidate other bull seals.

Southern elephant seals spend most of their lives at sea in Antarctic and sub-Antarctic oceans, where they consume a diet that consists of cephalopods and fish.

They breed on islands situated on both sides of the Polar Front, on the Argentinian coast in Patagonia, and on the Antarctic mainland—though rarely. They occur in three discrete groups: a southwestern Atlantic population, centering on South Georgia, and including the Falkland Islands and Signy Island; a population in the southern Indian Ocean, based largely on Heard Island and Iles Kerguelen; and a population on Macquarie Island, south of Australia, and south of New Zealand. Non-breeding animals are regularly spotted along the coasts of Antarctica. One long-distance traveler was tagged in South Georgia and some time later spotted in South Africa—a massive swim of about 4,800 kilometers (3,000 miles).

➤ MISLEAD BY A NOSE
Only the male has the big inflatable
nose that gives elephant seals
their name. It grows as the male
matures, and reaches full size at
about eight years, but even then
is only fully enlarged during the
breeding season. When an
elephant seal snorts and grumbles,
his sounds are directed into the
resonating chamber of his mouth
and pharynx. This makes him
sound bigger and tougher than he
really is, and is intended to scare
off other bulls without resorting to
actual fighting.

The adult males are the
first to come ashore at the
beginning of the breeding
season, in late winter to early
spring (from late August to
September). The bulls stay on
the beaches for more than two
months, fighting to establish
and maintain dominance
hierarchies. Adult females
arrive a few weeks after the
males, in September or
October, and they remain for
four to five weeks. Within less
than a week of coming ashore,
the females give birth to the
pups that were conceived the
previous year. When ashore
the females form large
aggregations, known as
harems, which can contain more than 100 individuals,
although the average harem size is generally about forty.
Neither sex feeds while ashore, and both lose vast
amounts of weight.

There is generally plenty of space when the females
first come ashore, and so they begin in small, peaceful
harems, which are managed by the dominant bulls. Once
they have given birth, and more females come ashore,
the seals become much more aggressive. And then, as
the season progresses, more cows arrive and more pups
are born. The beaches become thronged with females

➤ NOT BUILT FOR WALKING
Young elephant seal pups are comparatively light and flexible,
but as they grow they quickly become wrapped in fat. Adult
elephant seals carry a lot of blubber, and have limited ability to
move on rough ground. When not in the water, they prefer to
remain near the sea, on gravel, sand or mud, and in flat tussock
grass areas near beaches.

▲ A WELL-DESERVED REST
Elephant seals are great divers, and of their time in the water, 93 percent is spent underneath it. The average dive duration for elephant seals is 22 minutes, the longest recorded dive being 120 minutes. The average depth of their dives is 400 meters (880 ft), but they regularly do down to 1 kilometer (²/₃ mile).

➤ HIGH ENERGY ACTIVITY
When elephant seals undergo a molt, they lose both hair and skin from all over their body at the same time. The energy cost of this process is very high because a lot of blood must be at the rapidly growing surface. The molt takes two weeks in adults and three weeks in sub-adults, and during this time they must fast.

A few days after mating, the females permanently abandon their pups and go back to sea to feed and regain the weight lost during the breeding season. At this time the cows are pregnant, but implantation and development of the embryo is delayed for four months, which ensures that the pups will not be born until eleven months later, when the mothers come back to the breeding beaches in the following season.

After their mothers have returned to sea, the pups congregate around the edges of the beaches. Together they learn to swim and hunt, first in nearby shallow waters and gradually further out to sea. They leave the beaches for the last time when they are two months old, usually a month after the departure of their mothers.

Bull Southern elephant seals retain their dominance, and control of their females, by fighting other males to maintain rank. When a challenger appears, he announces his arrival on the scene by uttering a huge roar in an attempt to intimidate his rivals with noise and bluster. If answering roars from the resident bull, known as the beachmaster, do not frighten off the intruder, a fight will ensue. The bull rears up in front of his enemy and strikes at his opponent's throat and neck with his teeth. The conflict is an awesome sight to behold and often leaves the opponents bloodied. However, the seals' thick necks, which are well covered in a layer of blubber, usually prevent any serious injury being inflicted upon either seal.

Early in the breeding season, big, experienced bulls dominate the beaches—a powerful bull being able to control a breeding territory that is home to as many as 60 females. As the harem grows larger, with the arrival of more females, a second male snatches the chance to mate with the new females as they come into estrus, and on very populated beaches, where more than 130 females may haul out, a third bull may be lucky enough to gain access to the newly arrived females.

A strong and dominant bull can keep his position at the top for the whole season, but this requires a great many battles, which are exhausting. Before the breeding period draws to a close, weaker males, particularly those that are older or younger, or have been injured, are likely to lose to a vigorous bull in his prime. However, as soon as the new bull takes up his dominant position, he, in turn, is certain to be challenged by other fresh males, each awaiting his chance to make a bid for the top spot.   LW

and pups fighting to claim space among the disorderly hundreds of animals. Meanwhile, the bulls fight constantly for dominance and attempt to keep control over as many cows as possible.

Newborns are suckled by their mothers for only 20 to 25 days, but during that period of time they gain 30–105 kilograms (70–230 lb), often tripling their birth weight. The cows produce as much as 5 liters (1¼ gallons) of milk each day and may lose up to a third of their body weight during their short nursing period. Right at the end of their nursing period, 17 to 22 days after giving birth, the cows come into estrus. Sometimes the bulls may weigh up to 25 times more than the pups. In their amorous enthusiasm, these large males scarcely seem to notice the pups, and can accidentally trample on the youngsters—some of whom may die.

▲ **DOMINANCE FIGHTING**
Fighting is the last resort of males, after a lot of posturing and roaring. Bulls rise up on their hindquarters and slam their chests into their opponents, reaching for vulnerable spots, such as the throat or head, to inflict a bite. They are able to do a great deal of harm to each other, but their thick necks, which are well covered in a layer of fat, are protected.

Whales, Dolphins, and Porpoises

Whales

ORDER: Cetacea

Previous pages

◀ INDIVIDUAL MARKINGS
A Southern right whale approaches. Biologists studying right whales use the unique patterns of their callosities—the white lumpy patches on the head—to identify individuals, and thereby to follow their movements and study their life histories.

Whales, dolphins, and porpoises belong to the Order Cetacea, from a Greek word meaning "large sea creature." And many are indeed large—no dinosaur ever reached the size of a Blue whale which, at 30 meters (100 feet) long, is the largest animal ever known to exist. Not all cetaceans are large, however; some dolphins reach only 1.2 meters (4 feet) but are still classified as whales. There are two sub-orders of whales: Mysticeti, the baleen whales, and Odontoceti, the toothed whales. Of the almost 90 known whale species, 11 are baleen feeders and 76 are toothed species, which are generally smaller than the baleen whales. Dolphins and porpoises are toothed whales.

All whales are carnivorous, but different species favor different types of food. In one of nature's paradoxes, the huge baleen feeders mainly live on zooplankton, filtering enormous quantities of these minute sea creatures from the water through comb-like plates in their mouths called baleen. Toothed whales use echolocation to hunt larger animals, such as fish and squid, and in Antarctic waters Orcas hunt warm-blooded prey such as penguins, seals, and even other cetaceans.

Whales inhabit all the world's seas, from shallow tropical waters to the deepest and coldest oceans. Their lives revolve around two major requirements—feeding, and breeding. The feeding season for most baleen whales is determined by the annual cycle of productivity in Antarctic waters. During the summer feeding season, sea ice is at its minimum extent, and zooplankton including krill are most abundant and easily found. During winter, when ice covers the sea, krill are more difficult to find, and the majority of baleen whales head north to warmer tropical or temperate waters to breed. The reason for this is still not certain, but it is thought likely that warm water allows calves to channel their energy into growth in early life, when their insulating blubber is thin. However, it is now known that some Minkes and Orcas, at least, remain in sea ice during winter, and may breed there. Because warm seas tend to be food-poor, most baleen whales have evolved a cycle of feast and famine, storing enough energy in blubber to fuel them during their winter migrations.

Living in water

Whales are mammals, and like all mammals they must breathe air. But because they have adapted over tens of millions of years to a totally aquatic existence, they differ from land mammals in many ways. Two of the important differences are how they breathe, and—a related problem—how they give birth in water.

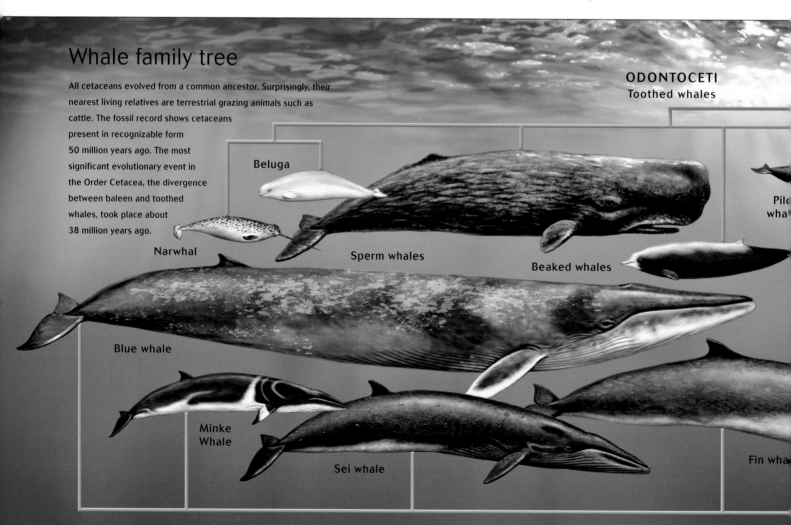

Whale family tree

All cetaceans evolved from a common ancestor. Surprisingly, their nearest living relatives are terrestrial grazing animals such as cattle. The fossil record shows cetaceans present in recognizable form 50 million years ago. The most significant evolutionary event in the Order Cetacea, the divergence between baleen and toothed whales, took place about 38 million years ago.

ODONTOCETI
Toothed whales

Beluga

Narwhal

Sperm whales

Beaked whales

Pilot whale

Blue whale

Minke Whale

Sei whale

Fin whale

Most mammals breathe through their mouth as well as their nostrils. As an adaptation to aquatic life, however, whales have an airway that is separate from their mouth, and their nostrils (the blowholes) are on the top of the head for efficient breathing when surfacing. This also means that baleen whales can breathe while simultaneously filling their mouths with water and food; which may be particularly important for suckling calves. Unlike humans, whales are voluntary breathers: they hold their breath and choose when to breathe. The breath is often exhaled explosively—the blow—followed by a very rapid inhalation before diving again. Newborn calves are often helped to the surface for their first breath, assisted in some species by "aunts," who accompany the mother during birthing.

In other adaptations to an aquatic lifestyle, whales have virtually no hair, no sweat glands, no external ears, no external male sex organs, and no hind limbs. Although they do not have gills or scales as fish do, they have developed some fish-like characteristics—a streamlined, torpedo-like form, for example. Unlike fish (which have vertical fins and tails that move from side to side), however, whales have horizontal tail fins, called flukes, that move up and down. Cetaceans have evolved to move through water with minimal turbulence compared with most other mammals.

Temperature control

Being warm-blooded, whales need to maintain a constant body core temperature. Water quickly draws heat from warm bodies, and some species which migrate to polar waters have to contend with sea temperatures as low as –2°C (28°F). To cope with this, they carry a warm coat of blubber, a mixture of fat, oil, and connective tissue, which lies under their skin. In some species, blubber may make up as much as 30 to 40 percent of total body weight, and is the energy store which fuels long winter migrations,

▲ SEASONAL MOVER
A Southern right whale near the Auckland Islands, a winter breeding area where mating and calving take place. In summer, Southern rights migrate southward to rich feeding grounds in the Southern Ocean.

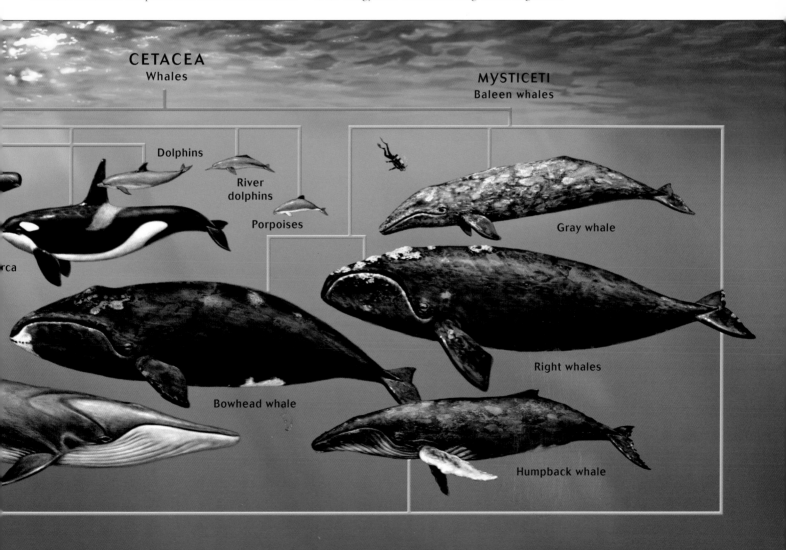

CETACEA
Whales

Dolphins

River dolphins

Porpoises

Orca

Bowhead whale

MYSTICETI
Baleen whales

Gray whale

Right whales

Humpback whale

▲ SURVIVAL SKILLS
Insulating blubber enables the Minke whale to live throughout the year in ice-packed Antarctic waters. The smallest of the baleen whales, Minkes spend their lives locating krill swarms beneath the ice, while at the same time playing a cat-and-mouse game of survival with their own predators—Orcas. Minkes can hold their breath for only 20 minutes or so, and may use echoes of their own calls to detect vital breathing holes.

including the production of calves and milk. Humpback whalers in New Zealand and elsewhere caught whales on their northward migration to breeding areas, while they were heavy with blubber, and ignored them on their southward migration at the end of winter, when they were thin and hungry. Many species that migrate to tropical waters to give birth may do so for the benefit of their infants: the calves have time to grow a layer of blubber before encountering colder waters for the first time. On the other hand, some species (including Orcas, Bowheads and Narwhals) live their lives in icy waters and bear their calves there, and these calves are quite capable of surviving these conditions.

Whales also have a heat exchange mechanism in the circulatory system that helps to regulate their body temperature. In mammals generally, warm blood travels to the extremities and cooler blood, stripped of its oxygen, returns to the lungs to be replenished. The whales' blood system is different; they have inter-woven inward and outward blood vessels. When they are cold, they can remove the heat from the blood traveling to the extremities and bring it back to the core to keep their vital organs warm. Conversely, when a whale becomes too hot, it is able to direct warm blood to its body surface, especially the fins. The colder water absorbs the heat, and the cooled blood then returns to lower the temperature of the internal organs.

A consequence of large size in whales is that larger bodies conserve heat more efficiently because of the comparatively low ratio of skin surface to body mass. Animals with a larger body mass have proportionally less surface area to lose heat through than small animals. One possible cause of the evolution of large size is to enable whales to spend more time in cold, food-rich waters.

Cetacean anatomy

Whales moved into the oceans millions of years ago, and they have evolved adaptations that are not seen in land animals. For example, they have no gall bladder or appendix, as humans do. They also do not have carotid arteries; instead, they have small blood vessels, the retia mirabilia, that carry blood to the brain and may have a role in

maintaining consistent pressure in the blood flowing to the brain, particularly during deep dives.

Some whales have become so large because living in water, they are not as subject to the constraints of gravity as terrestrial animals are. When large whales strand, they may be crushed under their own weight. The bones of whales are spongy and filled with oil, which probably increases their buoyancy, although most whales will sink when dead. The skeletons of whales are flexible in the vertical plane, but generally they lack sideways flexibility. Neck vertebrae fuse during maturation to support the head as it travels forward through the water. The head itself has become elongated during the course of evolution, to support the feeding structures such as baleen or an array of teeth. The front limbs have become flattened paddles, which are used for steering and control, yet their bone structure still shows their common mammalian ancestry in the form of five "fingers." Towards the tail, the vertebral column becomes more tapered and flexible, permitting the vertical movement used for locomotion. The flukes themselves are not supported by bone, but by muscle and firm, elastic connective tissue. Whales do not have any trace of leg bones, and the vestiges of the pelvic girdle (the hips) have dwindled to two small bones that are embedded in the body wall rather than attached to the skeletal structure. In land mammals these bones support the hind limbs, but in whales their only purpose is to support the muscles of the male's external reproductive organs, which are usually hidden in a slit that runs along the underside of the body.

Taking to the water

Prehistoric cetaceans are among the earliest recorded marine mammals, and fossil cetaceans have been found on all continents. Aquatic dinosaurs died out 65 million years ago, leaving oceanic niches to fill. Mammals took to the water, and cetaceans were already present 50 million years ago. There are fossils of animals with webbed hind feet—semi-aquatic precursors of whales—and there is evidence that whales descended from a wolf-like, hoofed land mammal with large teeth, also thought to be an ancestor of even-toed hoofed mammals such as cattle. (An alternative. but less likely theory holds that whales are more closely related to odd-toed hoofed mammals such as horses.) Cetaceans gradually became more efficient in water; by 38 million years ago, toothed and baleen whales had evolved their main characteristics, and 13 million years later different species had evolved within the two groups. About 15 million years ago, climatic cooling produced the nutrient-rich polar seas of today, and larger whales seem to have evolved as a result of these conditions. Modern groups of whales had largely emerged by approximately seven million years ago.

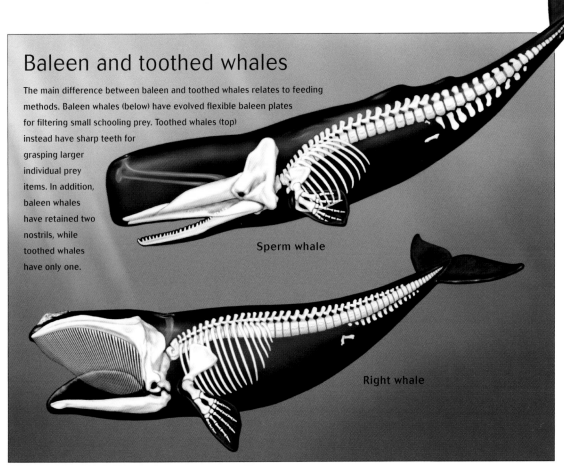

Baleen and toothed whales

The main difference between baleen and toothed whales relates to feeding methods. Baleen whales (below) have evolved flexible baleen plates for filtering small schooling prey. Toothed whales (top) instead have sharp teeth for grasping larger individual prey items. In addition, baleen whales have retained two nostrils, while toothed whales have only one.

Sperm whale

Right whale

▼ TO DARK AND ICY DEPTHS With flukes raised, a Sperm whale begins a deep vertical dive to feed on squid. Sperm whales reach phenomenal depths—some 3,000 meters (9,850 feet) or more—and can remain submerged for more than two hours. However, most dives are much shallower and shorter. Whales use loud clicks both to communicate with each other and to echolocate their elusive, fast-moving prey.

Baleen whales

Only about 15 percent of whale species are baleen whales—but what they lack in numbers they make up for in size. Even the smallest baleen whale, the Pygmy right whale, reaches 6 meters (20 ft), and the mighty Blue whale grows to about 30 meters (100 ft).

Unlike the toothed whales, which have only one blowhole, baleen whales have two blowholes protected by a raised splashguard, or coaming, that prevents water from entering the blowholes while they are breathing.

Baleen whales are named for the comb-like baleen plates that hang from their upper jaws, which are used to strain their prey from seawater. The baleen whales are divided into two groups: rorquals (the groove-throated whales) and right whales. Right whales have enormously long baleen, and continuously strain prey as they swim, while most rorquals rapidly engulf huge mouthfuls of food, extending their throat capacity by ballooning out their throat pleats, creating a huge sac containing water and food. The muscular pleats are then contracted, squeezing water through the filtering baleen while their prey remains inside.

In order to catch enough food, these whales need enormous jaws that can provide enough room for the filtering baleen. This requirement has resulted in major modifications to the shape of the baleen whale skull, which is very different to that of the toothed whale. The head of some species, such as right whale, may be as much as a third of its body length. Each species of baleen whale has a different color, thickness, and number of baleen, depending on feeding habits. Coarse baleen filters out larger prey, whereas finer baleen filters smaller prey. Right whales, which generally hunt very small prey, have long, very fine baleen as well as a very bowed upper jaw to accommodate the length of baleen.

While baleen whales are known for feeding on zooplankton, they also devour small fish, particularly schooling species like herring, cod, mackerel, capelin, pilchards, and sardines. These species, or bottom fish like cod and sand lance—and even the occasional squid—make up the bulk of baleen whale food in areas, especially in the northern hemisphere, where small schooling fish are abundant. However, in the cold southern waters of Antarctica, zooplankton, particularly krill, is the whales' principal food source.

The distribution of krill—and of baleen whales—is determined by the dynamic oceanographic processes which influence phytoplankton production. These minute sea plants—and krill—live under and among sea ice during winter, and as the ice melts back during spring

▲ BREATHING EASY

A rorqual surfaces, showing the twin blowholes characteristic of baleen whales. Nostrils on the top of the head enable these whales to breathe while mostly submerged. Many fast-moving species also have a raised splashguard (seen here), so they can breathe while swimming, even in rough seas.

Hearing under water

Sound travels about four and a half times faster in water than it does in air, and much further; vision, on the other hand, is poor, especially in deep or murky water. So it seems likely that hearing plays an important role for species that live mainly under water, conveying information rapidly and from great distances.

As whales' internal ear bones are well developed, hearing is probably important to them. However, whales may not hear as humans do; that is, through their ear canals. In toothed whales a waxy plug blocks a narrow ear canal that does not appear to be connected to the eardrum, which suggests that their hearing mechanism is not like that of humans. Scientists suspect that sounds may be transmitted to the whales' internal ears through soft tissues, possibly the fatty deposits in their chins, or through their skull bones. It may be that under water the tissues or bones receive and transmit sound more effectively than would the ear canal.

The internal ears of a whale are protected by foamy surroundings that reflect sound, and this may form a sophisticated stereophonic sound system that allows whales to use echoes to pinpoint a mate, an enemy, or their prey.

and summer, migratory baleen whales follow its retreat to feed. While it was long considered that the southern seas were largely empty during winter, recent studies have shown that phytoplankton, krill, and Minke whales (which are small enough to live in the open cracks in the ice), are present year-round.

Whale talk

All cetaceans produce sound for communication, and at least some for echolocation. Toothed whales, which are known to echolocate, generally use higher frequencies than do baleen whales, which are not yet proven to echolocate. Higher frequencies travel only short distances, while very low frequencies may travel hundreds or even thousands of kilometers. Some dolphins produce sounds which are too high-pitched for the human ear, while the deep rumbles of Fin and Blue whales are below our hearing range. In between, most species produce sounds that can be heard by the human ear. Many of us are familiar with the haunting songs of Humpback whales, or the whistles and clicks of bowriding dolphins.

Size matters

Female baleen whales are larger than males; a mature female Blue whale, for example, is 2 meters (6 ft) longer than a male. This sexual dimorphism is thought to be linked to the migratory habits of the whales: females, which fast during long winter migrations, must have the energetic reserves not only to travel from feeding grounds to breeding grounds while fasting, but also to produce and suckle a calf while doing so. The greater size of females enables more energy supplies to be carried, in the form of blubber.

Whale "bone"

During the nineteenth century, when commercial whaling was at its peak, baleen was in great demand. Though marketed as whalebone, baleen is not actually bone but a fibrous material similar to fingernails. Because it is both stiff and flexible it was an ideal material for the manufacture of umbrella ribs, corsets, skirt hoops, watch springs, and brushes. Happily for the world's whale population, plastics have replaced baleen in modern industries.

▼ MASSIVE MOUTHFULS
Two Humpback whales surface feed on krill. Both whales have rolled onto their right sides and, with mouths wide open, they lunge through their prey. The rear whale's bristly fringe of baleen is clearly visible. Humpback whales often form pairs in feeding grounds and may cooperate to herd and capture prey.

Right whales

SUBORDER: Mysticeti

FAMILIES: Balaenidae, Neobalaenidae

Right whales were so named by whalers because they were the "right" whale to hunt—they inhabited coastal waters, they were slow swimmers, they floated when dead, and they provided valuable baleen and high quality oil. All these factors contributed to the mass slaughter of right whales in the nineteenth and early twentieth centuries.

A right whale has a deeply curved jawline to provide room for its long, slender baleen plates. This feature—and its large head, which is up to a third of its body length—make it easy to identify. It uses its fine baleen to filter copepods, krill, and other small crustaceans from the water, generally preferring to skim feed near the surface with its mouth open. Unlike the rorquals, right whales have no throat pleats, and so cannot expand their throats to maximize a mouthful when feeding. They are quite stocky, have no dorsal fin, and are usually almost completely black-brown, except for white patches called callosities around their head, and white belly patches. Only Southern right whales and the two northern right whale species grow callosities, and so their presence is a positive identification of a right whale.

Callosities are raised, horny layers of skin that may be heavily infested with barnacles, parasitic worms, and whale lice (cyamid amphipods). They are present from birth, and males have more than females; possibly they function as

weapons against other males. Early whalers gave names to the various patches—the largest, on the tip of the upper jaw, they called the bonnet. Other patches were described as the beard, the eyebrow, and the lip patch, and any further patches were called islands. But each right whale has a unique pattern, and can be recognized by its callosities, rather as humans can be by their fingerprints.

The five right whale species—the Bowhead, the North Pacific right, the North Atlantic right, the Southern right and the Pygmy right—occupy the earth's cooler seas. The Bowhead is restricted to Arctic waters, and the two northern rights to their respective ocean basins. The Southern right and Pygmy right whales share southern cool temperate waters, with Southern rights penetrating further south into polar waters.

▲ BREACHING
A Southern right whale breaching. It is believed the main reason for breaching is communication with other whales—the thump can be heard more than a kilometer (²⁄₃ mile) away. Breaching may also be to remove parasites or dead skin, to observe objects above the surface, or simply to express exuberance. Whales may breach when threatened by a predator, another whale, or a boat that approaches too closely.

Parasites

Larger whales all collect various external parasites, including diatoms, whale lice, and barnacles. Many species wear a thin brownish-yellow layer of diatoms—a type of algae—over their skin during the polar summer, and scientists have used the presence of a particular diatom species to infer migratory movement of whales from feeding to breeding areas.

Cyamid amphipod crustaceans, better known as whale lice, are found on most large whales, but especially on Gray, Humpback, and right whales. These small creatures congregate on parts of the whale's body where water flow is such that they are unlikely to get washed off—for example, in the throat grooves, on the margins of the lips, and particularly on callosities, the rough white patches unique to right whales. They feed mainly on shedding skin, but they also rapidly move in on flesh wounds. Surprisingly, whale lice cannot swim, and must move to a new host by jumping from one animal to the next when they are touching. Whale lice often blanket callosities surrounding

sensory hairs, and so little is understood about their relationship with their host that it is uncertain whether they help or hinder the function of these hairs. They may interfere with the functioning of a sensory hair by blocking or bending it, but it has also been suggested that they may help the whale to locate plankton swarms by becoming excited when a food source is in the offing.

Many different types of barnacles colonize whales, but they do not appear to do any serious harm to their host. Acorn, stalked, and pseudostalked barnacles all attach themselves to favorite species and feed on the microscopic life associated with whale skin, or on drifting plankton. Slow movers such as right whales, Grays, and Humpbacks grow more barnacles, (and carry more cyanids) than faster moving species; in cool waters, a Humpback may accumulate as much as half a tonne of barnacles. It must be a relief to the whale when it moves to warmer waters and most of these passengers drop off.

Southern right whale

EUBALAENA AUSTRALIS

FAMILY: Balaenidae
LENGTH: female 14 m (45 ft); male 13.7 m (44 ½ ft)
WEIGHT: female 23 tonnes, male 22 tonnes
STATUS: approximately 7,000, increasing; IUCN: Lower Risk—
    conservation dependent   CITES: Appendix I

The Southern right whale is often consigned to the same species as the North Atlantic whale (*Eubalaena glacialis*), but the two species never mix, and recent research makes it clear that the groups probably separated as much as two million years ago.

    Southern right whales range from Antarctic to temperate waters. The species reaches the northern end of its distribution during winter, which they spend off the coasts of the southern parts of Africa, South America, New Zealand, Australia, and sub-Antarctic islands. During their breeding season in winter, Southern right whales come very close to land, preferring shallow protected bays. They then travel south to feed in the plankton-rich sub-Antarctic waters during summer. Southern rights have shorter migration routes than most other baleen

▲ SKIMMING THE SURFACE
Southern right whales spend most of their time near the ocean surface where krill and other small crustaceans swarm. When feeding, the whale opens its enormous mouth and swims slowly, continuously filtering prey through its very long baleen as it moves along. Among the fattest of whales, the thick blubber coat that warms Southern rights in cold waters also equips them for their buoyant life at the surface.

Courtship and mating

No whale species mates for life, and both males and females will have several partners. In the first stage of courtship, the female signals that she is in estrus and ready for a mate—possibly by releasing hormones in her urine, but this is speculation. Male Humpbacks signal their availability by singing, but once the female has chosen him, he falls silent, perhaps to avoid attracting the attention of other males who will attempt to oust him through aggressively competitive behavior such as fluke slapping. The mating pair are less rough with each other, but they do engage in flipper slapping, which is believed to help synchronize the pair. When right whales mate it is common for a second or even third male to be present.

▼ **MOTHER AND CALF**
A Southern right whale and her white calf. The calf is not an albino, but is probably a male with a recessive gene for pale skin (white female calves are not known). The color will darken to a milk-chocolate brown during its first year of life, but never to the brown-black of its mother.

whales, because they do not migrate to tropical waters. As their numbers are increasing, however, there is evidence that their migratory range is expanding into both warmer and colder seas—there have been increased sightings along the Antarctic sea ice edge in recent years. It is thought that they may now be re-occupying habitat that they have avoided since the whaling era when their numbers were so drastically reduced.

Right whales are comparatively slow swimmers, seldom exceeding 8 kilometers per hour (5 mph). While they are often regarded as placid animals, they are capable of energetic displays such as lobtailing (slapping their flukes on the water), breaching (leaping from the water), or other social behaviors. Calves may splash about boisterously, breaching or lobtailing repeatedly, and may pester their mothers for a feed. On the other hand, Southern right whales may use subtle changes in posture to signal information. For example, they will sometimes warn an approaching whale simply by shifting their body

so that their flukes—their main defensive weapon—face the intruder. This symbolic action is usually sufficient to deter further unwanted approach. Other baleen whales inflate their throat pleats aggressively, or let loose a stream of bubbles to drive other whales away. A further weapon in the whale's armory is explosive blowing in order to signal alarm or annoyance when another whale—or even a boat—comes too close.

Right whales have an extensive repertoire of sounds, which include a loud bellow that is sounded when their blowholes are above water, and which can be heard several hundred meters away. Southern rights also produce low-frequency sounds, which have been recorded during travel, courtship, and play, but these are not prolonged and repetitive like Humpback songs. However, they do make grouped sounds called belches or moans, ranging from 50 Hz to 500 Hz, for up to a minute. The purpose of these different noises is not yet known, nor is it understood why Southern right whales appear to be most vocal at night.

Pygmy right whale

CAPEREA MARGINATA
FAMILY: Neobalaenidae
LENGTH: to 6 m/20 ft
WEIGHT: up to 3–3.5 tonnes
STATUS: unknown but not rare
    IUCN: Lower Risk—least concern   CITES: Appendix I

Pygmy right whales appear to be part way between a right whale and a rorqual. They have a comparatively slim build, more reminiscent of rorquals than right whales, and shorter baleen than the other right whales. They also have a dorsal fin (not present in other right whale species), a comparatively small head for their body size, and four fingers instead of five. Another feature they have in common with rorquals is throat grooves, but whereas rorquals have many, they have only two, and other right whales have none. However, they do have the characteristic bowed upper jaw of right whales, and very similar coloration, being dark brown or black on top, with a slightly paler underbelly.

Very little is known about this animal, which is clearly not very closely related to other right whales. The Pygmy right whale is the smallest of the baleen whales, and is thought to be mostly pelagic, ranging through the cold and temperate waters of the southern oceans. It has very fine bristles in its baleen, which suggests it prefers very small prey such as copepods. While Pygmy rights were long thought to be solitary animals, they have been sighted in large, sociable groups, confounding yet another long-held belief about a cetacean species. They are not believed to carry out seasonal migrations, perhaps remaining in temperate waters year-round. Available data on strandings suggest that younger animals are more likely to roam near the coasts of South America, South Africa, Australia, and New Zealand.

△ INSEPARABLE
The close bond between Southern right whale mothers and calves made them easy targets for nineteenth-century whalers. Newborn calves will not leave their mother's side, but after a few months they will begin to venture further afield, in preparation for weaning at 12 months. Whale watchers can once again observe Southern right mothers and their calves at breeding grounds, as the species slowly recovers from near-extinction in the 1800s.

Whale footprints

Whale watchers sometimes follow whales by tracking their "footprints"—areas where a whale has surfaced and left behind disturbed water with an oily residue. A whale's blow looks like spouting water, but it is really water vapor condensing in the warm air breathed out by the whale. With the air, the whale exhales an oily substance that is probably the foamy mucus from its sinuses. The oil settles behind the whale, leaving the water slightly greasy. In calm water this may remain visible for several minutes.

Rorquals

SUBORDER: Mysticeti
FAMILY: Balaenopteridae

▽ DRAWN BY CURIOSITY
Inquisitive Humpbacks investigate the photographer's boat in the placid waters of the Antarctic Peninsula. The whale on the left is spyhopping (raising its head above the surface to get a clearer view), while its companion prepares to dive under the boat. Humpbacks are among the most curious of whales; if they have no pressing business—such as feeding, courtship, or travel—they will often spend considerable time around and under vessels.

Rorquals are the most abundant and diverse group of baleen whales. Their name comes from the Norse word for a tubed or furrowed whale, which refers to the grooves, or pleats, that run the length of the throat and which allow it to expand when the whale is feeding. With their mouths open wide, rorquals engulf huge mouthfuls of water containing prey, then pump out the water while retaining their prey inside their filtering baleen, which is much shorter than that of right whales. The rapid expansion of throat pleats to accommodate their food may give rorquals such as Blue whales the appearance of giant tadpoles when seen from the air. Otherwise, most rorquals are slender and fast-moving, except the stocky, slower Humpback, which is unique in many respects, including its exceptionally long flippers and its head tubercles.

◁ CUTTING THROUGH WATER
Minke whales in the Antarctic sometimes behave like giant dolphins, racing in to join a ship and riding at its bow or in its wake. Minkes are thought to be one of the fastest whales as can be seen from this photograph—a sheet of water is jetted out to each side as the Minke's head re-enters the water at speed.

There are six species in the Family Balaenopteridae. The Minke, Blue, Fin, Sei, and Bryde's whales form the genus *Balaenoptera*, while the Humpback has its own genus, *Megaptera*. The size difference among Balaenoptera is huge: an adult Minke, at 10 meters (33 ft), is only fractionally larger than a newborn Blue whale, which may grow to 30 meters (100 ft). Most rorquals migrate seasonally between warm waters, where they breed and calve, and cool temperate to polar waters, where feeding occurs. The exception is Bryde's whale, which is found only in tropical and warm temperate waters. During the nineteenth and twentieth centuries, the predictability of the timing and destinations of rorqual migrations made it easy for whalers to locate aggregations of the larger species, which were heavily hunted as a result.

Rorqual life cycles

Whalers believed that the various species migrated at slightly different times of the year, with Blue whales leaving Antarctic waters first, followed by Fins and Humpbacks, with the Sei whales being the last to depart. While they are migrating, rorquals may segregate according to both age and sex. In Humpbacks, lactating females leave first, followed by immature animals, adult males, non-breeding adult females, and finally pregnant females. Blue whales and Minkes tend to feed near the edge of the sea ice, but Fins and Seis generally feed further north. Humpbacks may be found between the ice and ice-free offshore areas.

The yearly cycle of rorquals is centered around migration. Conception and birth generally fall within the three to four months that the whales spend in warm waters. Rorquals usually mate when in warm waters during winter, and then migrate to their polar or cool temperate feeding grounds where they spend three to six months feeding. With summer over, they migrate back to the tropics where, eleven to twelve months after mating, females give birth to single young. After about three months, the newborns accompany their mothers on the long migration south, nursing as they travel. Whale calves learn to feed themselves in the rich southern feeding grounds, where there is plenty of prey on which to hone their hunting skills. Some species, such as Blue whales, abandon their young at this point, after only seven months of life, while others, including Humpbacks and Southern right whales, do not do so until they have returned to their wintering grounds and the young are nearly one year old. In later life, calves often winter in the area they were born.

▲ MOTHER CARE
A Humpback cow and her new-born calf at ease in the clear waters of a breeding area. Suckling its mother's fat-rich milk, the calf will grow at a phenomenal rate. Within a few months it will be large and strong enough to begin the migration to Antarctic feeding grounds. However, the strain of calving migrations on the mothers is so great they may lose up to 40 percent of their body weight.

Temper tantrums

The bond between a whale mother and her calf is a strong one, and parent and offspring regularly nuzzle and touch each other. Mother whales will place themselves between their young and any perceived threat, such as a boat or an Orca, to protect their young. But life is not easy for mothers of young whales. Should they refuse their calves a feed, they may be subjected to behavior that any parent would recognize: head butting, fluke slapping, and very large temper tantrums. Whale mothers occasionally have to resort to using their pectoral fins to restrain unruly young.

▲ BREATHING SPACE
Minke whales in Antarctic
waters are superbly adapted
to spending their lives in heavy
pack ice. They feed on krill,
constantly moving between
breathing holes that open and
close under the influence of
winds and currents. This whale
has raised its snout through
brash ice, which is no real
obstacle, but sometimes
breathing holes are so
small that Minkes must rise
through the ice vertically to
draw breath.

Minke

BALAENOPTERA BONAERENSIS

OTHER NAMES: Lesser rorqual, Piked whale, Pikehead
LENGTH: female 10 m/33 ft, male 8 m/26 ft
WEIGHT: up to 13 tonnes
STATUS: northern hemisphere: 60,000–80,000
   southern hemisphere: 300,000–1 million
IUCN: Lower Risk—near threatened   CITES: Appendix I

Minkes are found in all of the world's oceans, and they
are equally at home in the Arctic, the Antarctic, and the
tropics. While many inhabit the open ocean, they are just
as capable of living in the heaviest of Antarctic sea ice.

Growing to only about 10 meters (33 ft), the Minke
whale is the smallest of all the rorquals. It is very sleek in
shape and its head is sharply pointed. Its body is dark gray
on top, graduating through various shades to a pale underside, and it often has a whitish patch on its flipper. At birth,
it is only 2.5 meters (9 ft) long, which is about the same
size as some of the smaller toothed whales, such as the
Pygmy sperm and Dwarf sperm whales. Unlike the larger
rorquals, who breed only every two or three years, there is
some evidence to suggest that Minkes probably have an
annual breeding cycle.

Minkes mainly eat krill, but will also eat fish and the
occasional mollusk. In Antarctic waters they consume a
larger proportion of krill than other Minke populations,
due to the plentiful supplies of krill in the Antarctic
ecosystem. Sightings of Minke whales in winter sea ice
suggest they are perfectly capable of feeding in the
winter, when krill is thought to be relatively difficult to
catch. They sometimes herd prey into tight aggregations
before engulfing them in great mouthfuls. At times,
hundreds of Minkes may gather to feed on very large
aggregations of krill. Minkes are sometimes drawn
to ships, racing in from a distance to swim ahead or
alongside of the ship, before quickly losing interest and
drifting off.

Because they are relatively small, Minkes were the last
of the Antarctic rorquals to be hunted, whalers preferring
larger whales that netted better returns per catch, and
therefore increased profits. Now Minkes are pursued by
the Japanese whaling industry for "scientific whaling," and
commercial whaling may soon resume in the Antarctic.
Norway has already resumed limited commercial whaling
in its waters. Another threat which Minkes and other
whales may face in the future is increased exploitation of
Antarctic krill as a food source or as fertilizer.

Do Minkes echolocate?

Toothed whales use clicking sounds, generated by forcing air between their blowhole and their larynx, to "sound out" their environment, but baleen whales lack the physical features required for this form of echolocation. However, it is possible that they have developed their own unique form of echolocation. Some rorquals utter repeated sounds, at a frequency that may be effective for detecting large, slow-moving objects such as fish schools and krill swarms.

Sei whale

BALAENOPTERA BOREALIS

OTHER NAMES: Pollack whale, Short-headed sperm, Japan finner, Sardine whale
LENGTH: female 17 m/56 ft; male 15 m/50 ft
WEIGHT: 20–25 tonnes
STATUS: unknown population, thought to be greatly reduced from original numbers
IUCN: Endangered   CITES: Appendix I

Sei whales are a worldwide species that occupy cold, temperate, and tropical waters—although some tropical sightings have been confused with the tropical species, Bryde's whale (*B. edem*). The two species are of similar size and appearance, both being dark gray on top and paler below. Some southern hemisphere Sei whales are slightly larger than their northern hemisphere counterparts.

Sei whales will devour almost anything they can find, but they seem to prefer smaller species such as copepods and amphipods. They will consume larger crustaceans such as krill, as well as fish and squid, but they are basically adapted to hunt smaller species. One of the favorite techniques of these whales is skimming, where they swim along the surface, twisting through the water and skimming zooplankton into their open mouths. When feeding like this, they do not take individual mouthfuls of water, but instead constantly push water through the baleen by the pressure of swimming. This technique is also used by right whales.

Seis are often found away from the coastal edges of the seas, in the open ocean, where they gather along oceanic fronts in pods of three or more. While most rorqual species have been observed to usually form groups of ten or less, it may be that small group size is a consequence of whaling, and that rorqual groups may have been larger before the days of mass whaling. However, on feeding grounds, group size may be a result of prey availability and patchiness: for example, Humpback whales are not usually found in groups larger than two, despite a considerable population increase in recent decades, because this appears to be an optimal number of animals feeding on

small, scattered prey swarms. It remains to be seen whether rorqual pod sizes will become larger as populations increase with conservation measures.

Fin whale

BALAENOPTERA PHYSALUS

OTHER NAMES: Common rorqual, Finner, finback, Razorback
LENGTH: female 26m/85ft, male 21m/69ft
WEIGHT: up to 80 tonnes
STATUS: northern hemisphere population estimated at 40,000; southern hemishere: 5,000–30,000
IUCN: Endangered   CITES: Appendix I

Fin whales are dark gray above, with swirling patterns known as chevrons on their backs, and white underneath. Although they are the largest whale after the Blue whale, they are slimmer, and not particularly heavy given their size. It is the only whale with a distinctive asymmetrical body coloration, with its right lower jaw being white, while the left side is dark. The baleen plates also follow this pattern. This coloration may be an advantage when feeding as it possibly helps to confuse their prey.

Fin whales range from polar to tropical waters. There are clearly discrete populations which are highly migratory, and many individuals have been identified returning to mate in their mother's breeding grounds. Fin whales are

⌄ **WARM-WATER COUSINS**
These Dwarf minke whales are best known from the Ribbon Reefs north of Cairns, Queensland, where tourists are able to swim with them. Only recently declared a separate subspecies, Dwarf minkes are thought to inhabit mostly temperate to tropical waters. However, they are sometimes seen further south in polar waters, the haunt of their larger cousins, the Antarctic Minke whales.

▷ **SLEEK AND FAST**
A Sei whale, recognizable by its tall, hooked dorsal fin, surfaces to breathe. Midway in size between a Blue whale and a Minke and one of the most diverse feeders of the baleen group, Sei sometimes exploit copepod swarms in cool-temperate or sub-Antarctic waters, and at other times they move south into icy waters to feed on krill. Little is known of their social organization, migrations, or breeding grounds.

normally seen in pods of six to ten, but they are also often found in pairs. The size of the group may be related to the amount of food in the area, or to courtship and mating.

The Fin whale likes fairly deep water, and is usually found outside continental shelves. However, there have been recent sightings of Fin whales in waters less than 200 meters (655 ft) deep. Faster than most rorquals, the Fin whale can swim at 37 kilometers per hour (23 mph) for short periods; however, it can maintain a steady pace of 22 kilometers per hour (14 mph) for weeks. This active whale also dives for longer than some species, and will submerge for up to 30 minutes—particularly when feeding. Dives may reach 500 meters (1,640 ft), which is exceedingly deep for baleen whales. The Fin whale prefers a wider range of prey species than the Blue whale. In the northern hemisphere it consumes krill and fish, such as Capelin and herring.

▼ TELLTALE BLOW

The blow of the Blue whale is unmatched in height and power. Keen-eyed Antarctic whalers could spot these blows from a great distance, a factor that contributed greatly to the near extinction of Blue whales during the early twentieth century. Now researchers, not whalers, scan the waters, hoping to learn more about this magnificent species and so help protect it into an uncertain future.

▼ WHY SO BLUE?

Seen from space, the earth is blue. Seen underwater from above, a Blue whale is also blue. This is puzzling, as the true skin color of a Blue whale, when viewed above the water surface, may vary from silver to almost black. Yet once they slip beneath the waters there is an instant transformation, as if a light is switched on.

Blue whale

BALAENOPTERA MUSCULUS

OTHER NAME: Sulfurbottom

LENGTH: female 31 m / 102 ft, male 30 m / 100 ft

WEIGHT: 100–120 tonnes, possibly up to 150 tonnes

STATUS: unknown population, but probably less than 10,000

IUCN: Endangered   CITES: Appendix I

The Blue whale is probably the largest animal ever to have lived; *Brachiosaurus*, the biggest known dinosaur, reached a length of only 23 meters (75 ft), compared with the Blue whale's 30 meters (100 ft). The largest accurately measured Blue whales were both females. The longest had a body length of 33.6 meters (110 ft), and the heaviest weighed 190 tonnes and measured 27.6 meters (90 ft). Larger animals live longer than smaller ones, and the Blue whale is no exception; some individuals are thought to have lived for 110 years.

Blue whales are a mottled slate-gray, but the reason for their name becomes apparent when they submerge and their skin appears a pale, luminous turquoise-blue. In Antarctic waters their skin even takes on a yellowish-brown hue from the diatoms that grow on them. Blue whales may be confused with other rorquals, but their main distinguishing features are their size, the position of their dorsal fin, their generally paler color, and their spotted appearance.

Although it is so large, the Blue whale is sleek and slender. Like most rorquals, it is streamlined and is a comparatively fast swimmer, having a torpedo-shaped body with no unnecessary appendages that could drag. In both males and females, a slit along the underbelly contains and protects the sexual organs when not in use.

In their internal structure, the testes of a mature Blue whale male are identical to other mammals but may be as long as 75 centimeters (25 in) and as heavy as 37 kilograms (82 lb), and the penis can be as much as 3 meters (10 ft) long and 30 centimeters (12 in) in diameter. This largest of whales produces very large babies—about 7 meters (23 ft) long.

The Blue whale is a cosmopolitan species, living in all oceans, and ranging from polar to equatorial waters. When in Antarctic waters, it will follow the ice edge seeking krill swarms. It goes deeper into the Antarctic ice than all other baleen whales, except Minkes, and like most migrates from the poles to the tropics at summer's end. Blue whales seem to gather in low numbers, which may be the effect of slow recovery from large-scale whaling.

Several breeding stocks of Blue whales may mingle in the polar regions while feeding, but they part company as they return to their various breeding grounds. Some Blue whales may live in warmer waters and not migrate to the poles at all. While some Blue whales are feeding in temperate waters, such as those off southeast Australia, others are feeding along the ice edge much further south.

Blue whales are regarded as fast swimmers, and have been timed at up to 48 kilometers (30 mph) when alarmed. They are relatively shallow divers, but feeding dives to 360 meters (1,180 ft) have been recorded, and they are capable of submerging for about 20 minutes. When they surface, their blow can reach a spectacular 10 meters (33 ft). When traveling, they may blow eight to fifteen times over a five-minute period before diving again.

Blue whales, the largest of all animals, prefer one of the smallest for their food, specializing almost entirely in euphausiid crustaceans, commonly known as krill. To satisfy the needs of its huge body, an adult Blue whale must catch a total weight of around 3.6 tonnes of krill each day. Blue whales' energy requirements are so huge that, unlike other rorquals, they probably search out tropical breeding areas where food is abundant.

Like many rorquals, the Blue whale uses the gulping technique to feed, expanding its throat pleats to maximize the capacity of its mouth, and then simultaneously closing its mouth, contracting its throat pleats and raising its tongue to force the water out while retaining the food. A single mouthful of food and water may weigh as much as 45 tonnes. Gulp feeding may occur at any depth, depending on the prey, and the whale may modify the technique to concentrate the food, forcing prey back against the surface of the water, an iceberg, or land.

Humpback whale

MEGAPTERA NOVAEANGLIAE

LENGTH: 15 m/50 ft

WEIGHT: generally 25-30 tonnes; up to 36 tonnes

STATUS: possibly 20,000–30,000

IUCN: Vulnerable   CITES: Appendix I

Humpbacks are much heavier in build than other rorquals and are slow swimmers, and as a result, they were heavily hunted by whalers. Though only 15 meters (50 ft) long, they can weigh up to 40 tonnes. A distinctive feature is their long pectoral fins—much longer than those of any other rorquals, reaching about one-third of body length. These fins make them very maneuverable and, despite their stocky appearance, they are graceful acrobats underwater.

While their coloration can vary from all black to all white, they tend to be dark above and white below. The undersides of their tail flukes, and often their flanks too, are pigmented with distinctive patterns which enable photo-identification of individuals. They hunch their back when they dive—hence their common name—and there are a series of lumps, or tubercles, around their head. They are one of the few whales to grow any hair, and young Humpback whales may have a single hair growing in the middle of each tubercle.

▲ MAKING A SPLASH

A young Humpback whale breaches, twisting as it breaks free of the water, so that its muscular back, and not its soft underparts, bears the brunt as it crashes back to the surface. Its expandable throat pleats are visible, as are the numerous barnacles on its chin and pectoral flippers. Breaching is a learned behavior: young calves may sometimes be seen breaching in synchrony with their mothers.

▶ A WHALE WATCHING
A Humpback whale eyes the photographer, displaying the curiosity typical of this species. Two unique features of the Humpback are visible here: the extremely long flippers and the raised tubercles on the head. Humpbacks have fewer throat pleats than other rorquals, but their function remains the same—to distend when feeding to allow the whale to engulf huge quantities of water and food.

▲ PHOTO-ID
A diving Humpback whale shows its distinctively pigmented tail flukes. In the 1970s researchers realized that these pigment patterns are unique for most individuals, and change little over time. This enabled the start of a new era of whale research in which it was possible to study the life histories and migrations of individuals. Catalogs of photographs of animals are now kept for most of the oceans of the world.

Humpbacks are born in the tropics, about 4.3 meters (14 ft) long, in the late autumn or early winter. They must feed and grow fast to be ready for their first migration. Like most baleen whales, the mother and calf is the basic unit of Humpback society, which tends to be fluid. Groups form and split according to the abundance of food, or the dynamics of the breeding season when groups of adult males compete boisterously, and violently, for the attention of receptive females. In feeding areas, they are usually found in ones or twos, reflecting the scattered, patchy nature of the krill they feed on.

The song of the Humpback

Humpbacks are well known as the "singing whales." They sing complex, repetitive songs, each of which may last for 10 to 15 minutes. The individual sounds or "units" of Humpback song have been described as yaps, snorts, chirps, groans, whooos, eees, and ooos. Units are organized into "phrases," which, in turn, are organized into "themes"— and the themes are organized into songs. Songs may be repeated for hours, and have been heard at distances of at least 31 kilometers (20 miles).

Mature males are the only ones known to sing, and song is most often heard in their breeding grounds. Each breeding population has its own song; over time, mostly during the winter, new phrases are added and elaborated and old ones omitted. At any one time, all the whales in a group sing much the same song, but that song is constantly evolving, for reasons unknown.

Humpback songs are thought most likely to be a display to attract females. They may serve to synchronize ovulation in females, and to mark out temporary "territories" of competing males. Singers often stop singing when they are joined by another whale, probably a responsive female.

Migration

Many baleen whales undertake long migrations from cold feeding waters to warmer breeding grounds. Humpbacks travel from near the poles to tropical waters. The longest migration route of any mammal is that of some Antarctic Humpbacks that feed on the western side of the Antarctic Peninsula and migrate north across the equator to breeding grounds off Colombia. Antarctic Humpbacks breed in the tropical waters off all southern continents, taking about three months to commute from Antarctica at a speed of about 2–5 kilometers (1–3 mph). Their leisurely migration is often punctuated with rests, courting, and occasional feeding. How the whales find their way over such vast distances is not known for certain, but several possible navigational aids have been suggested. They may use the sun and stars as guides, or natural features such as land masses, and wind and swell patterns. Like birds, they may have magnetic navigation systems that detect anomalies in the earth's magnetic field. Or perhaps they "hear" the shape of the sea floor by listening to sounds made by other marine life or echoes of their own sounds.

Toothed whales

▼ DANCING ON WATER
This Dusky dolphin displays
its distinctively striped flank
as it porpoises to breathe.
Research shows that leaping
clear of the water—or
porpoising as it is called—is
the most efficient way to
breathe while moving at
speed because it momentarily
helps the animal to avoid the
turbulence associated with
surface travel. Dusky dolphins
and other species are
sometimes visible from a
distance as a huge mass of
foaming white water, as
thousands porpoise together.

The best known of all whales must be Moby Dick, the great white Sperm whale immortalized in the eponymous novel by American writer Herman Melville. Moby Dick was a giant male Sperm whale, and it is this whale's tall, blunt profile that often features in children's books and drawings of whales. The Sperm whale is the largest of all the toothed whales and can reach 20 meters (65 ft) in length. However, its vast and impressive size and shape are unique among toothed whales. The dolphins, porpoises, and beaked whales that make up the bulk of toothed whale species are generally much smaller, some of them barely more than 1 meter (3 ft), and most of them under 5 meters (16 ft). They also range in habitat from steamy jungle rivers to ice-covered seas, and many of them live their lives rarely seen by humans.

Toothed whales hunt larger prey than do baleen whales, feeding on fish, mollusks, birds, and mammals, indeed, almost any animal they can catch—up to and including baleen and other toothed whales. Each toothed species has its preferred hunting patterns and prey: some home in on cephalopods such as octopus, squid, and cuttlefish; others specialize in grubbing up bottom-dwelling worms and crustaceans. Dolphins use their conical teeth to grab fish and swallow them whole, whereas porpoises have shearing teeth to rip prey apart before swallowing it.

Like baleen whales, toothed whales have a smooth, torpedo-like body but, due to their smaller size, they are far more flexible and agile than their more ponderous baleen cousins, and are far less likely to be encrusted with barnacles and whale lice. Males generally grow larger than females, which is the opposite to baleen whales. This sexual dimorphism may extend to include differences in other features; for example, the size and shape of the dorsal fin. As well as using their teeth to grasp prey, toothed whales may use them for fighting, and sexually mature animals of both sexes are often savagely striped with scars inflicted by the teeth of other whales.

Head pieces

Most toothed whales have a melon—a large, lens-shaped pocket of fatty oil at the front of their head. The melon is more highly developed in species that feed in very deep

Supportive species

Most toothed whales are gregarious, spending their whole lives in tightly knit social groups. These groups are critically important to the whales, who seem unable to function in isolation. Many mass whale strandings may occur because a group refuses to abandon a sick or dying individual.

In the heyday of whaling, this grouping behavior worked against whales; a whole pod would surround an individual wounded by whalers, who would then harpoon all of them one by one with very little effort. Sperm whales, in particular, were known for this remarkable nurturing behavior, surrounding injured or weak individuals with their flukes—their most effective weapon—facing outward in a protective pattern called the marguerite formation, so-named because it looks like the petals of a flower.

▲ BATTLE-SCARRED
The flukes of this Sperm whale show tooth rakes, probably caused by an Orca or by one of its relatives, such as a False killer whale or a Pilot whale. Whales of many species show similar scarring. It is likely that Orcas nip individuals to "test" them. A spirited response sends the Orca into retreat. Feeble resistance, however, may be the cue for a more determined attack.

water where light does not penetrate, or in muddy river water where visibility is poor. It is thought that the melon acts as a lens to focus echolocation sounds, which are generated inside the skull, into a beam of varying width and intensity. Returning echoes are received through oil-filled channels in the lower jaw. Echolocation is a sense that is dimly understood—humans have nothing comparable. What is known, is that it is used for hunting prey, for navigation, and possibly for investigating the emotional state of companions, or what their last meal was.

The jaws of many toothed whale species project in a beak shape in front of the melon, a feature that is very highly developed in both beaked whales and most dolphins. Unique among vertebrates, toothed whales have only one external blowhole. They originally had two nostrils, but these have merged within the whale's head, and a single nasal passage reaches the skin.

All toothed whales have teeth at some stage, but these teeth are often vestigial or highly modified, particularly in the beaked whales which may have just one pair of teeth in the lower jaw. This raises the question of how these animals with minimal dentition are able to secure their prey. The answer is that they use a sudden, powerful suction once they are near enough to their prey for it to be effective. At the other end of the scale, those species with functional teeth may have more than 260 of them. These are usually simple, conical pegs, but some species have evolved highly specific tooth shapes that are of little practical use in normal feeding. Male Narwhals are the most dramatic example of this development. They have one normal tooth and one extremely long, spirally twisted tusk, and adult males have been seen using these to "joust" with each other. Many beaked whales have similar but less spectacular dental equipment.

Echolocation

Toothed whales share with some bats the remarkable faculty of echolocation—the emission of sounds that bounce off surrounding objects and provide the animal with a "picture" of its environment. The sound that whales use is a click, loud enough to generate an echo, and short enough to be repeated rapidly to detect any change in the environment. Clicks may be widely spaced when seeking prey but become much more rapid as the item of interest is approached. Species that require a clear image of their immediate surroundings employ high frequencies, effective at short range, while species that need to scan the more distant environment use lower frequencies.

Bottlenose dolphins can echolocate to a distance of 750 meters (2,500 ft)—possibly further. Even in very murky conditions, dolphins can use echolocation to catch their prey, avoid predators, and map the sea floor. Echolocation operates in a very narrow band, so the animal must swing its head to scan a broad area, but captive dolphins can distinguish different types of metal in balls, and can pick hollow metal balls from solid ones.

Toothed whales seem to echolocate by forcing air through the passages between blowhole and larynx. The skull may operate as a parabolic reflector, and the melon, which can change shape, may focus the sound beam to a chosen distance. The echo is registered by oil-filled sinuses in the whale's lower jaw, or possibly in some species, by the teeth.

It appears that whales echolocate only when something interesting—such as food—is around. Sperm whales echolocate constantly during feeding dives, though they are occasionally quiet. They usually echolocate at a steady rate, but sometimes the click speed accelerates rapidly. This is probably because the whale has located prey and is seeking more detailed information about its size, shape, speed, and direction.

It is possible that echolocation is such an effective faculty that whales can use it to "X-ray" other whales in order to examine their stomach contents, or even in order to detect mood. It is also possible that some toothed whale species could use an extremely loud echolocation click to stun their prey into immobility.

Dolphins and porpoises

SUBORDER: Odontoceti

FAMILIES: Delphinidae, Phocoenidae

Oceanic dolphins—family Delphinidae—constitute one of the world's most successful groups of whales. More than 30 species are distributed across almost every oceanic environment, as well as many of the world's rivers. Some have quite limited geographic ranges, while others have broad or even worldwide distribution. Dolphins occupy both coastal and open ocean habitats, and both shallow and deep waters. In deep water, they may hunt at great depths or pursue surface-feeding fishes. Dolphins range in size from small river dolphins, at 1.5 meters (5 ft) long to the Orca, or Killer whale, at almost 10 meters (32 ft). Generally, dolphins are the fastest and most agile of cetaceans, and most species have a long, beak-like snout and a sleek, streamlined body.

Fundamentally, dolphins are social animals. They often hunt cooperatively to herd schools of prey, and they join forces to attack their prey and to avoid predators. Under the right conditions, and with a rich source of prey, pelagic dolphins may gather in hundreds, even thousands. These large gatherings string out over distances of several kilometers, swimming in smaller sub-groups as they move between feeding grounds, but quickly reassembling into a tight group when the opportunity to feed arises.

Porpoises—family Phocoenidae—are generally smaller and stockier than dolphins, with a more rounded body and no real beak. They have a long, shallow, triangular or rounded dorsal fin, and spade-shaped teeth. All porpoises hunt small fish and cephalopods, using echolocation and sight. They are not as gregarious as dolphins, occurring in much smaller groups. Dall's porpoise, a North Pacific species, is thought to be the fastest of all cetaceans, though they are preyed on by Orcas—and by humans.

Dolphin characteristics

Dolphins exhibit a wide range of color patterns, which to some extent reflect their way of life. Animals that live in turbid riverine or coastal waters, and form small groups, are often relatively drably patterned; whereas animals that hunt in the open sea in larger groups are often more strikingly patterned. This patterning may help to deceive prey, to avoid predators, or to aid in recognition.

Most dolphins have many functional teeth in both their upper and lower jaws. These teeth are designed for grasping prey and preventing it from escaping before being swallowed. Dolphins are capable of swallowing surprisingly large prey. Most species prefer to hunt either fish or squid, and fish-eating dolphins often have narrower heads than squid-eating dolphins. Some species are more generalist feeders, taking whatever food presents itself. The bottlenose dolphin is a good example of this.

Some dolphin species, such as Striped or Hourglass dolphins, are oceanic nomads with no fixed territory, constantly roaming in large groups over vast distances in search of seasonally available prey, and foraging along oceanic fronts or upwellings. At the other extreme, river dolphins and coastal Bottlenose dolphins, for example, form much smaller groups and are quite restricted in their range: they may inhabit a single river or bay, or short stretch of coast.

Dolphins make sounds ranging from low-pitched chirps to ultrasonic clicks, which they use for echolocation, and possibly to stun their prey. Sounds at higher frequencies carry only short distances under water, but most dolphin species have no need to communicate over long distances as their social cohesion keeps group members fairly close to each other.

A social species

Each dolphin species has its own pod structure. Many have a flexible, open-group policy where individuals come and go freely and often move from group to group. Others live in very stable groups of related animals, with very little change over periods of years. Previously it was thought that several species formed harems. Now we know that many toothed whale societies are matrilineal; that is, the older females form a stable nucleus, around which younger adults and immatures gather. Adult males in many species form their own groups, usually approaching female groups only for mating. In other species, males remain with their mother's group for life, and mating occurs between such groups when they meet. Promiscuous mating is the general rule for dolphins, and male–female and male–calf bonds are weak; the bonds between female and calf, and between female and female are usually much stronger.

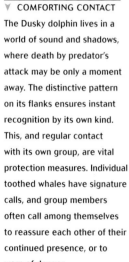

▼ **COMFORTING CONTACT**
The Dusky dolphin lives in a world of sound and shadows, where death by predator's attack may be only a moment away. The distinctive pattern on its flanks ensures instant recognition by its own kind. This, and regular contact with its own group, are vital protection measures. Individual toothed whales have signature calls, and group members often call among themselves to reassure each other of their continued presence, or to warn of danger.

DOLPHINS

FAMILY: Delphinidae

Orca

ORCINUS ORCA

OTHER NAMES: Killer whale, Blackfish, Grampus

LENGTH: female 8.5 m/28 ft; male 10 m/33 ft

WEIGHT: female 7.5 tonnes; male 11 tonnes

DISTRIBUTION AND HABITAT: worldwide; found in all ocean habitats, from the tropics to the polar ice, but most common near rich food sources in high polar latitudes

STATUS: unknown population, but considered locally abundant

  IUCN: Lower Risk—conservation dependent

  CITES: Appendix II

APPEARANCE: These are the largest of the dolphins, and they are huge by dolphin standards; their size, and their black and white panda-like pattern, make them impossible to miss. Mature males are almost twice as large as adult females, and they have an exceptionally tall dorsal fin, and very large fins and flukes.

The Orca, or Killer whale, is one of the most successful dolphin species, with populations occurring in all the world's seas and oceans. Orcas have 44 conical teeth with which they seize their prey. They are frequently referred to as Killer whales, which is somewhat undeserved because every other cetacean species also kills to eat. And, they have never been known to attack people in the wild, even when provoked. Unfortunately, the reverse is not true: Orcas are often killed because of their propensity for

▲ DETERMINED HUNTER
An Orca in its element—the Antarctic sea ice, where it hunts penguins, fish, seals, and whales. Orcas have been seen rushing at ice floes where seals are resting, creating a bow wave that rocks the floe and tips the hapless prey into the water. They are formidable predators capable of complex, coordinated attacks on animals ranging in size from sardines to Blue whales.

stealing from human fisheries. To such a hunter, taking fish from longline hooks is as easy as picking grapes.

An adult Orca needs to consume an average of 68 kilograms (150 lbs) of seafood each day. Orcas usually hunt in packs, and will show a remarkably disciplined division of labor, such as when attacking large whales. Some populations have discovered how to hunt seals resting on shore by surfing in and out on waves—this is probably the closest any modern whale gets to going on land. No matter what the target, Orcas are efficient, cooperative hunters. Depending on their location, they will eat whatever is available, including fish, squid, sharks, stingrays, shellfish, seabirds, dugongs, seals, dolphins, porpoises, and baleen whales up to, and including, Blue whales. No predator is known to prey on Orcas, however.

▲ FEMALE AT THE CENTER
This Orca can be identified as an adult female by the size and shape of its dorsal fin. An individual whale can be recognized by the unique shape of its saddle—the patch of white pigment behind the dorsal fin. Adult females are at the center of Orca society. The smallest stable groups consist of an adult female and her calves, some of them with calves of their own.

Off the coast of British Columbia and Alaska, there are two different types of Orcas: residents, which inhabit a home range and eat mostly fish; and transients, which range much more widely, and favor marine mammals as prey.

The basis of Orca society is a matrilineal group consisting of a mature female and her offspring, and often their offspring. Such groups of resident animals form a series of progressively larger groups, up to about 60 individuals. Transients form much smaller groups of up to four, consisting only of a female and her immediate offspring. So far there have been no studies of social organization in Antarctic Orcas, but there is evidence that the two types exist there as well.

Females are physiologically capable of bearing a calf every three years, but they normally produce one only every eight years. The young measure 2 meters (7 ft) at birth, and calves remain in their birth pod with their mother for many years. Females generally live for 80 to 90 years and bear about five calves; males have a shorter lifespan of about 60 years.

◄ CARING ROLE
Long-finned pilot whales are one of the few mammalian species, other than humans, known to undergo menopause. While mothers make deep feeding dives that may last an hour or more, females past calving age act as aunties, caring for the young at the surface. They are even capable of suckling the calves.

◄ SELDOM SEEN
Southern right whale dolphins rarely come close to land and are never seen for long at sea. They roam the Southern Ocean in search of squid and small fish, porpoising along in a welter of white water in a motion similar to bouncing. Because they have no dorsal fin, at a distance they can be confused with penguins.

◄ BOLD AND CURIOUS
The dramatically marked Hourglass dolphin is sometimes mistaken for a small Orca. Ranging further south than other dolphins, almost to the edge of the sea ice, it is one of the least studied cetaceans. Hourglass dolphins tend to accompany vessels and may bowride for hours.

Commerson's dolphin
CEPHALORHYNCHUS COMMERSONII
OTHER NAMES: Southern dolphin
LENGTH: 1.4 m/4 $\frac{1}{2}$ ft
WEIGHT: 50 kg/110 lb
DISTRIBUTION AND HABITAT: The southern oceans off both the southern coasts of South America and into the Atlantic Ocean.
STATUS: unknown population
 IUCN: Data Deficient   CITES: Appendix II
APPEARANCE: These dolphins are dramatically black and white, with a black head, a black dorsal fin, and a black tail. A white patch runs from just behind the head down to the belly. They have no melon, and a very short beak.

Long-finned pilot whale
GLOBICEPHALA MELAS
LENGTH: female 5 m/16 ft; male 6 m/19 $\frac{1}{2}$ ft
WEIGHT: female 1.8 tonnes; male 3.5 tonnes
DISTRIBUTION AND HABITAT: North Atlantic Ocean and southern oceans generally; possibly all cold–temperate waters.
STATUS: probably common
 IUCN: Lower Risk—conservation dependent
 CITES: Appendix II
APPEARANCE: Long-finned pilot whales have a square, bulbous head, and a strongly sickle-shaped dorsal fin. The back and flanks are black, and their chin and belly are a grayish-white that darkens slightly as they mature.

Southern right whale dolphin
LISSODELPHIS PERONII
LENGTH: 1.8 m/6 ft (possibly larger)
WEIGHT: 60 kg/132 lb
DISTRIBUTION AND HABITAT: An offshore species, known to inhabit southern hemisphere cold–temperate waters and Antarctic waters south of Argentina.
STATUS: unknown population, but probably common
 IUCN: Data Deficient   CITES: Appendix II
APPEARANCE: These dolphins are black on top and have a large, wide, white belly patch running up their flanks, including their flippers, head, and tail. They have no dorsal fin.

Peale's dolphin
LAGENORHYNCHUS AUSTRALIS
OTHER NAME: Whitesided dolphin
LENGTH: 2 m/6 $\frac{1}{2}$ ft
WEIGHT: 115 kg/253 lb
DISTRIBUTION AND HABITAT: Cold to temperate waters off the southern coasts of South America and the Falkland Islands.
STATUS: population unknown
 IUCN: Data Deficient   CITES: Appendix II
APPEARANCE: Peale's dolphins have round snouts and short beaks. They are black or dark gray on the dorsal surface, with a light gray stripe on the flanks and a fine white line above the gray patch that enlarges to a wider pale stripe.

Hourglass dolphin

LAGENORHYNCHUS CRUCIGER

OTHER NAME: Whitesided dolphin

LENGTH: 1.6 m/5 $^1/_4$ ft

WEIGHT: 100 kg/220 lb

DISTRIBUTION AND HABITAT: Southern hemisphere; Antarctic and cold–temperate waters.

STATUS: unknown population, presumed abundant.

  IUCN: Lower Risk—least concern   CITES: Appendix II

APPEARANCE: These dolphins are black or dark gray and white in color, and derive their name from the hourglass-shaped patch on their black flanks. They have rounded snouts and short beaks. Often grouped with *L. obscurus*.

Dusky dolphin

LAGENORHYNCHUS OBSCURUS

OTHER NAME: Whitesided dolphin

LENGTH: 1.6 m/5 $^1/_4$ ft

WEIGHT: 100 kg/220 lb

DISTRIBUTION AND HABITAT: Southern hemisphere; Antarctic and cold–temperate waters.

STATUS: unknown population

  IUCN: Data Deficient   CITES: Appendix II

APPEARANCE: Similar to *L. cruciger*, with which they are often grouped. They have a rounded snout and short beak; complex coloration—pale below, dark gray above, with a distinctive pale forked blaze running forward on the flanks, and a pale trailing edge to the dorsal fin.

PORPOISES

FAMILY: Phocoenidae

Spectacled porpoise

AUSTRALOPHOCAENA DIOPTRICA

LENGTH: 2 m/6 $^1/_2$ ft

WEIGHT: 85 kg/190 lb

DISTRIBUTION AND HABITAT: A Southern hemisphere species, possibly circumpolar, occupying cold–temperate waters off South America and the Falkland Islands, South Georgia, Heard Island, Macquarie Island, and the Auckland Islands.

STATUS: unknown population;

  IUCN: Data Deficient   CITES: Appendix II

APPEARANCE: Spectacled porpoises have black backs and white bellies, and a distinctive black outline around their lips and eyes—hence their common name.

Burmeister's porpoise

PHOCOENA SPINIPINNIS

LENGTH: 1.8 m/6 ft

WEIGHT: 70 kg/155 lb

DISTRIBUTION AND HABITAT: Ranges from Brazil southward into sub-Antarctic waters; prefers coastal habitats.

STATUS: uncertain, but possibly as low as 500 animals.

  IUCN: Data Deficient   CITES: Appendix II

APPEARANCE: Very dark black-gray on top; lighter below.

▲ **WORKING TOGETHER**
Dusky dolphins, like most dolphins, are gregarious animals, sometimes forming temporary groups of up to several thousand, although they are usually found in much smaller, more permanent groups of 15 or more. When they locate schools of fish, individuals summon others with exuberant leaps, as cooperative herding increases the chance of a good feed for all. Then, well-fed, they lounge around in small groups, relaxing and socializing.

Sperm whales

SUBORDER: Odontoceti

FAMILIES: Physeteridae,
Kogiidae

There are two families of sperm whales. The Sperm whale—the largest of all toothed whale species—belongs to the family Physeteridae; two other species, which are anatomically similar but much smaller, the Dwarf and Pygmy sperm whales, belong to the family Kogiidae. These two are as small as dolphins but several characteristics make it clear that they are more closely related to the Sperm whale than to any of the smaller toothed whales.

The most significant shared feature of these three species is the spermaceti organ—a highly specialized melon—which gives sperm whales their distinctive blunt profile. They also share highly unusual nasal passages. In all three sperm whales the blowhole is located far to the left side of the head, but in the two smaller species it is at the back of the head, close to the normal position, whereas in the great Sperm whale it is at the front of the melon. All three species have functional teeth in their small lower jaw, and all are very deep divers, feeding on squid that live at great depths.

Sperm whale

FAMILY: Physeteridae

PHYSETER MACROCEPHALUS

OTHER NAMES: Catchalot, Spermaceti whale

LENGTH: female 13 m/43 ft; male to 20 m/65 ft

WEIGHT: female up to 20 tonnes; male up to 50 tonnes

DISTRIBUTION AND HABITAT: Worldwide; observed in all the world's seas and oceans from the Antarctic through cold and temperate zones to tropical waters.

STATUS: unknown population, probably abundant
IUCN: Vulnerable CITES: Appendix I

APPEARANCE: Sperm whales are very dark gray-brown with a wrinkled appearance over much of their body, and characteristic white scarring from deep-water squid.

A species at risk

Before species conservation became a matter for concern, adult male Sperm whales were hunted almost to extinction for the commercially valuable spermaceti oil from their melon and ambergris from their gut, and the huge males of the species are no longer found. Whalers took the biggest mature males and left females and immature males. When these young males reached breeding age (about 27), they had not acquired the social skills necessary to reproduce, so the birthrate of the species—already quite low—fell even further. The young males seem to have been learning, though, because the number of Sperm whale births is increasing.

One of the largest extant whales, the Sperm whale has a dorsal hump, rather than a dorsal fin. Its most recognizable feature is its huge, rectangular head, which may be up to one third of its body length, and which is most fully developed in mature males. The blow of the Sperm whale has a forward angle of 45°, a feature which readily identifies it at sea. It has no teeth in the upper jaw and a relatively short, narrow lower jaw. The teeth in the lower jaw are the largest functional teeth known in the animal world, and may be up to 25 centimeters (10 inches) long. Sperm whales are one of the most sexually

dimorphic whale species: a particularly well-grown male may weigh three times as much as a mature female and can grow to twice the length. As well as being much smaller than the males, female Sperm whales have a less well developed spermaceti organ—the huge melon that gives the male such a distinctive square shape. In a male Sperm whale, the large melon can be up to one-quarter of the animal's total body length.

The largest brain on earth

Behind the spermaceti organ, the Sperm whale's huge skull contains a brain that can weigh more than 9 kilograms (20 lb). In comparison, the weight of a human's brain is only 1.3 kilograms (3 lb). Relative to its body size, the Sperm whale's brain has probably been as large as this for more than 30 million years, whereas humans reached their present brain capacity only 100,000 years ago. The whale's brain is very convoluted, with a highly developed cerebral cortex. This has been cited as evidence of high intelligence. However, intelligence is extremely difficult to compare across cultures, and is probably impossible across species. While many scientists believe that the size of a sperm whale's brain may be necessary for the animal to deal with the complex echolocation information that it

must process, it is not possible to rule out intelligence of a different order than our own.

The skull of a Sperm whale is unusually asymmetrical for a mammal. The spermaceti organ in its forehead is filled with a huge quantity of waxy oil, and nasal passages run between air sacs at the front and back of this organ. Spermaceti oil was the most highly prized product of the nineteenth century whaling industry (except for ambergris, an aromatic waxy deposit also obtained from sperm whales). It was used as a fine machine oil, and is still unrivaled as a lubricant for missile inertial guidance systems.

There are several theories about the function of the spermaceti organ. According to the buoyancy-control theory, a Sperm whale can remain motionless at a chosen depth, or swim vertically up or down, by regulating the temperature, and thus the density, of the waxy contents of the spermaceti organ, heating it with body-warmed air and blood or cooling it with sea water drawn into the nasal passages. Another theory holds that these whales use their spermaceti organ for long-range echolocation. It is believed that air moves through the right nasal passage, generating sound at the entrance to the forward air sac, and is then cycled back along the left nasal passage, possibly amplified en route.

Sperm whales use low-frequency clicks for echolocation or communication. Clicks may be arranged into organized patterns known as "codas," which are thought to mediate social interactions between animals that come together after a separation. They also produce "creaks"— very fast click trains rising in frequency—which may be used to track prey at close range. Whereas the sounds of Sperm whales appear relatively simple when compared with the songs of Humpback whales, their giant brains are probably capable of processing information of a richness and complexity that we cannot imagine.

Hunting

Sperm whales normally swim at 10 kilometers per hour (6 mph), but can reach speeds of up to 30 kilometers (20 mph). In the summer, mature males migrate slowly to polar waters where they live as solitary hunters, diving in search of squid to depths unmatched by any other mammal. They may ambush or pursue their prey through the depths before surfacing. Sperm whales eat mainly squid and octopus, but they will also opportunistically grab any deep-water species, including sharks and fish. They have been tracked to depths of 2,500 meters (8,000 ft), and there is evidence that they dive even deeper after prey.

▽ DEEP-SEA DIVER
A Sperm whale shows its characteristic blunt head, short underslung lower jaw, dorsal hump, and wrinkled skin. Sperm whales breathe, travel, rest, and socialize at the surface, but the serious business of feeding is carried out far below, in total darkness and under unimaginable pressure. Even Sperm whales with broken or malformed lower jaws can feed successfully, suggesting that prey (mainly squid and octopus) are swallowed by suction.

Dwarf and Pygmy sperm whales

FAMILY: Kogiidae

Adult Dwarf and Pygmy sperm whales feed on deep-water cephalopods, fish, and crustaceans, while the younger members of these species hunt similar prey from shallower waters. Smaller and younger *Kogia* species are unable to hunt successfully in deep waters, but by the time they are mature and have reached their full size, they have learned to plumb deep waters for their prey. Both whales have a largely offshore distribution and keep to small groups, ranging from two (usually mother and calf) to a maximum of ten. They are generally quiet and inactive while on the surface of the water.

When threatened by predators, Dwarf and Pygmy sperm whales exhibit a unique protective mechanism that involves releasing a cloud of reddish-brownish feces, and then thrashing their flukes to spread the cloud as much as possible. They then make their escape into the murk. Curiously, this tactic is also used by squid on which the whales prey.

Pygmy sperm whale

KOGIA BREVICEPS

OTHER NAMES: Short-headed, Small or Lesser sperm whale

LENGTH: 2.8 m/9 ft

WEIGHT: 400 kg/880 lb

DISTRIBUTION AND HABITAT: Worldwide; from cold and temperate waters to the tropics.

STATUS: unknown population, probably widely distributed. IUCN: Lower Risk—least concern CITES: Appendix II

APPEARANCE: The head of a Pygmy sperm whale is blunt, rather like that of a shark. They share some Sperm whale features, but have a dorsal fin.

Dwarf sperm whale

KOGIA SIMUS

OTHER NAMES: Owen's pygmy sperm whale, Rat porpoise

LENGTH: 2.7 m/9 ft

WEIGHT: to 270 kg/600 lb

DISTRIBUTION AND HABITAT: Worldwide, from colder and temperate to tropical waters.

STATUS: unknown population, probably widely distributed. IUCN: Lower Risk—least concern CITES: Appendix II

APPEARANCE: This species is difficult to distinguish from the Pygmy sperm whale, and was once consigned to the same species. It is now thought to be smaller, with a rounder nose and a larger dorsal fin.

▲ AVOIDING THE BENDS
Before a feeding dive, Sperm whales lie at the surface blowing repeatedly—roughly one blow for every minute they will be submerged. Dives can last well over an hour. Although the whale plunges to extraordinary depths, it does not get the bends because it does not breathe oxygen under pressure. Its lungs empty soon after diving, and it relies instead on oxygen stored in blood and muscle.

They usually patrol waters that are too deep for light to penetrate, and totally blind sperm whales have been captured in perfect health, and with food in their stomachs. Clearly these whales must have evolved a highly effective technique for detecting and catching their prey without the aid of vision, and echolocation is the obvious answer.

It is the stuff of legend that Sperm whales devour giant squid. In fact, most of their squid victims are relatively small—1 meter (3 ft) or less. But body scars suggest that these whales do regularly engage in battle with larger specimens, and a squid 10.5 meters (35 ft) long has been retrieved from the stomach of one Sperm whale.

Family life

Female and young Sperm whales remain in nursery groups of some 20 to 40 whales, which are usually related through the mature females, and stay in temperate and tropical waters. These whales have very low reproductive rates, calving in summer or autumn and producing one offspring every four to six years. The young are 4 meters (13 ft) at birth. They are weaned into juvenile pods, but females may suckle their last offspring for up to fifteen years. As the whales mature, females rejoin their nursery groups, and males form bachelor herds that contain fewer numbers as the whales grow, until really large males tend to be very widely spaced, though they may still communicate with a group. During breeding, each nursery group is joined by up to five mature males, about one male for every ten females. Recent research suggests that males stay with a nursery group for only a few hours and then depart in search of another group where the females are ready to mate. Males compete for females, and only 25 percent are successful.

Sperm whales feed mainly on squid, including giant squid. They hunt around the edges of continental shelves, oceanic islands, and underwater seamounts, and are often found in water at least 1,000 meters (3,250 ft) deep. Sometimes they follow their prey into shallow coastal waters where they are prone to stranding.

Beaked whales

FAMILIY: Zipiidae

The scientific name for beaked whales comes from the Greek word *xiphos*, meaning sword. Beaked whales are small to medium species. They have a single pair of throat grooves, unlike most toothed whales, which do not have any. They are a very successful group of whales, and there are around twenty species, most of which are less than 6 meters (20 ft) long.

Beaked whales are generally deep-water animals, and are very reclusive. Populations could be quite large, especially for Arnoux's beaked whale and the Southern bottlenose whale, which live in Antarctica in summer; however, they are seldom seen, and scientists know very little about their behavior. On occasions, they have been spotted in groups of fifty or more, which suggests that they are gregarious. Beaked whales feed on squid, and while males have one or two tusk-like lower pairs of teeth, females have no real teeth. Males seem to use these teeth for fighting amongst themselves.

Arnoux's beaked whale

BERARDIUS ARNOUXII

OTHER NAME: New Zealand beaked whale

LENGTH: 10 m/33 ft

WEIGHT: 10 tonnes

DISTRIBUTION AND HABITAT: Circumpolar throughout the southern oceans.

STATUS: unknown population, probably common in Antarctica
  IUCN: Lower Risk—conservation dependent
  CITES: Appendix I

APPEARANCE: These whales are blue-gray or brownish, with slightly darker flippers, flukes, and back. Older males tend to be paler and can be heavily scarred. This is one of the few beaked species with functional teeth.

Southern bottlenose whale

HYPEROODON PLANIFRONS

OTHER NAMES: Flower's/Flat-headed/Antarctic bottlenose whale, Pacific beaked whale

LENGTH: 7.5 m/25 ft

WEIGHT: 8 tonnes

DISTRIBUTION AND HABITAT: Throughout the southern hemisphere, particularly in Antarctic waters, but occasionally reaching into the tropics.

STATUS: unknown population, thought to be common in Antarctica
  IUCN: Lower Risk—conservation dependent
  CITES: Appendix I

APPEARANCE: Southern bottlenose whales are an unusual metallic gray on top and paler below, and they may be heavily scarred. Females are larger than males.

Cuvier's beaked whale

ZIPHIUS CAVIROSTRIS

OTHER NAME: Goose-beaked whale

LENGTH: female 7.5 m/25 ft; male 7 m/23 ft

WEIGHT: 3 tonnes

DISTRIBUTION AND HABITAT: Worldwide, in cold to tropical waters; believed to live in small, deep-water pods, feeding on squid and deep-sea fish.

STATUS: population unknown
  IUCN: Data Deficient CITES: Appendix II

APPEARANCE: Cuvier's beaked whales can be very light yellow-brown or gray-blue. The head is slightly concave and their beak is less developed than in most beaked whales.

Mesoplodon species

COMMON NAMES: Gray's beaked whale, Scamperdown whale (*M. greyi*); Hector's beaked whale (*M. hectori*); Strap-toothed whale (*M. layardii*)

LENGTH: *M. greyi*—6 m/20 ft; *M. hectori*—4 m/13 ft; *M. layardii*—5 m/16 ft

WEIGHT: *M. greyi*—5 tonnes; *M. hectori*—2 tonnes; *M. layardii*—3.5 tonnes

DISTRIBUTION AND HABITAT: Cold and temperate southern hemisphere waters, in singles or groups up to three; deep divers, submerging for up to two hours.

STATUS: population unknown
  IUCN: all Data Deficient CITES: Appendix II

APPEARANCE: These whales are seldom seen, and hard to distinguish. Color patterns of some species are unknown, but they are generally dark gray above and paler below, often with extensive scarring. They are squid eaters with few functional teeth, and may capture their prey by intense suction. LW

▼ MYSTERY WHALE
Little is known of the Southern bottlenose whale. Although it is believed to be reasonably common in the Southern Ocean, it is rarely seen, and is difficult to identify unless observed at close range. It dives deep to feed on squid, but its surface behavior is inconspicuous. These enigmatic animals are thought to migrate to warm waters during the winter months.

Seabirds

Seabirds

Previous pages

◄ **ATTRACTING A MATE**
Young Wandering albatrosses
(*Diomedea exulans*) engage in
complex courtship dances that
include outstretched wings,
bill pointing, and preening.

Of the nearly 10,000 bird species on earth, only about three percent are considered seabirds, meaning that they obtain virtually all their sustenance from the sea. Some 70 percent of the earth's surface is covered in water, and this is the world of seabirds. Many seabirds and a few land birds make their home in Antarctica or on the many sub-Antarctic islands scattered around the Southern Ocean. To stand on the deck of a ship in the Drake Passage, the waterway that separates the southern tip of South America from the Antarctic Pensinsula, is to marvel at the birds that wheel and dive above the waves. Humans cannot imagine a life away from *terra firma*—so how do these birds survive on an apparently featureless ocean?

At home at sea

Unlike marine mammals, no seabirds have evolved the ability to reproduce without coming ashore somewhere to find a mate, build a nest, and lay eggs. Most seabird species breed only once each year (some only every other year), spending four to five months in the process. Tied to land for less than half of each year, seabirds are truly at home when at sea. Most seabird species are relatively long-lived and take a long time to reach sexual maturity, so that once a young bird fledges and strikes out on its own, it may be years before it alights on land again. There is no fresh water at sea, and food, while abundant, is often found in widely separated patches or hidden deep beneath the surface. Finding ways to meet these challenges has given rise to the diversity of seabirds in the world today.

Several groups of birds have adapted to life at sea,

◄ **THE NEXT GENERATION**
A White-capped albatross
(*Diomedea cauta*) gently preens
its small chick. Once the egg
hatches, the parents tend the
chick for about three weeks.
During that time the male and
female take turns going to sea
for one to three days in search of
food while the other feeds,
broods, and guards the chick.

▲ BREEDING COLORS
In the midst of the breeding season, the Imperial shag (*Phalacrocorax atriceps*) sports a bright blue eye-ring and yellow caruncles at the base of its bill. It builds a substantial nest from seaweed and pieces of grass cemented together with guano.

▼ DISTINGUISHING TRAITS
A Northern giant petrel (*Macronectes halli*; top) and a bloodstained Southern giant petrel (*M. giganteus*; bottom) scavenge an elephant seal carcass. The Northern giant petrel has a characteristic red tip on its bill; only Southern giant petrels have an all-white form.

and there is no single story of seabird evolution. Of the flying birds around the world, 113 species are Procellariiformes, or tube-nosed birds, expert fliers that roam huge expanses of the oceans to find food; they are an old group with great diversity that evolved in the southern hemisphere. Descended from ancestral Procellariiformes, another 50 species of seabirds are the Pelecaniformes, the pelicans, gannets and boobies, tropicbirds, cormorants and shags, and frigatebirds; these tend to live in temperate or tropical habitats, and stay closer to shore than the tube-nosed seabirds. Another 107 species come from a larger group of waterbirds, the Charadriiformes; these are the skuas, gulls, terns and noddies, skimmers, and auks. Primarily northern hemisphere birds, they have spread throughout the world's seas. Several other scattered species can also claim the title seabird: a few species of ducks and geese of the order Anseriformes, and grebes (Podicepediformes) and loons (Gaviiformes).

Seabirds wander more than land birds, and transequatorial crossings have occasionally led to the rise of closely related species pairs, with one species in the north and the other in the south—for example, Arctic/Antarctic terns, Lesser black-backed/Kelp gulls, Great/Antarctic skuas, and Northern/Southern fulmars. Some of these pairs are literally sister species while others are the result of convergent evolution.

Eliminating salt

Seabirds face a problem not usually encountered by land birds: when they spend months at sea, there is no fresh water for them to drink. It was known for many years that bird kidneys were less efficient than human kidneys, and no one understood how birds dealt with excess salt. Anatomists knew that all seabirds have two small glands lying in a small groove above their eyes, but it was not until the late 1950s that animal physiologist Knut Schmidt-Nielsen found the explanation: the function of these glands is to get rid of excess salt. The salt glands are ten times more efficient at removing salt than bird kidneys. Salt is picked up by the circulating blood, and as it passes through the salt glands it is excreted as a highly saline solution that drains into the nasal cavities. In most birds this solution drips from the nostrils to the end of the bill, and the characteristic head-shaking of most seabirds is to get rid of these drips of salt. Many kinds of land birds possess rudimentary salt glands, but they are tiny compared with those of seabirds.

Albatrosses

ORDER: Procellariiformes
FAMILY: Diomedeidae

Albatrosses have an almost mythic significance for sailors: early venturers into the Southern Ocean believed that to shoot an albatross would bring bad luck, and biologist Robert Cushman Murphy said: "I now belong to the higher cult of mortals, for I have seen the albatross." Despite their remote locations, commercial interests have imperiled albatrosses for many years. In the late nineteenth and early twentieth centuries, many species were brought to the verge of extinction by hunters for the feather trade. Today the threat is long-line fishing; thousands of albatrosses are hooked and drowned when they try to seize baits from the hooks of long-line fishing operations.

The earliest fossil of the family Diomedeidae dates from about 50 million years ago, in the Late Eocene, and has been identified as an albatross or petrel. The first fossil that is clearly an albatross comes from about 32 million years ago, when the seas were changing and there was an important diversification in the bony fish and squid. Most modern albatrosses are squid-eaters, and it is believed that the increasing diversity and abundance of squid sparked the development of more albatross species. Albatrosses evolved in the southern hemisphere and then spread to all the world's oceans, but they have always been more diverse in the southern hemisphere. Today, only three species live in the northern Pacific Ocean and none in the North Atlantic, although there are fossils from a former resident there. Fossil history indicates that there were probably never any species larger than the Wandering albatross of today, but several extinct species were smaller than the smallest extant albatrosses.

The distinctive feature of albatrosses is their size: the Wandering albatross, with a wingspan of up to 3.5 meters (11½ ft), is the largest flying bird in the world. Albatrosses belong to two genera: the 12 *Diomedea* species, and the two smaller and darker *Phoebetria* species. The Wandering and Royal albatrosses are popularly known as the great albatrosses, and the smaller *Diomedea* species as mollymawks. A recent study lists 24 species, and argues that many subspecies should be regarded as species in their own right.

Built for gliding

Most albatrosses have typical seabird plumage—a dark back and upper wings, and a light body and under-wings. The great albatrosses, born conspicuously darker than adults and slowly growing more and more white feathers, may take up to seven years to develop full adult plumage, whereas the two *Phoebetria* species remain dark as adults. Albatrosses have long wings relative to their body size, and their vertebrae and upper wing bones are hollow yet strong, reinforced with struts to reduce their body weight and make gliding easier. Even with these adaptations their long wings are awkward, and they are

▲ **ECONOMICAL FLIGHT**
A Royal albatross (*Diomedea epomophora*) hangs in the wind near its breeding colony. By taking advantage of the constant winds, albatrosses use no more energy on their long foraging flights than they do sitting quietly on their nests.

◄ **MATING FOR LIFE**
Royal albatrosses (*D. epomophora*) normally take several seasons of courtship dancing to choose a mate—a very important decision. They will stay with the same mate throughout their long lives unless one of them fails to return to the breeding site.

► **A CROWDED COLONY**
Black-browed albatrosses (*D. melanophris*) are more social than other albatrosses and crowd into denser colonies. Windswept head-lands give them quick access to the ocean, where they collect squid and fish to bring back to their growing chicks.

Following ships

Seabirds are opportunistic and optimistic, and they have learned
that a ship usually means food. Most ships discharge kitchen scraps
and sewage into the sea, so a ship's wake often rewards a following
bird. Fishing boats especially attract large flocks of seabirds,
because they use bait and throw unwanted catch overboard, as
well as the offal of the catch that they keep and clean on board.
But even non-fishing ships attract seabirds. Many smaller seabird
species live on minuscule prey such as krill and amphipods, and the
currents created by ships' propellers throw small crustaceans and
fish to the surface, attracting birds that rely on these prey.

Ships also create wind currents that are kind to birds.
Albatrosses and petrels soar on the almost constant winds of the
Southern Ocean, riding the eddies and updrafts generated by the
swell of the open ocean. There may be vast distances and no
obstructions, but wind swirling around and over a moving ship
creates powerful updrafts and currents that can assist seabirds.
The birds take advantage of the windward flow, where the wind rises
over the ship, and on the leeward side they travel a little further
away to avoid the dead air near the ship.

not very maneuverable. They soar majestically over the waves, seldom flapping their wings. Albatrosses are clumsy and slow on land and, without sufficient wind-speed, they must land on the ocean. They are the most oceanic of the seabirds and cover vast distances. Even during the breeding season, when they must return regularly to their nest to feed chicks, they may travel 8,000 kilometers (5,000 miles) in a week in search of food. Modern satellite-based studies show that on their long journeys, albatrosses take curving paths around the ocean to avoid both calm regions and areas of the highest wind speed.

Most albatross breeding colonies are on sloping exposed hillsides of isolated islands at high southern latitudes, where there are persistent strong winds to help the birds take off. The three largest species breed only every other year, but many of the smaller species return to their breeding sites to mate each year. Most albatrosses are sexually mature at four to six years of age, but they do not normally attempt to breed until they are seven to nine years old, and the largest species may wait until they are ten or eleven. Courtship is protracted, taking perhaps two to four years of visiting the breeding colony to form a pair bond with a bird of similar age. Once paired, albatrosses usually stay together for life; if one fails to return, its mate will try to find a new partner, but this may take two or more seasons.

Albatrosses have a marvelous repertoire of courtship displays. They may stand motionless with wings outspread, or point skyward with their bills, or perform an exaggerated head-swaying walk. They clack their bills together, and when they eventually form a pair bond they usually preen each other vigorously. The Sooty albatross and the Light-mantled sooty albatross also perform one of the most beautiful duet flights imaginable; in perfect synchrony they swoop above the cliffs where they will nest, turning and banking in an aerial *pas de deux* that consolidates the pair bond and prepares them for breeding. Southern albatrosses build a substantial nest of mud, grass, and moss, returning year after year and adding to it to create a tall pedestal or a massive mound. Northern species build a more rudimentary nest, a simple scrape in the sand surrounded by twigs. The Waved albatross of the Galapagos Islands makes no nest at all, but lays a single egg onto the ground and rolls it around during incubation.

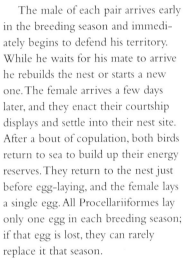

◄ **FEEDING FRENZY**
Discarded bycatch or offal from fishing boats can cause a feeding frenzy. Here, White-capped alba-trosses (*Diomedea cauta*) and Northern giant petrels (*Macronectes halli*) compete vigorously for scraps. Smaller species cannot compete in such hectic pile-ups so they stay away from the central mêlée.

▶ **BEGGING FOR FOOD**
After the end of the brooding stage, parents feed their chick every three or four days. When the parent returns and approaches with food, the hungry chick eagerly bites at the bill of the parent to beg for food. When the parent opens its bill, the chick reaches inside and across the mouth of the parent to receive a meal of regurgitated fish and squid.

The male of each pair arrives early in the breeding season and immedi-ately begins to defend his territory. While he waits for his mate to arrive he rebuilds the nest or starts a new one. The female arrives a few days later, and they enact their courtship displays and settle into their nest site. After a bout of copulation, both birds return to sea to build up their energy reserves. They return to the nest just before egg-laying, and the female lays a single egg. All Procellariiformes lay only one egg in each breeding season; if that egg is lost, they can rarely replace it that season.

The parent birds share incubation and feeding duties. The male takes the first incubation stint while the female goes to sea to feed. Her trip may last a few days or a few weeks, during which she may travel up to 15,000 kilometers (9,300 miles), and when she returns the male goes foraging. Observations suggest that males travel south and forage around the Polar Front, whereas females tend to fly north, where they are more likely to encounter the hazards of long-line fishing.

The parents take turns to bring food back to the nest in their stomachs and regurgitate it to the chick. When the adult returns, the chick pecks and bites at the adult's bill for food, the adult begins to regurgitate, and the chick maneuvers its bill crosswise inside the adult's bill so that the food is transferred directly into the chick's mouth. The parents feed the chick until it is considerably heavier than they are—up to 166 percent of typical adult weight. The extra weight is mostly fat to provide the energy to grow feathers, and to help the chicks survive their first days alone at sea.

Albatrosses either alight on the sea or plunge into the ocean surface in quest of squid, fish, krill, and other crustaceans, but they cannot dive very far below the surface. Many albatrosses feed in the early morning and the evening because squid, their favored prey, rise to the surface at those times.

Wandering albatross

DIOMEDEA EXULANS

LENGTH: 107–135 cm/42–53 in

WINGSPAN: 250–350 cm/98–138 in

STATUS: about 21,000 breeding pairs; 100,000 birds

IUCN: Vulnerable   CITES: not listed

WANDERING ALBATROSS

Distribution
Breeding sites

▼ **WAITING PATIENTLY**

A fledgling Wandering albatross awaits the next meal. Parents stop feeding their chick after about 270 days. The fledglings usually stay around their nests until the breeding males arrive and claim territories for the upcoming season. When the juveniles approach them to beg for food, the males aggressively drive them away.

The Wandering albatross is the largest of the albatrosses. As a juvenile its feathers are brown but as it matures, it becomes progressively whiter. First the body whitens and then the upper wings, starting with white spots on the mid-wing close to the body and slowly whitening along the wings to cover the entire wing except the tips. The tails retain some black.

The breeding cycle of Wandering albatrosses occupies just over a year, and the birds spend the year following breeding at sea. This means that at best they can raise one chick every two years. Wandering albatrosses breed in loose aggregations on gentle hills with good exposure to the

▼ **COURTSHIP DANCE**

Starting at age three, young Wandering albatrosses return to the islands where they were reared to find a mate. But it may be four to six seasons before they successfully find a partner and breed.

wind. The eggs are incubated for 78 to 79 days before the fluffy white chick hatches. Chicks stay in the nest and are fed by both parents for an average of 246 days before they are ready to go out on their own. Towards the end of the nestling period, the older fledglings leave the nest and walk around, exercising their wings.

Wandering albatrosses are found all around the Antarctic in the open ocean, away from all land except for their own nesting areas. They range over the entire Southern Ocean as far north as 27°S and as far south as 67°S—a short distance from the ice edge. The most abundant form nests throughout the sub-Antarctic. Although traditionally considered a single species containing five subspecies, recent genetic studies argue

that the Wandering albatross is actually five distinct species, each with unique characteristics in its plumage. The four recognized new species nest on Gough Island, Ile Amsterdam, and the Auckland and Antipodes islands, and each island population has its own home range.

Wandering albatrosses feed primarily on squid and fish, which they seize once they land on the water surface. They eat more carrion than other albatrosses, and although they are habitual ship-followers, they tend to hover further from ships than Black-browed albatrosses and other petrels.

Studies on South Georgia have found that female Wandering albatrosses tend to fly north to forage, whereas the males head south. The northward-flying females are more likely to encounter long-line fishing boats, and so are more likely than males to get caught and drown. The chicks of drowned adults do not survive because a single parent cannot provide enough food for a growing chick. Breeding among paired adults can also be threatened by the excess loss of females: widowed males searching for a new mate may pester already paired females, and the confusion and harassment sometimes causes breeding females to fail.

Royal albatross

DIOMEDEA EPOMOPHORA

LENGTH: 107–122 cm/42–48 in

WINGSPAN: 305–350 cm/120–138 in

STATUS: about 10,000–20,000 breeding pairs

IUCN: Vulnerable   CITES: not listed

Closely related to the Wandering albatross, the Royal albatross is very similar in appearance. Immature Royal albatrosses, however, are mostly white, and only young juveniles have black on their tails. Royal albatrosses also have distinctive black lines along the cutting edges of their upper bill, although these can only be seen at close range. The adults of the southern subspecies retain the dark tops of the wings.

Royal albatrosses are most common in the Pacific, but can be seen throughout the Southern Ocean. They are divided into two subspecies (or species in the new taxonomy)—the Northern royal (*D. e. sanfordi*), which breeds on the South Island of New Zealand and on the Chatham Islands, and the Southern royal (*D. e. epomophora*), which breeds on Campbell Island and the Auckland Islands. In the non-breeding season the two subspecies mingle at sea.

The breeding behavior of the Royal albatross is exactly like that of the Wandering albatross, although the precise timing may differ according to the island where they breed and the conditions that they find there. Like Wandering albatrosses, Royal albatrosses feed extensively on squid, which they take by surface-seizing. Unlike Wandering albatrosses, they are not habitual ship-followers, but they readily join other petrels to feed on offal discarded from fishing boats.

▲ DEVOTED PARENTS

Sitting on its egg high on Campbell Island, this Royal albatross and its mate will tend the egg constantly for 79 days. Males usually incubate for a few days longer than females, but each sits for 4 to 13 day shifts, patiently waiting—and fasting—in all kinds of weather.

ROYAL ALBATROSS

This species is difficult to identify at sea and may be circumpolar.

0°

Tristan da Cunha   Gough I

South Georgia   Sth Sandwich Is   Bouvetøya   Prince Edward Is   Is Crozet

Falkland Is   Sth Orkney Is

Heard I   Is Kerguelen

90°W   South Pole   90°E

Balleny Is   Polar Front

Campbell I   Macquarie I   Auckland Is
Antipodes Is
Bounty Is
Chatham Is

180°

Distribution
Breeding sites

▲ EGG LAYING

A female Royal albatross lays its egg in mid-November. The egg takes about 40 days to form in the female, but laying typically lasts just three minutes.

Yellow-nosed albatross

DIOMEDEA CHLORORHYNCHOS

OTHER NAME: Yellow-nosed mollymawk

LENGTH: 71–81 cm/28–32 in

WINGSPAN: 200–256 cm/79–101 in

STATUS: 80,000–100,000 breeding pairs

IUCN: Lower Risk–near threatened

CITES: not listed

Distribution
Breeding sites

YELLOW-NOSED ALBATROSS

▼ RESTING AT SEA

A Yellow-nosed albatross swims calmly on the sea. The usual image of albatrosses is soaring on the wind, but they regularly alight on the water to feed or rest. They have no trouble taking off as long as there is a wind blowing.

▼ MUTUAL PREENING

Allopreening, where one bird preens another, is an important social behavior among albatrosses, particularly for strong pair bonds. While adults that have chicks are at sea finding food for their growing chicks, failed breeders and non-breeders continue to seek mates or reconfirm their pair bond.

The Yellow-headed albatross, with its white body and dark wings, looks like the other mollymawks, but its gray head is lighter than that of the Gray-headed albatross, and it has a whitish cap. Its underwings have more white than those of the Black-browed and Gray-headed albatrosses but not as much as those of the White-capped albatross. It is more slimly built than other mollymawk species and its long, thin bill is black, with pronounced yellow lines above and below.

There are two subspecies, which the revised taxonomy defines as species in their own right. The eastern subspecies (*D. c. bassi*) breeds in the western Indian Ocean around Prince Edward Island, and Iles Amsterdam and Crozet; the western subspecies (*D. c. chlororhynchos*) breeds around Gough Island and the Tristan da Cunha group of islands in the South Atlantic. These albatrosses feed mainly on squid, but they also take crustaceans and offal. Around fishing boats, they typically grab a morsel of offal or bait and fly away from the other scavengers to consume it.

Yellow-headed albatrosses start their annual breeding cycle in August and September, earlier than other southern hemisphere species. They nest singly or in loose groups on fairly flat ground, in nests of mud and grass. Parents incubate the single egg for 71 to 78 days, and then feed the chick for about 130 days, when it fledges. After the breeding season they disperse around the South Atlantic and Indian oceans between latitudes 15 and 50°S. They occur along the coasts of South Africa and Argentina, and south of Australia to New Zealand. In recent years they have been seen along the coast of North America as far north as Canada.

Black-browed albatross

DIOMEDEA MELANOPHRIS

OTHER NAME: Black-browed mollymawk

LENGTH: 83–93 cm/32–36 in

WINGSPAN: 240 cm/94 in

STATUS: 550,000–600,000 breeding pairs; more than 2 million birds

IUCN: Lower Risk–near threatened   CITES: not listed

The Black-browed albatross is probably the most common of the albatrosses and one of the most recognizable. Its body is white with a black saddle between its wings, its upper wings are black, and its underwings are white with a broad black margin. It has a yellowish bill, and a white head with clearly marked black eyebrows. From a distance, and at rest, it can be mistaken for the Kelp gull, but its soaring flight, in elegant contrast to the flapping motion of the Kelp gull, identifies it as an albatross. It feeds on crustaceans—mainly krill—as well as squid and fish, but it also devours carrion, joining other petrels at fishing boats to scavenge the offal.

Black-browed albatrosses fly over the whole of the Southern Ocean, and have been spotted close to shore as well as far from land. Their range extends along the cold-water currents of the west coasts of South America and Africa, and in high southern latitudes they are regularly seen along the Antarctic Peninsula in the Gerlache

White-capped albatross

DIOMEDEA CAUTA

OTHER NAMES: Shy albatross, Mollymawk
LENGTH: 90–99 cm/35–39 in
WINGSPAN: 220–256 cm/87–101 in
STATUS: 800,000–1,000,000 birds
    IUCN: Lower Risk–near threatened   CITES: not listed

The White-capped albatross is the heaviest and most thickset of the mollymawks. The undersides of its wings are almost entirely white, with a thin border of black, and it has characteristic black "thumbprints" on the underwing near the body. Its head is a very pale gray, and its dark eyes and eyebrows make it look stern. Its bill is light gray with a yellowish tip.

There are three subspecies of *Diomedea cauta*, distinguished by the coloration of the head and bill. The Tasmanian shy albatross (*D. c. cauta*) breeds around Tasmania and the Auckland Islands; the Chatham albatross (*D. c. eremita*) breeds on the Chatham Islands; and Salvin's albatross (*D. c. salvini*) breeds on Iles Crozet in the Indian Ocean, and on Snares Island and Bounty Island in the South Pacific.

WHITE-CAPPED ALBATROSS

▲ BONDING

A White-capped albatross preens its mate. Most albatrosses have chewing lice in their feathers, but there is no indication that preening can alleviate the problem. Instead, it is a social behavior used by paired adults.

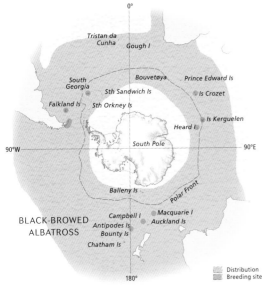

BLACK-BROWED ALBATROSS

Distribution
Breeding sites

Strait, where there may be considerable ice. They breed annually, the females laying a single egg in a substantial nest of mud and grass. Breeding colonies can consist of as many as 100,000 pairs of birds, and are often shared with shags and penguins.

Little is known of their migratory patterns after breeding, but there is a general movement northward in the southern autumn. Stragglers are regularly seen in the North Atlantic, and there have been over 40 recorded sightings in Britain.

> **COURTSHIP ON THE WING**
The two species of sooty albatrosses use aerial displays for some of their social interactions. Here a male Light-mantled sooty on the ground calls to a female flying overhead. She will fly over several males and perhaps visit a few of them before settling on a prospective mate.

▲ **SLOW AND STEADY**
Gray-headed albatrosses (*D. chrysostoma*) are biennial breeders. Their chicks grow slower and are fed longer than other mollymawks, but they do have consistently higher fledging success.

White-capped albatrosses normally breed annually, starting in September, and their breeding cycle is the same as those of other mollymawks. They eat mostly squid, as well as fish, crustaceans, and offal from fishing boats. Despite one of their common names—Shy albatross—they often feed with other seabirds, and they have been seen following whales. They compete successfully with the birds that crowd around fishing boats.

After the breeding season, White-capped albatrosses are seen throughout sub-Antarctic latitudes in the Pacific and Indian oceans, with the exception of the Chatham Island subspecies, which is rarely seen far from its island home. Unlike other albatrosses, White-capped albatrosses are frequently seen over continental shelves, or even close to shore.

Gray-headed albatross
DIOMEDEA CHRYSOSTOMA
OTHER NAME: Gray-headed mollymawk
LENGTH: 81–84 cm/32–33 in
WINGSPAN: 180–220 cm/70–87 in
STATUS approximately 500,000 birds
IUCN: Vulnerable   CITES: not listed

Except for its gray head, which gives it a hooded appearance, this species is very much like the Black-browed albatross. Its bill is black with yellow lines along

the top and bottom, although this is difficult to see in flight. Its flight patterns, food, and lifestyle are similar to those of the Black-browed albatross, and the two species often share colonies, although the Gray-headed albatross prefers steeper slopes. The structure and timing of the reproductive cycle are also similar, except that Gray-headed albatrosses feed their chicks for three or four weeks longer, so that if they succeed in raising a chick they need to breed only every second year.

Gray-headed albatrosses prefer colder water than most other albatrosses, and they stay well away from land except at their breeding sites. They eat mostly squid, and they follow fishing vessels less than other albatrosses. Like most albatrosses, they are at risk from long-line fishing operations, competition from local squid fisheries, and introduced animals at some of their breeding islands.

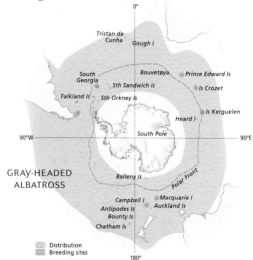

GRAY-HEADED ALBATROSS

0°
Tristan da Cunha
Gough I
South Georgia
Bouvetøya
Prince Edward Is
Sth Sandwich Is
Is Crozet
Falkland Is
Sth Orkney Is
Heard I
Is Kerguelen
90°W
South Pole
90°E
Balleny Is
Polar Front
Campbell I
Macquarie I
Antipodes Is
Auckland Is
Bounty Is
Chatham Is
180°

▢ Distribution
▢ Breeding sites

Sooty albatross

PHOEBETRIA FUSCA

OTHER NAME: Dark-mantled sooty
albatross

LENGTH: 84–89 cm/33–35 in

WINGSPAN: 203 cm/80 in

STATUS: 80,000–100,000 birds

IUCN: Vulnerable   CITES: not listed

Light-mantled sooty albatross

PHOEBETRIA PALPEBRATA

LENGTH: 78–79 cm/30–31 in

WINGSPAN: 183–218 cm/72–86 in

STATUS: about 150,000 birds

IUCN: Lower Risk–near threatened

CITES: not listed

The two albatrosses that
constitute the genus *Phoebetria*
can be distinguished from other
albatrosses by their narrow,
pointed wings and long, wedge-
shaped tails. Returning to shore
only during their breeding
period, they nest in loose
colonies among vegetation
on cliffs or steep slopes. If they fail in their breeding
attempt one year, they can try again in the following
year, but if they succeed in raising their chick, they
can only breed every other year. The breeding areas of
the two species overlap in some areas, and where this
happens they lay their eggs at different times. On Iles
Crozet, for example, the Sooty albatross breeds two
weeks earlier than the Light-mantled sooty albatross, and
on Marion Island one month later. They feed mainly on
squid, but they also take fish, crustaceans, and carrion,
usually penguin carcasses.

Sooty albatrosses are uniformly dark, with
a partial white eye ring and a yellow line
along the cutting edge of their lower
bill. Little is known of their feeding
habits, but they probably surface-
feed at night. They are found
throughout the southern Atlantic
and Indian oceans from Argentina
to Tasmania, and they nest on
Gough and Prince Edward islands,
on Iles Crozet, and on the islands of
the Tristan da Cunha. Sooty albatrosses
on Nightingale Island in the Tristan da
Cunhas are under threat from humans;
each year as many as 3,000 chicks are taken
from a major colony there, and the population
cannot withstand such heavy exploitation.

Identical to the Sooty albatross except for an ash-
gray collar and mantle and a blue line on the
lower mandible, Light-mantled sooty albatrosses prefer colder water than do the
Sooty albatrosses. They nest around
New Zealand, on Iles Crozet and
Kerguelen, on the Prince Edward
islands, and on South Georgia.
These albatrosses like to breed
further inland than the Sooty
albatrosses, but during the non-
breeding season they are much
more widespread than the Sooty
species and occur throughout the
Southern Ocean.

▼ ELABORATE RITUALS

**Light-mantled sooty albatrosses court with alternating
aerial and ground-based displays. At the nest, courtship
proceeds with a series of behaviors such as mutual preening,
open-mouthed thrusting, tail fanning, and bowing.**

SOOTY ALBATROSS

0°
Tristan da Cunha
Gough I
South Georgia
Bouvetøya
Prince Edward Is
Sth Sandwich Is
Is Crozet
Falkland Is
Sth Orkney Is
Is Kerguelen
Heard I
90°W
South Pole
90°E
Balleny Is
Polar Front
Campbell I
Macquarie I
Antipodes Is
Auckland Is
Bounty Is
Chatham Is
180°
Distribution
Breeding sites

LIGHT-MANTLED SOOTY ALBATROSS

0°
Tristan da Cunha
Gough I
South Georgia
Bouvetøya
Prince Edward Is
Sth Sandwich Is
Is Crozet
Falkland Is
Sth Orkney Is
Is Kerguelen
Heard I
90°W
South Pole
90°E
Balleny Is
Polar Front
Campbell I
Macquarie I
Antipodes Is
Auckland Is
Bounty Is
Chatham Is
180°
Distribution
Breeding sites

▲ LARGE BABIES

Light-mantled sooty chicks are guarded and fed
by their parents for the first 19–21 days. They
grow up to 1.4 times heavier than their parents,
but by the time they make their first flight they
dwindle to 91 percent of typical adult weight.

Fulmars

ORDER: Procellariiformes
FAMILY: Procellariidae

▶ SOCIABLE BIRDS
Cape petrels are habitual
ship-followers, often flying in
very close over the bow or
alongside ships. For this
reason, and because they
are so easy to identify, they
are a favorite of ship travelers
in the Southern Ocean.

This family consists of a diverse group of petrels—small to medium-sized seabirds, dark in color, with long wings that enable them to use the wind to fly great distances. The nostrils on the top of their bill join into a single tube with a visible septum separating them. All petrels lay only one egg at each breeding attempt, but incubation times are quite variable. They often leave their egg unattended while they forage. Because of these breaks in incubation, petrel embryos withstand cold better than those of most other birds. The period from hatching to fledging also varies due to changes in local feeding conditions and the parents' ability to raise their chick.

Family origins

The first group of petrel-like birds diverged from the divers, or loons (Order Gaviiformes) about 50 million years ago, and soon after the Diomedeidae (albatrosses) diverged from the petrels. The first fossil evidence of the Procellariidae dates from 40–50 million years ago and is considered to be from a shearwater. They probably originated in the southern hemisphere, later spreading northward. Petrels are now found in all of the world's oceans, but the southern hemisphere still supports many more petrel species than the northern hemisphere.

This group presents many taxonomic puzzles that are exacerbated by lack of information about their breeding behavior and ecology. However, the family can be split into four main groups based on anatomical and ecological differences such as their flight patterns, feeding habits, and bill shapes; they are the fulmarine petrels, the prions, the gadfly petrels, and the shearwaters.

Fulmarine petrels

Ranging from the Arctic to the Antarctic, the fulmarine petrels are the most diverse of this group. All fulmarine petrels have relatively long, broad wings, but they vary considerably in size and plumage. There are two sets of sibling species—the Northern and Southern giant petrels, and the Northern and Southern fulmars—and three further species: the Snow petrel, the Antarctic petrel, and the Cape petrel. The giant petrels fly with slow, powerful wingbeats, while the smaller species have faster wingbeats; all are good gliders. Fulmarine petrels feed on squid and crustaceans, mainly surface krill. The giant petrels, fulmars, and Cape petrels are ship-followers.

Prions

Prions are small petrels of the genus *Pachyptila,* and the Blue petrel—the only member of the genus *Halobaena.* (The Blue petrel has been placed in each of the four Procellariid groups, but its coloration and other structural features link it with the prions.) Prions are similar in appearance and habits, sharing a buoyant, erratic flight. All are a light blue-gray with white underparts, dark primary feathers, and upper wing coverts forming an M-shape on their upper wings. All have black tips on their wedge-shaped tails, except for the Blue petrel, which has a white tip. Some breeding sites are threatened by predation by introduced rats, cats, and pigs, and commercial krill harvesting may limit food supply .

Gadfly petrels

Traditionally this group comprises 23 to 34 *Pterodroma* species and two *Bulweria* species, but relationships within the group are not understood. Several of the *Pterodroma*

◀ UNGAINLY ON LAND
The Broad-billed prion, seen here clambering towards its burrow, is ungainly on the ground but agile on the wing. With its large bill and well-developed filtering system, it is particularly adapted to skimming the sea suface for food, usually crustaceans.

species are known only from a few specimens taken long ago and only recently rediscovered. Gadfly petrels have long, wedge-shaped tails and relatively short wings. In flight they flap and glide, rising in high, swooping arcs, and they tend to bend their wrists while flying. They rarely land on water, but pluck squid, crustaceans, and fish from the surface; they are not ship-followers, and they do not dive for food. Most nest in small colonies or scattered pairs, but because of their isolated breeding areas and their secretiveness, the breeding sites of several species are unknown. Many colonies are threatened by predation and harassment by introduced species.

Shearwaters

Shearwaters are a very old and complex group consisting of four species in the genus *Procellaria*, some 15 to 20 in *Puffinus*, and two in *Calonectris*—the latter two species rarely venturing into Antarctic waters. The range of many shearwaters is only partially in Antarctic or sub-Antarctic waters, and several migrate to northern hemisphere waters. They are medium-sized to large petrels with long bills, long, narrow wings, and short, rounded tails. The smaller shearwaters alternate bursts of rapid wingbeats with stiff-winged banking glides near the ocean surface, dipping into the troughs and rising over the crests of the waves. The flight of the larger species is more relaxed and deliberate. Their plumage is dark above, and brown or white below. They feed on fish, crustaceans, and squid on the surface or from shallow dives. Only the four *Procellaria* species commonly venture into Antarctic waters, and none of the species has a wholly Antarctic distribution.

ARMED AGAINST ATTACK
Southern fulmars nest in the open on cliff ledges and steep slopes. They can protect themselves and their chicks from attacks, often by skuas, by spitting or spraying sticky stomach oil onto attackers. They will even spit on people who make them nervous by approaching too close.

FAIRIES OF THE SOUTH
Apsley Cherry-Garrard, a biologist on R.F. Scott's British Antarctic Expedition of 1910–13, called Snow petrels "the fairies of the south," a poetic name for one of the most beautiful Antarctic birds.

OCEAN SCAVENGERS
A Northern giant petrel feeds on a dead penguin in shallow water. The giant petrels, especially the males, are the top scavengers throughout the Southern Ocean. They dominate all other petrels and skuas at a carcass.

▲ ANTARCTIC VULTURES
Because of their size and
strength, giant petrels are the
only birds that can gain access
to the larger carcasses of
seals. While feeding, the
dominant bird threatens all
others with outspread wings
and erect, fanned tail. Once
he takes his fill, the next
dominant giant petrel will
get its turn feeding.

Southern giant petrel

MACRONECTES GIGANTEUS

OTHER NAME: Southern giant fulmar
LENGTH: 86–99 cm/34–39 in
WINGSPAN: 185–205/73–81 in
STATUS: approximately 36,000 breeding pairs
    IUCN: Vulnerable   CITES: not listed

The Southern giant petrel, with its close relative, the
Northern giant petrel, is the largest of the Procellariidae.
These species are sometimes mistaken for small
albatrosses, but they do not glide as much as albatrosses,
their bodies are heavier and more hunched, and they
have thicker bills because of the size of their tubular
nostrils on the top of their bill.

Southern giant petrels have heavy, yellow-
green bills with pale green tips. Most adults
are brownish gray, with dirty white heads,
necks, and upper breasts; juveniles are
sooty black with a few white flecks on
the head. A few adults have white
plumage scattered with occasional
black spots, and the juveniles have
the same coloration as the adults.
Their circumpolar range extends
south to the edge of the Antarctic
ice and north into the subtropics.
They breed on exposed hilltops
and plains on sub-Antarctic islands
throughout the region in small, loose
colonies, building nests of small stones
or grass. Breeding starts in October, with
both parents taking turns to incubate a single

egg for 55 to 66 days. After about four months the
chicks fledge and leave the breeding site to roam the
Southern Ocean for six or seven years until they are
mature and ready to breed.

Giant petrels are probably Antarctica's most
important scavengers, feeding on carrion, especially
the carcasses of seals, penguins, and petrels. They can be
aggressive, and will drown or batter to death such birds
as Cape petrels or immature albatrosses. They also
surface-feed on squid, krill, and fish. They are avid
ship-followers in quest of refuse or offal, and many are
caught on the hooks of long-line fishing boats. Partly
for this reason, populations appear to be dwindling, but
this apparent decline may also be due to their extreme
sensitivity to disturbance at their nest sites by humans.

Northern giant petrel

MACRONECTES HALLI

OTHER NAME: Northern giant fulmar
LENGTH: 81–94 cm/32–37 in
WINGSPAN: 180–200 cm/71–79 in
STATUS: 7,000–12,000 breeding pairs
    IUCN: Lower Risk–not threatened   CITES: not listed

The coloration of this species resembles that of the
dark-phase Southern giant petrels, although they are
never as light on the head and neck as their Southern
cousins. There is no white form. They are slightly
smaller than the Southern species, but the physical
feature that distinguishes them positively from the
Southern giant petrel is their bill, which is horn-colored
with a pinkish or dark red tip. Like the Southern species

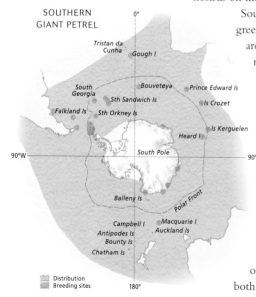

SOUTHERN
GIANT PETREL

Tristan da
Cunha
Gough I
South
Georgia
Bouvetøya
Prince Edward Is
Sth Sandwich Is
Is Crozet
Falkland Is
Sth Orkney Is
Is Kerguelen
Heard I
South Pole
90°W
90°E
Balleny Is
Polar Front
Campbell I
Macquarie I
Antipodes Is
Auckland Is
Bounty Is
Chatham Is
0°
180°
Distribution
Breeding sites

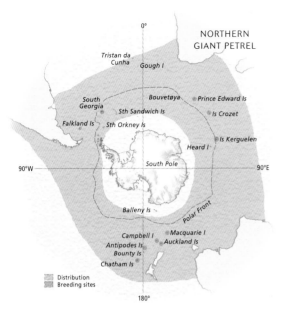

NORTHERN GIANT PETREL

Tristan da Cunha
Gough I
South Georgia
Bouvetøya
Prince Edward Is
Sth Sandwich Is
Is Crozet
Falkland Is
Sth Orkney Is
Is Kerguelen
Heard I
South Pole
90°W
90°E
Balleny Is
Polar Front
Campbell I
Macquarie I
Antipodes Is
Auckland Is
Bounty Is
Chatham Is
180°

Distribution
Breeding sites

◄ KEEPING CLEAN

Vultures in warmer and drier climates have no feathers to soak up blood on their heads and necks. Giant petrels cannot afford to forgo insulation around their heads, but they wash long and often to prevent their head and neck feathers from becoming matted with blood.

▼ PROMINENT NOSTRILS

The Southern fulmar is a good example of the tube-nosed seabirds whose nostrils develop into prominent tubes on the top of the bill. Studies show that the tubes facilitate a well developed sense of smell. The clear drop of liquid hanging from the bill is the highly saline excretion from the salt glands.

they have a circumpolar distribution, but they generally stay north of the Polar Front, breeding mostly on South Georgia and Iles Crozet and Kerguelen, and in the New Zealand region. They nest alone or in small colonies, and their breeding season starts in August. Interestingly, the sexes prefer different diets, the males feeding on the carrion of seals, penguins, and seabirds, and the females mostly going to sea to catch live prey, especially krill and squid.

Southern fulmar

FULMARUS GLACIALOIDES

OTHER NAMES: Silver-gray fulmar, Antarctic fulmar

LENGTH: 46–50 cm/18–20 in

WINGSPAN: 114–120 cm/45–47 in

STATUS: abundant, but no accurate estimates

IUCN: not listed   CITES: not listed

The Southern fulmar is the largest petrel after the two giant petrels, easily recognized by its pale gray upper parts and white underparts. Its wings have a dark trailing edge, a white flash at the base of the primaries, and variable amounts of black at the wingtips. Southern fulmars are circumpolar, and normally frequent cold

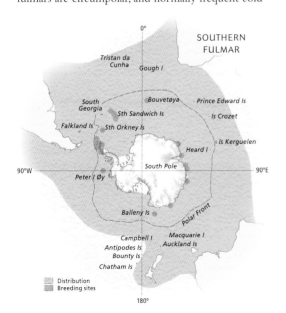

SOUTHERN FULMAR

Tristan da Cunha
Gough I
South Georgia
Bouvetøya
Prince Edward Is
Sth Sandwich Is
Is Crozet
Falkland Is
Sth Orkney Is
Is Kerguelen
Heard I
South Pole
90°W
90°E
Peter I Øy
Balleny Is
Polar Front
Campbell I
Macquarie I
Antipodes Is
Auckland Is
Bounty Is
Chatham Is

Distribution
Breeding sites

180°

water areas on the edges of the pack ice. Outside the breeding season they range widely around Antarctic waters. They feed mainly on crustaceans (krill) and fish (Antarctic silversides), with varying amounts of squid, carrion, and offal, depending on the location. They are primarily surface-feeders, but sometimes make shallow dives.

They nest in colonies on sheltered ledges on cliffs and steep slopes. Breeding sites are restricted because of the birds' need for cliffs, but there are huge colonies on the South Sandwich and South Orkney islands. Breeding begins in November, and the parents take turns incubating the single egg for about 46 days. The chick fledges after 48 to 56 days.

ANTARCTIC PETREL

Tristan da Cunha
Gough I
0°
South Georgia
Bouvetøya
Sth Sandwich Is
Prince Edward Is
Falkland Is
Sth Orkney Is
Is Crozet
Is Kerguelen
Heard I
90°W
South Pole
90°E
Balleny Is
Polar Front
Campbell I
Macquarie I
Auckland Is
Antipodes Is
Bounty Is
Chatham Is
180°

Distribution
Breeding sites

▼ INLAND NESTERS
Antarctic petrels usually feed in open water around the pack ice, but they nest on cliff ledges, sometimes more than 200 kilometers (120 miles) from the coast.

Antarctic petrel

THALASSOICA ANTARCTICA

LENGTH: 40–46 cm/16–18 in
WINGSPAN: 101–104 cm/40–41 in
STATUS: many millions of birds
IUCN: not listed   CITES: not listed

This is one of the most numerous of the Antarctic birds. Like the Cape petrel, for which it is often mistaken from below, it has a white underbody and wings and a black face and throat. However, its upper side is quite distinctive. Its head, body, and the front half of its wings are dark brown to dark gray, while the back half of its wings are white, as is its tail except for a black tip. The female is slightly smaller than the male.

Antarctic petrels are difficult to observe because they stay mostly within the boundaries of the pack ice,

although some do fly as far north as the Polar Front in winter. They are truly Antarctic birds, feeding in open water near pack ice and favoring areas with icebergs. They are sometimes found in association with other seabirds and whales, and will sometimes roost in large numbers on icebergs. They nest on cliff ledges and begin breeding in November. About 35 nesting colonies have been identified, some as far as 250 kilometers (155 miles) inland, but the known colonies cannot account for their estimated numbers. They feed mainly on krill, and to a lesser extent on fish and squid, which they take by surface-seizing, although they also plunge-dive and dip for food.

Cape petrel

DAPTION CAPENSE

OTHER NAMES: Pintado petrel, Cape pigeon
LENGTH: 38–40 cm/15–16 in
WINGSPAN: 81–91 cm/32–36 in
STATUS: several million birds
IUCN: not listed   CITES: not listed

One of the most familiar of the fulmarine petrels, the Cape petrel has an unmistakable upper wing pattern of white splotches on black, and the larger white patches on its back and tail are often flecked with black. Its tail is white with a broad black terminal band. Underneath it is white with a black head and throat.

Cape petrels have a circumpolar distribution that ranges from the subtropics to the edge of the Antarctic Continent. In winter they stay far from shore, and from April to August they leave the Antarctic and move north into the sub-Antarctic. During the breeding season, however, they stay closer to their Antarctic home, and forage mostly in inshore waters. In November they

> **THE SMALLEST PETREL**

The Snow petrel is the smallest of the fulmarine petrels and the ecological equivalent of the Ivory gull (*Pagophila eburnea*) of the Arctic. These petrels forage in leads in pack ice and nest in crevices in cliffs or scree. They nest further inland than any other petrel, and to combat the cold their downy feathers cover the base of the bill and extend well down the legs.

CAPE PETREL

Tristan da Cunha
Gough I
Bouvetøya · Prince Edward Is
South Georgia
Sth Sandwich Is · Is Crozet
Falkland Is · Sth Orkney Is
Heard I · Is Kerguelen
90°W
Peter I Øy
South Pole
90°E
Scott I
Balleny Is · Polar Front
Campbell I · Macquarie I
Antipodes Is · Auckland Is
Bounty Is
Chatham Is
180°

☐ Distribution
☐ Breeding sites

return to their nesting cliffs to breed, building simple nests on small ledges. Adults take turns to incubate the single egg for 41 to 50 days. The young fledge in 47 to 57 days, and those that survive have a life expectancy of 15 to 20 years.

Snow petrel

PAGODROMA NIVEA
LENGTH: 30–40 cm/12–16 in
WINGSPAN: 75–95 cm/29½–37 in
STATUS: possibly several million birds
IUCN: not listed   CITES: not listed

Snow petrels have pure white plumage, dark eyes, and a dark bill. They forage in areas with pack ice; with their white camouflage they seem to materialize out of nowhere, flitting by with a dancing flight and then disappearing into the distance. They dip-feed on the wing, taking mainly krill with some fish, squid, and carrion. They also land on the surface and feed by surface-seizing or shallow diving. They are circumpolar, nesting extensively on the Southern Continent and on some of the islands in high latitudes. Some remain

around the Continent all year, but the main influx of breeders arrives from mid-September to early November.

Snow petrels, along with South polar skuas, breed further south than any other birds in the world, in some of the harshest conditions on earth. They nest in small to large colonies in crevices in cliffs or among boulders on scree slopes, mostly near the coast, but sometimes as much as 345 kilometers (215 miles) inland, at elevations as high as 2,400 meters (7,870 ft). In the bitter weather they must dig the snow out of the crevices where they nest. Incubation averages 45 days, and the young are brooded for eight days after hatching; the parents then go to sea to bring food back to their chick. Chicks leave the colony at 42 to 50 days old.

▲ **ALMOST ADULT**

Already sporting the checkered wings of the adult, this Cape petrel chick still has some nestling-stage down. Soon he will be off on his own for six years before returning to his natal area to breed.

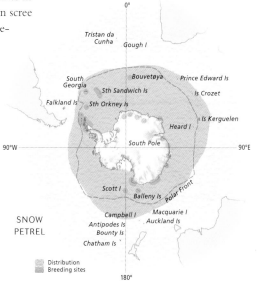

Tristan da Cunha
Gough I
Bouvetøya · Prince Edward Is
South Georgia
Sth Sandwich Is · Is Crozet
Falkland Is · Sth Orkney Is
Heard I · Is Kerguelen
90°W
South Pole
90°E
Scott I
Balleny Is · Polar Front
SNOW PETREL
Campbell I · Macquarie I
Antipodes Is · Auckland Is
Bounty Is
Chatham Is
180°

☐ Distribution
☐ Breeding sites

Narrow-billed prion

PACHYPTILA BELCHERI

OTHER NAMES: Slender-billed prion, Thin-billed prion

LENGTH: 25–26 cm/10½–11 in

WINGSPAN: 56 cm/22 in

STATUS: more than one million breeding pairs

IUCN: not listed   CITES: not listed

The Narrow-billed prion is one of the more distinctive of the prions. It has the blue-gray body that is typical of all prions, but the black on the end of its tail is just a small central spot instead of a broad terminal band. Its slender bill, black eye line, and clearly marked white stripe above and behind the eye give it a striped facial pattern that may be distinguished in flying birds at sea.

These birds are gregarious. They feed mainly at night on small crustaceans—mostly amphipods, but they will also take some krill and small fish, using their thin bill to pick prey from the water individually. Although feeding at sea for most of the year, during the breeding season they feed inshore close to their breeding colonies. They breed in burrows on the Falkland Islands, on some islands off southern Chile, on Iles Crozet and Kerguelen, and possibly on Macquarie Island and South Georgia.

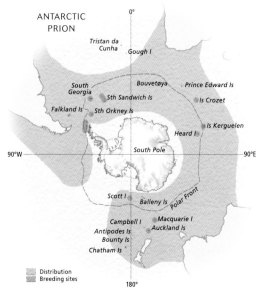

NARROW-BILLED PRION

Tristan da Cunha
Gough I
South Georgia
Bouvetøya
Prince Edward Is
Sth Sandwich Is
Is Crozet
Falkland Is
Sth Orkney Is
Is Kerguelen
Heard I
90°W
South Pole
90°E
Balleny Is
Polar Front
Campbell I
Macquarie I
Antipodes Is
Auckland Is
Bounty Is
Chatham Is
180°

Distribution
Breeding sites

Antarctic prion

PACHYPTILA DESOLATA

LENGTH: 25–27 cm/10–11 in

WINGSPAN: 58–66 cm/23–26 in

STATUS: many millions of birds; abundant and widespread

IUCN: not listed   CITES: not listed

The Antarctic prion is a fairly large bird, with the widest distribution of all the prions. It can be identified by the broad black tip on its tail and its relatively extensive

dark blue-gray collar. It feeds mainly on krill and other small crustaceans, although it will also take some fish. Its bill is adapted for filter feeding, and it has developed a specialized behavior, called hydroplaning, to take advantage of this feature: it scurries along the surface of the water into the wind, with its body resting on the water but partially supported by the wind flowing under its outstretched wings. Holding its mouth open at the surface, or with its head entirely submerged, it filters its prey from the water as it swims, sometimes swinging its head from side to side.

Antarctic prions range from latitude 40°S to the ice pack, and are circumpolar except for a gap in the southern Pacific. They breed on islands from New Zealand westward to South America, nesting in burrows in very dense colonies. They begin breeding in November—later if there is too much snow on the breeding grounds. After their chick fledges, at 45 to 55 days, all the birds leave the colonies and disperse northward over a broad area.

Despite the large numbers of Antarctic prions, they have problems in some of their breeding areas from introduced rats, cats, and pigs. Large-scale harvesting of krill may cause concern in the future.

Broad-billed prion

PACHYPTILA VITTATA

LENGTH: 25–30 cm/10–12 in

WINGSPAN: 57–66 cm/22½–26 in

STATUS: several hundred thousand birds

IUCN: not listed   CITES: not listed

The Broad-billed prion has a fairly restricted range, nesting on Gough Island and the islands of the Tristan da Cunha group in the Atlantic, on Marion Island, and on the sub-Antarctic islands off the coast of New Zealand. It is highly gregarious, but rarely follows ships. Broad-billed prions are difficult to identify in flight, but at close quarters their large size and massive bill distinguish them from other prions. They live on small planktonic crustaceans and, like Antarctic prions, they feed by hydroplaning, using their specialized bill to filter

ANTARCTIC PRION

Tristan da Cunha
Gough I
South Georgia
Bouvetøya
Prince Edward Is
Sth Sandwich Is
Is Crozet
Falkland Is
Sth Orkney Is
Is Kerguelen
Heard I
90°W
South Pole
90°E
Scott I
Balleny Is
Polar Front
Campbell I
Macquarie I
Antipodes Is
Auckland Is
Bounty Is
Chatham Is
180°

Distribution
Breeding sites

> **DISTINCTIVE PLUMAGE**
Unmistakably a prion, with its bluish back and the bold "M" pattern formed by black primaries and diagonal black stripes across the top of the wings, this Antarctic prion also has a strong black eye line and an extensive dark collar.

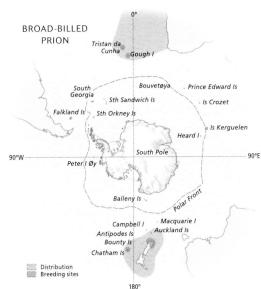

BROAD-BILLED PRION

Tristan da Cunha
Gough I
0°
South Georgia
Bouvetøya
Prince Edward Is
Sth Sandwich Is
Falkland Is
Sth Orkney Is
Is Crozet
Is Kerguelen
Heard I
South Pole
90°W
90°E
Peter I Øy
Balleny Is
Polar Front
Campbell I
Macquarie I
Antipodes Is
Auckland Is
Bounty Is
Chatham Is
Distribution
Breeding sites
180°

▲ PERFECT CAMOUFLAGE

A Fairy prion drops gently to the surface of the water. Prions have no difficulty taking off from the surface, even in calm conditions. The blue-gray of their backs makes ideal camouflage against the sea.

◄ PROTECTIVE MEASURES

A Broad-billed prion emerging from its burrow. Skuas are their major predator on land because they can dig prions right out of their nests. The concrete mouth of this burrow may provide some protection.

food from the water. The weather at their breeding sites is comparatively mild, and breeding begins early, in July or August. They nest in burrows, occasionally sharing with another pair of birds. The parents incubate the egg for 50 days, and then spend a further 50 days raising the chick.

Fairy prion

PACHYPTILA TURTUR

LENGTH: 23–28 cm/9–11 in

WINGSPAN: 56–60 cm/22–23 ½ in

STATUS: several million birds

IUCN: not listed   CITES: not listed

Little is known of the distribution of Fairy prions, but they are probably circumpolar. They have the standard

prion plumage, with few distinguishing features. Highly colonial, they breed on most sub-Antarctic islands in burrows on cliffs and rock falls, or in grassland with limited vegetation. Breeding begins in September, and they incubate their egg for about 55 days. They feed their chick for 43 to 56 days, and then all the birds leave the colony, probably flying to subtropical waters off the coasts of Australia and South Africa.

FAIRY PRION

Tristan da Cunha
Gough I
0°
South Georgia
Bouvetøya
Prince Edward Is
Sth Sandwich Is
Falkland Is
Sth Orkney Is
Is Crozet
Is Kerguelen
Heard I
I Amsterdam
South Pole
90°W
90°E
Balleny Is
Polar Front
Campbell I
Macquarie I
Antipodes Is
Auckland Is
Bounty Is
Chatham Is
Distribution
Breeding sites
180°

Fulmar prion

PACHYPTILA CRASSIROSTRIS

LENGTH: 24–28 cm/9½–11 in

WINGSPAN: 60 cm/24 in

STATUS: 100,000 breeding pairs

IUCN: not listed   CITES: not listed

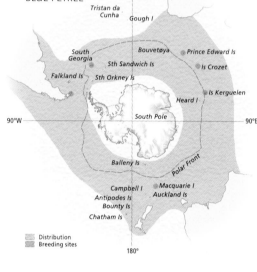

One of the least known of the prions, the Fulmar prion is so similar in appearance to the Fairy prion that it is sometimes classified as a subspecies of the Fairy prion.

The Fulmar prion has a limited range and breeds on Heard, Auckland, Snares, Chatham, and Bounty islands in October, when it occupies rock crevices and cracks in coastal cliffs and boulder slopes, sometimes alongside albatrosses. It is largely sedentary, and some adults roost in their nests all winter. They are affected by the presence of cats, rats, and pigs on some of the islands on which they breed.

will also dive and plunge. They breed in colonies on sub-Antarctic islands that have soil and vegetation, nesting in long burrows that they excavate in soft soil under tussocks. They begin breeding in September and raise their chicks for 43 to 60 days. Many of the adults are sedentary, staying close to their breeding sites in the winter and even visiting the sites intermittently, but the young birds disperse more widely, traveling throughout the range for several years before they are sexually mature and ready to breed.

▲ DAYLIGHT COURTSHIP

Unlike other prions, who shun the daylight hours, Fulmar prions regularly perform courtship displays during the day while perched in the open. Predatory skuas are uncommon in many of their breeding areas, but prions will mob a skua if it dares come near the nest area.

Blue petrel

HALOBAENA CAERULEA

LENGTH: 26–32 cm/10–12½ in

WINGSPAN: 62 cm/24 in

STATUS: several million birds

IUCN: not listed   CITES: not listed

In appearance this species resembles the prions, except for its white tail tip. Its flight is a little slower and less erratic because it is slightly larger. It has a typical circumpolar distribution and flies over all the seas from the temperate latitudes along the coasts of Africa, Australia, and South America to near the Antarctic Continent. A marine pelagic bird, it does not venture into the pack ice in the far south.

Blue petrels feed mostly on crustaceans in some parts of their range, but take more fish in other parts. They take food mainly by surface-seizing and dipping, but

▲ SMALL DIFFERENCES
With its species-identifying white band visible on the end of its tail, a Blue petrel takes off from the surface. These petrels are usually classified with the prions, though they are slightly larger and have a narrower bill.

Great-winged petrel

PTERODROMA MACROPTERA

OTHER NAMES: Gray-faced petrel

LENGTH: 41 cm/16 in

WINGSPAN: 97 cm/38 in

STATUS: about 400,000 breeding pairs

    IUCN: not listed   CITES: not listed

The Great-winged petrel is uniformly dark with a stubby bill and black feet; some have a patch of light gray feathers around the base of their bills, hence their other common name, Gray-faced petrel. They are almost circumpolar in the sub-Antarctic, being absent only from the western Atlantic. They feed mainly at night on squid, which they may track down by their bio-luminescence. They start breeding in April in scattered pairs or small colonies on Iles Kerguelen, the south coast of Australia, and New Zealand and its islands. At the end of the breeding period, Little penguins or one of the shearwater species often move into the burrows after the Great-winged petrels have left.

▲ A PRECARIOUS EXISTENCE
Great-winged petrels are still taken legally by the Maori in New Zealand for personal use, but a former commercial harvest in the Australian region has been phased out. The most important current problems for this species are introduced cats and rats that prey upon them, and rabbits that evict them from their burrows.

◄ SHARING NEST SITES
The boulder-strewn shores of Bounty Island furnish nesting areas for several species. A juvenile Salvin's albatross (*Diomedea cauta salvini*) occupies a partially built nest, while nearby an Erect-crested penguin (*Eudyptes sclateri*) is swollen up as it begins its molt, and a Fulmar prion (*Pachyptila crassirostris*) has just emerged from its nesting crevice.

▲ AWAY FROM THE COLONY
With an unmistakable white body, dark underwings, and a black eye-patch, the White-headed petrel is easily recognized at sea. These petrels usually nest in large colonies but, like most of the gadfly petrels, they are often seen alone when flying at sea away from the nesting area.

White-headed petrel

PTERODROMA LESSONI

LENGTH: 40–46 cm/16–18 in

WINGSPAN: 109 cm/43 in

STATUS: about 100,000 breeding pairs

IUCN: not listed   CITES: not listed

This bird's white head and underbody contrast with dark underwings. The white of the throat and head becomes gray on the mantle, darkening towards the tail, and a broad, dark eye line gives them a masked appearance at sea. They range throughout the sub-Antarctic from about latitude 30°S to the pack ice in the south. They breed on most sub-Antarctic islands except South Georgia, but are less common in the South Atlantic. They start breeding in October, dispersing around the Southern Ocean in May.

Kerguelen petrel

PTERODROMA BREVIROSTRIS

LENGTH: 33–36 cm/13–14 in

WINGSPAN: 80–82 cm/31–32 in

STATUS: several hundred thousand birds

IUCN: not listed   CITES: not listed

Kerguelen petrels are uniformly dark on the upper parts and slightly lighter below, with plumage so shiny that it appears to have white or silvery patches on it. The birds' large head and steep forehead give them a hooded appearance.

Kerguelen petrels are abundant throughout their range, and are circumpolar right from the sub-Antarctic region down to the pack ice. The birds begin to breed in August, and the chicks fledge by the end of December.

Soft-plumaged petrel

PTERODROMA MOLLIS
LENGTH: 32–37 cm/12 ½–14 ½ in
WINGSPAN: 83–95 cm/32 ½–37 ½ in
STATUS: tens of thousands of birds
IUCN: not listed   CITES: not listed

The Soft-plumaged petrel is mostly dark above and white on the breast and belly. The crown, nape, and upper back are usually slate-brown, as is the partial or complete collar around the upper breast. The lores, chin, and throat are white. From a distance the underwings look dark, but at close range white highlights can be seen, and the upper wings are marked with a dark "M" like that of the prions. They are found around the sub-Antarctic except in the Pacific Ocean, on the islands around New Zealand westward through the Indian and Atlantic oceans to the South American coast. They breed on most sub-Antarctic islands within their range except South Georgia and the Falklands. They commence breeding in September, and the chicks fledge in March.

Mottled petrel

PTERODROMA INEXPECTATA
LENGTH: 33–35 cm/13–14 in
WINGSPAN: 74–82 cm/29–32 in
STATUS: 10,000–50,000 breeding pairs
IUCN: Lower Risk–near threatened   CITES: not listed

The Mottled petrel is distinguishable from other *Pterodroma* species by its white underbody and tail, with a contrasting dark belly. The tail, throat, and chin are usually white, the rest of the lower body dark. The underwings are mostly white, with a dark line along the lower leading edge of the wing out to the white primaries. The upper body is gray, and the face is white with various amounts of gray mottling. They range south from about the Drake Passage to Prydz Bay, Antarctica, and northward through the central part of the Pacific into the Bering Sea. They breed in October, forming dense colonies, and use burrows or rock crevices on the islands around New Zealand. During the breeding season adults probably feed in the south near the Antarctic pack ice. After breeding they quickly fly north to the Bering Sea to feed. They are vulnerable because they suffer from predation by introduced mammals and wekas (New Zealand wading birds with a thievish and aggressive disposition).

White-chinned petrel

PROCELLARIA AEQUINOCTIALIS
LENGTH: 51–58 cm/20–23 in
WINGSPAN: 134–147/53–58 in
STATUS: probably several million birds
IUCN: Vulnerable   CITES: not listed

The White-chinned petrel is a dark, heavily built species with white chin markings that may be small and inconspicuous, or may extend in a curving line up the cheeks to partially encircle the eye, giving the bird a spectacled appearance. The bill is thin in relation to the heavily built head and neck, and is normally ivory in color with some black on the top ridge. They are circumpolar, breeding on most sub-Antarctic islands; however, most—several million birds—breed on South Georgia. They begin breeding about October in burrows on grassy slopes or boggy ground, and disperse widely around the Southern Ocean after breeding. They gather in thousands around ships after offal and bait, and are sustaining severe losses by getting hooked and drowned while trying to steal baits.

Gray petrel

PROCELLARIA CINEREA
OTHER NAMES: Brown petrel, Pediunker
LENGTH: 48–50 cm/19–20 in
WINGSPAN: 115–130/45–51 in
STATUS: poorly known; probably in 100,000s of birds
IUCN: Lower Risk–near threatened   CITES: not listed

The Gray petrel has a whitish underbody contrasting with ash-gray underwings. It has a dark cap and a light mantle that darkens towards the tail and towards the primary feathers on the wings. Gray petrels return to their breeding colonies in February/March, which is unusual for a southern petrel. They incubate the egg for 52 to 61 days and then feed the chick for 110 to 120 days. They are usually solitary at sea, but will feed in larger flocks in productive spots. They are common around New Zealand and its islands, and they also breed on Gough Island and the Tristan da Cunha group. They are not seriously affected by commercial fishing, but have suffered high losses from predation. Introduced wekas on Macquarie Island probably extirpated them there, and in other places rats take many chicks.

▲ SEPARATE SPECIES
The Soft-plumaged petrel is one of the gadfly petrels, a group of different species that are sometimes classified together. Indeed, gadfly species are very similar, but the fact that they are distinct is demonstrated by four species nesting on Gough Island without interbreeding.

▼ CONSUMMATE DIVERS
White-chinned petrels are heavily built birds, often with such inconspicuous white markings that they look to be all dark. They are particularly vulnerable to depredation around fishing boats because they are better divers than most other petrels and will dive after long-line baits, often pursuing the baits far below the surface.

◄ MIGRATORY HABITS
The Sooty shearwater is the most widespread and abundant of the shearwaters. Its breeding range is restricted to the southern hemisphere, but after breeding it crosses the equator and spreads across the world's oceans.

Sooty shearwater

PUFFINUS GRISEUS

OTHER NAMES: Shoemaker petrel, Cape hen

LENGTH: 40–51 cm/16–20 in

WINGSPAN: 94–109 cm/37–43 in

STATUS: several million birds

IUCN: not listed   CITES: not listed

Sooty shearwaters are dark brown with slender bodies, long bills, and long, backswept wings. Their only markings are variable amounts of light brown, gray, or white on the underwing. They have the largest range of all shearwaters, and are found throughout the world's oceans except the Arctic basin and the tropical portions of the Indian Ocean. They generally prefer cold off-shore water, where they feed on shoaling fish and squid. They readily dive from the wing and are good underwater swimmers.

Their breeding starts in October, and they nest in burrows in large, dense colonies, primarily on New Zealand's sub-Antarctic islands: Snares Island alone has 2,750,000 pairs. The incubation period is 53 to 56 days, and chick rearing takes another 86 to 109 days. After breeding they travel great distances around the globe, migrating into the North Pacific and the North Atlantic. In the North Pacific, thousands are killed by entanglement in gill nets. Even so, they are abundant, and the chicks have historically been a source of food, fats, and oil for the New Zealand Maori. The Sooty shearwater is currently the only seabird that is commercially harvested in New Zealand; 250,000 young are taken each year.

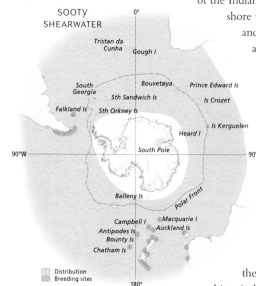

SOOTY SHEARWATER

0°
Tristan da Cunha
Gough I
South Georgia
Bouvetøya
Prince Edward Is
Sth Sandwich Is
Is Crozet
Falkland Is
Sth Orkney Is
Is Kerguelen
Heard I
90°W
South Pole
90°E
Balleny Is
Polar Front
Campbell I
Macquarie I
Antipodes Is
Auckland Is
Bounty Is
Chatham Is
180°

▢ Distribution
▢ Breeding sites

➤ MASS FEEDING
A flock of Sooty shearwaters mass to feed on small shoaling fish, squid, and crustaceans. When the food is concentrated, seabirds can congregate in impressive numbers to take advantage of the abundant—but usually short-lived—bounty.

Great shearwater

PUFFINUS GRAVIS

LENGTH: 43–51 cm/17–20 in

WINGSPAN: 100–118 cm/39–46½ in

STATUS: total population too large to count

IUCN: not listed   CITES: not listed

One of the easiest shearwaters to recognize at sea, the Great shearwater has pale underparts, with a white neck and throat, and a dark cap. White bands across the base of the tail and the back of the neck emphasize the dark cap. The dark feathers on the back and top of the wings have light edges, giving the bird a scaled appearance. They are colonial and nest in burrows or in rock crevices. Breeding starts in October and the pairs

incubate the egg for 53 to 57 days. The chick fledges at about 105 days, and then the birds travel throughout the Atlantic Ocean as far north as Greenland, Iceland, and Scandinavia. They feed mainly on fish, squid, and fish offal, following trawlers in large numbers and plunge-diving from high above the surface. They also follow their prey under water.

Great shearwaters breed only in the Falklands (a few hundred pairs), the islands of the Tristan da Cunha group, and Gough Island. Each year a few thousand adults and about 50,000 chicks are harvested from Nightingale Island by Tristan locals, and there is some concern that the current exploitation is too severe, despite the large overall population.

Little shearwater

PUFFINUS ASSIMILIS

LENGTH: 25–30 cm/10–12 in

WINGSPAN: 58–67 cm/23–26 in

STATUS: little data; widespread and locally abundant

IUCN: not listed   CITES: not listed

The Little shearwater is very small, dark above, and white below. The entire underside of its wings and body is white, and the white of the throat extends to the base of the bill and around the back of the eye. Little shearwaters fly low to the waves, with a flutter and glide pattern. Even in higher winds they "shear" less than other species. They dive and swim well and will investigate ships, but are not ship-followers. They are nearly circumpolar in subtropical and sub-Antarctic waters, but also occur in the Azores and Cape Verde Islands and the surrounding ocean. Adults appear to be sedentary; Little shearwaters that range widely are mostly juveniles wandering the ocean before they reach breeding age. Because they breed in such widely separated places, the breeding season is highly variable. Normally they breed in the local summer period. They are colonial, nesting in burrows or rock crevices among boulders, incubating the egg for 52 to 58 days, and caring for the chick for a further 70 to 75 days. The main threats are the human harvesting of eggs and predation by introduced species.

Storm-petrels

ORDER: Procellariiformes
FAMILY: Hydrobatidae

Storm-petrels probably diverged quite early from the other petrels. There are two subfamilies, the Hydrobatinae in the southern hemisphere, and the Oceanitinae in the northern hemisphere, but all storm-petrels probably originated in the southern hemisphere and colonized the north.

About 18–20 centimeters (7–8 in) long, storm-petrels have shorter and broader wings than other petrels. They dart and weave at wave level in agile and restless flight. In a strong wind they may glide slowly, with wings raised in a V-formation and legs dangling. They often dip their feet into the water as they fly, so that they seem to be scurrying along the surface. They have highly developed salt glands, long, tubular nostrils, and a keen sense of smell for detecting food. Most are black to dark brown on their upper parts, with some gray or white on the face, rump, tail, flanks, or belly. Their flight feathers are mainly dark because of a high concentration of the pigment melanin, which toughens their feathers to withstand salt and wind.

Although small, storm-petrels are usually found far out to sea except during the breeding season, and they are relatively long-lived, with life spans of six to twenty years. They inhabit all the world's seas except in the highest northern latitudes. They nest in remote, predator-free areas in cliff crevices, boulder fields, or scree slopes, coming to their colonies only at night except in high southern latitudes with long hours of summer daylight. Colonies range in size from a few dozen nests to tens of thousands scattered over large slopes.

Storm-petrels are monogamous unless they repeatedly fail to reproduce. Both sexes incubate the egg in stints of two to eight days, although they may leave the egg unattended for long periods of time. The chick fledges after seven to eleven weeks. Chicks that survive the first couple of years join with flocks of non-breeders that visit the breeding colony for several years before becoming mature enough to breed themselves.

▼ **WIDELY TRAVELED**
Perhaps the most abundant seabird in the world, the Wilson's storm-petrel breeds only in Antarctica but migrates throughout the world's oceans after breeding. These birds cannot rest on the wing and actively flap and glide very close to the water. Despite their relatively small size, they perform one of the longest migrations of any bird species.

WILSON'S
STORM-PETREL

Distribution
Breeding sites

Wilson's storm-petrel

OCEANITES OCEANICUS

LENGTH: 15–19 cm/6–7½ in

WINGSPAN: 38–42 cm/15–16½ in

STATUS: many millions of birds

IUCN: not listed   CITES: not listed

Wilson's storm-petrel is all dark with a simple white band across its rump, and sometimes lighter bars above and below the wings. In flight, its legs project beyond its tail. It prefers cold waters above continental shelves, feeding mostly on crustaceans such as krill, on floating carcasses, and on fish offal and rubbish from ships. It is a transequatorial migrant, with one of the longest recorded migrations. It breeds only on sub-Antarctic islands and around the Antarctic Continent, but in the non-breeding season it flies to the North Atlantic, to the Gulf of Alaska in the Pacific Ocean, and throughout the Indian Ocean—amazing journeys for such small birds.

▼ **RELYING ON THE BIGGER BIRDS**

Gray-backed storm-petrels often attend the feeding sites of other birds—especially if there is a carcass floating in the water. As they are quite small, they cannot pull the carcass apart for themselves, but there are plenty of scraps to scavenge.

Other storm-petrels

Three other storm-petrels breed in Antarctic and sub-Antarctic waters. The Black-bellied storm-petrel (*Fregetta tropica*) is widespread in the Southern Ocean. It has the typical dark back and white rump, but the white extends to most of its belly and breast, with a thick black line bisecting the belly. It breeds on islands and isolated offshore stacks to avoid predation by introduced cats, rats, and mice. In the non-breeding season it travels throughout the southern hemisphere, and may even reach the northern parts of the Indian Ocean.

The Gray-backed storm-petrel (*Garrodia nereis*) has a dark back, a white belly, and a gray band on its rump and tail. It occurs throughout the Southern Ocean, but is more common around New Zealand and the nearby sub-Antarctic islands in the Indian Ocean and western South Pacific. It stays closer inshore over the continental shelf than other storm-petrels, probably visiting deeper waters only when it disperses.

The White-faced storm-petrel (*Pelagodroma marina*) takes its name from the prominent white pattern on its face. Its back is gray-brown, and its underside mainly white. Six subspecies are scattered on remote island groups throughout the Atlantic, southern Pacific, and Indian oceans. Although they are very common—with about a million in New Zealand and its outlying islands alone—they are vulnerable to destruction of their breeding areas by livestock, and to exploitation by fishers. In 1970, over 200,000 were found dead on Chatham Island, their legs entangled in the filaments of *Distomium filiferum* (large flatworms).

▲ **BLACK STRIPE**

In seeming contradiction to their name, Black-bellied storm-petrels appear to have a white belly when seen on the wing. However, though it is hard to see in flight, they do have a thick black line down the center of their belly.

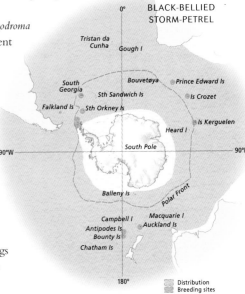

BLACK-BELLIED
STORM-PETREL

Distribution
Breeding sites

Diving petrels

ORDER: Procellariiformes
FAMILY: Pelecanoididae

There is sparse fossil evidence of the origins of diving-petrels, but they probably evolved in isolation somewhere around southern South America or Australia, and then spread around the Southern Ocean. Whereas most tube-nosed seabirds are supremely adapted to an aerial life on the open ocean, diving-petrels have evolved in a different direction; they have adapted to flying under water, and could be considered flying penguins. To adapt to swimming, diving-petrels developed heavy, stocky bodies with short, rounded wings. Like those of most aquatic birds, their feathers are dark above and light below, with very little pattern. Their short, black bills have a small elastic pouch like that of a pelican for food storage—hence their family name, Pelecanoididae. To counter buoyancy in the water, their bones are denser than those of their flying cousins.

These adaptations act against them when they fly, so they must work very hard when in the air. Most petrels glide on wind currents, flapping their wings intermittently, but diving-petrels are a constant whir of wings, and their bones are so dense that they would drop like stones if they stopped flapping. Their wings beat so quickly that they appear as a blur alongside a small bullet-shaped body. They normally fly short distances, and very close to the surface. They do not skip and dance over the waves like a shearwater, but fly directly into waves and pop out on the other side.

Diving-petrels are strictly marine, but because they need so much energy to fly, they are most often found inshore rather than far out to sea. Most tend to stay close to their breeding areas, although the South Georgia and Peruvian diving-petrels are sometimes seen further offshore. At sea they are usually seen singly or in small groups, roosting on the water when not actively feeding. Typically, they dive rather than fly to escape approaching ships.

Diving-petrels begin visiting their breeding areas well before breeding starts. They nest in burrows that they dig on sloping ground, sometimes in colonies of thousands of pairs. Normally they come ashore under cover of darkness to avoid the gulls and skuas that prey on them. They spend the night inside their burrows, calling during the night. During the breeding season, there are always many birds flying overhead. Once paired, the females spend some time feeding intensively at sea, then lay a single egg. Both

sexes take turns to incubate the egg for seven or eight weeks in daily shifts—the shortest incubation shift of any of the tube-nosed seabirds—because their feeding areas are so close to their breeding grounds. The parents feed the chick for about eight weeks before it goes to sea on its own. Chicks that survive their very difficult first year join the flocks of immature birds wheeling over the breeding colony in the following year.

The South Georgia diving-petrel and the Common diving-petrel—two of the most common sub-Antarctic diving-petrel species—are both small, dark birds with white bellies and no distinctive plumage characteristics, and it is virtually impossible to distinguish them at sea. Occasionally, however, diving-petrels become stranded on a ship during the night, providing the opportunity to examine one in the hand for the only reliable means of identification, the bill shape.

South Georgia diving-petrel

PELECANOIDES GEORGICUS
LENGTH: 18–21 cm/7–8 in
WINGSPAN: 30–33 cm/12–13 in
STATUS: more than 4 million breeding pairs
IUCN: not listed   CITES: not listed

The South Georgia diving-petrel has a wide-based bill that tapers in a continuous curve to a point. It breeds on most sub-Antarctic islands; on Iles Kerguelen, where its range overlaps with that of the Common diving-petrel, the South Georgia species forages further from the island. It feeds mainly on planktonic crustaceans, especially krill and amphipods, as well as some small fish and squid. There are more than 2 million pairs on South Georgia, and their breeding areas tend to be in gravelly substrate.

Common diving-petrel

PELECANOIDES URINATRIX
LENGTH: 20–25 cm/8–10 in
WINGSPAN: 33–38 cm/13–15 in
STATUS: more than 4 million breeding pairs
IUCN: not listed   CITES: not listed

The Common diving-petrel's bill, the feature that distinguishes it from the South Georgia diving-petrel, looks long and narrow because the sides are parallel along the length of the bill until near the tip. The species is found throughout the sub-Antarctic region. The Kerguelen diving-petrel (*P. u. exsul*), one of six subspecies of the Common diving-petrel, ranges from South Georgia east to the Antipodes Islands. It is most abundant around Iles Kerguelen, where there may be as many as one million breeding pairs.

▼ BUSY BIRDS
Common diving-petrels seemingly stop only when they arrive at their nest burrows at night. At sea they are a constant whir of wings as they dive through the waves and pursue their prey. They will rest on the water, but dive quickly out of sight when a boat approaches.

Cormorants

ORDER: Pelecaniformes
FAMILY: Phalacrocoracidae

Cormorants are medium to large aquatic birds with webbing across all four toes. They are excellent divers and swimmers but rather awkward on land. Their bodies are heavy and elongated, and they have long necks, heads, and bills. Like pelicans, cormorants have a pouch of loose skin on the throat, called a gular pouch. They use it for holding food and for thermoregulation. On a warm day it is common to see cormorants fluttering their throats to increase the evaporation on the inside of their gular pouch.

Because they are heavy, cormorants swim quite low in the water, sometimes with just their head and neck raised. They normally stay submerged for only a few seconds while foraging, but they can dive for more than 90 seconds and reach depths of 25–50 meters (65–165 ft). Like all seabirds, they have oil glands to waterproof their feathers, but their feathers also have a special structure that allows their plumage to become waterlogged to adjust their buoyancy for diving. They quickly shake off most of the water when they return to land, but it is common to see cormorants standing on low rocks or on the shoreline with their wings outstretched to dry off their feathers.

Cormorants do not rest on water, so they require dry perches. Because they need to dry their feathers, they are almost always within sight of land, and most do not migrate. Many species are also restricted to relatively shallow water because they forage for small, bottom-dwelling fish and invertebrates. Some species follow coastlines seasonally in search of food, but all the island forms are entirely sedentary.

Most cormorants are very dark brown to black, with a metallic green or blue sheen to the feathers. All the Antarctic and sub-Antarctic shags have some white on their necks and bellies. As the breeding season approaches, mature adults develop white, hairlike plumes and brightly colored facial patches: these are important during court-ship, but are lost after breeding. Juveniles are

> **BLAST-OFF**
> Antarctic shags, like most shags and cormorants, are not strong flyers. They nest in ledges on cliffs or on the edges of flat bluffs overlooking the water so that they can launch directly into the air.

> **SUB-ANTARCTIC SHAGS**
> Shag colonies are always near water. The birds are ungainly on land, with huge feet and an upright stance. They usually live on cliffs or ledges, but this colony of Auckland Island shags (*Phalacrocorax colensoi*) occupies a flat beach.

◄ JUVENILE PLUMAGE
It takes a juvenile Antarctic shag three to four years after it fledges before it becomes sexually mature and develops the bright blue eye ring and the yellow caruncles of the breeding adult.

uniformly drab brown with brown eyes, developing the typical adult plumage and eye color as they mature.

Cormorants breed in busy, noisy colonies, often sharing their breeding sites with other species—gannets, boobies, or penguins. Their colonies are on cliffs, low ledges, rocky slopes, or even on sand. Each year, after attracting a female, the male cormorant goes in search of seaweed, mud, or grasses. He flies back with his bill filled with nesting material, and the female plasters it to the nest with mud or excrement. Nests are reused year after year, and many become quite large. Females lay two to four eggs. Both sexes take their turn on the nest, and chicks hatch in laying order after 23 to 25 days. Then the real work begins: with two to four chicks on each nest, all raising their heads and calling for food, the din is unrelenting.

◄ VULNERABLE SPECIES
The Campbell Island shag (*Leucocarbo campbelli*) is restricted to Campbell Island and the associated small outcrops nearby. There were an estimated 2,000 nests in 1975, and the species is considered vulnerable. Because they nest on ledges and cliffs, they suffer little impact from introduced mammals such as rats, cats, and sheep, but skuas manage to take some of their eggs.

Imperial shag
PHALACROCORAX ATRICEPS
OTHER NAME: Blue-eyed shag
LENGTH: 68–76 cm/27–30 in
WINGSPAN: 124 cm/49 in
STATUS: about 10,000 breeding pairs
IUCN: not listed   CITES: not listed

The Imperial shag is widespread in the southern part of South America, the Falkland Islands, Prince Edward Islands, Iles Crozet, Heard, and Macquarie islands. It has a large white patch on its back, which is visible only in flight. During the breeding season it develops a brilliant cobalt, blue eye ring and bright yellow-orange caruncles at the base of its bill. Some populations also develop fine filoplumes that accentuate their upper cheeks or eyebrows.

Antarctic shag
PHALACROCORAX BRANSFIELDENSIS
LENGTH: 75–77 cm/29–30 in
WINGSPAN: 124 cm/49 in
STATUS: 12,000 breeding pairs
IUCN: not listed   CITES: not listed

Antarctic shags are a little larger than Imperial shags, although similar in appearance. They are found on the Antarctic Peninsula year-round, except during short foraging flights to open water. Living in such a cold climate, they do not hold their wings out to dry and possess an extra-thick coat of down under their contour feathers. They normally congregate in small groups, often on the edges of penguin colonies.

Kerguelen shag
PHALACROCORAX VERRUCOSUS
LENGTH: 65 cm/25½ in
WINGSPAN: 110 cm/43 in
STATUS: 6,000–7,000 birds
IUCN: not listed   CITES: not listed

The smallest of the blue-eyed shags is the Kerguelen shag. It is restricted to Iles Kerguelen, and is the only shag of the islands of the southern Indian Ocean that has remained a separate species; all others were classified with the Imperial Shag. The Kerguelen shag feeds in shallow water within about 6 kilometers (4 miles) of the coast of Kerguelen. It normally breeds in small colonies of 330 nests, but some colonies have expanded to more than 400 nests.

► A BUSY COLONY
Imperial shags nest in crowded, sometimes very large, colonies. They lay two to five (usually three) eggs in substantial nests made of grasses, seaweed, and excrement. Once the chicks begin to hatch, these large colonies are a scene of bustling activity with arriving and departing parents trying their best to keep their hungry chicks fed.

Waterfowl

ORDER: Anseriformes
FAMILY: Anatidae

▼ A SHELTERED LIFE
Small populations of several species of waterfowl have made their home on sub-Antarctic islands. The Auckland Island teal (*Anas aucklandica*) is mostly found in coastal waters and sheltered bays, where it feeds on aquatic invertebrates. It usually feeds in shallow water by probing or dabbling near the shore.

Most members of the family of ducks, geese, and swans are found in temperate and tropical environments; only four species, all belonging to the genus *Anas*, are regarded as resident in sub-Antarctic and Antarctic environments, but a few other species visit during the southern summer.

Waterfowl are direct and rapid flyers, the smaller ducks having more rapid wingbeats than the larger geese and swans. On land they move slowly and awkwardly, but with surprising speed if disturbed. All have webbing connecting the three front toes, and are excellent swimmers. They feed on plant and animal material, partly by foraging on land but mostly by dabbling— upending in shallow water and reaching for the bottom. All except for the Magpie goose shed all their flight feathers at once, soon after breeding, which renders them flightless for a few weeks.

Waterfowl are territorial during the breeding season, but quite gregarious at other times. They nest on the ground, usually near water, in grass lined with down from the female's breast, hidden in tall grass or shrubs. Females do most of the incubation, and most are drab brown providing good camouflage. They often stay on their nest when approached, relying on their camouflage for protection. If geese and swans attack smaller ducks, the ducks burst from their nests in a flurry of wings.

Gray duck

ANAS SUPERCILIOSA

OTHER NAMES: Black duck, Brown duck
LENGTH: 58 cm/23 in
WINGSPAN: 89 cm/35 in
STATUS: about 10,000 breeding pairs
  IUCN: not listed   CITES: not listed

The Gray duck is common in Australia, and is also found north into Malaysia, Indonesia, and other Australasian islands, as well as on Macquarie, Chatham, Snares, Auckland, and Campbell islands. Drab brown, with a scaled appearance to its feathers, it has a striking facial marking of black and buff stripes. Unlike many of the *Anas* ducks, females are very much like the males in coloration, although a little smaller. Macquarie Island Gray ducks are timid, and active mainly at twilight. They pair over an extended period between fall and spring, the timing of their breeding varying across their range.

Yellow-billed pintail

ANAS GEORGICA

OTHER NAME: South Georgia pintail
LENGTH: 43–55 cm/17–22 in
WINGSPAN: 71 cm/28 in
STATUS: subspecies *A.g. georgica*: less than 2,000 birds
  IUCN: not listed   CITES: not listed

Generally brown with a scaled appearance, the Yellow-billed pintail has a chestnut crown, a long, brownish tail, and a bright yellow bill marked with black. It normally feeds on emergent vegetation and small invertebrates in shallow water by dabbling, but it also dives frequently, especially for favorite prey such as fairy shrimp. Of the two living subspecies, *A.g. georgica* is resident and sedentary on South Georgia, where it likes to live around ponds surrounded by tussock grasses. On South Georgia, flocks form in winter though pairs may stay together. They begin breeding in early spring, and unmated males will court paired females in an attempt to lure them away from their current mates. Females lay up to five eggs, incubating them until they hatch, and the male helps to raise the chicks.

Kerguelen pintail

ANAS ACUTA EATONI

LENGTH: 35–45 cm/14–18 in
WINGSPAN: 65–70 cm/25–30 in
STATUS: 5,000–10,000 breeding pairs
  IUCN: Vulnerable   CITES: not listed

Resident in a number of sub-Antarctic islands, this is a subspecies of the Northern pintail (*A. acuta*), one of the

▲ **CARNIVOROUS DUCK**
This Yellow-billed pintail (*Anas georgica georgica*) is restricted to South Georgia. These birds are common around freshwater lakes and protected bays, and generally feed on vegetation, algae, and aquatic plants. On South Georgia, however, where there are few aquatic insects, Yellow-billed pintails are carnivorous and feed on carcasses of seals along with other scavengers.

most common ducks throughout its large range. Like many island species, the Kerguelen pintail is a little smaller than its more widespread northern relative, but similar in appearance. It is not officially threatened, but in southern regions feral cats are taking their toll.

Speckled teal

ANAS FLAVIROSTRIS

OTHER NAME: Yellow-billed teal
LENGTH: 35–45 cm/14–18 in
WINGSPAN: 76 cm/30 in
STATUS: 40–50 birds on South Georgia
  IUCN: not listed   CITES: not listed

In 1971, a small breeding population—a mere 40 to 50 birds—of Speckled teal was discovered in Cumberland Bay on South Georgia. Whether these teal were naturally or intentionally introduced is unknown. An adaptable species, it is abundant throughout its range—southern South America and the Falklands—occurring in freshwater, brackish, and marine habitats. Still restricted to Cumberland Bay on South Georgia, it is grayer and less reddish than South Georgia's very similar Yellow-billed pintail and has more pronounced spotting on its breast, which helps to distinguish the two species.

▲ **A LARGER PINTAIL**
The Yellow-billed pintail (*Anas georgica spinicauda*) found on the Falkland Islands, is also found throughout southern South America. It is larger by a third than the South Georgia subspecies, and is partially migratory. Yellow-billed pintails are found from sea level to 4,600 meters (15,000 ft) in habitats such as lagoons, freshwater lakes, and flooded meadows.

Sheathbills

ORDER: Charadriiformes
FAMILY: Chionidae

▼ NOT FUSSY EATERS
Snowy sheathbills, the garbage collectors of Antarctica, eat anything small enough for them to swallow, including spilled food, eggs of any other species, excrement, carrion, and mucus from seal's noses. They are restricted to the relatively milder regions around the Antarctic Peninsula, and are omnipresent in penguin colonies.

Sheathbills are medium-sized, pigeon-like shorebirds that belong to the same order as skuas, gulls, and terns. They probably evolved around the tip of South America from a plover-like ancestor. Sheathbills are the only bird family found entirely within the Antarctic and sub-Antarctic regions. Today, there are just two species in the family, both of the genus *Chionis*. They are different in appearance, especially around the face, but very similar in habits.

Both species inhabit coastal areas and rely on the sea for some of their foraging, feeding on algae, limpets, and other small invertebrates in the intertidal zone. They are the ultimate scavengers of Antarctica, eating virtually any form of organic matter. They forage among penguin and cormorant colonies for unattended eggs or small chicks, dropped food, the expelled stomach linings of penguins, and the feces of other birds and seals. They congregate around concentrations of seals in the early spring, feeding on placentas and the inevitable seal carcasses, picking at scabs and wounds on adult seals, and even feeding on the nasal mucus of elephant seals.

They are the smallest of the Antarctic predator/scavengers, which limits their predatory activities to the smallest of penguin chicks and eggs; it also means that they can move about the colonies more freely than larger predators without arousing aggression. They are expert at stealing food from penguins, cormorants, and albatrosses by leaping against adults who are feeding their chicks and quickly scooping up spilled food. They are very agile, easily dodging penguins and seals that lunge for them as they forage. Their fearlessness also brings them into close contact with humans at settlements and scientific bases, where they scavenge absolutely anything they can find. Unfortunately, they sometimes eat dangerous items like grease from machinery or small batteries.

Sheathbills nest in crevices or under protective overhangs near penguin colonies, and time their own breeding to coincide with that of the penguins. They attempt breeding at three years old, but most fail. Once they succeed in breeding, they remain faithful to their territory and their mate, and are strongly territorial during the breeding season. They normally lay two or three eggs in a rough nest of tussock, moss, algae, and old bones, built in crevices or under an overhang to protect them from the weather and marauding skuas. Once the chicks hatch in January, parents deliver food by the billful, but do not regurgitate it. When the chicks become independent, at about 50 days, the adults gather in small flocks to forage and roost.

Snowy sheathbill

CHIONIS ALBA
OTHER NAMES: American sheathbill, Greater sheathbill, Pale-faced sheathbill, Yellow-billed sheathbill
LENGTH: 34–41 cm/13–16 in
WINGSPAN: 75–80 cm/29½–31½ in
STATUS: about 10,000 breeding pairs
   IUCN: not listed   CITES: not listed

With their thick, entirely white plumage and sturdy bodies, Snowy sheathbills are well adapted to the cold Antarctic environment. Their greenish bill sheaths and pink facial caruncles make their faces look untidy. They live and breed around the Antarctic Peninsula and the islands of the Scotia Arc. Many migrate to Tierra del Fuego and Patagonia in winter, but on Signy Island and the Falklands some stay near their breeding areas throughout the winter.

Lesser sheathbill

CHIONIS MINOR
OTHER NAME: Black-faced sheathbill
LENGTH: 38–41 cm/15–16 in
WINGSPAN: 74–79 cm/29–31 in
STATUS: 6,500–10,000 breeding pairs
   IUCN: not listed   CITES: not listed

This species has a black sheath and caruncles, and pink eye-rings. Leg color varies from pink to black, depending on the subspecies. Four separate subspecies occur on sub-Antarctic islands in the Indian Ocean, one resident on Iles Kerguelen, one on Iles Crozet, one on Marion and Prince Edward islands, and one on Heard and McDonald Islands. In the breeding season Lesser sheathbills nest in association with penguin colonies. King penguins provide a host colony all year so that the sheathbills can scavenge around their colonies all winter, but Rockhopper and Macaroni penguins abandon their colonies in winter to forage at sea, leaving the sheathbills to fend for themselves by foraging for terrestrial and marine invertebrates. Sheathbills without winter host colonies have a lower reproductive success rate.

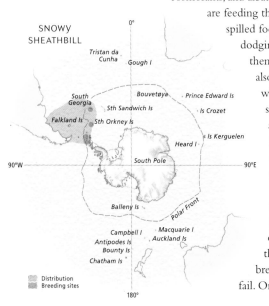

SNOWY
SHEATHBILL

Tristan da Cunha
Gough I
0°
South Georgia
Bouvetøya
Sth Sandwich Is
Prince Edward Is
Falkland Is
Sth Orkney Is
Is Crozet
Heard I
Is Kerguelen
90°W
South Pole
90°E
Balleny Is
Polar Front
Campbell I
Antipodes Is
Bounty Is
Chatham Is
Macquarie I
Auckland Is
Distribution
Breeding sites
180°

Skuas

ORDER: Charadriiformes

FAMILY: Stercorariidae

This family is divided into the larger *Catharacta* genus and the smaller, more agile *Stercorarius* genus, sometimes known as jaegers. Both groups probably originated in the northern hemisphere, splitting from the gulls about 10 million years ago, but the jaegers have remained wholly Arctic while *Catharacta* skuas are found in high latitudes in both hemispheres.

Skuas have heavier bodies than their gull relatives, and stronger, more hooked bills, reflecting their more predatory lifestyle. Their plumage is predominantly dark, with conspicuous white patches on the top and bottom of their wings, and they employ powerful wingbeats in flight. With strong, hooked claws on their toes, as well as full webbing between them, they are true seabirds of prey. Like the land birds of prey—the raptors and the owls—they have hard, scaly plates called scutes on their legs, and females are larger than males by 11 to 17 percent, depending on the species.

Skuas may live for 38 years, and are very faithful to both their breeding territories and their mates. They are migratory during the winter, arriving at breeding areas in early spring, just as the penguins are returning to their colonies. They are mature by age three, but cannot normally claim a territory until they are six or seven. Most pairs produce two eggs. If both eggs survive, the first egg hatches two or three days before the second. The older chick will pick fights with the younger, sometimes killing it but more often preventing it from getting enough food. The younger chick will either die of starvation or stray into other territories in search of food—usually to be killed and eaten by adult skuas. Parents work hard to raise their chicks, but the dangers of living around other predatory skuas and the trampling feet of penguins take their toll, and their reproductive success is usually very low.

Two *Catharacta* skuas breed in Antarctic or sub-Antarctic regions. Antarctic skuas nest on most of the sub-Antarctic islands, on cold temperate-zone islands south of New Zealand, and on the northern part of the Antarctic Peninsula. On the peninsula they overlap with South polar skuas, which breed exclusively around the coast of Antarctica. Where the species overlap they regularly interbreed, even though Antarctic skuas normally will not allow South polar skuas to nest around penguin colonies, forcing them to forage almost exclusively at sea for fish and krill. Further south, away from the overlap with Antarctic skuas, most South polar skuas nest in loose colonies near penguin or petrel colonies, on which they prey. Outside the breeding season they mainly fish for themselves by plunge-diving at sea.

Antarctic skua

CATHARACTA ANTARCTICA

OTHER NAME: Brown skua

LENGTH: 63 cm/25 in

WINGSPAN: 120–160 cm/47–63 in

STATUS: 13,000–14,000 breeding pairs

    IUC: not listed   CITES: not listed

Antarctic skuas are large, powerful birds with heavy bodies and a fierce demeanor. They are dark brown with yellowish streaks in their neck and nape feathers. Most

▲ FAITHFUL MATES

A South Polar skua stands alert on its territory. These birds breed only on the Antarctic Continent and nearby islands, but migrate far into the northern hemisphere. Some pairs are known to have stayed together and nest at the same site for 20 years.

▼ GUARDING THE HOME

A pair of South Polar skuas perform the "long call" display to protect their territory and their food. The display tells all other nearby skuas that they are at home and ready to fight for their territory.

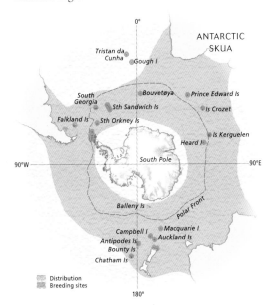

ANTARCTIC SKUA

Tristan da Cunha
Gough I
South Georgia
Bouvetøya
Prince Edward Is
Sth Sandwich Is
Is Crozet
Falkland Is
Sth Orkney Is
Heard I
Is Kerguelen
South Pole
90°W   90°E
Balleny Is
Polar Front
Campbell I
Macquarie I
Antipodes Is
Auckland Is
Bounty Is
Chatham Is

0°
180°

Distribution
Breeding sites

➤ **PRECOCIOUS CHICKS**
Within just a few hours of hatching, South polar skua chicks are able to walk on their own around the territory. If danger threatens, the parents call out and the chick usually crouches down among the rocks.

▼ **PREDATORY BIRDS**
This Brown skua (*Catharacta antarctica lonnbergi*) from Auckland Island shows the rich brown typical of the larger species. Skuas are related to gulls, but are much more predatory and have very strong bills with pronounced hooks on the tip. Many Brown skuas, particularly on the sub-Antarctic islands around New Zealand, nest in trios or larger groups, with one female and multiple males attending the territory.

◄ **SAFE HAVEN**
South polar skuas normally lay two eggs, but accidents caused by aggressive penguins and predation by other skuas often cause pairs to fail before their eggs hatch. The chicks must be protected and brooded regularly, but not constantly, for the first couple of weeks.

populations are fairly small, but bigger populations occur on larger islands that harbor more prey species. They are generalist predators who adapt to local conditions, and are also scavengers whose numbers tend to increase around human settlements.

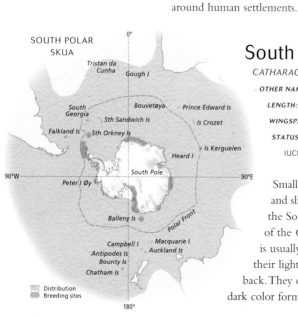

SOUTH POLAR SKUA

Tristan da Cunha
Gough I
South Georgia
Bouvetøya
Prince Edward Is
Sth Sandwich Is
Is Crozet
Falkland Is
Sth Orkney Is
Is Kerguelen
Heard I
South Pole
Peter I Øy
Balleny Is
Polar Front
Campbell I
Macquarie I
Antipodes Is
Auckland Is
Bounty Is
Chatham Is

0°
90°W
90°E
180°

Distribution
Breeding sites

South polar skua

CATHARACTA MACCORMICKI

OTHER NAME: McCormick's skua
LENGTH: 53 cm/21 in
WINGSPAN: 130–140 cm/51–55 in
STATUS: 5,000–8,000 breeding pairs
IUCN: not listed   CITES: not listed

Smaller than the Antarctic skua and slightly more agile in flight, the South polar skua is the grayest of the *Catharacta* skuas; even so, there is usually a clear contrast between their lighter nape and their darker back. They occur in pale, medium, and dark color forms.

A pirate species

Herbert Ponting, photographer on Scott's British Antarctic Expedition of 1910–12, called skuas "the Buccaneers of the South." They regularly prey on storm petrels and other petrels, as well as on the eggs and chicks of penguins and other seabirds. They protect their territories savagely, even diving at and striking human intruders. Skuas are also klepto-parasites, attacking seabirds bringing food to their mates or chicks, and stealing the food. This behavior is not unique to skuas, but in this species it is developed to the extreme.

Gulls

ORDER: Charadriiformes
FAMILY: Laridae

Gulls probably diverged from a shorebird ancestor about 65 million years ago. Fossil records suggest that they originated in northern Europe and Asia, and then spread throughout the world. Nowadays they occur on all coasts and in most inland areas. Gulls are fairly uniform in shape, with heavy bodies, long wings, and webbed feet. Most have a rounded tail and a stout bill that is either straight or slightly hooked. They are generally white-bodied, with a darker mantle that varies from black to a pale silvery-gray. The wingtips of most species are black, with white patches near the tips. Juveniles typically have brown plumage speckled with black or buff, and molt within a few months into their first winter plumage, which may resemble adult winter plumage or may be somewhere between the juvenile and adult plumages. Some species take four years or more to develop full adult plumage.

Most gulls breed in colonies and form strong pair bonds, usually heterosexual and monogamous, but some species in some colonies may include a small proportion of female pairs—perhaps because of a lack of breeding males. Both females mate with already paired males and lay fertile eggs, and as a result their clutches are often twice as big as usual. Within their colonies gulls are territorial and defend their own nests, but the entire colony will attack any intruder approaching the colony.

An infinitely adaptable species

Most gulls live on sea coasts, but some species live well inland and nest near fresh water. Their salt glands grow or shrink as required: the salt glands of inland gulls are very small during the breeding season, but when the birds must migrate in winter to coastal areas, these glands enlarge to cope with food that contains more salt.

Gulls live in a wide variety of habitats, and are omnivorous and opportunistic feeders. They eat virtually anything, and are familiar visitors to refuse tips around human habitations, and they can obtain their food in a variety of other ways as well. On the wing they catch food such as insects or food dropped by others. They can hover above water and drop down to pick items off the surface, or land on the water to surface-feed, and at times they will plunge-dive to seize food from below the surface. Most species also patrol shores and tidal areas to glean invertebrates and scavenge whatever they can find; more than half the food they eat is snatched up from the ground.

▲ WINGED BEGGARS
Young Kelp gulls begin to grow their juvenile plumage very soon after hatching, and take their first flight at about 45 to 60 days. The parents continue to feed them for another couple of weeks, so the juveniles harass their parents in the air, chasing them and begging for food.

▶ NEST BUILDERS
Kelp gulls normally nest in colonies but will sometimes nest as individual pairs. Each pair builds a substantial nest from whatever vegetation and materials are available. Both parents contribute to the building, but males do most of the work. Colonies are near water on sloping ground or rocky ledges.

Kelp gull
LARUS DOMINICANUS
OTHER NAMES: Dominican gull, Southern black-backed gull
LENGTH: 54–65 cm/21–26 in
WINGSPAN: 128–142 cm/50–56 in
STATUS: more than 1,085,000 breeding pairs;
10,000–20,000 in Antarctica and sub-Antarctic regions
IUCN: not listed   CITES: not listed

The only gull species commonly seen in Antarctica, Kelp gulls are most abundant in New Zealand, although they are also found on coasts throughout the southern hemisphere. In Antarctica they occupy the coasts and islands of the Antarctic Peninsula throughout the year; in winter they may move away from the coast in search of open water, but they do not migrate north into South America. They are large, white-headed gulls, with a black mantle and upperwings and completely white tails. Their straight, yellow bill has a red spot at the tip of the lower section.

Kelp gulls eat mollusks, fish, worms, arthropods, reptiles, amphibians, birds, and even small mammals. They also eat carrion and scavenge on rubbish dumps, and sewage. The Antarctic gulls specialize in the Antarctic limpet (*Nacella concinna*), which they seize from rocky shores at low tide or glean from the rocks as they swim parallel to the shore. They swallow the limpets whole, digest the flesh, and regurgitate the shells in piles resembling stacks of miniature dinner plates. Heaps of discarded limpet shells are often found on high rocks well away from the water, where the gulls have roosted.

▶▲ EYES PEELED FOR FOOD

Kelp gulls are generalist predators and scavengers. Their diet varies depending on location and availability. They are experts at swooping down to pluck food from the surface of the water—whether it be a single krill or small fish that has strayed too near the surface, or a tidbit from a penguin being devoured by a Leopard seal.

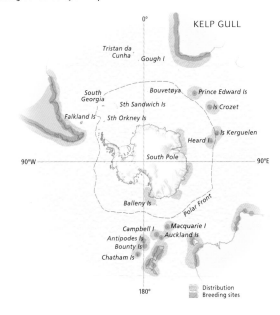

KELP GULL

Distribution
Breeding sites

Classifying gulls and terns

Gulls and terns are closely related, terns probably deriving from a primitive gull-like ancestor. Recently, however, DNA studies suggest that that they should be grouped with the skuas (Stercorariidae) and the skimmers (Rynchopidae) into the Laridae family, each current family retaining its distinction but at the subfamily level. Gulls are an enigmatic group; even with powerful modern classification techniques, hybridization is a persistent source of taxonomic confusion, and the new classification is not universally recognized. The classic separation of gulls, terns, and skuas into separate families is used here.

Terns

ORDER: Charadriiformes

FAMILY: Sternidae

> **WINTER QUARTERS**
Arctic terns travel huge distances on their migration. They breed exclusively in the Arctic during the northern summer, but during their non-breeding season migrate far into the Southern Ocean—even reaching the Antarctic Continent around the Antarctic Peninsula. There they occupy coastal and ice habitats like those of the high Arctic.

Like gulls, terns are also found all over the world. All Antarctic terns belong to the genus *Sterna*, except the Brown noddy (*Anous stolidusis*). Brown noddies are mostly tropical breeders, but there is a thriving population on Tristan da Cunha in the South Atlantic.

Terns are generally small to medium-sized, with white or light gray underparts, darker gray upper parts, and a black cap when in breeding plumage. They have long, forked tails and slender wings, making them more maneuverable and acrobatic flyers than gulls. Terns forage by swooping down to pick food from the surface or plummeting into the water. They hover briefly, slam into the water, and then surface and take off, holding a small fish or crustacean in their bill. They steal food from each other and from other species, but conversely they often lose their food to larger gulls or skuas.

Like gulls, terns nest in colonies and may occupy the same colony site for many years. They protect their nesting sites by attacking intruders; in some colonies, thousands of birds will soar into the air and then swoop on a fox—or a human—who ventures too close.

American coast, but—unlike Arctic terns—they do not migrate into the northern hemisphere. Their favorite prey is small fish and krill, which they obtain by plunge-diving.

Kerguelen tern

STERNA VIRGATA

LENGTH: 33 cm/13 in

WINGSPAN: 71 cm/28 in

STATUS: 2,400 breeding pairs

    IUCN: Lower risk–near threatened   CITES: not listed

One of the most endangered of terns, this species lives and breeds only on Iles Kerguelen, Iles Crozet, and Marion and Prince Edward islands in the southern Indian Ocean. The largest population—about 2,000 pairs—lives on Iles Kerguelen, where it is threatened by introduced cats. The species resembles the Antarctic tern, but is a little smaller and duskier. Its gray under-parts are a little darker than those of the Antarctic tern, and its bill is also darker. In the breeding season it feeds largely on insects in the uplands, but for the rest of the year it reverts to the typical tern prey of small aquatic animals.

Antarctic tern

STERNA VITTATA

OTHER NAME: Wreathed tern

LENGTH: 35–40 cm/14–16 in

WINGSPAN: 74–79 cm/29–31 in

STATUS: About 50,000.birds

    IUCN: not listed   CITES: not listed

This species, which is the most common tern in Antarctica, nests on most sub-Antarctic islands, as well as along the coasts of the Antarctic Peninsula, New Zealand, and some areas of South America and South Africa. It is smoky gray on its upper parts with a white rump, a black cap on its gray head, and a red bill. Its underparts are mostly white, sometimes with streaks of gray.

Antarctic terns breed in November–December, nesting in colonies in rocky areas near the shore or just inland. Around the Antarctic Peninsula they usually establish their colonies in flat gravel areas away from the beach. If left undisturbed they come back to the same site year after year, but they can be flexible: on Tristan da Cunha, they once nested on sandy beaches, but shifted their colonies to ledges on small offshore outcrops when rats were introduced. During the non-breeding season they move to ice edges to forage; they may migrate northward as far as the South

ANTARCTIC TERN

Tristan da Cunha · Gough I · Bouvetøya · Prince Edward Is · South Georgia · Sth Sandwich Is · Is Crozet · Falkland Is · Sth Orkney Is · Is Kerguelen · Heard I · I Amsterdam · South Pole · Balleny Is · Polar Front · Campbell I · Macquarie I · Antipodes Is · Auckland Is · Bounty Is · Chatham Is

90°W 0° 90°E 180°

Distribution
Breeding sites

KERGUELEN TERN

Tristan da Cunha · Gough I · South Georgia · Bouvetøya · Prince Edward Is · Sth Sandwich Is · Is Crozet · Falkland Is · Sth Orkney Is · Is Kerguelen · Heard I · South Pole · Balleny Is · Polar Front · Campbell I · Macquarie I · Antipodes Is · Auckland Is · Bounty Is · Chatham Is

90°W 0° 90°E 180°

Distribution
Breeding sites

Arctic tern

STERNA PARADISAEA

LENGTH: 33–36 cm/13–14 in

WINGSPAN: 76–85 cm/30–33 in

STATUS: about 500,000 breeding
pairs; IUCN: not listed
CITES not listed

ARCTIC
TERN

Tristan da
Cunha   Gough I

South
Georgia   Bouvetøya   Prince Edward Is

Sth Sandwich Is   Is Crozet

Falkland Is   Sth Orkney Is

Is Kerguelen

Heard I

90°W   South Pole   90°E

Balleny Is   Polar Front

Campbell I   Macquarie I
Antipodes Is   Auckland Is
Bounty Is
Chatham Is

180°

Distribution
Breeding sites

The Arctic tern is a common, black-
capped tern that breeds exclusively in the
high latitudes of the northern hemisphere. It
resembles the Antarctic tern, but has longer wings, a
longer, more slender bill, and a more delicate appearance. The underparts are lighter, with more white in the
cheeks. In the northern winter it flies halfway round the
world to the Antarctic to feed in the Antarctic pack ice
or around the islands of the Antarctic Peninsula—one
of the longest of bird migrations. Having bred in the
northern summer, the Arctic tern has acquired its
non-breeding winter plumage by the time it reaches
Antarctic waters. This
means that it is easy to
distinguish between Arctic
and Antarctic terns during
November and December
because Antarctic terns are
wearing their breeding
plumage. Later in the
southern summer, juvenile
Antarctic terns are easily
confused with winter-
plumage Arctic terns.   GM

▽ INTRUDER ALERT

Most terns, including the Antarctic tern, nest in dense colonies
on flat ground. When an intruder appears, the entire colony
rises up to mob it. They are agile and swift flyers, and will
dive at the intruder and peck at it aggressively.

➤ TIRELESS FLYERS

Found mainly along the
Antarctic Peninsula and
the sub-Antarctic islands
surrounding the continent,
Antarctic terns seem to be
tireless flyers, although they
do alight to rest sometimes.
They feed during the day on
crustaceans and small fish,
which they mainly take by
plunging into the water from a
few meters above the surface.
Occasionally they will take
prey from the surface by
dipping while on the wing.
They never land on the
surface to take prey.

Penguins

Penguins

ORDER: Sphenisciformes
FAMILY: Spheniscidae

Previous pages

◄ **IDENTIFYING COLORS**
Most penguin species are easily identified by the coloration around the head and neck, and the brightly painted King penguin is no exception. Even the mandibular plate, part of the lower beak, is pink or orange in the adult.

With their sway-backed upright posture, smart black-and-white coloration, and endearing waddling gait, penguins are perhaps the most distinctive of birds. Many people think of penguins as strictly Antarctic, but in fact only four species—the Emperor, the Adélie, the Chinstrap, and the Gentoo—breed on the Antarctic Continent proper, while seven others live and breed south of the Polar Front that defines the far southern oceans. Because penguins are flightless and must feed in cool water, they cannot travel across the warm waters of the tropics to inhabit the northern hemisphere. Only a few groups of Galapagos penguins (*Spheniscus mendiculus*) breed north of the equator.

Grouped scientifically into about 17 species, penguins are highly specialized, non-flying, marine birds, ranging in size from the Little penguin, at 1 kilogram (2¼ lb) in weight and 40 centimeters (16 in) in height, to the Emperor penguin, which weighs up to 38 kilograms (84 lb) and reaches 115 centimeters (45 in) in height. All have a dark back and a white front, and each species has a unique pattern, in some cases including orange or yellow, around the head and chest. Possibly

the basic dark and pale coloration, repeated with variations, provides both protection from predators and camouflage while hunting: from above, penguins on the ocean surface are dark against the dark water, but viewed from below, they are light against a light sky.

Male penguins are generally heavier than females and have larger, more powerful bills and longer flippers. In most species the difference between the sexes is hard to distinguish by eye, but behavioral observations do help. Scientists can sex Macaroni penguins, but find distinguishing male and female Gentoo and Chinstrap

> BALANCING ACT
King penguins do not use nests to care for their eggs, instead they tuck them up on their feet and warm them against a bare patch of skin on their lower belly. Because of this balancing act, a King penguin with an egg will only move a few meters, and even then only if disturbed.

penguins more difficult, although they have achieved a 95 percent success rate. Adélies are even more difficult to sex by eye, with scientists succeeding only 85 percent of the time. Interestingly, penguins from high and low latitudes are more difficult to sex than mid-latitude species, suggesting there may be some environmental factor contributing to the presence (or absence) of sexual dimorphism. One reason may be the need for males to compete for nesting space and females; larger males would win more battles and so be more success-ful. A further theory is that if the two sexes are different sizes they can effectively exploit different food resources, but this has never been shown to occur regularly.

◄ A FEEDING EXPEDITION
Emperor penguins gather at the edge of the fast sea ice at Cape Crozier, about to set out onto the thinner ice for a feeding trip. It is springtime, and parents are taking it in turns to feed their rapidly growing chicks.

Keeping warm—and cool

Unlike most other seabirds, penguins have lost the ability to fly, so that they do not need to minimize their weight. On the contrary, they have developed heavy blubber reserves as insulation and for long-term energy storage, an adaptation usually seen only in mammals. Their feathers have also modified from flying equipment into waterproof insulation, thickening the birds up and shaping them to hold air against the body. Penguins may not be able to take to the air, but they do "fly" through water, using their legs as rudders and their rapidly beating wings for propulsion. Swimming compresses their feathers and gradually forces out the insulating layer of air, so that a swimming penguin leaves a fine trail of bubbles.

Polar penguins are well insulated, and can remain in the water almost indefinitely without losing too much heat. But they are designed to retain body heat, and when conditions are too warm they have extreme difficulty cooling down. They shed heat through the undersides of their flippers and through their legs and feet, as well as by fluffing feathers and panting. On a warm day, penguins may lie on their stomachs, flippers raised, feet waving in the air. In the heat of summer, bigger chicks that are still too young to go to sea and dissipate heat into the water have an extremely difficult time.

▼ ICY CONDITIONS
The very best nesting sites are those that become snow-free first, allowing for an early start to breeding. However, even the best locations suffer the occasional surprise snow-storm, even in the middle of summer. Very young chicks are not well insulated, but older ones are quite capable of sitting out a snow fall.

▲ HOW FAST?

Despite some enthusiastic estimates of speeds of up to 50–60 km per hour (30–37 mph), small to medium-sized penguins such as the Adélies seen here actually swim at about 5–10 km per hour (3–6 mph), or up to twice that in short bursts.

Fishing the southern oceans

All penguins are carnivores but the various species prefer different prey, ranging from small plankton to large fish and cephalopods. These prey types are all fairly active, so they are difficult to catch and must be swallowed whole before they can escape. Bird beaks do not have teeth, but penguin mouths have "teeth" made from keratin. They are small hooks, rather like an extremely rough cat's tongue, that project backwards inside their mouth and prevent their prey from darting to freedom at the last second.

It is estimated that there are at least 23 million breeding pairs of the 11 species of Antarctic penguins living and fishing within the region south of the Polar Front. Scientists have calculated that they must consume at least 539,000 tonnes of squid, 3.6 million tonnes of fish, and 13.9 million tonnes of crustaceans—mostly krill—every year. The great bulk of this fishing takes place near South Georgia and the South Sandwich Islands. This is an immensely productive area, and it supports 24 percent of all the breeding penguins that live within the Polar Front. It is the principal location for penguin consumption of marine resources. South Georgia alone provides 23 percent of all food taken by penguins within the southern region. This massive feeding frenzy is due largely to Macaroni penguins, which make up 94 percent of the penguins in the area. The South Sandwich area accounts for another 19 percent of the prey consumed in the southern oceans, in this case almost entirely by Chinstrap penguins.

New feathers for old

A bird's feathers get damaged by the wear and tear of daily life, and must be replaced approximately once a year. Most birds lose and replace their feathers one by one, molting gradually over a period of time, but for penguins the story is different. Each year, penguins undergo a rapid complete molt, in which all their feathers are replaced at once. First the new feathers grow, pushing out the old feathers, which remain attached to the tips of the new. Eventually the old feathers are shed in batches, creating a veritable snowstorm of feathers around the penguin colonies, and then the new feathers grow further and thicken up before the penguin is ready to go to sea again. The process usually takes about three weeks, and during that time every bit of the penguin's energy is focused on renewing its feathers.

Without feathers, penguins have no insulation or waterproofing, so a molting penguin cannot go to sea to feed—it would freeze to death. From when they begin

The way home

Most penguins seem to navigate by the sun, and are equipped with an internal clock to adjust for the time of day. Experiments have shown that if penguins are captured and then released far from their colony, the weather is crucial to the operation of their homing instinct. In sunny conditions they will immediately head in exactly the right direction, but in overcast conditions it takes them longer to orient themselves and begin the trek to their colony. If it is very cloudy, they simply cannot work out what direction to take, and their homeward journey is erratic. They will wander in the wrong direction for hours, then orient themselves correctly during brief clear spells but lose their way again when it clouds over. Evidently penguins know which direction they want to take, but can find it only by the position of the sun in the sky.

to molt until their new feathers have grown in, penguins are very exposed to the elements and are at risk of freezing or starving, or both. In order to conserve energy, they move around very little while they are molting; they simply stand on shore sheltering from the weather, and use their blubber reserves to grow new feathers. Even so, the molting phase results in massive weight loss—a molting penguin can lose more than half its body weight in just over three weeks.

Preening, in which birds use their bill to stroke their feathers from base to tip, is essential for penguins. Penguins coat their feathers with an oil secreted by the uropygial gland, located at the base of the tail, which waterproofs the feathers. They squeeze oil out of their uropygial gland with their bill, then run the oil through their feathers. To put it on their head and neck—the hard-to-reach places—they put the oil on the edge of a flipper and rub their head over the flipper area. In some species, partners will preen each other's head and neck; this is called allopreening. While allopreening may help to spread oil, it is considered to be mainly a nurturing

behavior. Emperor penguins, which do not seem to have external parasites, have not been observed to allopreen. Adélies, Gentoos, and Chinstraps—the brush-tailed penguins—are also exclusively Antarctic species, and they do not allopreen, but King penguins, which breed in warmer conditions, do. Allopreening is generally more common in warmer conditions, where there are more parasites, but it is thought to be primarily a social behavior that strengthens the pair bond.

▲ VARYING DIET
Temperatures along the Antarctic Peninsula fluctuate over time and affect the types of prey species available. Adélies have been observed to change their diet to cope with changed prey availability, and may do so regularly.

> FIRST MOLT
By its first winter a King penguin chick (left) can weigh almost as much as an adult. Over the winter, the chick loses weight, but when spring returns and fishing improves, it regains weight. It also loses its stringy chick down, molting into immature feathers, which are similar to adult feathers but not so bright.

Sharing resources

Good breeding locations on the sub-Antarctic islands and along the Antarctic shoreline are few and far between, and during summer, when all penguins must come ashore to reproduce, they gather in very large numbers. During the summer season several species often share a general area, and its food supply, until they can reproduce, molt, and disperse throughout the southern oceans again. To accomplish this sharing of a limited resource, the penguins employ different feeding strategies and target different prey. Usually, one species fishes inshore, very close to land, while other species go farther out to sea to feed. If there are three or more species in the area, offshore feeders specialize in the types of food they hunt, partitioning the prey by both size and type. Foraging depth may be one of the deciding factors in this separation of resources, with larger species of penguins, capable of longer, deeper dives, taking larger prey items.

Breeding in colonies

Most penguin species prefer to live in large colonies of thousands, or even hundreds of thousands. To attract breeding partners from among these vast crowds, penguins put on attention-seeking courtship displays including calling and highly ritualized posturing. Penguins are highly imitative birds, and when one couple begins a courtship ritual, all the neighboring birds tend to copy them. Experiments playing courtship sounds to pairs of Royal and Little penguins found that it increased the rate of courtship and mating by shortening the time between pairing and egg laying. In further experiments, the cacophony of colony sounds was recorded and played back to the whole colony and the result was an increase in synchronicity of egg laying. This may partially explain why larger penguin colonies tend to be more successful. The acoustic background of a colony influences its seasonal reproductive rate and synchrony, with larger colonies more synchronous than smaller ones, and more successful. Synchronized egg laying produces many chicks all at the same time, so imitative mating and laying behavior is an effective survival technique, as it floods an area with a large number of chicks all hatched at the same time. This gives each individual chick a better chance of surviving the local predators, and the shorter the period with chicks vulnerable to predators, the fewer chicks can be taken.

➤ A CROWDED LIFE

Successful King penguin colonies like this one at Salisbury Plain, South Georgia, are not as chaotic as they might seem. Breeding adults usually take over the center or rear of the colony, separate from roosting and unoccupied birds—which remain at the edge. Chicks tend to remain close to each other, near where their parents left them, and during winter adults and chicks tend to segregate.

▼ DETECTIVE WORK

Examination of the stomach contents of Adélies in high summer suggests that their diet is mainly krill and a bit of fish. Analysis of the guano at some Peninsula colonies, however, has revealed that Adélies must also eat squid at some time when they are ashore.

Penguins must provide all the protection they can for their eggs. They incubate their eggs in nests of rocks, tail feathers, bones, and sometimes plant material, but these are not good insulators, and do not provide much protection. To protect their eggs and keep them warm, penguins tuck them into a brood patch, which is a bare patch of skin surrounded by thick feathers low between their legs. During incubation this patch becomes engorged with blood vessels located at the surface of the brood patch. The vessels are heated by the adult's body core and keep the egg warm. When a penguin stands up or swims between bouts of incubating, the stomach feathers close up over the patch and prevent the belly from losing too much heat.

PENGUIN PROCESSION

"They are extraordinarily like children, these little people of the Antarctic world, either like children, or like old men, full of their own importance and late for dinner, in their black tail-coats and white shirt fronts—and rather portly withal."

Apsley Cherry-Garrard

THE WORST JOURNEY IN THE WORLD

DISPUTING RIGHT-OF-WAY

Tensions are high during the breeding season, and every small territorial incursion is taken as an invitation to battle. Breeding penguins will attack any other penguin that enters the small space surrounding their nest, and may even chase them a small distance to see them off, but they seldom inflict any serious physical damage.

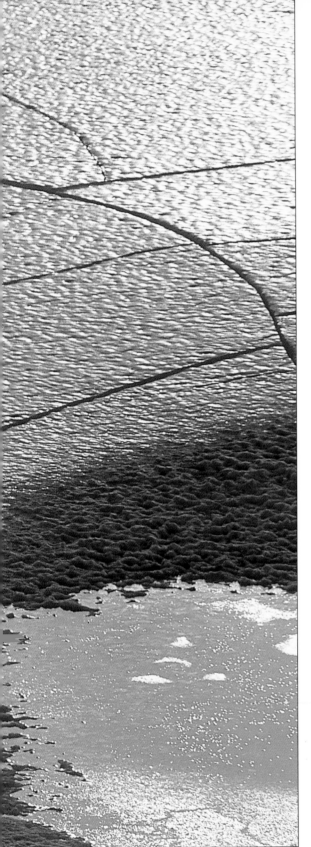

NATURAL FISH HOOKS
Penguins have no teeth, and
must catch and swallow their
food whole, using only their
beak. This is not quite as
difficult as it would appear
because they have backward-
pointing spines inside their
mouths that can grip and hold
onto prey.

THE WOOLLY PENGUIN
Early explorers were pleased
to discover the many species
of penguins in the southern
oceans, but their enthusiasm
occasionally outstripped
their skill as naturalists. King
penguin chicks like this one
were described by at least
one author as an entirely
separate species—the
Woolly penguin.

Wingless divers

▲ RUSHING TO MATE
King penguins are usually four years old when they first begin to breed, and after successfully raising a chick, the female is often ready to mate again a month sooner than the male. This means that successful couples don't usually re-mate.

Wingless divers, genus *Aptenodytes*, are the largest living penguin species. There are only two members of this group, the Emperor penguin and the King penguin. Both these stately birds stand at an impressive height of more than 90 centimeters (35 in). The Emperor penguin has a bold splash of golden yellow on the sides of the head, neck, and chest, and the King penguin is similar but more brightly marked with a rich orange color.

The two species are quite closely related, but they live in vastly different conditions. Emperors, the most southerly of penguins, live and breed in the harshest of environments, in the fast ice of the Antarctic Continent, whereas King penguins inhabit the comparatively mild sub-Antarctic islands north of the Polar Front. Neither species builds a nest, even though King penguins live in an area where nest-building materials are readily available. Emperor penguins, on the other hand, raise their young deep in the southern ice, where there is not even a scrap of rock to build a nest.

Bringing up baby

King penguins hatch in summer and reach their parents' body weight in about two months, but when winter sets in and prey becomes scarce, they are not regularly fed and so lose weight. The next spring, prey returns, the parents can feed the chicks again, and their weight recovers to about 80 percent of the adult weight before they fledge, at approximately 13 months old. Emperor penguins hatch in mid-winter, four months later than Kings, and mature steadily over five months. They fledge in December with a weight and beak size 40–60 percent that of adults—but with feet 80 percent of adult size. Presumably they need large feet to walk from their birthplace in the fast ice to the open ocean, where they can begin to hunt independently.

Each strategy has its pay-offs. King penguins' two-stage rearing means that energy is not wasted on weak youngsters that do not survive winter, but chicks that do make it through have good survival prospects. The Emperors' shorter rearing period means that they can breed annually and produce more offspring, balancing out the fact that many chicks perish long before reaching the ocean.

➤ MUTUAL ADMIRATION
Emperor penguins could not survive the bitter Antarctic winter without cooperation, and the early socialization of chicks is part of this overall strategy for survival. Emperor penguin parents will often shuffle together, face-to-face, so that the growing chicks on their feet can get to know each other.

➤ RECOGNIZING THEIR OWN
Parents returning from fishing expeditions locate their chicks by sound. King chicks are taught to recognize the characteristic rise and fall of their parents' voices from a very young age. The chicks respond with a high-pitched, three-noted whistle.

Emperor penguin

SPECIES: *Aptenodytes forsteri*
LENGTH: 100–130 cm/40–50 in
WEIGHT: 38 kg/84 lb

STATUS: at least 500,000;
population thought to be stable
IUCN: not listed   CITES: not listed

Emperor penguins have blue-gray backs shading to a black tail, and their underparts are white flushed with yellow, strengthening to deeper yellow curving ear patches. Uniquely among penguins, the chicks have beautiful, pale gray body down and a black head with a white eye mask.

Emperor penguins live in 40 known colonies south of latitude 65°, and there are three regional populations, in East Antarctica, the Ross Sea, and the Weddell Sea area. They spend their lives in the pack ice, avoiding the open water beyond, and breed only on the permanent ice attached to the Southern Continent and nearby islands. Emperors are the only penguins that breed on ice and snow rather than exposed land. In order to feed, they must visit open water, and if open water is not available they dive into the breathing holes maintained by seals and into narrow cracks in the continental ice. They feed on fish, and Emperors have been recorded

▲ SURVIVING THE COLD
Emperors have large, rounded bodies that help them to combat the cold. They are well insulated, mostly by a thick layer of waterproof feathers which overlap like tiles on a roof. A layer of fat under the skin provides further insulation, and doubles as a food reserve.

▶ CHICKNAPPING
So strong is the brooding instinct in Emperor penguins that chickless adults will try to steal any chick that leaves its parent for even a few seconds. Up to a dozen adults may try to claim a chick by pushing it under their belly feathers with their pointed bill, often injuring or even killing it in the process. Without a mate to share brooding and feeding, the foster parent must eventually head for the sea to feed, leaving the chick to die.

diving for as long as 18 minutes to depths of more than 400 meters (1,300 ft). On most dives, Emperor penguins spend two and a half to nine minutes holding their breath under water.

Emperors have many ways of dealing with their rigorously cold environment. Large bodies retain heat more efficiently than small ones because their surface area is small in relation to their weight, and the Emperor is the largest of all modern penguins; it is twice as heavy

as the King penguin, its warmer-climate relative, even though there is little difference in height. All the Emperor's extremities are reduced in size, limiting the heat that can be lost through them. Its head is small in relation to its body, and its flippers and bill are 25 percent smaller than those of other penguins. It also has a highly developed blood countercurrent mechanism— a heat exchange system designed to retain warmth in the body. It is twice as developed as the King's system, and three times as effective as the Adélie's. The Emperor's body is designed to withstand temperatures of −10°C (14°F) before it must use body energy to keep warm, whereas King penguins must draw on body energy at a temperature of only −5°C (23°F). Like all penguins Emperors have a certain amount of insulating blubber, but their main form of insulation is their feathers, which are dense and tightly packed, and lock together to trap air in the layer of down against the skin.

Family life

Living as far south as any animal in the world, Emperors must reproduce in indescribably savage conditions. They have little time or energy to spare on courtship, and pair bonds form quickly. The male lays his head on his chest and calls out, sometimes repeatedly, and a female responds by standing in front of him. Both slowly raise their heads, stretching very tall, and freeze into immobility for a few minutes. Then they relax, and either pair or part.

Unlike many penguin species, Emperors do not go fishing to stockpile energy before laying their eggs. After pairing for six weeks, the female lays one egg, and the male immediately takes it from her feet, puts it on his own feet, and covers it with his brood patch to keep it warm. In this position the egg is sheltered by a fold of skin, and the male rocks back onto his heels to minimize the amount of ice touching his feet.

▲ **SURVIVAL RATES**
Life as an Emperor penguin chick is very harsh, and many do not survive. Only 65 percent become independent from their parents, and of those that do fledge, less than 20 percent manage to survive their first year. If they make it through a few years, their survival rate improves dramatically to 95 percent in adulthood—and then they will probably live another 20 years.

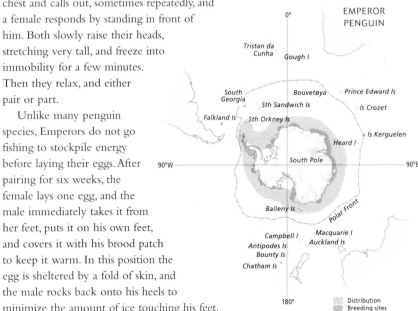

EMPEROR PENGUIN

Distribution
Breeding sites

Going into a huddle

Emperor penguins are the only adult penguins that huddle, and forming these tight groups is essential for survival during the cold winter. It is believed that huddling reduces the loss of body weight by 25–50 percent. In large colonies, up to 5000 birds will form one huddle, grouping together in as little space as possible. The huddle moves as a whole, all the birds keeping their backs to the wind. In a fantastically complex group dance, the windward birds move down the flanks of the huddle, then into the middle. Eventually, they are pushed out to the edge and into the worst of the weather, before moving down the flanks and back to the middle. This circulating pattern ensures that all the birds get some time in the warmth, and no bird should freeze to death.

Meanwhile, the female has done her part for the time being, and immediately heads for the sea to fish.

The males incubate the eggs throughout the winter in constant darkness, surviving by huddling together; the temperature often drops to –60°C (–76°F), with winds up to 180 kilometers per hour (110 mph). After nine weeks the chicks hatch and the males give them a small feed. Shortly afterwards the females come back and set about persuading the males to give up their charges; eventually the males give in and hand the chicks over to their mates. If a chick falls to the ice, another penguin may steal it—or, if left alone, it may freeze to death in the extreme cold.

Now, at last, the males are free to make for the sea to fish. They will have been fasting for four to six months and will have lost about half their body weight by the time the females return to claim their offspring, but they still usually have to walk about 100 kilometers (60 miles) across the ice to reach open water and a much-needed meal. Males never die of starvation while caring for their chicks. If the female is late in returning, the male simply abandons the egg or the chick and makes for the sea to feed before it is too late for him to survive. Abandoning a chick may seem harsh, even counter-productive—but the male will live to breed again, an older, wiser, and presumably more effective parent the next time around.

After the chicks have begun to hatch, two lines of Emperors snake between the colony and the sea, one coming and one going. The chicks need constant feeding, and the parents take turns providing meals for about seven weeks. After that, the growing chick needs more food than a single parent can provide, so the chicks huddle together for warmth while both parents go to sea at the same time to fish.

 BEGGING FOR FOOD
When a young Emperor penguin chick is hungry, it stretches as tall as it can, waves its head from side to side, and begs for food. After much pleading, the parent regurgitates an oily mess of fish, squid, and krill into the chick's throat.

A WARM PLACE
The temperature inside the brood patch—the bare patch of skin on the parent's belly that keeps the precious egg warm over winter and now does the same for the young chick—is over 30°C (86°F), even though the air outside is a chilly –30°C (–22°F).

King penguin

SPECIES: Aptenodytes
   patagonicus
LENGTH: 95 cm/37 in
WEIGHT: 13 kg/29 lb

STATUS: approximately 2 million;
   most populations increasing
   IUCN: not listed   CITES: not listed

The King penguin is the second largest living penguin species. Adults weigh around 13 kilograms (29 lb) and average 95 centimeters (37 in) in height. Females look identical but are a little smaller and slighter than males. Unlike most penguins, Kings have big splashes of color in their feathering. They are one of the more colorful penguins, with a beautiful spoon-shaped patch of bright orange on each side of their neck, fading into a paler orange upper chest. Their lower chest is white, and the rest of the penguin is a dark blue-gray with a silvery sheen, shading to black lower down the body. There are two subspecies, *Aptenodytes patagonicus patagonicus*, which lives on South Georgia and the Falklands, and *A. patagonicus halli*, from Prince Edward, Crozet, Kerguelen, Heard, and Macquarie islands.

▲ SMALLER AND LIGHTER

The King penguin is smaller and much lighter than its more southerly cousin, the Emperor, and it has slightly longer flippers. In addition, its elegant bill is thinner and longer than that of the Emperor, whose shorter bill exposes a smaller area to the weather. The feathers that extend over the base of the Emperor's beak are a further adaptation to its more extreme environment.

Recovery of a population

King penguins are large birds that gather in great numbers, and in the past they were ruthlessly hunted for their oil. Their numbers were seriously reduced, and some colonies were completely exterminated. However, with the end of sealing and whaling, most King populations seem to be growing at a yearly rate of 5–15 percent, and recently vacant areas are rapidly being colonized by nearby populations. There are now large, secure colonies at most breeding locations. Future risks to King penguins are largely limited to those related to the impact of humans on their small islands. An accidentally introduced disease, pest, or predator could inflict great harm on a large, closely packed colony in a very short period, and with human activity around the sub-Antarctic islands increasing, the risk of such introduction is increasing.

King penguins range widely in the southern oceans but stay clear of the pack ice, which they leave to their closest relative, the Emperor penguin. Five of the seven breeding populations live on islands just south of the Polar Front, but they prefer to fish waters just north of the Polar Front, where the surface temperature is around 4.5°C (40°F). They are also often found fishing over the slopes of the continental shelf. They stay close to their breeding colonies during summer, but are somewhat pelagic, particularly in winter, and juveniles are regularly spotted several hundred kilometers from the nearest colony.

Wooing and wedding

King penguins acquire their adult feathers just after their second birthday. They spend the next two seasons practising courtship behavior, and mate for the first time at four years old. A male who is ready to mate returns from feeding at sea and advertises his availability by calling and stretching to his greatest height. If he succeeds in attracting a female, the pair shake their heads together—this is perhaps a displacement activity to dispel any potential for conflict—and occasionally call. Then one of the pair, usually the male, struts off in a head-swinging "attraction" walk, and the new potential partner follows behind in a similar fashion. Swinging the head from side to side shows off the orange neck patches, and is believed to encourage following behavior. If a third penguin gets in the way of the forming pair during this ritual walk, there may be aggressive behavior such as wing flapping, mock beak stabbing, even fighting.

The next stage is "high-pointing," where the couple stand face to face, stretch slowly to their full height, and freeze for five or ten seconds before relaxing. The birds then sing antiphonally before bowing, and finally mating. After laying, they take turns to incubate the egg and brood the chick, strengthening their pair bond with each changeover by repeating the joint calling and bowing.

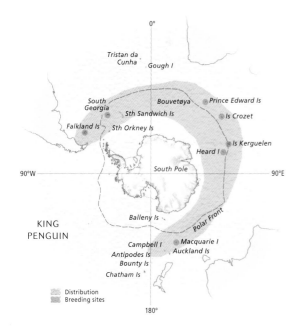

KING
PENGUIN

Distribution
Breeding sites

Once formed, a pair remains monogamous while raising the chick, but long-term pair bonds are uncommon: only about four percent of King penguins mate with the same partner in the following breeding season. Possibly re-bonding is infrequent because females stop feeding their chicks, molt, feed, and are ready to mate again a full month earlier than the males. But King penguins do try to get back to the same breeding place: one third return to the exact location of their previous territory, and almost 95 percent return to within a few meters of where they incubated their last chick.

King penguins gather in vast numbers to reproduce in a staggered three-year cycle, gathering on the plains and shelving beaches of their sub-Antarctic island homes. Pairs first reproduce early in the southern spring, laying one large egg in November and incubating it by balancing it on their feet, protected by a fold of skin. They do not build nests; instead, each pair's territory is defined by, and limited to, pecking distance, although territories do shift a little as the penguins shuffle slowly along with the egg resting on their feet. The parents take turns at two to three week incubation periods, until the egg hatches in late January.

▲ COURTSHIP

Mating is a long process usually instigated by the male after much displaying and parading. Once the male attracts a female, both birds begin head bowing, and after much strutting, and joint neck stretching, the penguins finally mate. At this stage, intrusion by a third party will result in an aggressive response of pecking and polysyllabic trumpeting.

The chick does not fledge until the following November at the earliest—but first it must survive the winter. Young chicks are at risk both from the elements and from predatory giant petrels and skuas, so at first one parent stays with the young while the partner goes fishing. As in the incubation period, the parents take turns at fishing and guard duty until the chicks are so big that they need two parents to hunt and bring back enough food. At this stage the chick joins a group of chicks known as a crèche. The crèche provides both protection from predators and some warmth through huddling together. Parents returning from the sea find their own offspring by calling and recognizing its voice. By the end of summer a healthy chick will have reached a weight of 11–12 kilograms (24–27 lb), but as winter approaches food becomes harder to find, so the parents can feed the chick only sporadically. Chicks lose about a third of their body weight over the lean winter period, but in spring, when prey becomes plentiful once more, the chicks regain most of their lost weight before molting into juvenile plumage.

The three-year breeding cycle

Breeding success has its price for King penguins. Healthy chicks hatched early one year will not fledge until November, so that its parents cannot produce another egg until much later in the following breeding season. The second offspring usually hatches in March, and the parents feed it as much as possible before prey becomes scarce, but inevitably this later-born chick weighs less at the beginning of winter than a youngster hatched in January, so it is less likely to survive. If a late-born youngster does live, its parents feed it again the following spring and it fledges very late in the season. Any youngsters born after a successful late-born chick are always too young to survive winter, and the parents breed on an early cycle the following year.

Swimming and diving

Penguin bodies are designed for efficiency in water. Both Emperors and Kings average slightly less than 9 kilometers per hour (6 mph) while fishing, although Kings can reach speeds of 12 kilometers per hour (8 mph). Not surprisingly, King penguins with small chicks spend only a fifth of their time swimming at 9 kilometers per hour, whereas parents feeding voracious larger chicks must work this hard for over a third of their time.

King penguins generally dive to around 50 meters (160 ft), but

▼ LARGE NUMBERS
▼ LARGE NUMBERS
Unlike most penguins, Kings were once hunted for oil, and several colonies were reduced or destroyed. They are now protected in most places, and are increasing in number and expanding their range to new areas around the southern tip of South America, a few spots in the Falkland Islands, and on Heard Island.

dives as deep as 300 meters (1,000 ft) have been recorded. Emperor penguins dive to an average of about 300 meters but often reach close to 400 meters (1,300 ft). These are not great depths compared with those plumbed by the sperm whale, which regularly dives to 300–750 meters (1,000–2,500 ft) and can go as deep as 3,000 meters (10,000 ft), but penguins can perform very impressive dives for their size. Emperor penguins have a probable aerobic dive limit of about four minutes, but two single dives have been recorded of approximately 22 minutes—more than five times that limit—to depths of between 40 and 60 meters (130 and 200 ft). In both cases the penguins were breaking a long fast.

▲ WALKING HOME

King penguins seem to avoid climbing, preferring large level areas on plains and in valley mouths, especially where there is protection from the wind. While their colonies may be close to shore, many penguins still have a long way to walk if they have left a mate or chick at the rear of a large colony.

➤ PUBLIC BATHS

Like most penguins, Kings like to scrub down after spending time on shore, where they are likely to get covered in guano and mud. When they first go to sea, they will spend some time rolling and splashing in the water, cleansing their feathers of any gathered grime, before leaving for their hunting areas.

Brush-tailed penguins

▲ CHANGING POSITION
To receive regurgitated food from its parent, a very young chick must stretch upward, usually thrusting its whole head into the parent's beak. Older chicks often just open their beaks while a parent tips a load of food in.

Like the Emperor penguin, the three brush-tailed penguins of the genus *Pygoscelis*—Adélie, Chinstrap, and Gentoo—live and breed in the Antarctic proper. They are fairly large birds, standing taller than all but the Emperors and the Kings, and their feathers are monochromatic black and white. Their genus name comes from two Greek words, meaning roughly "elbow-legs"—a reference to the shape of their leg bones, which in fact applies to all penguins—but the description "brush-tailed" may derive from their way of cocking their tail while swimming on the surface, so that it stands up from the water like a brush.

All these penguins nest on the land around the Antarctic Continent and its islands, colonizing headlands and hills exposed as the snow retreats with the coming of summer. Exposed land is very limited, but is essential to the success of a colony. The penguins build their nests from small rocks uncovered as the snow melts. It is not much of a nest, but better than nothing: meltwater drains away below it, and the rocks provide shelter from the wind. To shape the nest, one partner lies down and kicks backwards with one foot, scraping out a cup-shaped hollow among the stones.

The females lay two eggs and the parents incubate them in turns, lying along their bowl-shaped rocky nest with the eggs tucked into their insulating brood pouch. Given good weather and good fishing both chicks will fledge, but normally only one survives.

Precious stones

Brush-tailed penguins value stones so much that when they return from the sea they often bring one to their partner. They carry stones for great distances, and even steal them from the nests of their rivals.

They choose a stone with great care, and may replace one with an apparently better one. They often pick up stones from outside the nest rim and move them to a more central location to "tidy" around the nest. Neighboring couples respond in kind, and boundary disputes may follow.

> **GENTOO FEEDING CHICK**
This penguin chick is getting a feed
of krill from its parent, who has just
returned from fishing with a stomach
full of food, most of which will go to
the chick. Gentoos prefer to hunt
close to shore—their diet is mostly krill
from south of the Polar Front, but is
often more varied further north.

▼ **ANCIENT SITES**
Some penguin rookeries have been in
the same location for thousands of
years, especially those on the Antarctic
Continent. Rookeries like this one on
the Windmill Islands are about 3,300
years old, but some Ross Sea rookeries
have been operating for 13,000 years.
In contrast, the Peninsula's oldest known
colony is 644 years old.

Adélie penguin

SPECIES: Pygoscelis adéliae
LENGTH: 71 cm/29 in
WEIGHT: 5 kg/11 lb

STATUS: at least 2.6 million pairs;
10 million immature animals
IUCN: not listed   *CITES:* not listed

> ➤ LONG TREKS
Adélie penguins, despite their small size, will travel long distances to colony sites. They create paths by trekking across long, snow-covered slopes towards exposed ground. Very steep paths, rocks, and even cliff faces seem no obstacle, but rough ground, where the bird's short legs become a disadvantage, does seem to take its toll.

Adélie penguins, named for the wife of the French Antarctic explorer Dumont d'Urville, are everybody's idea of the proto-typical penguin. They have a clearly demar-cated and simple but strong black-and-white coloring, with a blue-black back and head and a white chest and throat. Their sole touch of color is a small patch of dark orange on the base of their bill, which is partially feathered, the black feathers making it look shorter than it really is. Around their eyes Adélies have vivid white rings, which the young develop long before they grow their adult plumage. This arresting feature is a vital part of the penguin's armory; eyes down and crest erect, it exposes even more white in the sclerae of its eyes while it performs its three main displays of aggression: the direct stare, the fixed and alternate one-sided stare, and the crouch. The startling contrast of black and white constitutes an unmistakable threat.

Adélies are rarely found north of 60°S, and always remain entirely south of the Polar Front, living on the Antarctic Continent and nearby islands. They prefer to live within the pack ice, but small numbers colonize the South Shetland, South Orkney, and South Sandwich islands. Adélie rookeries will form anywhere that the penguins can reach, as long as it is free of ice: they will colonize ridges, hillsides, scree slopes, peninsulas, knolls, beaches, and moraines, provided that they are ice-free. Sometimes the route to the rookery is very steep, but this does not deter the penguins as long as there are no large steps, which the birds cannot negotiate with their short legs. Adélie colonies may be very large, with populations of 20,000 to 30,000 birds common. There are seldom more than 100,000 penguins in one area, but occasionally a single large colony may be home to more than one million birds.

Adélie penguins are shallow divers, and like all other penguin species they feed by pursuit diving, pecking out their food as they swerve from side to side under water. Adélies prefer to eat euphausiids, but will also take fish, amphipods, and cephalopods. Their diet varies according to what is available and where they are; for example, the diet of the Adélies around Princess Elizabeth Land and Wilkes Land is largely fish, while that of the Ross Sea populations is about half crustaceans, and that of the Terre Adélie penguins may include a high proportion of cephalopods. Evidence suggests that Adélie penguins breed most successfully when krill is locally abundant.

All in the timing

Timing is crucial for successful Adélie breeding, and the level of pair-bonding seems to depend on the latitude of the colony: the further south the colony, the more important timing is to reproductive success and the less likely a penguin is to wait around for last year's partner to turn up. At Casey station, (about 66°S), 80 percent of Adélies rejoin their former partners, whereas at Cape Bird, 11 degrees further south, only 56 percent reunite with the previous year's partner. Adélies at Cape Bird have a shorter breeding season, so they cannot spare the time to wait for last year's partner to return. Instead, they begin as soon as possible, with a new mate.

➤ CONSUMMATE DIVERS
Penguins are heavier than other birds
and not at all suited to flying—their
bones do not have air spaces, their
air sacs are reduced, and they have
short feathers. On the other hand,
their added weight gives them an
advantage when diving, as they are
only slightly lighter than water.

The gathering of the clans

Adélie penguins are gregarious, and often gather together to molt on floes and bergs in the pack ice. Large gatherings can often be found on the lee side of a hummock or pressure ridge, sheltering until they can take to the water and fish again. Before going to sea, they congregate at the water's edge, calling back and forth to birds already in the water, until finally one penguin jumps. Many of the foremost birds follow this lead immediately, and then the birds at the back of the group move to the front and start calling to those already in the water—perhaps as a protective device to limit the risk of being taken by Leopard seals. When an Adélie comes from the sea in to land, it surfaces about 20 meters (65 ft) from the beach, looks around, then swims in and launches itself 2 meters (6 ft) out of the water, gripping the ice or rock with its long toenails to prevent itself from falling back in. Possibly this is protective behavior—the penguins steering clear of the land–sea interface where they are most vulnerable.

At the onset of winter Adélies disperse northward from the pack ice, the younger birds leaving the colonies first. Many depart before molting, moving into the ice, close to krill supplies, and sometimes returning to their breeding sites when storms open pathways through the ice. But the birds usually spend autumn and early winter in pack ice up to 650 kilometers (400 miles) north of the Antarctic Continent, and juveniles sometimes winter even further north than adults.

Homeward bound

In spring—October in southernmost locations, but earlier in warmer, more northerly sites—Adélies go back to their breeding colonies. The first to arrive are the older, more successful individuals, up to eight years old; the last are the two-year-olds, who are about to breed for the first time. But everywhere the time of home-coming is governed by the ice, and first-time breeders may be very late when the pack ice is heavier and more persistent than usual. Males arrive about four days before females, and the middle of the colony houses the more successful breeders—generally older, more mature couples. In this part of the colony reproduction is usually highly synchronized, with successful pairs returning to rebuild their nests in exactly the same location year after year.

Adélies cannot lay eggs until their colony surfaces are clear of snow, so once it has found a suitable location a colony stays put. After the female has laid, the males take the first incubation shift of two weeks and then the parents swap. Either parent may be on the nest when the eggs hatch. After hatching, the parents usually brood alternately, taking two-day shifts each for three weeks until the chick is large enough to be left on its own. Then both parents must go to sea in order to satisfy their off-spring's voracious appetite.

▼ POLLUTED WATERS
Signs of persistent organic
pollutants are showing up in
sub-Antarctic and Antarctic
penguins, mostly in the form
of PCBs (polychlorobiphenols)
and pesticides. Some residues
are linked to compounds
currently or recently in use
nearby, but others appear
to have come from much
further afield.

ADELIE
PENGUIN

Tristan da Cunha
Gough I
South Georgia
Bouvetøya
Prince Edward Is
Falkland Is
Sth Sandwich Is
Sth Orkney Is
Is Crozet
Is Kerguelen
Heard I
90°W
South Pole
90°E
Peter I Øy
Balleny Is
Polar Front
Campbell I
Macquarie I
Antipodes Is
Auckland Is
Bounty Is
Chatham Is
0°
180°

Distribution
Breeding sites

Chinstrap penguin

SPECIES: *Pygoscelis antarctica*
COMMON NAMES: Ringed, Bearded, Antarctic, or Stone-cracker penguin
LENGTH: 72 cm/30 in
WEIGHT: 3.8 kg/8½ lb
STATUS: 6.5–7.5 million pairs; some populations are increasing rapidly
IUCN: not listed   *CITES:* not listed

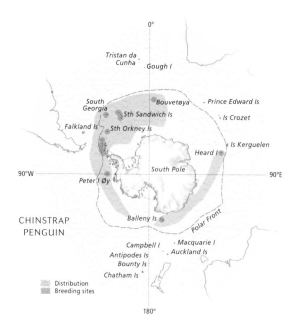

CHINSTRAP PENGUIN

Distribution
Breeding sites

Chinstrap penguins have a white throat, chest, and under parts, and a black body and head. Their bill is black, and a distinctive thin black line runs from ear to ear through the white of their throat. It is this band that gives them their main common name, Chinstrap, but these penguins were frequently seen by early Antarctic explorers, and were also named for their habits (hence "Stone-cracker") and their habitat (hence "Antarctic").

There are eight regional populations of Chinstrap penguins, all based inside the Polar Front; 99 percent of all Chinstraps live within the southern Atlantic Ocean. They inhabit both the sub-Antarctic and the Antarctic proper, preferring light pack ice. They breed on ice-free land on the Antarctic Peninsula and the islands south of the Polar Front, largely within the Scotia Arc. Chinstraps are very agile climbers, and their favorite nesting sites seem to be rocky slopes, headlands, rough foreshores, and high cliff edges.

These penguins are specialist feeders and eat mostly crustaceans, especially krill. Chinstraps dive to less than 50 meters (160 ft) most of the time, spending 40 percent of their time almost on the surface at a depth of less than 10 meters (32 ft), and 90 percent of their time at less than 40 meters (130 ft). Their maximum dive depth seems to be 70 meters (230 ft), but they probably reach

▼ MIGRATION

The Antarctic Peninsula is currently warming, and this seems to be one of the factors that is encouraging Chinstraps, like those shown here, to extend their range further south. Colonies found at the southern end of their Peninsula distribution limit did not exist 50 years ago.

Studying stress

The study of penguins is fascinating, but the disturbance of breeding penguins by human activity is the subject of constant concern and research. Initial investigations looked at obvious signs of distress, such as fleeing the nest and abandoning young at the approach of humans, or at the success or failure of reproduction. Currently, less obvious marks of distress, such as increased heart rates, are being examined and compared with the effects of normal stresses—for example, the disturbance shown by a nesting bird at the approach of a skua or other scavenger—but the effects and their implications are difficult to separate from normal fluctuations in reproductive success. One ingenious new way of measuring stress is a heart-rate monitor in the shape of a fake penguin egg, which is placed under experimental penguins.

this depth only when they are desperately trying to find food for their demanding chicks. Chinstrap dives have been timed; they generally feed for about two and a half hours, covering only 5 kilometers (3 miles) over a five-hour trip. Their shallow feeding dives last only about one and a half minutes each, with just over 30 seconds between dives. One reason for the current research interest in Chinstrap eating habits is that these penguins seem to have been very successful recently. Their range has expanded and their numbers are rising, perhaps because there is more krill in the southern oceans with the reduced number of krill-eating whales, or possibly because of the reduction in winter ice in the waters where Chinstraps hunt.

Noisy breeders

Chinstraps are extremely noisy, and it is surprising just how cacophonous a colony can be. All their calls are louder than those of the other brush-tailed species, and they also seem to call more frequently than other penguins. Away from the nest, Chinstraps are very gregarious and very inquisitive. Of all penguin species, they are the least likely to back away from an encounter and the most renowned for attacking humans who encroach on their territory—hence their reputation as feisty and aggressive.

While they are breeding Chinstraps stay close to their home colonies, traveling a maximum of 100 kilometers (60 miles) in the chick-rearing phase, and perhaps slightly further when the eggs are incubating. During this period they concentrate their hunting energies around nearby ice floes, presumably in search of the krill that is likely to be found nearby. After reproduction, when Chinstraps molt, some choose to wait it out just inland of their colony, while others seem to prefer to molt at a greater distance.

Retreating from the ice

In winter, as the temperature falls and the pack ice starts to expand northwards, Chinstraps move into the open waters north of the pack ice. They leave the Antarctic Peninsula and nearby King George Island in early April and do not return until spring (late October). But they stay together, and over the winter large groups congregate in the open water north of the pack ice. Little is known about where Chinstraps spend their non-breeding season, but they probably attempt to follow the krill and other organisms on which they feed. Chinstraps have been spotted a staggering 3,200 kilometers (2,000 miles) from their breeding site, but this figure may be based on individual birds that are extending the range of the species, rather than on the movements of the main population.

▲ CHASING FOOD
Chinstrap parents returning with food often lead older chicks on a chase through the colony, making them scramble after them for some time before finally stopping to feed them. This behavior possibly serves to separate weaker chicks from stronger ones when there is only enough food for one chick.

▶ SPECIALIZED FEATHERS
Penguins' feathers are highly specialized for insulation. The very short, broad, and flat rachis (center shaft) reduces the overall length of the feather and allows it to lie flat against its neighbor, and the closely spaced barbs, especially near the rachis and tips, provide good interlocking. There are also small, downy afterfeathers, which form a second layer of insulation.

Gentoo penguin

SPECIES: Pygoscelis papua
COMMON NAME: Johnny penguin
LENGTH: 75 cm/30 in
WEIGHT: 5.5 kg/12 lb

STATUS: more than 300,000 pairs
IUCN: Lower Risk—near threatened
CITES: not listed

➤ SETTING UP

Gentoos will nest with Adélies or Chinstraps, but generally prefer lower ground than the other two brush-tailed species when nesting in the same area. Not surprisingly, they prefer to form colonies away from elephant seals and flood zones, and on the Peninsula usually site their colonies less than 100 meters (330 ft) from the shore.

▼ BEING RESOURCEFUL

Suitable building materials for nest building are often very scarce, so supplies will be obtained from any source. If small stones are not available, Gentoos may utilize coal, small bits of wood, and other debris found near some Peninsula stations.

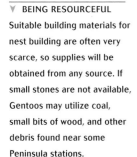

Gentoos have a triangular white flash above each eye, white flecks on the crown of their head, and a bright orange bill and feet. They live slightly further north than other brush-tailed species, remaining close to their breeding islands all year in sub-Antarctic and Antarctic regions right around the South Pole, but they move no further south than 65°S, shunning the most southerly pack ice and the Southern Continent itself, except near the Antarctic Peninsula. Most breed within the Southern Ocean, but about 25 percent breed north of the Polar Front. Taxonomists identify two Gentoo subspecies: the southern group, *P. papua ellsworthii,* is fully migratory, following a clear path each year, and has smaller feet, flippers, and bill than the northern *P. papua papua,* which disperses from its breeding areas but does not migrate in the true sense.

When they are not breeding, Gentoos mostly stay around the winter pack ice. The further south they breed during summer, the more likely it is that the whole colony will move northward for the winter, but generally they remain near their colonies, some staying as far south as possible in nearby open water. Many leave their breeding site after molting to feed and regain weight, but they never totally abandon a colony, and populations build up again after a few months. Nevertheless, whole colonies of Gentoos have been known to move on from time to time, perhaps to escape parasite infestation: large numbers of ticks have been observed on Gentoos. They peck constantly at each other and often squabble with their Gentoo neighbors over nesting materials, but they do not come to blows with other penguin species, even when competing species are nesting close by.

Finding food

Most Gentoos feed during the day and return to their colony in the evening. They generally feed inshore, finding shallow-water prey when breeding. Early in the day, they go to sea *en masse*; later they take to the water individually. They tumble in and swim briefly, then surface and clean themselves of the accumulated muck of the colony by rolling about and wiping their bodies all over with their flippers. When they come in to land after fishing, they will

pause to inspect the landing site, and then surf in on a wave. They leap onto ice or rocks, then shake their head and preen before moving on to their nest. Gentoos living near Antarctica mostly consume euphausiids, but those that breed north of the Polar Front eat more fish. If need be, they will dive deeper than 165 meters (530 ft) in search of their prey, but their diving is often less than 20 meters (65 ft).

Breeding behavior

Gentoos breed on ice-free land on the sub-Antarctic and Antarctic islands, and the Antarctic Peninsula, and may form colonies some distance inland or directly on the coast. They have been observed to occupy slopes, flats, terraces, valleys, headlands, ridges, and cliff tops, but some scientists have suggested that underwater land-forms dictate these penguins' breeding sites, so that they can always breed close to the shallow waters that are their most profitable hunting areas.

Around 90 percent of Gentoo penguin couples locate their former mate in the next breeding season, but, unlike the Adélie, Gentoos often move to new nest sites, presumably because they have a longer breeding season, so it pays them to wait for a successful partner to come back before beginning to breed. When Gentoos pair at the start of the breeding season, the male collects stones, moss, and dirt, depending on what is available, and starts to build the nest. After pairing, the male devotes his attention to foraging for nesting materials and leaves the actual building to his mate. Once the eggs are laid, individual incubation shifts last from about 24 hours to four days. At the northern end of their range, Gentoos start to lay in winter or very early spring, but on South Georgia they do not lay until the end of October, and on the Antarctic Peninsula they

may not lay until mid-November or later. These times may vary depending on snow cover, with the breeding season being delayed if there is heavy snow, and possibly getting under way early if the seasonal snowfall is unusually light.

Gentoos mostly choose elevated nesting sites at a safe distance from the breeding grounds of the elephant seals that share their regions. Most lay two eggs. In good years both chicks survive, but this is rare, except on the Peninsula, where usually both chicks survive or neither. Survival rates for chicks can almost always be traced directly to food supplies during the breeding

period. Gentoos feed very close to their nest sites, and eat about every 24 hours—much more frequently than the other brush-tailed penguin species. This daily feeding during mating, laying, and incubation seems to result in Gentoos being less likely than other brush-tailed penguins to abandon their chicks.

Young Gentoos first venture from the nest at about 20 days of age, and by 29 days most have joined other youngsters in a crèche, leaving both parents free to go to sea to fish. The youngsters fledge and are ready to go to sea at around 80–100 days, but at first they often spend more time on the beach than in the water.

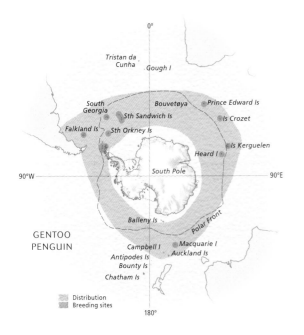

GENTOO PENGUIN

Tristan da Cunha
Gough I
South Georgia
Bouvetøya
Prince Edward Is
Sth Sandwich Is
Is Crozet
Falkland Is
Sth Orkney Is
Is Kerguelen
Heard I
South Pole
90°W
90°E
Balleny Is
Polar Front
Campbell I
Macquarie I
Antipodes Is
Auckland Is
Bounty Is
Chatham Is
0°
180°

Distribution
Breeding sites

What's in a name?

Both the common and the scientific names of the Gentoo penguin are mysteries. Gentoo, its common name, means "pagan" in Hindustani—a word believed to be of Anglo-Indian/Portuguese origin and somehow related to stories about dancing girls, but its etymology is unclear. As for the Gentoo's scientific name, *papua*, the first stuffed specimens to reach England for describing were thought to be from the island of New Guinea, and were named *papua*, which was the Malay word for "curly." Nothing about this penguin merits the description "curly," so perhaps it is simply a reference to the geography of its supposed place of origin.

True divers

Penguins of the genus *Eudyptes* are sometimes called "crested" penguins, a reference to their jaunty yellow "eyebrow" plumes. They occupy various habitats in the southern hemisphere, though none is truly Antarctic, and some live well within the sub-Antarctic region. They range from 45 centimeters to 70 centimeters (18–28 in) in height, and all have red or red-brown eyes. All species except the Fiordland penguin spend up to five months of the year at sea.

Eudyptes penguins differ from other penguin species in several ways. Allopreening, where one penguin preens another, is common in this group of penguins, although it is rare in most other groups. Some Eudyptids live in small groups, or even single pairs, hidden away from their neighbors, whereas other penguin species tend to be extremely gregarious. They always lay two eggs; the second, laid about four days after the first, is as much as 70 percent heavier than the first, which is almost invariably discarded.

▽ A SPEEDY MOLT
Most penguins molt quickly during the short summer: non-breeders molt first—in the middle of the summer; failed breeders may begin molting as soon as they lose their egg or chick; and successful breeders molt last, after their chicks have fledged, and after they have replenished their energy reserves at sea.

▽ ROCKHOPPING
True to their name, rockhoppers nest among ledges, rocks and grass, and prefer to come ashore amidst large rocks and rough seas. On arrival, they scramble from the sea, hopping ahead of the incoming waves with both feet together. This preference for hopping is not shared by other members of the genus.

▷ MACARONI HAVEN
South Georgia is home to millions of Macaroni penguins, but unlike the island's King penguins, most of the Macaronis live and breed far from easy human access. The island provides many good colony locations in tall tussock grass and on steep rocky hill-sides, and the surrounding waters supply plentiful food.

Rockhopper penguin

SPECIES: *Eudyptes chrysocome*
COMMON NAMES: Crested,
Tufted, or Rocky penguin
LENGTH: 45–58 cm/18–23 in

WEIGHT: 2.3–3.4 kg/5–7½ lb
STATUS: 3.7 million pairs, but
in decline. IUCN: Vulnerable
CITES: not listed

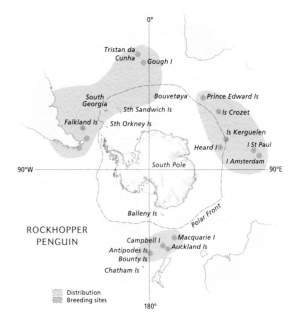

ROCKHOPPER
PENGUIN

Distribution
Breeding sites

The rockhopper is the smallest of the true divers, and the third smallest of all penguins. It has the classic black and white coloration, with a white chest, a black head and back, and an orange bill and feet. Its crest is a narrow band of yellow feathers above its eyes, thickening to a bushy dangling crest behind its eyes.

Rockhoppers are agile climbers: keeping both feet together, they scale cliffs quickly, gripping the rocky surface with their strong, sharp toenails. They hop on more level ground as well, and are surprisingly quick—probably the fastest of the smaller penguins, especially on steep or broken ground. They may have been named Rockhopper for their habit of landing on rocks, and have been known to spurn perfectly good beaches with smooth landing grounds, choosing instead to leap out onto rocky, wave-swept cliffs. They are much more likely than other penguins to jump into the water feet first, and they also seem to enjoy bouncing down hills, apparently without inflicting any damage on themselves.

Rockhoppers are found in sub-Antarctic regions around the Pole, but only about 12 percent of them, at five different locations, breed within the Southern Ocean. Most of their population base is north of the Polar Front, and in winter, during the colder, non-breeding period, they disperse northward into the ocean. They have been known to stay at sea for long periods, and there are even tales that on long sojourns their feet become encrusted with barnacles. Vagrants have been found on the southern shores of all the southern continents.

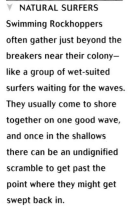

▼ NATURAL SURFERS
Swimming Rockhoppers often gather just beyond the breakers near their colony—like a group of wet-suited surfers waiting for the waves. They usually come to shore together on one good wave, and once in the shallows there can be an undignified scramble to get past the point where they might get swept back in.

These penguins are not large birds, so they do not consume large prey. Their diet regularly consists of the smaller crustaceans—amphipods, copepods, and isopods—and in some areas small squid. Because there are only a small number of them in the Southern Ocean, and because they are so small, they account for a mere one percent of the total energy consumed by all the penguins in the Southern Ocean.

Rockhoppers nest close to water, in among rocks and on ledges. They line their nests with rocks, or grass when they can find it, and on Ile Amsterdam there have even been reports of them digging burrows. Unusually for penguins living so far north of the Antarctic Continent, they have a remarkably precise breeding timetable—birds from the same site return to their nests on virtually the same day each year. Looser schedules, more attuned to each year's particular conditions, would be expected. Because Rockhoppers live so much further

north than many other penguins, they often share their rugged rookeries, which may be in very unlikely and exposed spots, with albatrosses and shags. They have a strong attachment to their original nest sites, returning over and over to the identical spot, whereas in other species partner loyalty is normally stronger than the site bond.

In southern rookeries, Rockhoppers arrive between mid-October and early November, and lay two eggs between late October and mid-November. The first egg is always smaller than the second, and if it is not jettisoned before it has a chance to develop it will be lost during squabbles that seem to pose much less of a hazard to the second egg. The eggs are incubated for 32–34 days and hatch from mid-December through January. The chick is brooded for 19–25 days, and the young fledge in February or March, after which the adults go to sea to feed for about five weeks and then return to molt before finally heading off to sea, usually in April or May.

Like other penguins, Rockhopper couples alternate hunting expeditions with nest duty. When the time comes to change shifts, both partners stretch their bills towards each other and trumpet loudly. Neighboring penguins often join in, directing their calls towards the returning partner. As the relieving bird reaches the nest, both partners trumpet vertically, with open bills pointed upwards, lifting their flippers with the calls.

Despite their small size, Rockhoppers are aggressive birds, and will attack when they feel threatened. They signal mild aggression by turning their head and bobbing it up and down, with flippers raised. If this fails to intimidate the opposition, one bird jabs its open bill towards its opponent, and eventually the two birds lock bills. The conflict ends only when one penguin manages to seize the other by the nape of the neck. The victor may even drag its vanquished opponent along behind.

Close encounters

Scientists differ about the classification of Rockhoppers. Some believe that there are three geographically defined subspecies: *E. chrysocome chrysocome*, found in the Falkland Islands and off the coast of Chile; *E. chrysocome moseleyi*, found at Tristan da Cunha, Gough Island, Ile St Paul, and Ile Amsterdam; and *E. chrysocome filholi*, from the region around Iles Kerguelen, Prince Edward, Heard, and Macquarie islands, and the Auckland and Antipodes groups. Others contend that there are only two types, a northern race and a southern race, distinguished by differences in behavior, song, and bill length.

In any case, relationships between the different *Eudyptes* species seem to be very close. Rockhoppers occasionally interbreed in the wild with Macaroni, Royal, and Erect-crested penguins, and these matings have produced hybrid offspring, some of which have successfully reproduced.

▲ GOLDEN HAIRED
The Rockhopper's scientific name, *chrysocome*, means "golden haired," and so is equally suited to any of the penguins in this group. The golden crest is smaller in the Rockhopper than in the Macaroni or Royal, but the small tufts, which often extend directly sideways, are clearly visible and distinctive. Rockhoppers have powerful bills for their size, and clear red eyes that can be quite disconcerting.

Macaroni penguin

SPECIES: *Eudyptes chrysolophus*
LENGTH: 71 cm/30 in
WEIGHT: 5–6 kg/11–13 lb

STATUS: about 12 million pairs; some populations increasing, some in decline
IUCN: Vulnerable   CITES: not listed

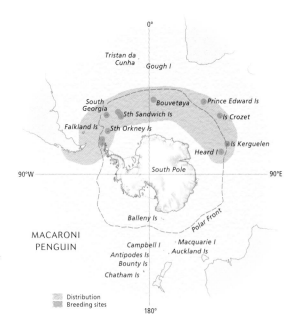

MACARONI PENGUIN

Distribution
Breeding sites

▲ INTER-SPECIES MATING
Macaronis are the most common penguins in the world, but even with around 6 million Macaronis of the opposite sex to pick from, some birds choose to mate with penguins from other species, including Rockhoppers.

Macaroni penguins are the largest of the *Eudyptes*. They are black and white, with an orange bill and feet, and a bright crest. The crest is dark yellow or orange, growing back along the brow line from the center of the bird's forehead, drooping over the edge behind the eyes, and hanging rather long at the back of the head. These birds were named for their resemblance to London dandies who went in for foppish dress and hairstyles associated with Italian culture.

Most Macaronis live south of the Polar Front, in sub-Antarctic and warmer Antarctic waters. They occupy colonies along the Scotia Arc and on South Georgia, and a few pairs nest with other species in the South Sandwich Islands. They often drift for long distances north and south of their colonies, but always remain north of the pack ice.

There are about 12 million breeding Macaroni pairs within the Southern Ocean, making up more than 50 percent of all breeding penguins south of the Front. This is the dominant penguin in the Southern Ocean, in terms of both biomass and consumption of resources, accounting for millions of tonnes of crustaceans and fish from the Southern Ocean each year.

Macaronis are noisy, pugnacious birds. They like to establish their rookeries on gently sloping ground, not too close to the water, with a pathway up along smooth, easily traveled stream beds. Small colonies often share rookeries with Chinstraps or Gentoos.

Breeding Macaronis reach their colony between September and November. The older birds come ashore earlier than the younger ones, and stay longer. The males arrive first and set about selecting or repairing a nest site, and the females come in eight days later. Then the birds pair, laying their eggs between October and November. Like the other *Eudyptes* penguins, Macaronis lay a very small first egg, which is usually lost after the second egg is laid four to six days later. The female takes the first incubation shift and the male takes the second, both about 12–14 days long. Both parents preside over hatching, which takes place some time between mid-November and early January. The male broods the chick for the first two to three weeks while the female makes several hunting trips to the sea to feed herself and the chick. The chicks fledge after about 65 days, and then both parents go to sea to feed for a few weeks, returning to the beaches in March to molt.

◄ ENOUGH TO GO ROUND
Penguin colonies make high demands on local waters for food. On the Prince Edward Islands, four species rely on the local food supply when they are breeding. Kings, Rockhoppers, and Macaronis eat 70 percent pelagic fish, whereas Gentoos hunt out benthic shrimp and fish.

Royal penguin

SPECIES: *Eudyptes schlegeli*
LENGTH: 70 cm/28 in
WEIGHT: 5–6 kg/11–13 lb
STATUS: 850,000 pairs
IUCN: Vulnerable
CITES: not listed

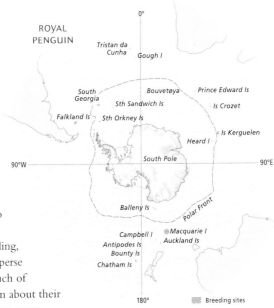

ROYAL PENGUIN

Many scientists believe that the Royal penguin is a subspecies of the Macaroni, the only difference being the coloration of the face, chin, and throat. Royals have a white or pale gray face, and there may be some black in the throat, but they have the massive head plumes and the very large bill characteristic of Macaronis. Royals are largely, but not entirely, limited to Macquarie Island, south of Australia. They are big for crested penguins, and extremely robust. Like most penguins, they walk by alternating their feet, but sometimes they hop along with both feet together, like their Rockhopper relatives.

Even though both sexes have very large bills, Royals are one of the easiest penguin species to sex when seen in pairs: the males have larger bills, and the females are seldom as white-cheeked as the males—more often gray.

Whereas most penguins are breeding by three to six years of age, Royals have a prolonged adolescence, and it may be 11 years before they breed. They breed in the same pattern as most *Eudyptes* penguins, in areas where vegetation is available. They excavate nests from sand or stones, and carry stones to their nesting site. In sandy areas, they also scoop sand into their bill and bring it to the nest site.

When they are not breeding, Royals either migrate or disperse to sea, where they spend much of their time, but little is known about their sea-going behavior or their sea range.

▼ **VULNERABLE SPECIES**

Royal penguins are isolated to breeding in a few large colonies on Macquarie Island and several nearby islets. Because they breed only in this one small area, if anything should disturb the population, all Royals would be at risk.

Other penguins

Some penguin species rarely venture into Antarctic or sub-Antarctic regions. Several of the crested penguins prefer to stay in warmer climes, as do the Yellow-eyed penguin, the Little penguin, and the four *Spheniscus* species, which hail from South America and South Africa.

▲ UNDER THREAT

Yellow-eyed penguin numbers are declining. These penguins prefer to live out of sight of other pairs—nesting close together reduces their breeding success—but fire and land clearing is forcing them together and reducing suitable nesting sites. Their numbers are also being affected by introduced animals.

▼ A RARE PENGUIN

Unlike other crested species, the Fiordland penguin breeds in single pairs or small groups scattered along long stretches of inhospitable coastline in southern New Zealand. Add to that a shy nature, and it is no surprise that, beyond knowing that they are rare, we do not have a clear idea of population size.

Eudyptes species

Three crested penguin species occupy New Zealand waters. Their crests are paler and less obvious than those of the Macaroni. Unusually for penguins, the small Fiordland penguin (*E. pachyrhynchus*) breeds in winter and is faithful to both nest site and partner. It is timid, living in small colonies or independent nests made of vegetable matter and stones in temperate rainforest close to the sea. It feeds in groups during daylight.

The Erect-crested penguin (*E.sclateri*) lives on islands south of New Zealand, while the Snares crested penguin (*E. robustus*) is found only on Snares Island. Uniquely, the Erect-crested penguin can raise and lower its crest. Little is known of these birds' movements at sea, but they are gregarious on land. They breed in large colonies in rocky breeding grounds.

Yellow-eyed penguin

This penguin (*Megadyptes antipodes*) is restricted to southern New Zealand waters. It has pale yellow irises, a yellow forehead and crown, and a long, narrow bill. In August it nests in isolation on grassy cliffs or in forests, feeding during the day and returning to land at dusk. It does not stray far from its breeding sites.

Little penguin

Eudyptula minor, also known as the Fairy, Blue, or Little blue penguin, is from Australia and New Zealand. By far the smallest of the penguins, at a mere 40 centimeters (16 in) high, it has a unique coloration—a pale, metallic blue-gray plumage with a white throat and chest. It is completely sedentary, neither dispersing nor migrating, usually feeding during the day and returning to its nest at dusk.

Spheniscus species

The four *Spheniscus* penguins are fairly large black and white birds with no crest and a striped chest and neck pattern. They are the least Antarctic of all the penguins, ranging from the equatorial Pacific Ocean to South Africa and South America, although the Magellanic penguin does reach the Falkland Islands. They nest in burrows or on rocky ledges and crevices, and their braying call has given them the sobriquet Jackass penguin.

The Jackass or African penguin (*S. demersus*) breeds along the coasts of southern Africa and is the only truly African penguin. It rarely strays far from land. It has a wide band of pink skin at the base of its upper bill and around its eyes, a broad white band running from above its eyes around its cheek to its breast, and a black band across its breast and down the side of its white underparts.

The Galapagos penguin (*S. mendiculus*) is unique in being equatorial, and unlike other penguins it molts more than once a year—before breeding as well as after. It gathers in small family groups in the waters around the Galapagos Islands. Its head is mostly black, with a thin white line curving from its eyes around its cheeks and onto its breast, where two thin black bands run across at flipper height.

The Magellanic penguin (*S. magellanicus*) is medium-sized and has the characteristic *Spheniscus* black-and-white bands across its face, neck, and chest, a black band around its white underparts, and another strong black band demarcating its head from its breast. It ranges through the Pacific and Atlantic coasts of southern South America and the Falkland Islands, occasionally appearing on South Georgia. It returns to its colonies around September, fledging its chicks and dispersing in March and April.

The Peruvian or Humboldt penguin (*S. humboldti*) lives off the Pacific coast of South America from Peru to Chile. It has a dark head, a pink patch at the base of its lower bill, and a single black stripe across its neck and breast. Peruvians are timid on land but gregarious in small groups at sea. They lay eggs year-round and do not migrate or disperse, although youngsters may drift both north and south before returning to breed. They nest in caves or burrows dug from sand and dirt a short distance from the ocean. LW

> SCIENTISTS' ASSISTANTS

Scientists in the Falkland Islands occasionally use Magellanic (seen here), Gentoo, and Rockhopper penguins as squid-catchers to help their research. About 50 percent of the food caught by penguins are types of squid that are not caught by either commercial fishers or research trawlers fishing the same areas.

▲ SWIMMING WITH PENGUINS

Penguins are so well adapted for a swimming lifestyle that it comes as a surprise to sometimes see them bumping into one another underwater—and not just when they are rushing, even when they are milling around.

➤ NOISY NEIGHBORS

A pair of Little penguins on their nest peer out from their home in a small cave. Little penguins commonly nest in burrows, caves, crevices, or under human structures along coastlines. Many New Zealanders can testify to how noisy a pair of penguins under a coastal house can be.

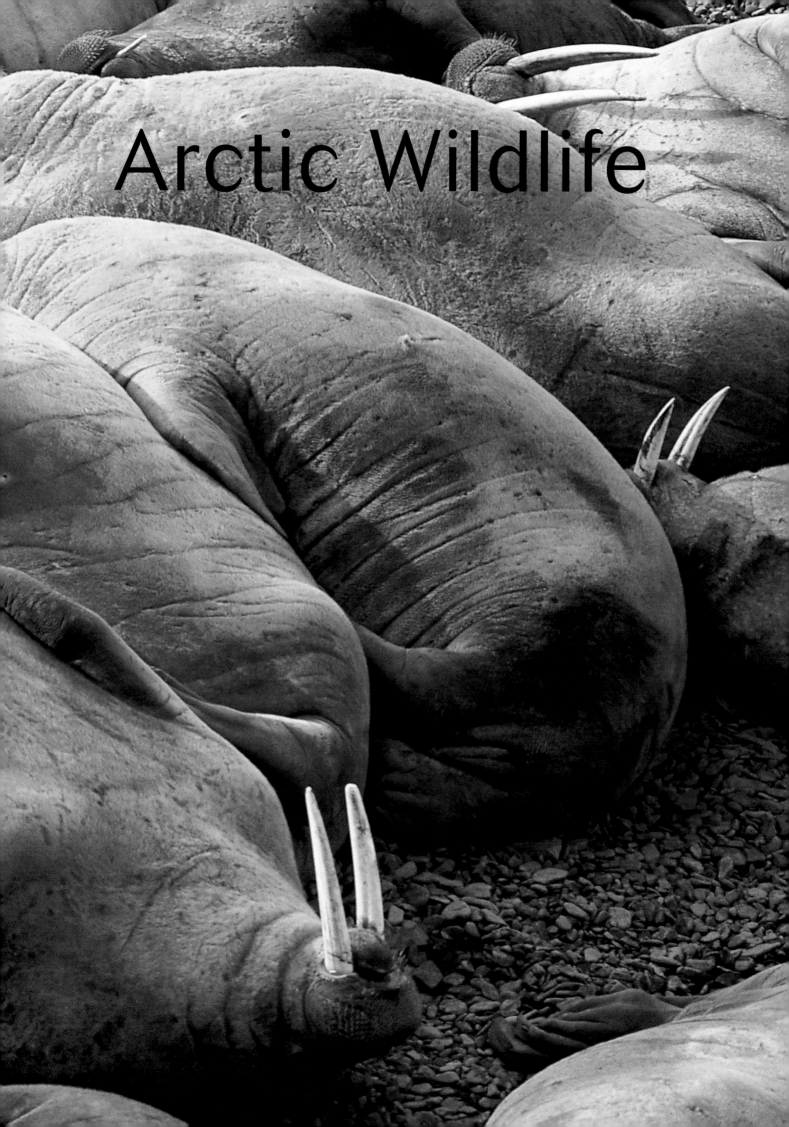

Arctic Wildlife

The Arctic region

Previous pages

◄ **CHANGING DISTRIBUTION**
The walrus is unique to the
Arctic. Five to ten million
years ago these large seals
were abundant in the Pacific
Ocean. Over time, they
became extinct in the Pacific,
but had meanwhile migrated
to the Atlantic and the Arctic
oceans, and flourished there.
Eventually they dispersed
across the northern seas
back into the Pacific.

Unlike Antarctica, the Arctic is an ocean surrounded by continents rather than a continent shielded by an ocean. This fundamental difference between the polar regions produces major climatic differences. The Arctic climate is generally cold, dry, and windy, but the northernmost frigid zone is open ocean, which retains heat and keeps the Arctic polar region warmer than the Antarctic polar region.

Arctic conditions

The Arctic contains the largest remaining wilderness areas in the northern hemisphere. Few species inhabit only the northern or southern polar region, and very few have successfully colonized both poles; those that have are plants and animals with extremely widespread distribution that have also colonized much in between. Generally, the Arctic and the Antarctic have recruited their species independently from local sub-polar life.

Arctic plants and animals are a combination of survivors from microenvironments unaltered by climatic change, and new colonizers from the south. There is no oceanic barrier for wildlife to pass in order to reach the Arctic; species are able to migrate across land that supports a gradual succession of plant types: tall trees give way to dwarf trees, then to low shrubs, low flowering plants and grasses, and finally to lichens and mosses. As the land is much warmer than in equivalent southern latitudes, forests grow inside the Arctic Circle.

Meeting of the seas

Several seas meet to form the main body of the Arctic Ocean, which consists of a deep central basin flanked by broad continental shelves. Compared with Antarctic seas, the Arctic seas have more multi-year ice, are closer to the Pole, and are generally much more ice-covered, with pack ice still covering almost half the Arctic Ocean in summer, when only the southern portion becomes ice-free. As a result, the major Arctic water masses are largely static. The upper layer of water remains frozen above a thin layer of richer water, largely from the Siberian rivers. Below that is a deep layer of warmer water from the Atlantic Ocean, and deeper still—below 800 meters (2,625 ft)—is very cold, very still water. The Arctic seas have large, powerful currents, of which the west Spitsbergen current within the Arctic Ocean is the biggest; it follows a deep trench and is also comparatively warm, above 3°C (37°F), and quite saline. As there is less land in the northern polar region, the Arctic has fewer icebergs than the Antarctic, but the northern ocean carries a large volume of river-borne material, both organic and inorganic, from the continents that fringe its southern borders.

Polar shores freeze over during winter. Expanding and contracting fast ice and the grinding physical

▲ **LAND OF LITTLE STICKS**
The great tracts of forest at
the southern edge of the
Arctic are known as taiga, a
Russian word meaning "land
of little sticks." From these
forests, a few huge rivers flow
into the Arctic, bringing to the
Arctic Ocean vast amounts of
fresh water, as well as
driftwood and other matter.

➤ **HARDY SCAVENGERS**
Kittiwakes are marine scavengers that have
learned to follow ships in search of food. These
small gulls, which inhabit most of the Arctic, are
more pelagic than most gulls, particularly in
winter. Only extreme conditions force them to
retreat to beaches and cliffs.

damage caused by moving ice prevent extensive Arctic shore flora and fauna developing. Even sub-Arctic shores are fit only for tough species, such as barnacles and periwinkles.

Polar water is generally oxygen-rich but nutrient-poor. Ice limits the mixing of waters, so very few nutrients reach the main body of colder waters. Consequently, most nutrients are concentrated in a small area, and the edges of the ice are rich in life for the four-month summer period.

Ring of life

The ice-covered areas of Arctic oceans have very little vegetative bloom in summer, and cannot support the vast biomass of krill that thrives in southern polar seas. However, where the more nutrient-rich Atlantic and Pacific waters meet the Arctic Ocean, there is a strong growth of plankton that feeds a huge number of small swarming fish, preyed upon by larger fish and seabirds. Much of this area is seasonally covered by pack ice, and many plants and animals make the ice edges, leads, and polynyas (areas of open water) their habitat. Highly specialized and adapted species of fish, seabirds, and marine mammals are so concentrated that the shelf seas here are called the Arctic ring of life.

Most life in the Arctic Ocean is plankton, and most plankton species are not endemic to the region, but reflect local temperate zones. Arctic plankton consists of diatoms, flagellates, and other green algae, with different species found in different regions. The far northern, central Arctic has limited plankton, with copepods predominating in summer. Krill, the mainstay of Antarctic zooplankton, does not dominate the Arctic region.

Squid—the major prey of male Sperm whales, and important food for Orcas, seals, and birds—are abundant and ecologically important to the Arctic, although very little is known about them. Polar squid are extremely fast and difficult to catch, and samples obtained for study differ significantly from the species usually found in the stomachs of squid-eating predators.

The Arctic supports a more diverse range of fish than Antarctica, although none of the main group found in the Antarctic—the Notothenioidei—lives in the Arctic.

The most important fish species in the Arctic is the Arctic cod (*Actogadus glacialis*), which feeds on larger plankton and lives in very cold waters. Salmon types are more common near freshwater inputs from the surrounding continental river runoff. These species, including Arctic char (*Salvelinus alpinus*), live and feed in the ocean, but travel up rivers to spawn. Further south, in sub-polar waters, fish tend to be related to local temperate species. Vast shoals of Capelin (*Mallotus villosus*) winter in deep water and surface during summer. Eelpout (*Zoarces viviparus*), Polar cod (*Boreogadus glacialis*), Skate (*Raja* species), Haddock (*Melanogrammus aeglefinus*), and Coalfish (*Pollachius virens*) also inhabit these low Arctic waters, along with a single shark species—the Greenland shark (*Somniosus microcephalus*).

> ▶ **MORE HOSPITABLE THAN THE SOUTHERN POLAR REGION**
> The diverse ecosystems of the northern polar zones are due to the warming effect of the Arctic Ocean, and to the surrounding continental landmasses that allow animals and plants to migrate over land rather than crossing inhospitable ocean.

Climate and vegetation

Northern polar regions can be divided into the central maritime basin and its peripheral land-masses. The whole northern region is dominated by anticyclones (high-pressure systems that rotate clockwise) in winter; in summer they lessen, and comparatively warm, wet oceanic air enters the region, which gives rise to tundra growth in southern areas and ice caps further north.

Climate

The interior of the Asian landmass has very cold winters and short, cool summers with low temperatures. The presence of the comparatively warm polar ocean means that the coldest point in the north is not close to the Pole, but in Siberia, near Verkhayansk. These central areas of the large northern landmasses are too cold to hold moisture, and so form large freezing deserts.

Growing conditions

A short growing season, meager winter light, and scant rainfall make it difficult for plants to survive in high Arctic lands. Temperatures are above freezing for only a couple of months each year. Arctic plants are very slow-growing due to the cold and lack of nutrients in the soil. The extreme cold stops matter decomposing and limits the rate at which new soil can form. Drainage is another problem. The top layer of soil melts under direct radiant heat in summer, but the ground below remains frozen. This permanently frozen underground soil, or permafrost, stops running water draining away from the surface soil, and creates bogs. Summer freshwater runoff from further south provides plenty of water, but may cause flooding or carry pollutants.

Topsoil changes make it very difficult for plants to take hold, and permafrost prevents them from putting down roots to anchor themselves against the winter winds. However, soil holds warmth at surface level, and conditions there are much better than at 1–2 meters (3–6½ ft) above or below ground level. These microclimates are very important to survival, and prostrate plants take advantage of the heat at ground level, where they are protected from snow damage. In cold conditions, rosette or cushion habits are common. Aerial shoots die back in winter, and new buds form under the ground.

Flora

The polar limit of the sprawling northern forests, known as the treeline, runs very close to the 10°C (50°F) summer isotherm (average temperature line). Taiga, a zone of stunted open forest, survives in this area despite the cold. Black spruce (*Picea mariana*), Dwarf birch (*Betula nana*), and willows (*Salix* species) resemble their southern cousins, but are smaller, shorter, and slower-growing. The taiga gradually gives way to tundra, where dwarf shrubs such as rhodo-dendrons and prostrate woody plants grow.

▼ **MILD ARCTIC CLIMATE**
Nordkapp, or North Cape, is high above the Arctic Circle in Norway, but it enjoys mild average temperatures ranging from −5°C to +10°C (23°F to 50°F). This is due to the North Atlantic current, which carries warm water from the Gulf Stream northwards into the Barents and Norwegian seas.

◄ **FIGHTING THE ELEMENTS**
The sparse tundra of the Svalbard island group supports more than 10,000 caribou. In harsh seasons, these numbers may fall because the caribou are trapped on small islands unable to move to better feeding grounds.

Shrubs, flowering plants, mosses, liverworts, lichens, and algae thrive on the treeless tundra. Flowering species include Purple saxifrage (*Saxifraga oppositifolia*), Moss campion (*Silene acaulis*), Mountain avens (*Dryas octopetala*), gentians (*Gentiana* species), Dwarf fireweed (*Epilobium latifolium*), Arctic poppy (*Papaver radicatum*), cotton grasses (*Eriophorum* species), Round-leaved sundew (*Drosera rotundifolia*), and Common butterwort (*Pinguicula vulgaris*). Further north, grass sedge tundra consists of grasses, sedges, herbs, and lichens, with bents, foxtails, anemones, gentians, and lichens and mosses dominant.

Tundra fades to polar desert to the north, with fewer species spread thinly. More barren than the tundra are the fellfields, where less than half the ground is plant-covered; then there are the Arctic barrens, where only slow-growing lichens survive. The problem in these northern barrens is lack of accessible liquid water, rather than cold; bogs and marsh are made up of acid sphagnum or peat moss and cotton grasses. The far north is usually locked in permafrost, and none of it is highly productive; large animals must graze huge areas in order to gain enough nourishment to survive.

Plants must manage the stresses caused by the huge seasonal changes in the amount of light. Most plants in the far north propagate by putting out runners and shoots, instead of relying on propagation by seed—a risky method where the the brief growing period may be cut even shorter. However, many plants also flower.

In the north, the production of flowers is based on the plant's response to the amount of light. Most require at least 12 hours of light to trigger the flowering process and rely on the almost continuous sunlight during the very short summer, when growth is extremely rapid. About 900 of the world's 250,000 or so flowering plants grow in the Arctic, compared with just two species growing in the Antarctic.

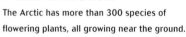

► CLOSE TO THE GROUND
The Arctic has more than 300 species of flowering plants, all growing near the ground. This narrow zone close to the ground provides favorable growing conditions—soil and rock warmed by long hours of summer sunlight, and shelter from blowing winds.

► A CARNIVOROUS SOLUTION
The Round-leaved sundew circumvents the difficulties of living in very poor soils by absorbing nutrients, especially nitrogen, from the insects that it captures on round, sticky glands raised as traps.

► FOLLOWING THE SUN
In Arctic conditions plants must flower and seed rapidly, before the end of the short summer. The Arctic avens, called *malikkat* ("the follower") by the Inuit, turns its flowers to the sun, the cup-shaped structure acting as a solar collector so no sunlight is wasted.

► SLOW-GROWING, LONG-LIVING
In the northern tundra, a smooth carpet of plant growth is usually a collection of many varieties of lichens and mosses, and a few tiny and hardy vascular plants like sedge and cotton grass. In the cold, dry, high Arctic, many plants are extremely slow-growing, and even tiny plants may be very old.

Marine mammals

Seal populations

Two species of eared seals and several true seal species live in the Arctic; Walruses belong to a separate family, Odobenidae, and are unique to this region, with one population in the Pacific and a separate one in the Atlantic. Seals are both predator and prey in the northern oceans. They consume fish, cephalopods, bivalves, and crustaceans, and are preyed upon by Greenland sharks, Orcas, Polar bears, and humans.

Eared seals

The Northern or Pribilof fur seal (*Callorhinus ursinus*), also known as the Alaska fur seal, is found in the northern Pacific Ocean and the Bering Sea, where it eats squid, herring, and other fish, depending on location and season. Not closely related to other fur seals, this species has a characteristically short muzzle. Youngsters, and females—which rarely exceed 1.5 meters (5 ft) and 50 kilograms (110 lb)—are generally silvery-gray above and reddish-brown below; adult males are black or dark brown with a thick neck and heavy mane. Males grow to 2.1 meters (7 ft) and weigh 250 kilograms (550 lb). Like southern fur seals, females mate about one week after giving birth; pups are born from June to August and suckle for approximately four weeks. Northern fur seals head south from October to December, with females and juveniles traveling further than adult males.

The Steller or Northern sea lion (*Eumetopias jubatus*) inhabits the northern reaches of the Pacific Ocean, hauling out onto the ice during winter. It is a light reddish-brown. The adult male's mane is less developed than that of other sea lions. Males may grow to 2.9 meters (9½ ft) and weigh a substantial 1,000 kilograms (2,205 lb); females reach only 2.4 meters (8 ft) and 270 kilograms (595 lb). Pups, born from mid-May through June, normally suckle for 8–11 months, but mothers sometimes nurse a yearling and a newborn pup together.

True seals

True or hair seals tend to be gray with characteristic spots and markings. Each type has its own niche in the ecosystem, which limits conflict for scarce resources. In Arctic and sub-Arctic regions most are just over 2 meters (6½ ft) long and weigh about 250 kilograms (550 lb), with little size difference between males and females. The exception is the male Gray seal (*Halichoerus grypus*) which may weigh up to 400 kilograms (880 lb). Gray seals occupy the northern Atlantic region, breeding on fast and pack ice as well as on beaches. They can be gray, silver, black, or brown.

Inhabiting the northern Atlantic and Arctic Oceans, the

Hooded or Bladdernose seal (*Cystophora cristata*) is a true ice species. The mating male intimidates its rivals by inflating a hood on its snout, forcing its nasal membrane out through one nostril—usually the left—to create a large red "balloon." Females nurse their young for less than four days, but they produce exceptionally rich milk, which is 61 percent fat, and feed their young every 25 minutes. Pups gain on average 7.1 kilograms (15½ lb) a day, 82 percent of which is fat.

The Bearded seal (*Erignathus barbatus*) is found among the ice and is circumpolar in distribution. Courting Bearded seals sing a one-minute song under water and perform a ritualized series of dives, sinking slowly in a loose spiral and then releasing a burst of bubbles before surfacing to repeat the dive.

Seals of the genus *Phoca* are the northern hemisphere's most abundant seal group, with different species tolerating varying degrees of cold. The word *phoca* comes from Sanskrit via Greek, and means "plump animal." These seals have comparatively small snouts, small nasal openings, and long first and second digits on their flippers. Males and females look very similar.

The truly ice-loving Ribbon seals (*P. fasciata*) live in the far northern Pacific Ocean and in the Bering Sea. They are a rich, dark, red-brown, with paler rings encircling their flippers, neck, and hindquarters.

Ringed seals (*P. hispida*) have a circumpolar distribution across the Arctic Ocean, where they frequent seasonally or permanently ice-covered waters, usually on and under stable fast ice. They are gray, with black spots ringed with paler gray. Ringed seals use their foreflippers to maintain breathing holes in the ice, and they dig lairs in snowdrifts with extensive tunnels, shared chambers, and extra spaces for pups.

Three populations of Harp seal (*P. groenlandica*) inhabit the pack ice of the northern Atlantic and Arctic oceans. Their colors vary with age and between individuals, but adult males are generally silver with black banding in the shape of a harp or horseshoe running across their back. Females gather in winter to give birth in comfortably spaced whelping patches on the ice.

The Spotted seal (*P. largha*) reproduces and molts along the edge of the pack ice and is largely circumpolar in distribution, but in summer it explores rivers some distance upstream. Pairs bond for more than a month, and family groups remain isolated for this period. The Spotted seal is usually pale gray with a sprinkling of darker spots, sometimes circled with paler rings.

The Common or Harbor seal (*P. vitulina*) is broadly distributed across oceans and seas, venturing as far north as Svalbard, generally in waters kept ice-free by strong currents. It preys on shrimp, fish, and octopus, generally diving for only three minutes at a time.

Inhabiting the Pacific Ocean as far north as southeastern Alaska, the Northern elephant seal (*Mirounga angustirostris*), or Sea elephant, is similar to its southern cousin (*M. leonina*), although in evolutionary terms the

▼ POLAR BEAR ALERT
An Antarctic seal resting out of the water does not have any major land-based predators to fear, but in the Arctic, Polar bears are a constant, if low-level, threat. When out of the water, this Bearded seal keeps a wary eye open, and an escape route in sight.

two species separated one to three million years ago. Mature northern males may have slightly larger noses than the southern species, but are a little smaller overall, reaching 4.2 meters (13¾ ft) and weighing approximately 2,300 kilograms (5,070 lb). Females rarely exceed 3 meters (10 ft) and 900 kilograms (1,900 lb). Northern elephant seals usually dive to between 350 and 650 meters (1,150 to 2,130 ft), feeding on bottom and mid-water fish and some squid. Recordings show they can dive to depths of 1,250 meters (4,100 ft) for up to 48 minutes, which means they may dive for longer and deeper than any other seal species. Northern elephant seals appear less happy in icy waters than their southern relatives.

Walruses

The massive, tusk-bearing walrus (*Odobenus rosmarus*) grows to 4 meters (13 ft) and can weigh 2,000 kilograms (4,410 lb). When warm, its gray-brown body flushes pink from the blood in a network of capillaries just beneath the skin. Walruses are bottom-feeders that use their whiskers, and perhaps their tusks, to dig for clams. Families live in pack ice and at the ice edge, hauling onto land when ice is unavailable. They prefer to winter inshore, mating underwater in midwinter with one bull servicing up to 15 cows; they disperse north in spring.

▲ SKIN AND FUR
Females and young walruses have fur about 1 centimeter (less than ½ in) long, but mature males have less hair, and can be identified by their nearly bare, knobbly necks. Walrus skin is very dense, generally 2–4 centimeters (¾–1½ in) thick, and quite inflexible. Its many folds often harbor lice.

▶ DAPPLED FUR
True seals are not common in tropical regions of the world. Most northern true seals have light and dark patterning, often spots, which may provide some camouflage from predators such as Orcas and polar bears.

Whale populations

Arctic waters support populations of toothed and baleen whales; of these, Humpbacks (*Megaptera novaeangliae*), Fin whales (*Balaenoptera physalus*), Minkes (*B. acutorostrata*), and Orcas (*Orcina orca*) inhabit both southern and northern polar waters. Only two toothed whale species are endemic to the Arctic—the Beluga (*Delphinapterus leucas*) and the Narwhal (*Monodon monoceros*).

In the past, whales were hunted indiscriminately for their meat, oil, whalebone, and ambergris. Even now, a few countries maintain reduced whaling industries. As a result, the Gray whale (*Eschrichtius robustus*) occurs only in the northern Pacific Ocean, the northern Atlantic Ocean population becoming extinct years ago. The other baleen whales endemic to the Arctic—the two northern right whale species and the Bowhead whale (*B. mysticetus*)—are both listed as endangered.

Toothed whales

Narwhals, or Unicorn whales, have no dorsal fin. Males grow up to 4.5 meters (15 ft) long and weigh about 1.6 tonnes, with females being slightly smaller. The young are born dark gray; later white patches develop, resulting in a mottled gray-white patterning. These whales live in family groups of up to 20, but gather in larger numbers close to shore for calving in midsummer. They then follow the ice edge south and may winter over in oceanic polynyas, but rarely venture south of the Arctic Circle. They have been observed using their foreheads in cooperative ice-breaking, and probably travel further north than any other whale—to as close as 320 kilometers (200 miles) from the Pole. Narwhals eat squid, cod, halibut, flatfish, and pelagic shrimp. They produce the most intense echolocation clicks of any toothed whale, and may use them to stun prey.

Belugas, also known as White whales or Sea canaries because they chirp loudly and whistle, are born gray and turn white by about five years of age. Males may grow to 5.5 meters (18 ft), females to 4 meters (14 ft); maximum weight in either sex seldom exceeds 1.5 tonnes. Belugas have no dorsal fin, but the ridge down their back enables them to break thin ice and surface after a freeze. These very sociable whales sometimes gather in thousands throughout Arctic and sub-Arctic waters, usually close to land in bays and fiords. Five populations follow the Bowheads north each summer, using the blowholes in the ice made by the larger toothed whales. In summer, Belugas also cluster at river mouths in Alaska, the western Arctic, Hudson Bay, and the northern rivers of Russia. Females often bring their young, still lacking insulating blubber, to the same river mouth each year, where the fresh water is up to 10°C (18°F) warmer. Although they are opportunistic

▼ THE TASTE TEST
Belugas, or white whales, are unique to the Arctic. These small whales, which grow to only about 5.5 meters (18 ft), gather every summer—sometimes in huge numbers—in river mouths. Careful feeders, they first visually inspect potential food before sucking it into their mouths. They have been known to spit out food they consider undesirable.

feeders, Belugas cannot swallow anything larger than about 4 kilograms (9 lb) because they cannot cut with their teeth. They consume Capelin, salmon, herring, Saffron cod, Arctic char, flatfish, shrimp, squid, marine worms, and bottom-dwelling invertebrates. Arctic cod are very important to them, especially at the far north of their range, where few other fish species are available.

Baleen whales

Gray whales, known locally as Mussel diggers, Hard heads, Devil fish, or Gray backs, live and feed in shallow waters—up to 100 in a group if the food supply is good—digging in the ocean floor with one side of their jaw for amphipods, polychaete worms, and mollusks. They have a narrow head and only two to four throat grooves, many fewer than those found in most baleen whales. Instead of a dorsal fin, Gray whales have a dorsal ridge with 8 to 14 humps. Barnacles and whale lice on their skin give them a mottled gray appearance. Females grow to 13 meters (46 ft) and weigh 35 tonnes; males are generally slightly smaller. Gray whales move very slowly, averaging only 8 kilometers per hour (5 mph) and rarely exceeding 13 kilometers per hour (8 mph).

Found only in four or five populations in the Arctic, Bowhead whales are often called Greenland right whales or Greenland whales. They are glossy black and callosity-free along their full length of 15 meters (49 ft), reaching a weight of 50–60 tonnes, and are insulated by blubber up to 30 centimeters (12 in) thick. Bowhead whales have an extra-long chin to house the longest, densest baleen of all the whales—up to 4.5 meters (15 ft) long, with 600 plates. They are surface-feeders, capturing very small crustaceans, including copepods and larval euphausiids. Bowheads travel north to 75° latitude in summer and south to only 55° latitude in winter. Large groups migrate north earlier in the season than other whales, individuals calling to each other during the journey—presumably to share information on leads and thin ice. Every so often, a whale will ram a blowhole in ice up to 22 centimeters (8½ in) thick with the pad of fibrous tissue on the top of its head, and take six or more breaths.

The Northern right whale, also known as the Black right whale or Biscayan right whale, is virtually identical to the Southern right whale. Separated one to two million years ago, each population feeds at its respective poles during summer and gathers in low latitudes for the winter breeding period. Northern right whales grow to 18 meters (59 ft) and weigh 40–80 tonnes. Small groups of three to ten feed on copepods and krill, usually in areas where seaweeds grow. They roll their tongues around their baleen to clean it, flicking balls of debris from their mouths.

▲ SORRY I MISSED YOU
Humpback whales that feed in rich Arctic waters are unlikely ever to meet their southern relatives. During the northern winter—when northern Humpbacks head south to breed in the tropical waters of the East China Sea, off Hawaii and Mexico, the west coast of Africa, and the Caribbean— southern Humpbacks are feeding in Antarctic waters.

➤ ZOOPLANKTON STEW
In summer the rapid increase in Arctic plankton brings whales, like this Humpback, north to feed. In the Arctic, zooplankton is made up of at least three-quarters large copepods, and the balance consists of a few euphausiid species (the Antarctic staple), many other crustaceans, small jellyfish, and even fish fry.

Arctic birds

Birds of the north

There are many types of seabirds in the northern oceans. Because the flora is more developed in the Arctic than in Antarctica, many land birds can also survive in the northern reaches of the continents. But birds that nest in the Arctic must contend with some skillful land-based predators. Polar bears can swim to raid birds' nests, and low sites become even more unsafe when the sea freezes and Arctic foxes have access as well.

Seabirds

Most Arctic seabirds raise their young on cliffs or in other inaccessible areas, and these protected sites support very high densities of birds during the summer breeding period. There are about 50 species of Arctic seabirds, including 14 different auks, guillemots, and razorbills, and 10 gulls and skuas. There are also terns, fulmars, and cormorants. The Northern gannet (*Sula bassana*) is the Arctic representative of the gannets. Like other gannets, it has a white body and a chunky bill, and has black wingtips and a straw-colored head.

Almost all seabirds breed during June and July and depart with their fledged offspring at the end of the warm season. The Northern fulmar (*Fulmarus glacialis*) raises its chicks in company with other fulmars, using holes in cliffs or burrows discarded by other birds. Northern fulmars are found in the northern Pacific, northern Atlantic, and Arctic oceans. They vary in color from pale gray or very dark gray to white. Always noisy scavengers, these birds follow ships on the off-chance of finding scraps.

Several gulls with a circumpolar distribution are unique to the Arctic. Of these, the Ivory gull (*Pagophila eburnea*) is probably the most obviously adapted. Pure white through all its life stages, it is well camouflaged in high Arctic pack ice. The Glaucous gull (*Larus hyperboreus*), which lives mainly north of the Arctic Circle, is a large bird with white and silvery-gray adult breeding plumage. The Black-legged kittiwake (*L. tridactyla*), a medium-sized, ocean-going gull is found throughout the north; it has a white body and gray wing tops, black wingtips, black legs, and a yellow bill.

Among northern-region skuas, the Long-tailed skua (*Stercorarius longicaudus*) of the high Arctic is the smallest and lightest. It disperses further south after breeding, and can be identified by a distinct dark cap, dark wings, pale gray mantle, and grayish-brown back. The Arctic skua (*S. parasiticus*) is stockier, with color variations from pale to darker brown.

The Arctic tern (*Sterna paradisaea*) is a beautiful bird with a black cap, grayish underparts, pale upper parts, and crimson legs and bill when in breeding plumage. These gregarious birds have circumpolar distribution across the Arctic and sub-Arctic regions, and breed in colonies numbering fifty to thousands of pairs. Medium-sized, about 30–38 centimeters (12–15 in) in length and weighing less than 1 kilogram (under 2 lb), they migrate to Antarctica in the northern winter, and therefore perhaps receive more summer sunlight than any other species of bird.

The Red and Red-necked phalarope (*Phalaropus fulicarius* and *P. lobatus*) are long-billed, long-legged birds that nest in freshwater pools and bogs in the northern regions before wintering on the open oceans. The females of both species are more brightly colored than the males. Also breeding singly in freshwater habitats, the duck-like Red-throated diver, or loon (*Gavia stellata*) winters in small groups on ocean shores. It swims with its gray head tilted upward, and can be identified by its red throat and the striped back of its long neck.

Permanent residents

There are eight species of non-migratory land birds in the Arctic. These birds tolerate the cold winters and breed in May and June. Their breeding success is very dependent on seasonal factors, including the amount of snow, the wind strength, and the temperature.

The Lapland bunting (*Calcarius lapponicus*), the Snow bunting (*Plectrophenax nivalis*), the Arctic or Hoary redpoll (*Carduelis hornemanni*), and the Common redpoll (*C. flammea*) are all seed- and insect-eaters, and move further south in winter when supplies begin to dwindle.

◁ **LONG-DISTANCE MIGRANT**
Arctic terns nest throughout the low and high Arctic in the northern summer, and in winter migrate south, some as far south as Antarctica—a journey that can range over more than 145° of latitude. Younger birds make rather shorter journeys, and gather along the southern reaches of the west coasts of South America and southern Africa.

◁ **MIGHTY TRAVELERS**
Northern gannets are large, heavy seabirds, often seen flying alone or in single file in small flocks. They live along the continental shelves of the Atlantic, with breeding areas ranging from eastern Canada to Norway, and their non-breeding distribution extends as far south as the Gulf of Mexico and western Africa.

> ADEPT FISHERS
Red-throated divers can often be found fishing along the fast-ice edge, usually singly or in pairs; but, like other birds, they will gather in large groups when the fishing is good. The Red-throated diver is the only one of four species of divers that can take off directly from the water—the others need a running start across the surface.

▼ SNOW CAMOUFLAGE
Most Arctic wildlife species wear some white as camouflage, but the Ivory gull is the only gull with all-white plumage. These black-legged, heavy gulls nest on both flat land and along cliffs. Like many colonial Arctic species, their breeding colonies are often under threat from Polar bears and Arctic foxes.

> FERTILIZING THE POLAR DESERT
Black-legged kittiwakes nest along coastlines well into the high Arctic, where their guano puts much-needed nitrogen into the local ecosystems. This organic material is critically important in the polar desert, where some plants may be largely dependent on it for nutrients.

The Rock ptarmigan (*Lagopus mutus*) and the Willow ptarmigan (*L. lagopus*) eat shoots as well as seeds.

The Arctic also supports various hunting and scavenging birds, including the Common raven (*Corvus corax*) and the Snowy owl (*Nyctea scandiaca*). Both eat insects, birds, small mammals, and any offal they can find—somehow managing to support themselves through the cold winter when food is scarce. Unusually for an owl, the Snowy owl hunts in the daytime, preferring lemmings to all other prey.

Since the Great auk (*Pinguinus impennis*) was driven to extinction in 1844, the Arctic has had no flightless birds. However, several flying seabirds occupy very similar ecological niches to the penguins of Antarctica. Included among these are Brunnich's guillemot or Thick-billed murres (*Uria lomvia*), Common guillemots (*U. aalge*), Razorbills (*Alca torda*), Atlantic puffins (*Fratercula arctica*), and Horned puffins (*F. corniculata*). All these species have dark upper parts and white underparts, and are quite stocky. However, the northern birds differ significantly from penguins in their ability to fly, and in their habit of nesting on cliffs and high ledges away from predators. It may be that this need to avoid land-based predators will prevent northern seabirds from ever evolving into fully specialized, non-flying, swimming birds.

▼ THE FACE OF LOVE
The remarkable faces of these Horned puffins are a breeding display; when not breeding the birds shed their brightly colored bill-sheath, together with the hardened, black, fleshy ornament that stretches above their eyes.

▲ THE GREAT AUK—EXTINCT AND PENGUIN-LIKE
Though similar in appearance and habit to the penguin, the Great auk is unrelated. Like penguins they were flightless, and they were hunted for food, fat, and feathers. Its last known breeding site was on Eldey Island, in Iceland, and it became extinct in 1844.

◄ SWIM FIRST, FLY LATER
Fledgling murres are reported to leave their crowded, rocky ledges before they can fly, joining adults in the water below. While most Arctic breeding birds fly north for high summer, murres remain in the north all year, close to their breeding grounds.

▶ OUT OF HARM'S WAY
Many Arctic nesting spots may appear safe, but when the sea freezes they become easily accessible to land-based predators such as Arctic foxes. Sheer rock spires that can be reached only by air, like this one in Siberia, give nesting birds the isolation and protection they need to raise their young.

Land mammals

Life in a cold climate

About 40 species of land mammals take advantage of northern habitats, and about 12 of these occupy the far north all year round. Many northern mammals have southern, temperate, and tropical relatives, but these polar cousins tend to be larger and have shorter limbs, and more fur and blubber. These adaptations, which help reduce heat loss, enable the animal to function better in the cold.

Only the large animals stay out in winter, their body surface to mass ratio allowing them to retain heat; some also huddle together for warmth during bad weather. The small land mammals, on the other hand, retreat under the snow. They do not hibernate, but shelter close to the ground nibbling plant roots, with no need to surface in inclement weather. This use of microclimates allows small animals to occupy niches much further north than they would otherwise be able to reach. Three types of lemming and nine kinds of voles live in the Arctic, sheltering under the snow for most of the year. Their short tails, rounded muzzles, and small ears and eyes are well adapted for the polar climate.

The hunters

The Arctic fox (*Alopex lagopus*) is gray-brown during summer—when it blends with rock and dry grasses—and largely white, with some gray, in winter. It raises a litter of pups in early summer, and may give birth again later in the year if food is plentiful. Arctic foxes forage singly or in pairs, preying on hares, birds, and rodents. Consummate scavengers, they will eat worms, eggs, nests, carrion, seal droppings, plankton, and other edible flotsam. They often trail larger predators, hoping to score leftovers, and also store food against future need.

The Gray wolf (*Canis lupis*) is the largest and heaviest of all wolves; males may weigh up to 80 kilograms (175 lb), and females 55 kilograms (120 lb). They are generally light brown or gray in color, with paler underparts; occasionally pure white and very pale individuals occur in populations in the far north. Wolves, which are the widest ranging land-based predators except for humans, generally move at night, but will travel during the day in cold weather. They often follow known pathways, returning to a rendezvous site from up to 200 kilometers (125 miles) away. Their home ranges are usually limited to defendable territories, but some northern wolf packs migrate with herds of caribou and musk oxen. They hunt by routinely testing the herds for weaker animals—with a success rate of less than 10 percent.

Gray wolf packs usually consist of five to eight individuals, increasing to about thirty when prey is plentiful. Packs generally consist of an adult pair and their offspring. When there is more than one adult pair in a pack, only the dominant pair mates, and constant aggressive behavior reinforces the hierarchy.

Grazers and browsers

The Arctic's grazing and browsing animals need ample space to satisfy their needs, and most of them travel huge distances in search of food.

Caribou (*Rangifer tarandus*) move through annual migrations in huge herds: the eight largest herds in North America range from 50,000 to 500,000 animals. After the spring migration, the herds split for calving, gather again after summer to migrate southward, and finally disperse over winter—generally to the forests just south of the Arctic, where the animals add trees to their standard diet of grass and lichen.

Wild North American caribou, which have the same species name as the partially domesticated and smaller European reindeer, vary from grays to browns, usually with paler underparts. Some are almost black or white, and Ellesmere Island has a subspecies that is pure white. Male caribou grow to 1.5 meters (5 ft) tall at the shoulder and weigh up to 250 kilograms (550 lb). They are the only deer species in which both sexes have antlers—the male's growing to about 1.3 meters (4 ft) long, and the female's to about half that length. Antler size determines dominance within the herd, and in September, during the breeding season, males try to control a group of five to fifteen females, or attempt to sneak into someone else's harem. Most mating occurs during October, with births in late May and early June—a single offspring is produced while the herd is migrating north.

The Moose (*Alces alces*), which prefers some tree cover, can often be found in marshy areas, browsing on shrubs, trees, and aquatic vegetation. Known as the Elk in Europe, the moose is the largest of the deer, having a maximum shoulder height of 2.3 meters (7½ ft) and weighing from 200 to 800 kilograms (440–1,760 lb). Moose have huge, heavy heads—the males crowned with palmate antlers—and broad, flat muzzles; their coats are dark brown in summer, turning paler in winter. Single or twin calves, born in spring, are able to browse by three weeks of age, and are fully independent at five months.

Herds of 12 or more Musk oxen (*Ovibos moschatu*) spend summer grazing the tundra, and winter on high ground where the snow cover has blown off, enabling them to dig for

▲ ORGANIC GARDENERS
Ground squirrels and other rodents hibernate through the winter and emerge to take advantage of the bountiful summer vegetation, storing much of it away to utilize over the following winter. The digging of dens and burrows is very beneficial to the Arctic, as it aerates the soil and mixes in organic material.

▼ SHEEP OF THE NORTH
Gregarious and extremely alpine in habit, Dall's sheep range over Alaska and northwestern Canada. They are closely related to the Bighorn sheep (found in the same region) and the Snow sheep of Siberia.

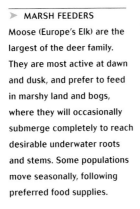

ON THE HOOF

Of the large grazers and browsers occupying the northern regions, the North American caribou is the only truly Arctic species, and it feeds on Arctic lichens, deciduous and evergreen leaves and twigs. The Wapiti, (pictured here) and its close relative the European red deer, are sub-Arctic species, exploiting the vast expanses of grassland that take over where the boreal forests flag.

MARSH FEEDERS

Moose (Europe's Elk) are the largest of the deer family. They are most active at dawn and dusk, and prefer to feed in marshy land and bogs, where they will occasionally submerge completely to reach desirable underwater roots and stems. Some populations move seasonally, following preferred food supplies.

◁ DECREASED RANGE
Well adapted to cold, Musk oxen live exclusively in the Arctic tundra. Musk oxen were once widely distributed across northern regions, but exploitation by Arctic peoples for their meat, hide, and horn has reduced their range to parts of Canada and Greenland, with introduced populations in the Eurasian Arctic.

vegetation that is still partially buried. Musk oxen have huge shoulders, and massive horns that meet in a flat plate on the forehead; males reach up to 1.5 meters (5 ft) at the shoulder, and weigh 400 kilograms (880 lb). Their long, matted guard hair, which is windproof and waterproof, protects them from the worst of the Arctic elements, and covers an extremely warm, dense inner layer of soft, downy hair. They calve from April through to summer; when wolves or other predators are nearby, the herd forms a circle around the young, their horns pointing outward.

Heavyweight hunters

The Grizzly or Brown bear (*Ursus arctos*), found through Eurasia and North America, grows to 1.5 meters (5 ft) at the shoulder, with adult males larger than females. Grizzly bears inhabiting the southern coast of Alaska may weigh up to 780 kilograms (1,720 lb), but average specimens weigh 200–300 kilograms (440–660 lb). Their shoulder hump—a sharp bend between snout and forehead—and longer claws and fur distinguish Grizzly bears from other bears; only some have lighter, grizzled tips to their brown fur.

Grizzlies prefer open areas such as tundra and feed in the morning and evening, eating mainly grasses, roots, and mosses in the spring; salmon and berries in the autumn; and insects, fungi, mice, ground squirrels, and occasionally larger animals. Hibernation begins between October and December; during this time heart rate and respiration slow down, but body temperature does not drop significantly, and animals can recover quickly. Cubs, one to four to a litter, are born from January to March while the mother is still hibernating. Bears wake between March and May, and mating occurs in spring and early summer. Although they are weaned at five months, cubs remain with their mother until their third or fourth spring.

The swimming bear

Semi-aquatic and circumpolar, the Polar bear (*Ursus maritimus*) is as much a marine mammal as a land dweller. Fully-grown males are two to three times larger than females, reaching about 1.6 meters (5 ft) at the shoulder and weighing 410–720 kilograms (900–1,600 lb). Like many polar species, Polar bears have plenty of blubber and fur to insulate them in very cold conditions. Young are born in January or February, when the mother is torpid in her den. The new family emerges during March or April, and cubs stay with their mother for two years. Males generally travel alone.

Polar bears are excellent swimmers and hunt between tundra, fast ice, and pack ice for Ringed and Bearded seals, but also relish Narwhal and Beluga calves. They also eat plants, birds, eggs, small mammals, and freshwater fish. They frequently wait quietly by the edge of a breathing hole in thick ice and, when an animal surfaces, attempt to sweep it out of the water with one swing of a front paw. Their patient vigil is worthwhile, as a 28-kilogram (62-lb) Ringed seal will satisfy a 230-kilogram (500-lb) Polar bear for about six days.   LW

Cycle of predation

When conditions are good, lemmings and voles breed in their burrows, producing up to six young per litter and up to four or five litters a year. At this rate, numbers increase rapidly, but then the population starves and collapses. This three- to four-year cyclic pattern has a major effect on their predators. With the drop in prey, Snowy owls fly south in search of other food, but Arctic foxes, unable to migrate as effectively, starve. Other species caught in the lemming cycle include skuas and weasels.

In the sub-Arctic, the Ruffled grouse (*Bonasa umbellus*), Snowshoe hare (*Lepus Americanus*), and Canadian lynx (*Lynx canadensis*) are involved in a similar predator–prey relationship, with a cyclic pattern of nine to ten years. Reasons put forward to explain this cycle include changes to external conditions, such as food supply, predators, or disease; internal changes, such as psychological stress; or changes to the adrenal and pituitary system, which breaks down under high-density living.

MEAT EATERS

Polar bears spend much of their time traveling on or in water, and may cover more than 20 kilometers (43 miles) a day in search of seals—their main food source. All bears will eat any type of food, but Polar bears are almost completely carnivorous. Other bears usually consume herbivorous diets with occasional protein feasts.

ARCTIC HEAVYWEIGHTS

Grizzly and Polar bears are the largest members of the bear family, and some big Grizzlies can challenge Polar bears for the title of heaviest bear. Grizzly bears are not social animals, and they usually establish solitary territories, which they will defend against other bears.

Part IV

POLAR EXPLORATION

It was not until the fifteenth century AD that the world began to gain accurate knowledge of the area beyond the Arctic Circle—and the Antarctic remained virtually unexplored until early in the twentieth century. Explorers of the eighteenth and nineteenth centuries sailed in frail wooden vessels into the harshest environment on earth, and gradually a true picture of the southern polar region emerged. National interests moved in and through heroic effort most of Antarctica was charted—though by no means conquered.

Early Explorers

1487–1900

First speculations

Previous pages

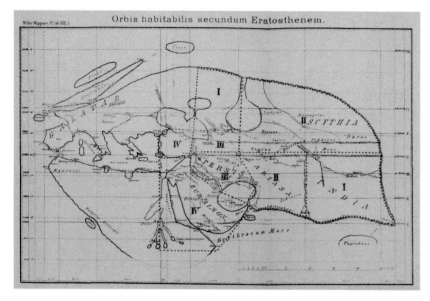

◄ PRIMITIVE TECHNOLOGY
A wooden sailing ship, held fast in ice—here one of Dumont d'Urville's—graphically represents the perils of early Antarctic exploration. In those days, no "motion of the hand [could] set the screw going [to] wriggle out of the first difficulty" encountered.

Antarctica is the only continent that, from the perspective of human thought, began as a sophisticated concept emerging from a series of deductions. In the sixth century BC Greek philosopher and mathematician Pythagoras calculated that the earth was round, and about a century later Parmenides divided the world into five climatic zones not unlike those that we know empirically today. He postulated frigid zones at the poles, a torrid zone at the equator, and temperate zones separating these uninhabitable extremes of heat and cold. In the fourth century BC Aristotle suggested that the landmass of the northern

Claudius Ptolemy, the influential Alexandrian geographer of the second century AD, added greatly to cartography with his refinement of grid lines of latitude and longitude to give every location fixed coordinates, northern orientation for maps, and various scales for different uses. However, his mistakes had an equally great effect. He agreed that there must be a southern land, but thought it linked the known continents and was fertile and inhabited. He also ignored the work of Eratosthenes, preferring that of the Greek astronomer Poseidonius, whose theories yielded a calculation of the earth at about 75 percent of its true size. Because of Ptolemy's errors, Christopher Columbus expected to find Japan where he encountered America, and James Cook set out to find the riches of the supposedly fertile Great South Land.

Orbis habitabilis secundum Eratosthenem.

▲ EARLY OUTLINES
One of the earliest maps of the world—as it was then imagined—drawn around 250 BC by astronomer and mathematician Eratosthenes. The lands around the Mediterranean are easily recognizable; the remoter regions of India and Scythia (China) are less familiar. This map also shows Parmenides' five climatic zones.

hemisphere must be balanced by a large landmass in the south; later this became known as *Terra Australis Incognita*, "The Unknown South Land." Aristotle gave it a name: at the time, the north lay under the constellation Arktos (the Bear), so he called the other end of the world Antarktikos ("opposite to the north"). Aristarchus, a third-century astronomer, was the first to expound that the earth rotated around the sun.

In 240 BC, Eratosthenes of Cyrene (now Egypt's Aswan)—who coined the word "geographica"—calculated the earth's circumference by comparing shadow angles in two distant locations at the summer solstice. It was a simple, ingenious, and remarkably accurate method. The units he used were imprecise, but even so, he overstated the circumference by only about 15 percent. By AD 200 philosophers such as Pomponius Mela had postulated the existence of a cold continent at the southern pole of a globe roughly the size that we now know it to be, spinning around the sun. It was the nearest to the truth that anyone would be for 1,500 years.

The dark ages
But after Ptolemy and long before Columbus and Cook, an age of ignorance set in. In AD 391 the fabulous libraries of Alexandria were destroyed and an era of dogma based on literal interpretation of the Bible began. Theological objections to the concept of an inhabited southern land proliferated. The Old Testament *Book of Isaiah* states: "It is He that sitteth upon the circle of the earth … that stretcheth out the heavens as a curtain, and spreadeth them out as a tent to dwell in." It was an utterance that gave rise to a perception of the earth as a flat disk. The New Testament *Gospel According to Saint Mark* records God's command to Christ's disciples: "Go ye into all the world, and preach the gospel to every creature." The disciples did not go to the South Land—therefore it could not exist.

The age of exploration
When the Christian city of Constantinople (now Istanbul) was occupied by the Turks at the end of the fourteenth century, fleeing scholars brought Ptolemy's works back to Italy. The *Geography* was translated into Latin in 1405 and hundreds of copies were printed by the end of the century. Europe started to look outwards.

▶ ROYAL PATRONAGE
Prince Henry the Navigator took no part in the voyages he sponsored between 1419 and 1460. Although seen traditionally as the force behind fifteenth-century Portuguese exploration, more new coastlines were actually found while his brother Pedro was regent (1440–49).

The Portuguese Prince Henry the Navigator (1394–1460) is regarded as the initiator of the great age of exploration; although he himself did not travel much, he was the patron of many voyages in Portuguese caravels that ventured far down the west coast of Africa. Henry wanted to reach India by sea for missionary and trade purposes, but the furthest his vessels sailed was to the coast of Sierra Leone. It was left to his compatriot Bartolomeu Diaz (c.1450–1500) to launch the age of Antarctic exploration.

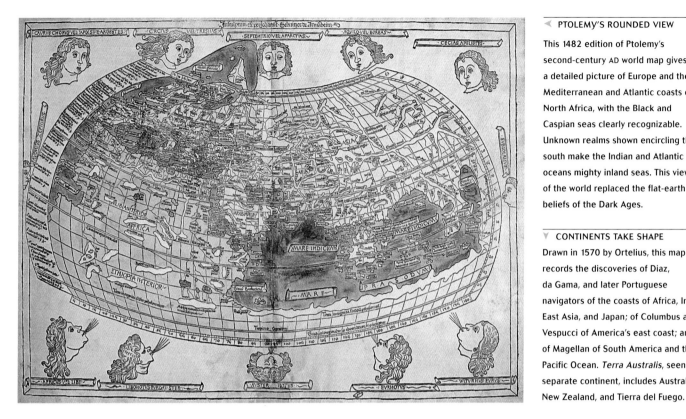

◄ PTOLEMY'S ROUNDED VIEW

This 1482 edition of Ptolemy's second-century AD world map gives a detailed picture of Europe and the Mediterranean and Atlantic coasts of North Africa, with the Black and Caspian seas clearly recognizable. Unknown realms shown encircling the south make the Indian and Atlantic oceans mighty inland seas. This view of the world replaced the flat-earth beliefs of the Dark Ages.

▼ CONTINENTS TAKE SHAPE

Drawn in 1570 by Ortelius, this map records the discoveries of Diaz, da Gama, and later Portuguese navigators of the coasts of Africa, India, East Asia, and Japan; of Columbus and Vespucci of America's east coast; and of Magellan of South America and the Pacific Ocean. *Terra Australis*, seen as a separate continent, includes Australia, New Zealand, and Tierra del Fuego.

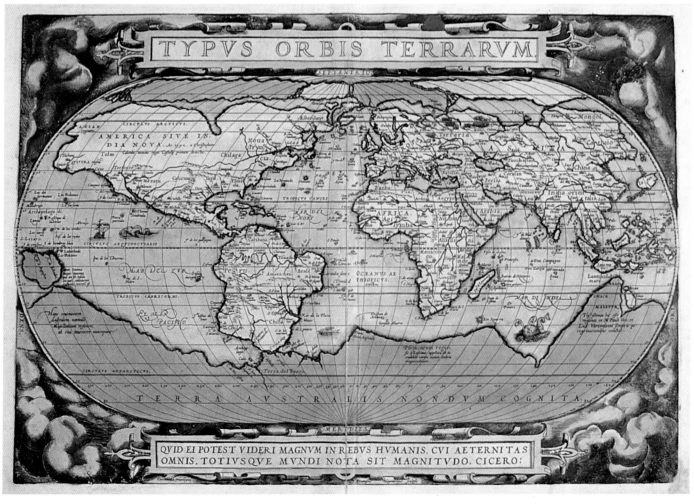

Early navigators

▲ VASCO DA GAMA

It took Vasco da Gama
23 days to sail from the
Cape of Good Hope to Calicut
on the west coast of India.

Most early exploration of Antarctica was a process of whittling down the fabled Great South Land as empirical knowledge replaced conjecture. The Portuguese explorer Bartolomeu Diaz took the first significant step when he sailed down the west coast of Africa in 1487. At the time, it was still credible that Ptolemy was right and that *Terra Incognita* filled the bottom of the world with a coast across the temperate zone, so that the Atlantic and Indian oceans were in effect inland seas. It was a brave venture by Diaz, because other ancient predictions stated that humans could not survive the "torrid zone" at the equator. In January 1488 he was alongside the coast of what is now South Africa when storms drove him out to sea and he sailed south for a few days. When he turned east, he found no coast, and made landfall only by sailing north again. He had rounded the bottom of Africa. He realized that he had discovered a sea route to India, and that

Africa, at least, was not joined to the Great South Land. When he returned there was talk of his leading a voyage to India, but it was another Portuguese explorer, Vasco da Gama (c.1469–1525), who sailed to India in 1497, thus proving that the Atlantic and Indian oceans were not landlocked. In March 1500 the Portuguese navigator Pedro Cabral led a fleet of 13 vessels to India, but to avoid the becalming waters of the Gulf of Guinea they sailed southwest far enough to see (and claim) the land now known as Brazil.

Next came Amerigo Vespucci, born in Florence in 1454 and employed by Spain, then by Portugal, and finally by Spain again. In 1501 he led a Portuguese expedition that reached Brazil in January 1502 and proceeded south to the River Plate. There is some dispute over the authenticity of Vespucci's papers, but it is likely that he sailed south along Argentina's Patagonian coast. He concluded that he had discovered

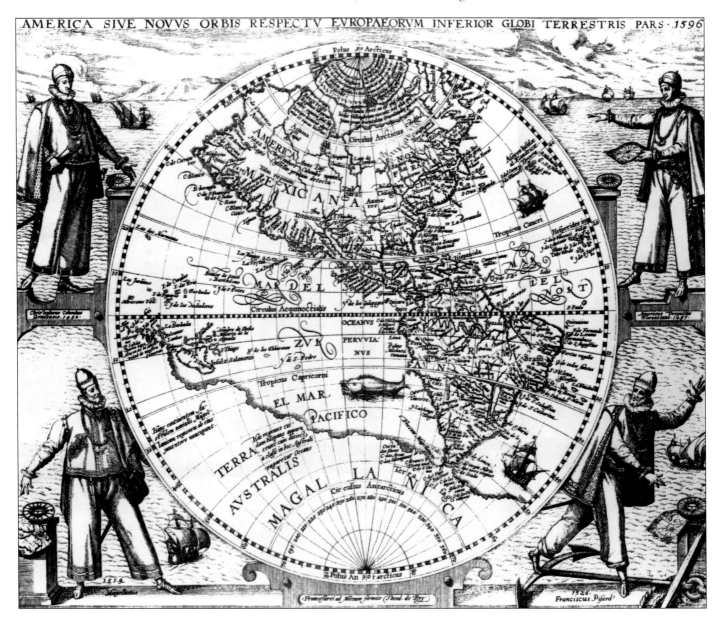

AMERICA SIVE NOVVS ORBIS RESPECTV EVROPAEORVM INFERIOR GLOBI TERRESTRIS PARS · 1596

◄ ATOP THE CORDILLERA
Clad in heavy armor, Balboa
and his men pushed through
thick jungle and crossed
the rivers and swamps of the
Isthmus of Panama to gaze
upon a mighty body of water
that after Magellan's expedition was to became known as
the Pacific Ocean.

not a new route to Asia but a whole new world that was
to become known as America—a name first used in
1507. Clearly the world was much larger than Ptolemy
had calculated, and this was confirmed when, on
25 September 1513, the Spanish explorer Vasco Núñez
de Balboa crossed the Isthmus of Panama and became
the first European to see the vast Pacific Ocean.

Ferdinand Magellan

Magellan's name is writ large in the annals of
exploration. Born in Portugal in 1480, he traveled east
as far as Malacca (in what is now Malaysia) with the
Portuguese fleet before seeking service elsewhere. He
persuaded the king of Spain to send him on a voyage
westward to prove that the spice islands—the Moluccas,
now part of present-day Indonesia—could be
considered part of Spain's western hemisphere of
influence. To accomplish this he needed to find a passage
around the bottom of South America, a difficult task
that took him much further south than he had

anticipated. His five ships left
Spain in 1519. From Rio de
Janeiro they sailed south, and
Magellan explored the River
Plate to see if it led through
to the Pacific—that would
have seemed reasonable, as
it was on about the
same latitude as
Africa's Cape of Good
Hope. He continued
southward, continually
seeking a westward
passage and continually being baffled. Finally,
he rounded the Cape of the Virgins (Cabo
Vírgenes) at 52°50′S and entered the strait
that now bears his name.

The Strait of Magellan threads a tortuous
and treacherous 525 kilometers (330 miles)
between the island of Tierra del Fuego and
the South American continent. Magellan sailed
through, but could not decide whether Tierra
del Fuego was the tip of a continent or an
island. It took 38 days to reach the ocean that
came to be known as the "Pacific" because it seemed so
peaceful. The fleet discovered that it was much larger
than expected: they had 14 weeks of near-starvation
before struggling into their next landfall at Guam.
Magellan was killed on 27 April 1521 in the Philippines,
but his surviving ships reached the Moluccas (and, in
1522, Spain), thus proving Magellan's thesis that the
spice islands could be reached by sailing either east or
west, and proving conclusively that the earth was round.

▲ NAUTICAL EXPERT
Merchant and explorer–
navigator Amerigo Vespucci
switched allegiance between
Spain and Portugal to support
his voyages to the New World.
He was the first European to
discover the Río de la Plata
(River Plate).

➤ THE EARTH IS ROUND
Magellan's circumnavigation in
1519–22 finally corrected Ptolemy's
drastic underestimate of the size
of the globe. This 1589 map detail
shows Magellan's ship, the Victoria,
forging bravely westward across the
vast Pacific Ocean.

Opposite page
◄ REACHING THE NEW WORLD
This map of the Americas was
drawn in 1590 by Theodor de Bry.
Eminent explorers associated with
the New World are featured in the
corners: Christopher Columbus,
Amerigo Vespucci, Ferdinand
Magellan, and Francisco Pizarro.

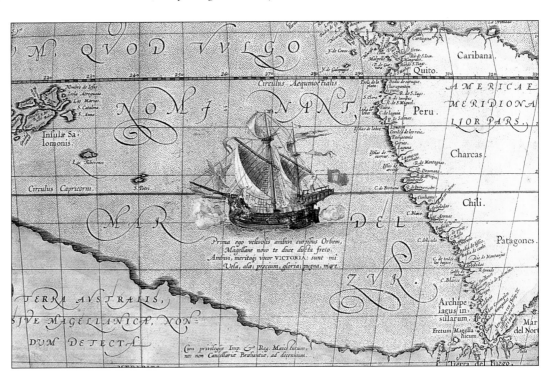

▲ DARING VOYAGE EARNS KNIGHTHOOD
Francis Drake returned from circumnavigating the globe in 1580;
only the second time this feat had been achieved, and the first
time by a British expedition. His knighthood, conferred on the
deck of *Golden Hind* by Queen Elizabeth I, was as much for
the commercial value of the voyage as for the geographical
discoveries he made.

◄ AWAITING DRAKE'S DISCOVERIES
This map of the Americas by Ortelius was published in 1570,
eight years before Drake proved there was a water passage
south of Tierra del Fuego. It shows Magellan Strait as separating
South America from *Terra Australis*. The 1630 revisions of this
map, however, show Tierra del Fuego as an island.

Sir Francis Drake

In 1572, the brilliant English navigator Francis Drake
(c.1540–96) led an expedition to raid Spanish
possessions in Panama. Here he had his first glimpse of
the Pacific and formed the intention "to sail once in an
English ship in that sea." In December 1577 he sailed
with five vessels with the goal of mapping the coast of

the South Land that was believed to lie below the
Pacific Ocean. He reached the entrance of the Strait of
Magellan in August 1578. Drake's chaplain, Francis
Fletcher, reported that the vessels sailed through a
"crooked strete," with mountainous land on either side
… "with such tops and spires into the aire and so rare
a height as they may well be accounted amongst the
wonders of the world …"

When he arrived at the Pacific Ocean in September,
Drake did not experience the same benign conditions
as Magellan. Rather, an "intolerable tempest" drove the
100-tonne *Golden Hind* southeast as far as 57°S. So, by

chance, and traveling backward, Drake discovered the place where the Atlantic and Pacific oceans "meete in most large and free scope" in what is now the Drake Passage. It was clear that Tierra del Fuego was an island, and that the tip of the Southern Continent must lie even further to the south.

In 1579 the Low Countries broke away from Spanish rule. It was the birth of a new seafaring nation, and the Dutch East India Company, founded in 1602, was to have a profound influence on exploration and trade for the next 200 years. Paradoxically, the next major development in exploration of the Southern Ocean resulted from an act of rebellion against the company. By the early seventeenth century it was illegal for ships other than the Company's to pass through the Strait of Magellan, so when the wealthy Amsterdam merchant Isaac le Maire organized a private expedition, he decided to seek out the southern seaway of which Drake spoke.

So in 1615 two vessels, the *Eendracht* and the *Hoorn*, commanded by Wilhelm and Jan Schouten respectively, set sail for Patagonia under the overall command of le Maire's son, Jacob. Fire destroyed the *Hoorn* during the voyage but on 29 January 1616 the *Eendracht* became the first ship to round Cape Horn, which Wilhelm Schouten named Kaap Hoorn after his birthplace in northern Holland. One of the world's most notorious nautical landmarks was now on the shipping maps.

By 1642, the Dutch colony of Batavia was well established and ships of the Dutch East India Company had explored part of the west coast of Australia. But no one knew whether Australia was part of the legendary Great South Land, so Abel Janszoon Tasman sailed from Batavia to explore the Indian and Pacific Oceans. Over 10 months he reached about 49°S and proceeded north to discover Tasmania, which he described as "not cultivated but growing wild by the will of God." He continued to New Zealand, which he assumed was the western shore of the southern

continent, and returned to Batavia having circumnavigated the continent of Australia without ever seeing it. The Great South Land had shrunk again.

Over the following century various national expeditions discovered land that they thought was the southern continent but turned out to be small sub-Antarctic islands. James Cook of the Royal Navy undertook the next major exploration of Antarctic waters.

Strange birds

In his passage through the Strait of Magellan, Francis Drake landed on an island, and Francis Fletcher described a "great store of strange birds which could not flie at all," their body size "less than a goose," that laid their eggs on the ground. Fletcher continued: "... in the space of one day, we killed no less than 3000 ... they are a very good and wholesome victuall ..." They were, of course, penguins, though no one on Drake's voyage used that name. It had been applied to the now extinct Great auks of the Arctic—there was a Penguin Island off Newfoundland by 1578— but the name seems to have been first used to refer to the familiar black and white birds of Antarctica by Francis Petty during a circumnavigation of the globe by Englishman Thomas Cavendish in 1586–88; Petty noted that "we trimmed our saved pengwins with salt for victual."

▶ WHERE OCEANS MEET
An "intolerable tempest" drove Drake south into the stormy passage between Tierra del Fuego and the Antarctic Peninsula that now bears his name. Here, viewed from a modern cruise ship, *Akademik Ioffe*, a Force 12 gale whips up surging seas in the narrow gap. How much more overwhelming must such mighty swells have seemed from the tiny *Golden Hind*.

James Cook

1772–75

When the Royal Society asked the Admiralty in 1768 to commission a voyage to the Pacific to observe the transit of Venus across the Sun, James Cook was their rather surprising choice of leader. Alexander Dalrymple who optimistically expected the great South land to begin south of the tropics and be populated with 50 million inhabitants had been lobbying to lead a voyage south. Cook, born in Yorkshire on 27 October 1728, was an experienced seaman and a self-taught mathematician and navigator, but was largely unknown. He was also commissioned to find *Terra Australis Incognita*. He left Plymouth in HMS *Endeavour* on 26 August 1768 and returned on 12 July 1771, having observed the transit, claimed New Zealand and the east coast of Australia for Britain, and mapped a great deal of the Pacific Ocean.

Cook's first voyage, successful though it was, did not discover the fabled Southern Continent. Naturalist Joseph Banks deduced from his observations: "That a Southern Continent exists I firmly believe," while Cook concluded that "as to a Southern Continent I do not believe any such thing exists unless in a high latitude." But he was determined to resolve the issue once and for all, and with his second venture a question more than two millennia old would be answered.

The second voyage

In July 1772 Cook, now a commander, set out with HMS *Resolution* and *Adventure* with instructions from the Admiralty for "prosecuting [his] discoveries as near to the South Pole as possible." First, though, he was to look for Cape Circumcision, mapped by French explorer Jean-Baptiste Bouvet de Lozier in 1739 and potentially the tip of the Southern Continent. The expedition was equipped with provisions for 18 months and the most up-to-date chronometer. Banks had fallen out with Cook and the Admiralty over such matters as accommodation for himself and his party of 17, which included two French horn players; instead, Cook had the erudite but demanding and difficult John Reinhold Forster, described by a Cook biographer as "one of the Admiralty's vast mistakes."

Cook sailed south from Cape Town but could not find Cape Circumcision. Exploring south and east, he realized that it could not have been part of a continent, and wrote dismissively: "I am of the opinion that what M. Bouvet took for land … was nothing but mountains of ice surrounded by field ice." On 17 January 1773 *Resolution* and *Adventure* became the first ships to cross the Antarctic Circle, but Cook ventured only a short distance inside the Circle before encountering heavy ice and retreating to the northeast, little realizing that he was only 130 kilometers (80 miles) from the Antarctic continent. Summer was over so he sailed directly to New Zealand.

Cook's two ships had lost touch in a storm off New Zealand's Cape Palliser, but Cook turned the *Resolution* south in November 1773. On 20 December he crossed the Antarctic Circle at about longitude 148°W. He had seen the first iceberg, but continued eastward through heavy ice. He turned north to explore ocean not covered on his previous voyage, but turned south again at 122°W—to the disappointment of his crew, who thought they were on their way to Cape Horn and home. Instead they crossed the Circle again on 26 January 1774. Over the next few days, they dodged north and south through thick fog and heavy ice. On 30 January

▲ COOK IN THE ANTARCTIC
James Cook made the first circumnavigation of Antarctica during the summers of 1772–75, crossing the Antarctic Circle four times. Returning to England he declared "no man will venture further south than I have done, and the lands which may lie to the South will never be explored."

◄ ISLANDS OF ICE
The painter William Hodges was employed by Cook to make a visual record of the expedition. His images provided the first glimpse of Antarctica's picturesque but hostile environment. Here, beside a huge, castellated "ice island," men in the ship's boats collect ice for water and shoot seabirds for food, while *Resolution* stands by.

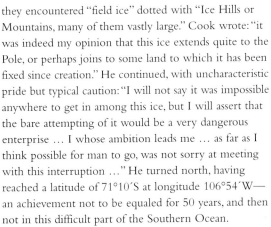

TIMELESS PEAKS, SOUTH GEORGIA

Heading back to England in January 1775, Cook discovered the island of South Georgia, which he named for the king and claimed for Britain. Possession Bay, with rearing jagged mountains behind, is a scene that has changed little since recorded then by William Hodges. However, although King penguins now preen undisturbed on the beach, Cook's reports of the island's abundant sea mammals unleashed a century of exploitation that virtually exterminated fur and elephant seals, and later whales, from these shores and waters.

they encountered "field ice" dotted with "Ice Hills or Mountains, many of them vastly large." Cook wrote: "it was indeed my opinion that this ice extends quite to the Pole, or perhaps joins to some land to which it has been fixed since creation." He continued, with uncharacteristic pride but typical caution: "I will not say it was impossible anywhere to get in among this ice, but I will assert that the bare attempting of it would be a very dangerous enterprise … I whose ambition leads me … as far as I think possible for man to go, was not sorry at meeting with this interruption …" He turned north, having reached a latitude of 71°10′S at longitude 106°54′W—an achievement not to be equaled for 50 years, and then not in this difficult part of the Southern Ocean.

Over the next few days Cook sailed eastward on both sides of the Antarctic Circle before winter (and pack ice) drove him back to warmer regions. Faced with impene-

trable ice, Cook and his crew willingly retreated, weary of the "dangers and hardships, inseparable with the navigation of the Southern Polar regions." He sailed south again from New Zealand in November 1774, heading for the polar spring thaw. For five weeks, *Resolution* sailed east toward Cape Horn, and on 28 December the ship rounded the Horn and continued eastward through the South Atlantic in search of the land reported by London-born merchant, Antoine de la Roche, in 1675. Cook found it—but far from where de la Roche had placed it. It had been considered a potential promontory of the Southern Continent, but at its southwestern point Cook realized it was an island. He named the point Cape Disappointment, and the island South Georgia.

From there Cook followed the 60° latitude through fog but in a sea clear of ice. On 27 January he encountered an iceberg, the harbinger of a lot of sea ice.

▲ IMAGES OF ICEBERGS
This painting, *Ice islands
with Ice Blink,* is attributed to
Wiliam Hodges, whose graphic
depictions of what he saw
on Cook's second voyage
were exhibited at the Royal
Academy. Some of his
paintings of the voyage
are displayed in London's
National Maritime Museum
at Greenwich.

He saw some islands and, tantalisingly, some mountain peaks, but in the fog and sea ice he could not discern whether they were located on islands or on the point of a large landmass. He optimistically named it Sandwich Land, after the first Lord of the Admiralty. Later he wrote of this land and ice: "I firmly believe that there is a tract of land near the Pole, which is the source of most of the ice which is spread over this vast Southern Ocean … I can be bold to say, that no man will ever venture farther than I have done and that the lands which may lie to the South will never be explored. Thick fogs, snow storms, intense cold and every other thing that can render navigation dangerous one has to encounter and these difficulties are greatly heightened by the inexpressible horrid aspect of the country, a country doomed by nature never once to feel the warmth of the sun's rays, but to lie forever buried under everlasting snow and ice."

Cook returned to England on 30 July 1775, three years and eighteen days after his departure, "In which time," he noted, "I lost but four men and only one of them by sickness." The first circumnavigation of Antarctica had been achieved. The quest for the mythical bounty of the fertile great Southern Continent was over: it did not exist. Cook wrote: "I had now made the circuit of the Southern Ocean in a high latitude … in such a manner as to leave not the least room for the possibility of there being a continent, unless near the Pole and out of the reach of navigation …"

The third voyage

Cook's third Pacific voyage once more took him into Antarctic waters. He sailed on the *Resolution* again, this time accompanied by the *Discovery*. The stated purpose was to return a native of Tahiti to his home, but the scope of the voyage was much wider: he was to find the Northwest Passage—the long-sought northern shortcut from Europe to Asia. He sailed via Cape Town to inspect some islands discovered by French captain Yves-Joseph de Kerguelen-Trémarec in 1772 in the southern Indian Ocean. Today they are called Iles Kerguelen, but Cook thought Desolation Islands a more apt name for them. He proceeded through the Pacific, along the west coast of America and through the Bering Strait to cross the Arctic Circle. On the voyage north he had discovered the Hawaiian Islands (which he named the Sandwich Islands), and when the Arctic winter set in he returned there. On 14 February 1779 he intervened when one of his boats was stolen. A fight ensued, and Cook was killed. It was a tragic end to a remarkable life.

By happy chance, William Wales, the astronomer on *Resolution*, went on to teach at the Mathematical School at Christ's Hospital in London. Young Samuel Taylor Coleridge was one of his pupils in the 1780s, and the schoolmaster's account of ice and albatrosses inspired the poet's later masterpiece, *The Rime of the Ancient Mariner,* which vividly and remarkably accurately evokes an environment he experienced only through hearsay.

Ne plus ultra

The contesting boasts of two men who were on the
Resolution when she reached Cook's southernmost
latitude on 30 January 1774 must have entertained
the crew on the long voyage home. English explorer
George Vancouver waited until the ship was ready
to tack about, and then climbed to the end of the
bowsprit to exclaim *Ne plus ultra* ["No further is
possible."] But Swedish doctor and naturalist Anders
Sparrman had a rival claim: "I went below [to his stern
cabin] … to watch … the boundless expanses of Polar
ice. Thus … I went a trifle farther south than any of
the others … because a ship … always has a little
stern way before she can make way on a fresh tack."

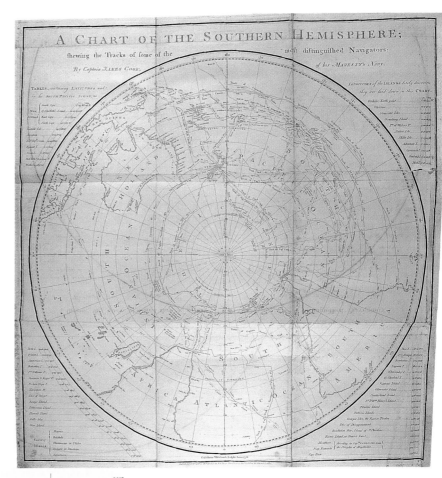

➤ COOK'S SECOND VOYAGE

George Forster, son of the expedition's naturalist, published this
map in 1777, only weeks before Cook's own "official" map of his
discoveries appeared. It shows the route of the expedition, and
details individual sightings by Cook in *Resolution* and by his
deputy Furneaux in *Adventure*. Cook's second voyage finally
established the true extent of *Terra Australis Incognita*—it had
to lie south of the Antarctic Circle.

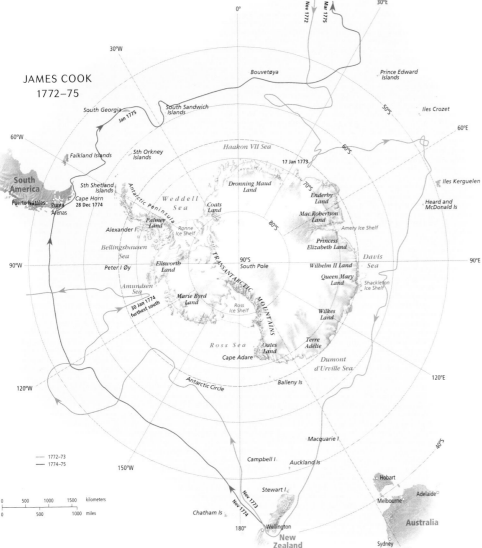

JAMES COOK
1772–75

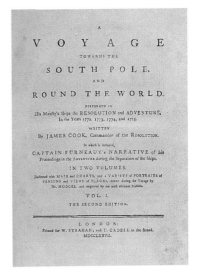

▲ MOMENTOUS WORK

The narrative of Cook's second voyage,
published in 1777, was evocatively but
accurately subtitled "Towards the South
Pole and around the World." It was an epic
that established the framework for virtually
all subsequent Antarctic exploration.

Thaddeus von Bellingshausen

1819–21

Cook died in 1779, and the world had to wait 40 years for another explorer to match his achievements. But in May 1819 the Russian explorer Thaddeus Thaddevitch von Bellingshausen took command of an expedition to Antarctica. Hugh Robert Mill, historian of the Heroic Age of Antarctic exploration, described this as "one of the greatest Antarctic expeditions … well worthy of being placed beside that of Cook … a masterly continuation of Cook, supplementing it in every particular, competing with it in none."

Born in 1778 and a naval cadet from the age of 10, Bellingshausen had served as fifth lieutenant on the first Russian voyage around the world, led by Admiral Adam Johann, Baron von Krusenstern from 1803 to 1806. When the Russian Antarctic expedition was being planned as part of a burgeoning nationalism, Bellingshausen was chosen to command the two ships, *Vostok* and the smaller and slower *Mirnyi*, with a total of 189 officers and crew. He was allowed very little time to prepare for this monumental undertaking—indeed, the expedition sailed in late July 1819, about two months after he was recalled to St Petersburg from survey work in the Black Sea. His instructions were succinct: to build on the explorations of Cook, whom the Russians greatly admired, and "to approach as closely as possible to the South Pole, searching for as yet unknown land, and only abandoning the undertaking in the face of insurmountable obstacles." The quest for the South Pole (rather than the Great Southern Continent) had begun.

A modest hero

Bellingshausen's expedition has often been overlooked in Antarctic history because his logs are lost and his unassuming personal record was available only in Russian until 1902, when a German translation appeared; there was no English version until 1945. Few visitors realize the historical significance of the name of the stretch of the Southeast Pacific to the west of the Antarctic Peninsula: the Bellingshausen Sea.

The two ships left Rio de Janeiro on 20 November 1819 for South Georgia, where Bellingshausen completed Cook's survey by mapping the southern coast. He then followed Cook's route to the South Sandwich Islands and, in better conditions than Cook had experienced, found that they were just more small islands, and that there were more of them than Cook had observed. Continuing eastward, he crossed the Antarctic Circle on 15 January 1820—his was the second expedition to do so—but did not even note this first crossing in his narrative. But while Cook sailed 24° of longitude within the Circle, Bellingshausen was to cover more than 42°.

▲ ICY FRINGE SIGHTED
Bellingshausen came much closer than Cook to the Continent. He sighted what was probably the Lazarevisen in February 1820 and found Peter I Øy in January 1821.

Map

THADDEUS VON BELLINGSHAUSEN 1819–21

30°W · 0° · 30°E · Feb 1821 · Dec 1819 · Bouvetøya · Prince Edward Islands · South Georgia · South Sandwich Islands · 50°S · Iles Crozet · 15 Jan 1820 · Haakon VII Sea · 60°E · 60°S · Falkland Islands · Sth Orkney Islands · Lazarevisen · 60°W · 60°E · South America · Sth Shetland Islands · Antarctic Peninsula · Weddell Sea · Dronning Maud Land · 70°S · Enderby Land · Iles Kerguelen · Puerto Natales · Ushuaia · Punta Arenas · Coats Land · Palmer Land · Mac.Robertson Land · Heard and McDonald Is · Alexander I · Ronne Ice Shelf · Amery Ice Shelf · 80°S · Princess Elizabeth Land · Bellingshausen Sea · Ellsworth Land · 90°S South Pole · Wilhelm II Land · Davis Sea · 90°W · Peter I Øy · 21 Jan 1821 · TRANSANTARCTIC MOUNTAINS · Queen Mary Land · 90°E · Amundsen Sea · Marie Byrd Land · Ross Ice Shelf · Shackleton Ice Shelf · Wilkes Land · Ross Sea · Oates Land · Terre Adélie · Cape Adare · Dumont d'Urville Sea · Vostok · Mirnyi · 120°W · Balleny Is · 120°E · 150°W · Macquarie I · Campbell I · Auckland Is · 40°S · Stewart I · Hobart · Melbourne · Adelaide · New Zealand · Australia · Oct 1820 · Sydney · 180° · 150°E

0 500 1000 1500 kilometers
0 500 1000 miles

➤ GRAND HARBOR
Deception Island in the South Shetlands group is a drowned volcanic caldera and one of the finest natural harbors in Antarctica. When Bellingshausen sailed through its narrow entrance (on the far left of this photograph) in January 1821, he found 18 British and American sealing ships at anchor, even though the island had only been discovered the previous year.

On 16 January he would have seen the continent if the weather had been fine, but he was sailing through snow when he observed "a solid stretch of ice running from east through south to west." Probably he was viewing an ice shelf at the base of Haakon VII Sea, which may at that time have extended far out to sea. Geographers regard continental ice as part of the land-mass—otherwise, the Antarctic Peninsula would be classified as an archipelago—so the first sighting of "Antarctica" may be credited to Bellingshausen. Some have contended that what he saw was pack ice, but his use of the Russian term for "continental ice" works against that argument.

Land in sight!

Bellingshausen's information for 5 February is more definite. On a day with good visibility he wrote: "The ice to the SSW is attached to cliff-like, firmly standing ice: its edges were perpendicular and formed bays, and the surface rose in a slope towards the south, over a distance whose limits we could not see from the cross-trees." He must have been looking at the Lazarevisen (Lazarev Ice Shelf). But he needed supplies, so he sailed

north to Port Jackson (now Sydney). On 31 October he sailed south again to complete his circumnavigation of the South Pole. However, like Cook before him and Shackleton to come, he encountered unseasonably heavy ice as he approached the continent. Thus he missed the Ross Sea and could not cross the Antarctic Circle until 14 December. But on 21 January 1821 he reached the southernmost point of his voyage—69°59′S—and discovered Peter I Øy, the first land ever seen within the Antarctic Circle. He also discovered and named Alexander I Land (now Alexander Island), stating: "I call this discovery 'land' because its southern extent disappeared beyond the range of our vision." It is in fact a large island, separated from the continent by a narrow channel but linked to it by ice. He sailed on to the South Shetland Islands, which he mapped, and on Teille Island (now Deception Island) he encountered American

▲ FROZEN FASTNESS
Peter I Øy, named by Bellingshausen for Russia's greatest czar, is one of the world's most inaccessible islands. The first landing was more than 100 years after discovery, and barely a dozen expeditions have since broached these forbidding shores. Only one has reached the interior—by helicopter.

BELLINGSHAUSEN'S SHIPS
Although here they appear identical, *Mirnyi* was not only a much smaller ship, but was also much slower, which frustrated Bellingshausen in *Vostok*. The creator of this image had probably not been south himself—unlike William Hodges, who painted scenes of Cook's voyages from personal experience.

ISLAND PROFILES
Bellingshausen's navigator drew these careful sketches of the islands that make up Cook's "Sandwich Land." They can be recognized from the same point of view today. Cook and Thule Islands are remnants of the rim of a large volcanic crater, and the caldera between them, though open, is of similar formation to Deception Island.

Bristol Island, 7 miles distant

Thule Island

Cook Island, 15 miles distant

STANDING THEIR GROUND
Bellingshausen called the South Sandwich Chinstrap penguins "ringed" or "common" penguins, and found that they pursued his officers aggressively.

and British sealing ships working the island group. On 25 January a young American captain, Nathaniel Palmer, came aboard; Palmer was later to lay dubious claim to have been the first to see the Antarctic Continent.

One of Bellingshausen's ships, *Vostok*, had been shipping water since leaving Port Jackson, and as winter approached Bellingshausen decided to turn north to Rio de Janeiro at the end of January, arriving there on 27 March. He overhauled the ships over the next month and sailed back to Kronstadt via Lisbon, completing the voyage on 4 August 1821. The end of his narrative is typically brief and factual: "We had been absent for 751 days. During that time we had been at anchor in different places 224 days and had been under sail 527 days. Altogether we had covered 57,073 ½ miles … During the course of our voyage we had discovered twenty-nine islands: two of these were in the Antarctic, eight in the South Temperate Zone, and nineteen in the Tropics."

He did not mention that only three men had died throughout the voyage—a record that Cook would have admired. And he had achieved the remarkable feat of circumnavigating Antarctica closer to the coast than Cook—so close that he was the first to see Emperor penguins, the largest and southernmost-dwelling of all penguins. But the voyage had suggested few commercial possibilities and was largely ignored in Russia; and because his charts and logs were not accessible to non-Russian speakers the voyage was overlooked by other nations. Bellingshausen rose to the rank of admiral during the next 30 years of his naval career and then was appointed governor of Kronstadt, the role he still held when he died in 1852.

The first sight and the first step

The South Shetland Islands, off the west coast of the Antarctic Peninsula, are central to the question of who was the first to "see Antarctica." The islands were discovered in February 1819 by sealer William Smith, whose report of his find triggered the island sealing boom. Smith was soon employed by the British Admiralty to survey the islands under the command of Edward Bransfield the following summer.

The first sight

On 30 January 1820 Bransfield and Smith saw and charted part of the Antarctic Peninsula, and named it Trinity Land (now Trinity Peninsula). Exactly two weeks earlier, Thaddeus Bellingshausen had probably seen the icy fringe of Antarctica, far to the west. However, he had glimpsed ice and they were looking at rocky mountains. In November 1820 Nathaniel Palmer, captain of the sealer *Hero*, sailed through the narrow entrance of Neptune's Bellows and into the spectacular caldera known as Port Foster. He was perhaps the first to do so; Bransfield and Smith had seen Deception Island on 29 January 1820, but did not investigate in the thick fog. Palmer met Bellingshausen in January 1821, and was later quoted as saying: "I informed [Bellingshausen] of … the discovery of land … and it was him that named it Palmers Land," but his only likely sighting is dated November 1820—10 months after Bransfield and Smith.

The first step

The honor of the first step onto the Antarctic Continent probably belongs to the crew of a boat from the *Cecilia*, an American vessel skippered by John Davis. On 7 February 1821 he recorded "… open cloudy weather and light winds a standing for a large body of land in that direction SE at 10 am close in with our boat and sent her on shore … I think this southern land to be a continent." The landing probably took place at Hughes Bay (64°13′S 61°20′W), and much pre-dates the other documented claim, that of the Norwegian businessman Henryk Johann Bull, who led a whaling expedition to the Ross Sea region in 1895. The area is now the Davis Coast, but the names of the crew who made the historic one-hour landing are unknown.

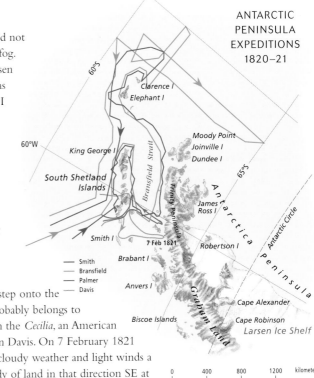

ANTARCTIC PENINSULA EXPEDITIONS 1820–21

50°W
60°S
65°S
60°W
Clarence I
Elephant I
Moody Point
Joinville I
Dundee I
King George I
Bransfield Strait
South Shetland Islands
Trinity Peninsula
James Ross I
Antarctica
Smith I
7 Feb 1821
Robertson I
Antarctic Circle
Brabant I
— Smith
— Bransfield
— Palmer
— Davis
Anvers I
Peninsula
Cape Alexander
Biscoe Islands
Cape Robinson
Larsen Ice Shelf
Graham Land

| 0 | 400 | 800 | 1200 | kilometers |
| 0 | | 400 | 800 | miles |

▼ ANTARCTIC PENINSULA FROM DECEPTION ISLAND
From Deception Island (foreground), the mountains of the Antarctic Peninsula can be seen across Bransfield Strait. British mariners Smith and Bransfield charted part of the Peninsula in January 1820, naming it Trinity Land. Ten months later American sealer Nathaniel Palmer explored Deception Island, and from there spied these icy peaks. He believed his was the first sighting of the Antarctic mainland.

James Weddell

1822–24

▲ FEARLESS SEALER
In 1819 James Weddell left the Royal Navy to become a sealing skipper. On his first, very successful, voyage to the South Shetlands, he also discovered the South Orkney Islands. In February 1823, seeking fresh seal populations, he pushed further south than anyone before him, into the icy sea that now bears his name. This feat was not matched until 1912.

Travelers to Antarctica still speak of James Weddell with awe. The Weddell Sea—one of the two great indentations into the Antarctic Continent—is a spawning ground of polar ice, yet in 1823 Weddell sailed further south than was conceivable at the time. His record stood for 18 years, until James Clark Ross ventured into the waters on the other side of Antarctica, and steam had ousted sail before anyone replicated Weddell's achievement in the same location.

Weddell was British, but was born in Ostend in 1787. He joined the Royal Navy at the age of nine, and then alternated between the Royal Navy and the Merchant Navy before taking command of the 160-tonne sealer *Jane* in 1819. He returned in 1821 with enough seal skins to buy the even smaller 65-tonne *Beaufoy* to partner *Jane*.

An inauspicious start

The voyage that set Weddell's mark on Antarctic history sailed down the Thames on Friday 13 September 1822. Directly after the inauspicious date, while still in English waters there was a collision and *Jane* was damaged. The vessels sailed to the South Orkneys but found few seals—and they were unlike any that Weddell had seen before. (He took the skins and skulls back to Robert Jameson of Edinburgh University, who declared a new species: the Weddell seal, *Leptonychotes weddelli*.)

But the immediate task was to find more seals, or more land where they might live. Weddell offered a reward of 10 pounds—most of an able seaman's annual wage—for the first sighting of land. By February 1823 he had concluded that such discovery would only be

made to the south. He pushed south into a prevailing wind through intense cold and a sea jagged with ice. It was a miserable voyage, but Weddell's judgment was partially vindicated on 16 February by a change in the weather: the wind shifted to the west, the ice disappeared, and flocks of seabirds were reflected in the calm sea. The next day they reached 71°34′S 30°12′W—no ship had ever been so far south.

A stroke of luck

In these ideal conditions Weddell pressed southward. On 19 February he wrote a paragraph that those who know the area still read with wonder: "In the evening we had many whales about the ship, and the sea was literally covered with birds of the blue peterel kind. NOT A PARTICLE OF ICE OF ANY DESCRIPTION WAS TO BE SEEN … had it not been for the reflection that probably we should have obstacles to contend with in our passage northward, through the ice, our situation might have been envied."

On 20 February 1823 a rising south wind forced Weddell to make a decision. At noon the ship was at 74°15′S 34°16′W and only three icebergs were visible, but winter and the polar night were closing in, and there was ice to be crossed in sailing north. Caution won the day, and Weddell fired the canon, raised the colors, and issued an extra allowance of rum. Weddell named his discovery George IV Sea, but in 1900 it was renamed the Weddell Sea.

Weddell was fortunate in his easy passage into the Weddell Sea—many who followed did not fare so well, even in larger, purpose-built ships. His feat was so difficult to replicate that skeptics doubted his veracity, and he had three of his crew swear under oath that the ship's logs were correct. But James Clark Ross, the next

◀ SKILL AND DARING
Taking advantage of uncharacteristically ice-free seas, Weddell's flagship, the brig *Jane*, followed by her smaller companion the cutter *Beaufoy*, reached their furthest point south on 17 February 1823. Although Weddell made the most of his opportunities, the voyage failed in its primary objective of finding new sealing grounds; nor did he discover new lands.

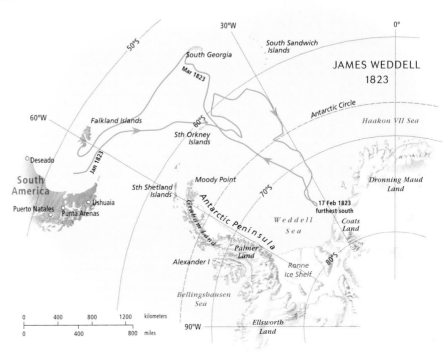

WEDDELL SEAL AT REST

On the South Orkneys Weddell discovered a new species of seal, which was named *Leptonychotes weddelli*. The encounter proved more significant to science than to commerce. Compared with the valuable fur seals found further north, Weddell seals have much poorer, thinner coats, of little economic worth.

AS SEEN BY WEDDELL

Weddell's drawing of a "Sea Leopard" clearly shows the mottled fur, the twinned rear flippers, and horizontal posture of a true seal. Weddell would have been more familiar with the more upright posture and separated rear flippers of the fur seal.

to hold the title of "furthest south" wrote generously that Weddell "was favored by an unusually fine season, and we may rejoice that there was a brave man and daring seaman on the spot to profit by the opportunity."

Jane and *Beaufoy* returned to England in July 1824. In *A Voyage Towards The South Pole*, published in 1825, Weddell wrote: "I have only done that which every man would endeavor to accomplish, who in the pursuit of wealth, is at the same time zealous enough in the cause of science to lose no opportunity of collecting information for the benefit of mankind." *Jane* had to be scrapped in 1829. It was a financial blow, and Weddell died in poverty in London in 1834, aged 47.

Sealers and whalers

It has been said that few did more to endanger Antarctic wildlife than James Cook, who, after his second southern voyage, told of islands teeming with seals and the Southern Ocean filled with whales. But Cook was not the first to note this profusion; earlier explorers, including Francis Drake, had also reported it.

Since at least the twelfth century Arctic peoples had used seal and whale oil for lighting and seal furs for clothing, and southern seals had long provided oil and clothing for the Patagonians. Wholesale slaughter of seals and whales by European profit takers, however, was a direct result of the demands of the nineteenth-century Industrial Revolution with its improved technologies.

In quest of bounty

European sealing probably began with Spaniard Juan de Solfo, who sailed to South America in 1515; he was killed and eaten by the natives, but his crew took seal skins back to Seville. In the eighteenth century Chinese furriers discovered how to remove the coarse outer hair of seal pelts, leaving the soft inner fur intact. Sealing became a maritime gold rush, beginning on the Falkland Islands in 1764, and rapidly spreading to South Georgia and the coast of South America.

In the late seventeenth century the piratical English navigator William Dampier called at the Juan Fernandez Islands, off the coast of Chile, and wrote: "seals swarm around … there is not a bay nor rock … but it is full of them." In 1797 throngs of sealers arrived, and by 1807 few seals remained. One vessel took off 100,000 skins, and the total harvest was estimated at three million. Forty thousand skins were shipped to London from the Falkland Islands in 1788, the year when the first British sealers reached South Georgia. In 1800, a New York sealer took 57,000 skins from the island, and in 1825 James Weddell calculated that 1.2 million fur seals had

▼ RETURN FROM THE BRINK
This fur seal pup at Cooper Bay, South Georgia, represents the gradual revival of fur seals after their virtual extinction in the early nineteenth century. Weddell calculated that by 1825, 1.2 million fur seals had been slaughtered on the beaches of South Georgia.

been killed on South Georgia. More than 120,000 skins had also been plundered from Australia's Macquarie Island.

Biscoe and the Enderby Brothers

As their quarry approached extinction, the sealers sought new hunting grounds—but important discoveries were made during those voyages. Nationalism had instigated the circumnavigations of Antarctica by Cook and Bellingshausen, but the third was an uneasy mixture of exploration and sealing. John Biscoe, at 36, was an ex-Royal Navy seaman employed by the Enderby Brothers, the firm that had sailed into Boston Harbor with a cargo of tea in 1773, thus triggering the "Boston Tea Party" and the American War of Independence. The company needed new ventures, so it switched to whaling and sealing. The Enderby tradition combined exploration with exploitation, and captains with education and naval experience were recruited to collect flora and fauna. The policy eventually bankrupted the company, but it left an enduring legacy of Antarctic exploration. Biscoe was to seek out new sealing grounds, but also to make discoveries in high southern latitudes. His vessels were the *Tula* and the much smaller *Lively*, sparsely equipped and provisioned, with 29 crew. They sailed from England in July 1830 and, often at an agonizingly slow 3 knots or less, crossed the Antarctic Circle in January 1831. On 28 January they arrived at their furthest south: 69°S at 10°43′E.

Biscoe ventured even further south than Bellingshausen, and in one way had better luck. He first definitely sighted land on 28 February 1831 at 66°S 47°20′E, when "several hummocks" resolved themselves into "the black tops

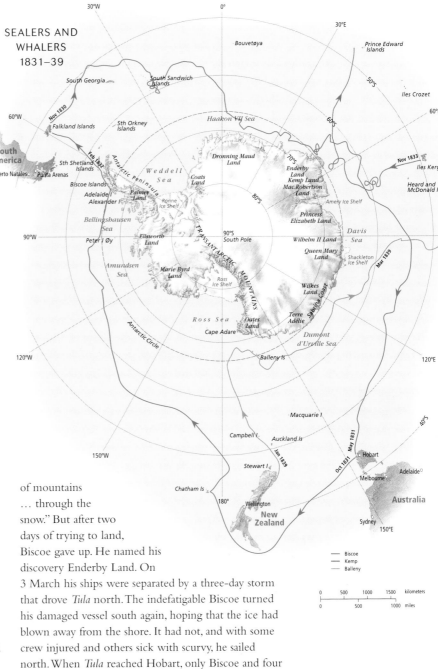

SEALERS AND WHALERS 1831–39

of mountains … through the snow." But after two days of trying to land, Biscoe gave up. He named his discovery Enderby Land. On 3 March his ships were separated by a three-day storm that drove *Tula* north. The indefatigable Biscoe turned his damaged vessel south again, hoping that the ice had blown away from the shore. It had not, and with some crew injured and others sick with scurvy, he sailed north. When *Tula* reached Hobart, only Biscoe and four others could stand, and two of the crew had died.

On 3 September Biscoe and *Tula* headed south again, but as they left port they met *Lively* at the mouth

> ### NO ESCAPE

Fur seals are agile rock climbers, but sealers pursued them remorselessly. The bodies were quickly stripped and the skins thrown down to the waiting boat, then taken to the ship to be salted to preserve them. This lithograph shows a sealing gang at work, on Beauchene Island in the Falklands, in the 1830s.

◄ PROFITABLE WORK
This watercolor painting by Augustus Earle (1793–1838) shows a sealer on one of the islands of Tristan da Cunha, stripping the skin from a young elephant seal in the process known as flinching or flensing. The tools of the sealer's trade were a sharp spear and a whetted knife.

▼ PERIL VERSUS PROFIT
This scene, painted by Thomas Buttersworth, c.1820, vividly shows the dangers of hunting whales from rowing boats in iceberg-strewn waters. These primitive techniques limited the whalers to hunting only right whales, which floated when killed and so could be hauled back to the ship.

of the River Derwent. *Lively*'s crew had a harrowing tale to tell. Seven of her crew of ten were dead, and the survivors had been marooned in Port Phillip Bay (now Melbourne) when the ship had drifted out to sea while they were foraging ashore; it had taken weeks to recover and refloat it. Undaunted, the two vessels headed southeast in October. After some unrewarding sealing in New Zealand, they continued slightly south of east to cross Cook's path and discover an island at 67°15′S 68°30′W; Biscoe named it Adelaide Island for England's Queen Adelaide, wife of William IV. In late February they arrived in the South Shetland Islands, having completed a circumnavigation of Antarctica and finally established that there was a large continent at the heart of the ice. Biscoe wrote: "I am firmly of the opinion that this is a large continent as I saw to an extent of 300 miles."

Biscoe needed to find seals before winter set in, but he failed, and *Tula*'s rudder was damaged in a late storm. He retreated to the Falklands, where *Lively* was wrecked. He wanted another season in Patagonia to make the voyage a financial success, but one by one his hard-pressed crew deserted. The bedraggled *Tula* sailed into London on 8 February 1833 with just 30 seal skins to show for a 30-month voyage. Biscoe received a gold medal from the Royal Geographical Society, went on to skipper craft in the West Indies and around Australia, and sailed south again as far as 63°S on a whaling venture from Sydney in 1838. In 1842 a public appeal in Tasmania raised money to send him back to England, but he died on the voyage.

Enduring legacy

In 1833 the *Magnet*, under the command of British sealer Peter Kemp, left London for Antarctica in quest of seals. Little is known of Kemp— even the date of his birth is uncertain—but he went to Iles Kerguelen and on to the Antarctic Circle, where he saw land to the south at about

58°E on Boxing Day. He called it Kemp Land, and then turned north for more sealing at Iles Kerguelen. He fell overboard and drowned on the return voyage, but the Kemp Coast retains his name, though it was not seen again until Douglas Mawson visited it almost a century later.

Even less is known about the life of another British sealer and whaler, John Balleny, but his name lives on in a group of five islands rising sheer from the water and lying across the Antarctic Circle below New Zealand. They were named for him by the British Admiralty, in a tribute to a significant voyage. The *Eliza Scott*, under Balleny, and the much smaller *Sabrina*, under Thomas Freeman, sailed from England in July 1838 via the Cape of Good Hope and Iles Amsterdam to New Zealand. On 1 February 1839 they reached 69°02′S at 172°11′E, but the way south was blocked by ice. But it showed that there was a possible route along this longitude: the way into the Ross Sea had opened a crack. On 9 February Balleny saw what are now the islands that bear his name, and he briefly—and damply—landed there on 12 February; it was the first landing below the Antarctic Circle. He followed a course some 500 kilometers (310 miles) south of Cook's voyage on the *Resolution*, and on 2 March he saw land to the south and named it the Sabrina Coast. It became a memorial to his ship: in a violent storm on 24 March a blue distress flare was the last that was seen of the *Sabrina* and her crew. The *Eliza Scott* returned to England via Madagascar and St Helena, bearing just 200 seal skins. Balleny's log and chart were paraded before the Royal Geographical Society, but the man himself faded into obscurity.

While early sealing brought wildlife populations to the verge of extinction, early whaling was less successful. The whaling industry's chance for mass destruction of species other than Southern right whales would have to wait for the invention of the explosive harpoon head and the factory ship. However, those few early whalers and sealers who pursued new lands with as much enthusiasm as they hunted their quarry left an enduring legacy in Antarctic exploration.

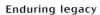

◄ SPOILS FOR OIL

Weddell reported that at the time of his visit to South Georgia the "sea-elephants were nearly extinct, not less than 20,000 tonnes of elephant-oil having been shipped to the London markets alone … since the reports by Captain Cook."

▲ PREY NO LONGER

The aggressive appearance of male Southern elephant seals failed to protect the species from being savagely hunted to near extinction. Now this species is protected, and any approach by humans is for peaceable scientific study only. Some elephant seals are tagged to record how far they swim: one from South Georgia has been spotted in South Africa—4,800 kilometers (3,000 miles) away.

Jules Sébastien César Dumont d'Urville

1837–40

It is due to Frenchman Dumont d'Urville that there is a sliver of French territory amidst the giant Australian claim to Antarctica. He called Adélie Land after his wife, and the Adélie penguin (*Pygoscelis adeliae*) is named for the land he discovered.

Dumont d'Urville was born in Normandy on 25 May 1790 to an aristocratic but poor family. He was an entomologist, a classicist, a botanist, a brilliant linguist, and a respected leader, though somewhat aloof. He joined the French Navy in 1807, was second in command of a Pacific voyage in 1822, and leader of another in 1826. Both voyages were on the same vessel, though for the second *Coquille* was renamed *Astrolabe*. He returned in 1829 and wrote his account of the voyage over the next five years.

When Dumont d'Urville proposed another Pacific voyage in 1835 he was 45—old by Antarctic leadership standards. His request was granted and he was ordered to go via the Southern Ocean with ambitious instructions to extend his explorations "towards the Pole as far as the polar ice will permit." There was a bonus for the crew if they reached 75°S—further south than Weddell's British voyage of 1823. Dumont d'Urville took two ships, the *Astrolabe* and the smaller *Zélée*, with a complement of 183 officers and crew. He sailed from

Toulon on 7 September 1837, was delayed by repairs before reaching the Strait of Magellan, but was off the coast of Tierra del Fuego by January 1838.

Ice and storms

Unhappily for Dumont d'Urville, and for his bonus-hungry crew, they struck unseasonably heavy sea ice at 59°30′S and had to dodge icebergs at 62°S. On 21 January they were trying to find a way through the pack ice, but by 25 January they had been forced to retreat northward toward the South Orkneys. A storm stopped them landing for fresh seal and penguin meat, and they sailed south again. Encountering pack ice in early February, Dumont d'Urville bravely elected to push into it. They stopped in a stretch of open water and the crew celebrated, but Dumont d'Urville who had been ill for weeks wrote: "I went to bed and could hear their rowdy celebrations … The rashness and imprudence of this move unfolded before my eyes … The only way out was by the way we had come in … was there not reason to fear … ice would completely block our exit?"

How right he was! Before midnight they were ice-bound, and it took five days of hard labor to break into open water. They continued westward, surveying and mapping around Graham Land and the South Shetlands before retreating to Chile, where they arrived ridden with scurvy. Dumont d'Urville wrote: "Either Weddell struck an exceptionally favorable season or he played on the credulity of his readers … As soon as just one other captain has penetrated only five or six degrees further than us in the same area will my doubts be allayed and Weddell right in my eyes."

Another try

When they reached Valparaíso on 25 May, two crewmen were dead and seven others had deserted. But a week later they set out across the Pacific, landing at many islands along the way, and arrived in Australia in March 1839. Early in 1840, after further Pacific exploration (and more sickness), they sailed south from Hobart. They found tabular icebergs at 64°S on 18 January. Their erratic compass suggested that they were close to the Magnetic Pole, but their bad luck with ice the previous season seemed about to be repeated. However, the next afternoon an officer climbed the mast and reported land ahead. After a frustrating few days becalmed, the ships reached the land on 21 January and launched a boat from each ship. On a tiny islet just off the mainland and barely within the Antarctic Circle, the excited boat crews landed, unfurled the tricolor, and collected some granite chips and a few unlucky penguins.

Dumont d'Urville named the area Adélie Land "to perpetuate … my deep and lasting gratitude to my devoted wife." In an endearingly French footnote,

▲ **RESPECTED LEADER**
Dumont d'Urville's voyage to the South Pole was his third major expedition for the French Navy. It was on this voyage that he made his most famous Antarctic discoveries, which earned him approbation and promotion to rear-admiral. His efforts led to the French claim to a slice of Antarctica.

▶ **HEAVY GOING**
Dumont d'Urville's ships, *Astrolabe* and *Zélée*, were naval corvettes, and not suited to pushing through heavy pack ice. In February 1838 both vessels were beset by ice. During the five weary days it took to break free into open water, the crew of *Astrolabe* took the opportunity to load supplies of ice for fresh water.

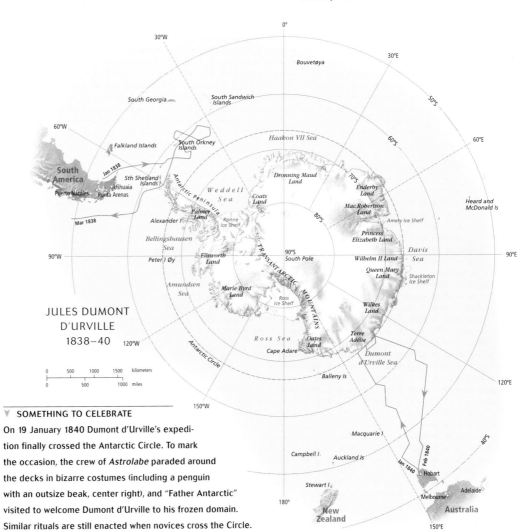

JULES DUMONT
D'URVILLE
1838–40

STAND-UP COMICS

Adélies are the most studied of all penguins.
The smallest but most numerous of the five
species that breed on Antarctica, they are
often found in colonies of several hundred
thousand. With their inquisitive natures and
endless squabbling over nesting territory,
they are the natural comedians of Antarctica.

SOMETHING TO CELEBRATE

On 19 January 1840 Dumont d'Urville's expedition finally crossed the Antarctic Circle. To mark
the occasion, the crew of *Astrolabe* paraded around
the decks in bizarre costumes (including a penguin
with an outsize beak, center right), and "Father Antarctic"
visited to welcome Dumont d'Urville to his frozen domain.
Similar rituals are still enacted when novices cross the Circle.

▲ TANTALIZING HORIZON
Despite infuriatingly calm weather that slowed their approach, on 20 January Dumont d'Urville's ships came in sight of a high, snow-covered land stretching from southeast to northwest. But they could not safely go ashore, and next day they landed on a small islet just offshore of the continent itself before heading north again.

an officer reported that they drank a bottle of Bordeaux wine that someone had brought along "to the glory of France." They continued westward along the coast for another couple of days, but a bad storm blew up and both ships were nearly lost. It was time to return to Hobart, but there was one more unexpected adventure. On 29 January a ship loomed out of the fog in the east—it was the *Porpoise* from the American expedition led by Charles Wilkes. The astonished French maneuvered to meet the newcomer, which changed course and departed without contact. Later, the secretive Wilkes claimed that the Frenchman turned and ran; Dumont d'Urville replied: "I would have been glad to give our

co-explorers the results of our researches … it seems the Americans were far from sharing those feelings."

Dumont d'Urville sailed north on 1 February, reached Tasmania's River Derwent on 18 February, and returned to Toulon on 6 November 1840 after exploring more of the South Seas. The voyage had aged him far beyond his 50 years, but his efforts were rewarded. He was promoted to rear-admiral and given the gold medallion of the Société de Géographie—their highest honor. But he had little time to enjoy his fame. In May 1842 he and his wife and son perished in a train accident—a tragic end to a family whose names would forever be associated with the frozen south.

◄ FRENCH CONNECTIONS
Bull kelp washed up on the shore of Adams island in the Auckland islands is called *Antarctica durvillea* for the French explorer Dumont d'Urville, and like several place names in Antarctica perpetuates his memory.

An art work salvaged

Ancient Greece, the first civilization to dream of a Southern Continent, also created great works of art. One was the incomparable statue of Venus de Milo, and it was Dumont d'Urville who acquired it for France. During a posting to the Mediterranean in 1819, he came across a peasant on the Greek island of Melos who wanted to sell an ancient statue. Dumont d'Urville pursued the deal and the bargain was sealed. The French delegation sent to finalize the sale was involved in a fight with bandits, and the arms of the statue broke off in the scuffle. The statue now stands in the Louvre Museum in Paris, and for his role in obtaining it Dumont d'Urville was promoted to lieutenant and awarded the Légion d'Honneur.

Charles Wilkes

1838–42

On 18 August 1838 the United States Exploring Expedition, nicknamed Ex Ex, left Chesapeake Bay commanded by the determined but irascible and largely inexperienced 40-year-old lieutenant Charles Wilkes. The fleet was led by *Vincennes*, a 780-tonne war sloop and the first United States Navy ship to circumnavigate the globe; the others were *Peacock*, *Porpoise*, *Sea Gull*, *Flying Fish* (the baby of the fleet, a 96-tonne New York pilot boat), and *Relief*, a capricious supply ship that was to prove a handicap. More than 400 men were on board, including nine scientists. At Orange Harbor, Tierra del Fuego, Ex Ex split into three groups: *Vincennes* and *Relief* were to survey the southern tip of South America; Wilkes, on *Porpoise* accompanied by *Sea Gull*, would sail south to try to outdo Weddell; *Peacock* and *Flying Fish* would explore southwest in an attempt to pass Cook's southernmost point.

Perilous ventures

Winter was approaching—it was already 26 February 1839—and Wilkes gave up a mere 10 days later and well short of the Antarctic Circle, defeated by impenetrable ice and inadequate clothing. On the retreat to South America *Porpoise* almost ran aground on Elephant Island, but the fog parted just in time for the hard-pressed crew to avert disaster. *Peacock* and *Flying Fish* were driven apart by a gale 48 hours after leaving port. Both ships endured great hardships. The valiant little *Flying Fish* had a particularly bad time; she disintegrated further in every storm, and the sailors were constantly wet and their quarters awash. When the thermometers broke, they suspended a tin cup of water and resolved to travel south until it froze. This they did, and on 22 March they reached 70°14′S—still a degree short of Cook—before accepting they would not outdo the British explorer and turned north.

Three days later the two vessels met, compared dismal experiences, and separated, *Peacock* for Valparaíso, *Flying Fish* for Orange Harbor.

It was a depleted expedition that regrouped in Valparaíso at the end of June: *Sea Gull* had been lost in a storm off Chile, and the cumbersome and unseaworthy *Relief* was sent home. The remaining four ships sailed to investigate the Pacific. Scientifically this Pacific leg was more successful than the Ex Ex's first Antarctic venture, but the relationship between Wilkes and his scientists and officers deteriorated; some even privately questioned his sanity. In Sydney in December 1839 officials were aghast at the terrible condition of the ships and the lack of equipment, heating, and waterproofing for their next sojourn in the icy south, and the question of Wilkes's sanity—or at least his judgment—became a matter of widespread concern. When Wilkes offered the naturalists the chance to stay in Australia and meet the ship in New Zealand in March 1840, they all accepted. A typical sentiment of those who did sail south on 26 December was this journal entry: "I do not suppose that a vessel ever sailed under the [United States] pendant with such a miserable crew … It will be a great wonder to me if we return from the southern cruise."

Claims and counter-claims

For his second Antarctic exploration Wilkes had been ordered to sail south from Tasmania as far as he could, to explore the continent westward to longitude 45°E, then to proceed to Iles Kerguelen. He substantially modified these instructions, sailing only as far as 105°E and then heading for New Zealand's Bay of Islands. He also entreated his captains to "avoid a separation as the lives of … the squadron may be jeopard[iz]ed by it."

> ▲ **CHARLES WILKES**
> Lieutenant Charles Wilkes (1798–1877) led the first United States expedition to Antarctica. Six ships set off in 1838, but only two survived the whole expedition, and none reached the continent, although lookouts on three ships reported sightings. However, the claim by Wilkes to have seen land just three days before Dumont d'Urville is regarded as highly suspect.

> ▶ **PROVIDENTIAL ESCAPE**
> In the expedition's first foray into Antarctic waters one ship sank, a second retired hurt, and in March 1839, Wilkes's own ship, *Porpoise*, only narrowly avoided being wrecked on the desolate, fog-shrouded shores of Elephant Island (left), while returning from the Weddell Sea. *Porpoise* survived to rejoin the remaining ships at Valparaíso. The depleted fleet then sailed via Sydney to the other side of the Antarctic.

Within 10 days *Flying Fish* and then *Peacock* were separated from the others and from each other. *Flying Fish* reached the ice but wisely turned toward New Zealand in early February. On 11 January 1840, *Porpoise* and *Vincennes* reached the "icy barrier" at 64°11′S 164°30′E and, proceeding to the west, fortuitously met up with *Peacock* five days later. That same day the lookouts on all three ships reported seeing land—but nothing was entered in any of the ships' logs. Visibility was good, but it remains doubtful whether they were seeing snowy land or a glacier, or merely trapped icebergs. Wilkes later wrote in his narrative for 19 January that he was "fully satisfied that it was certainly land," and named Cape Hudson after the skipper of *Peacock*—but there is no record in the ship's

◄ **LANDMASS A CONTINENT**

The scientific results, and suspect claims, of the Exploring Expedition's four-year odyssey appeared in this five-volume report, published in 1845. Despite inaccuracies in his charts, Wilkes had sailed further along, and closer to, the Antarctic coast than any previous expedition, and was thus the first explorer to declare it a continent.

log. His chart showed land 65 kilometers (40 miles) away, but later voyages established that the closest land lies 190 kilometers (120 miles) from his position that day. The first sighting of land entered in the ship's log was on 28 January.

On 24 January *Peacock* damaged its rudder in ice, narrowly escaping destruction, and limped back to Sydney and across to New Zealand. Meanwhile *Porpoise* and *Vincennes* pressed on to the west. Six days later *Porpoise* sighted Dumont d'Urville's *Astrolabe* through the fog. According to *Porpoise*'s commander, Cadwalader Ringgold, he intended to pass under the stern of the other vessel but when he saw it putting on more sail, "without a moment's delay, I hauled down my colors and bore up on my course before the wind." Apparently national egos are highly sensitive to any perceived slight, even in the wilds of Antarctica. *Porpoise* continued westward before setting course for New Zealand on 14 February 1840.

Vincennes also continued its coastal survey, skirting the edge of the pack ice under an increasing cloud of paranoia. Wilkes wrote: "I cannot help feeling how disgusting it is to be with such a set of officers (one or two I must except) who are endeavoring to do all in their powers to make my exertions go for nothing." On 21 February *Vincennes* reached what is now called the Shackleton Ice Shelf, which extends almost

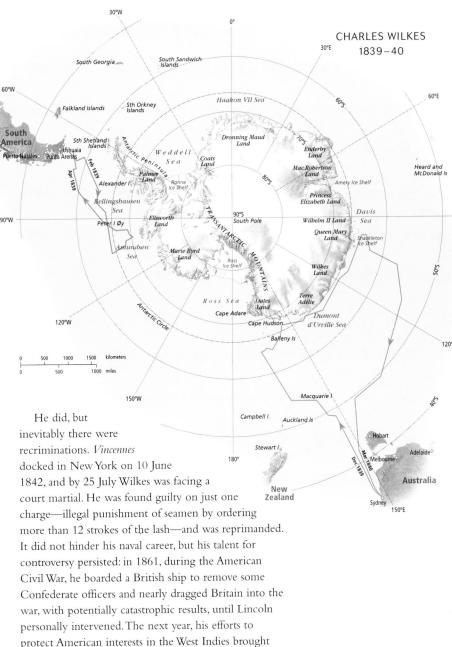

◄ ICEBERG LANDING

Wilkes claimed several sightings of the Antarctic coast during January and February 1840, but the closest he came to a landing was on this large iceberg, where the crew are shown enjoying themselves while collecting ice for water. The snowy hills behind may or may not have been the continent—many areas Wilkes marked on his chart as "land" were sailed over by later explorers.

CHARLES WILKES
1839–40

160 kilometers (100 miles) out to sea. With no time left for the long detour around Iles Kerguelen, Wilkes sailed for Sydney to report his success, and for a surprise meeting with *Peacock*. He had followed more than 2,700 kilometers (1,700 miles) of the pack ice, and could be justly proud of the achievement. He had also seen enough to confidently proclaim the landmass a continent—the first explorer to do so.

Wilkes was certain that he was about to announce the first discovery by a national expedition, rather than by whalers such as Balleny and Biscoe, of an Antarctic landmass south of the Pacific and Indian oceans. But word soon arrived from Hobart that Dumont d'Urville had not only seen land on the afternoon of 19 January but had landed on an island that day. Wilkes countered that he had seen land in the morning of that day, and later that he and others had first seen land on 16 January. These claims would later form the basis of a charge of "deliberate and wilful falsehood" at his court martial. Wilkes's cause was not helped when James Clark Ross returned from Antarctica to say that he had *sailed* across an area marked as solid land on the chart that Wilkes had generously given him.

Home at last

It was another two years before the expedition finally returned home, but only three months of the four-year voyage had been spent below latitude 60°S. During extensive Pacific exploration *Peacock* had been wrecked (with no loss of life) off the mouth of the Columbia River, and had been replaced by *Oregon*, and *Flying Fish* had been sold in Singapore. Wilkes deliberately issued time-wasting orders to delay *Porpoise* and *Oregon* so that he would arrive in New York on *Vincennes* several weeks ahead of them.

He did, but inevitably there were recriminations. *Vincennes* docked in New York on 10 June 1842, and by 25 July Wilkes was facing a court martial. He was found guilty on just one charge—illegal punishment of seamen by ordering more than 12 strokes of the lash—and was reprimanded. It did not hinder his naval career, but his talent for controversy persisted: in 1861, during the American Civil War, he boarded a British ship to remove some Confederate officers and nearly dragged Britain into the war, with potentially catastrophic results, until Lincoln personally intervened. The next year, his efforts to protect American interests in the West Indies brought claims of neutrality violations by several nations. He retired from the United States Navy with the rank of rear-admiral in 1866, and died in February 1877.

► DEEP-FROZEN

The United States commemorated Wilkes in a station it established in 1957. Wilkes Station was transferred to Australia in 1966, but snow build-up later forced it to be abandoned. Its derelict remains now present both nations with the twin problems of removal and environmental rehabilitation.

James Clark Ross

1839–43

▲ POLAR EXPERIENCE
James Clark Ross had already made six Arctic expeditions, and had located the North Magnetic Pole, when he sailed south in 1839 to find its southern equivalent. His was the first British naval voyage to Antarctica. Unlike the United States and French naval voyages of Wilkes and Dumont d'Urville, this expedition was built on decades of polar expertise.

When James Clark Ross set sail from England on 5 October 1839, Wilkes and Dumont d'Urville had already ventured into the Antarctic and had retreated to the north to wait out the winter. There was fierce competition among the Antarctic expeditions of the late 1830s, and Ross was spurred on by patriotism and personal ambition to surpass his rivals' achievements.

Ross was the most experienced Arctic navigator to visit Antarctica. Born on 15 April 1800, he joined the Royal Navy before his twelfth birthday. His most significant Arctic achievement was locating the North Magnetic Pole in 1831. This (and his Byronic appearance) won him great popular acclaim in Britain. A fellow officer described him as "the finest officer I have met with … He is perfectly idolized by everyone." When British scientists pressed for a voyage to find the South Magnetic Pole—the Magnetic Crusade—Ross was chosen to lead it.

His ships were the 372-tonne *Erebus* and the smaller *Terror*—bomb ships, built to withstand the recoil of mortars. These bleak names were not reflected in conditions on board; Ross ensured that they were waterproof, warm, and well provisioned. Ross commanded *Erebus*, and *Terror*'s captain was Irishman Francis Crozier, who had sailed the Arctic with Ross. Also on board was 22-year-old naturalist Joseph Hooker, who later brought London's Kew Gardens to their full Victorian glory. Just before the ships departed, the British sealer and whaler John Balleny brought news of land below the Antarctic Circle and a possible break in the pack ice along the 170° meridian.

Arriving in Hobart in August 1840, Ross received permission from the Governor of Tasmania to build the observatory "Rossbank" on a hill above the town. But there was bad news as well as good: Ross's national pride was pricked to learn that Dumont d'Urville and Wilkes had both spent their second Antarctic seasons seeking the Magnetic Pole. Ross sailed again on 12 November and crossed the Antarctic Circle on New Year's Day 1841, near the area that Balleny had suggested. He pushed south, sometimes through heavy ice, to break into open sea on 9 January. He had discovered the Ross Sea, the best ocean access to the South Geographic Pole. Ross's was the first expedition equipped with vessels suitable for ice work—but they were still sailing ships. Decades later Norwegian explorer Roald Amundsen wrote: "These men sailed right into the heart of the pack … It is not merely difficult to grasp this; it is simply impossible—to us, who with a motion of the hand can set the screw going, and wriggle out of the first difficulty we encounter. These men were heroes."

Ross was orientated towards the Magnetic Pole, and he was disappointed to encounter land instead of open water. He calculated that the Magnetic Pole was only 800 kilometers (500 miles) away, but clearly there was no direct sea route. But the discovery of new land was worthy of ceremony, so on 12 January 1841 Ross and Crozier landed on an island they later named Possession (71°52′S 171°12′E) and claimed and named Victoria Land for the young Queen Victoria.

Fire and ice

Proceeding southeast with the fading hope of finding a seaway to the Magnetic Pole, they had already passed Cook's furthest south on 11 January and passed

◄ CLAIMED FOR THE QUEEN
Ross sighted this mountainous coast soon after entering the open sea now named for him. He managed to land on an offshore island, and there, ankle-deep in guano from its raucous penguin rookery, took possession of all he could see. The most prominent mainland peak Ross named for Captain Edward Sabine, the leading authority on terrestrial magnetism and chief instigator of this voyage.

Weddell's 74°15′S on 22 January, when they reached 74°20′S and celebrated with a double ration of rum. Six days later they saw what Hooker described as "a fine volcano spouting fire and smoke." Ross named it Mount Erebus and its smaller neighbor Mount Terror. An active volcano 3,795 meters (12,450 feet) high and covered in snow was remarkable enough, but as they approached a low white line filled the horizon. Ross wrote that it was "a perpendicular cliff of ice between one hundred and fifty feet and two hundred feet above … the sea, perfectly flat and level on top, and without any fissures or promontories on even its seaward face." Ross marked it on his chart as a Barrier, as it killed the last hope of sailing to the South Pole; it is now the Ross Ice Shelf. Sailing 390 nautical miles (720 km) eastward they finally found a bay in the Barrier that a crewman described as the "most rare and magnificent sight that ever the human eye witnessed."

Ross's party could congratulate themselves. They had collected rocks and plants, taken the first Antarctic sea soundings (and established that the edge of the Barrier ice was afloat), found the Ross Sea, and, by rigorous measurement, established that the South Magnetic Pole was much further south than predicted. The two vessels tied up alongside "Rossbank" on 6 April 1841.

▲ FORGED IN FIRE

A volcano spurting a "dark cloud of smoke, tinged with flame" was astonishing in this land of ice and snow. Ross named it Erebus for his flag-ship, and called its smaller extinct neighbor Mount Terror.

▼ EASY HUNTING

Although the expedition had sheep—and a pet goat—on board, additional fresh meat was always welcome and helped ward off scurvy. So the men from both ships took every opportunity to kill the unsuspecting seals. Seal blubber was melted down to provide oil for lighting.

JAMES CLARK ROSS 1840–43

Southward again

The voyage south the following summer is better known for its tribulations and celebrations than for any ground-breaking achievements. The two ships sailed via Sydney and the Bay of Islands to cross the Polar Front (Antarctic Convergence) on 13 December. Coincidentally they again crossed the Antarctic Circle on New Year's Day, but this time both ships were locked in pack ice. An ice floe between the two ships became a dance floor to celebrate the holiday in style. Ross dressed up as a woman, and one officer later wrote to his sister, describing the festivities: "Captain Crozier and Miss Ross opened the ball with a quadrille … Ices and refreshments were handed round … You would have laughed to see the whole of us, with thick overall boots on, dancing, waltzing and slipping about … Ladies fainting with cigars in their mouths—to cure which the gentlemen would politely thrust a piece of ice down her back … a "lady" burnt the back of my hand with a cigar."

But this was the last light moment for weeks. The ships remained trapped as the ice drifted north, and on 19 January 1842 a gale ground the ice against the ships, smashing their rudders. Nevertheless, on 23 February they were again at the Barrier at 78°10′S—their furthest south, a record that would not be surpassed for 50 years. They were to winter on the Falklands, and the next day they set out on the long voyage around Cape Horn.

At first they saw few icebergs, but some emerged through the fog and snow of 12 March. Conditions worsened around midnight and Ross ordered *Erebus'* to heave to for the night. But even as sails were being furled an iceberg loomed dead ahead. Ross described how: "the ship was immediately hauled to the wind on the port tack … But just at this moment the *Terror* was observed running down upon us, under her top-sails and foresail; and as it was impossible for her to clear both the berg and the *Erebus*, collision was inevitable … Our bowsprit, foretopmast, and other smaller spars, were carried away, and the ships hanging together, entangled

▼ **TERRIFYING COLLISION**
To avoid a huge iceberg, which emerged out of the gloom in a storm, *Erebus* turned, only to find *Terror* "running down upon us, under her top-sails and foresail." *Terror* ploughed straight into *Erebus* and "the concussion when she struck us was such as to throw almost everyone off his feet" wrote Ross later. Locked together, the ships performed a hideous vertical dance that destroyed most of *Erebus*'s rigging.

▶ **LONG WALL OF ICE**
The pale light of the setting sun does little to soften the chilling bulk of the Ross Ice Shelf. Looking for an opening, Ross sailed 390 nautical miles (720 km) along these ice cliffs. At no point was the unbroken wall lower than 15 meters (50 ft), and all that could be seen from mast height was "an immense plain of frosted silver."

by their rigging, and dashing against each other with fearful violence, were falling down upon the weather face of the lofty berg under our lee, against which the waves were breaking and foaming to near the summit of its perpendicular cliffs." Half-naked seamen roused from sleep struggled desperately while Ross, according to various officers, "calmly gave the order to loose the sail … as if he were steering into any harbor."

In two strokes of luck the ships parted and Crozier found a gap through what had seemed a single berg. In its lee he was able to inspect the damage to *Terror*. *Erebus* was almost completely incapacitated and destruction against the iceberg seemed inevitable. The only chance was the intricate maneuver "stern board," in which a square rigger sails backward—this in a storm with the rigging and sails tied in knots. Few could hear Ross's calm commands in the chaos, but his do-or-die tactic was successful and *Erebus* joined *Terror* in the shelter of the iceberg. Amazingly, two days later the ships were able to continue to the Falklands, arriving on 6 April.

The final venture

Christmas 1842 was the ships' third one surrounded by ice as they approached the Antarctic Peninsula on their last southward voyage. Ross had wanted to extend Dumont d'Urville's exploration and perhaps follow Weddell's southward path, although he knew that no discovery could be expected along that track. But apart from some surveying around the eastern side of the tip of the Antarctic Peninsula, their main discovery was ice in all its permutations, cutting every access to the south. After six frustrating weeks they were finally forced to give up and resumed an eastward course. Even at 40°W, where Weddell had found open water, they found solid pack ice. However, the further east they proceeded the further south lay the edge of the pack, so they crossed the Antarctic Circle for the third time to reach 71°30′S 14°51′W on 5 March 1843. By 11 March they were north of the Circle, and on 2 September they saw England again after almost four and a half years.

▼ NO WAY THROUGH
As soon as he sighted the Ross Ice Shelf, which he named the Victoria Barrier, Ross realized that "we might with equal chance of success try to sail through the cliffs of Dover, as to penetrate such a mass." He described these icy white cliffs as "extending from its western extreme point as far as the eye could discern to the eastward."

The *Challenger* Expedition

1872–76

▼ RELEVANT RECORDS

Scientists aboard *Challenger* wrote meticulous reports with accurate illustrations that are still used by marine biologists today. Below is a series of radiolarians—tiny protozoans with complex skeletons, usually composed of silica. These creatures are named after HMS *Challenger* and the expedition's scientists.

For many years scientists believed that the deep oceans were "lightless and lifeless," but between 1868 and 1870 dredging off the Scottish coast disproved this belief, and the science of oceanography was born. The first voyage to explore the possibilities of the fledgling science was effectively a joint venture between the Royal Navy and the Royal Society. HMS *Challenger* sailed from Sheerness, Scotland, under the command of George Strong Nares (1831–1915) on 7 December 1872, and returned more than three years later, on 24 May 1876, with such a wealth of scientific data that it took from 1885 to 1895 to compile it into 50 volumes containing some 30,000 pages of information that would take decades to analyze.

The goals of the expedition were global and ambitious. Among other tasks, it was to "investigate the physical conditions of the Deep Sea, in the great Ocean basins … in regard to Depth, Temperature, Circulation, Specific Gravity, and Penetration of Light." *Challenger*, a military corvette with a displacement of 2,352 tonnes,

had been adapted for survey and scientific purposes, and carried six civilian scientists as well as 20 officers and 200 crew. The southward voyage was remarkably slow as they established the painstaking and time-consuming procedures for scientific dredging and sampling.

On 16 February 1874, *Challenger* became the first steamship to cross the Antarctic Circle, and it went on to reach 66°40′S at 78°30′E. The much-quoted steam-powered vessel claim is rather misleading, since the greater part of the voyage, both within and beyond the Antarctic Circle, was conducted under sail. Sublieutenant Lord George Campbell wrote that steam was used only "in a prolonged calm when, if we had coal to spare, we steamed slowly on our way." *Challenger* was not ice-strengthened; even using a combination of sail and screw, she barely survived her brief time among the icebergs.

Nevertheless, whether powered by its 1,200-horse-power engine or the 16,000 square feet of sails on its three masts, *Challenger* completed a remarkable scientific voyage. She sailed across the Atlantic to Brazil and then returned eastwards, calling at the Tristan da Cunha and Iles Kerguelen, and at Heard Island. The men saw their first iceberg only days after leaving Iles Kerguelen, and many more soon followed. *Challenger* remained in the ice for less than three weeks before continuing onwards to Melbourne, and then to Japan and the North Pacific. She did not go back to the Southern Ocean, but returned home by way of Tahiti, Chile, the Strait of Magellan, and Montevideo.

Extending the frontiers

Chief scientist Professor Wyville Thomson later reported that *Challenger* had made 362 observation stations during the voyage, "at intervals as nearly uniform as possible." Although the expedition's sojourn in the south was so short, it contributed greatly to Antarctic science, including establishing that there was a permanent cell of high atmospheric pressure over Antarctica. Earlier expeditions into the southern ice had observed rocks and debris trapped in icebergs drifting northward, but it was left to *Challenger* to dredge the floor of the Southern Ocean for the rocks released from these melting icebergs. The rocks she laboriously hauled to the surface with her 18-horsepower winch were quite different from the rocks of the sub-Antarctic islands, providing clear evidence of the existence of a distinct Southern Continent. South of 43°S, her dredging haul included about 400 animal species, more than 80 percent of which had never been seen before.

When *Challenger* steamed into Portsmouth in May 1876, she had covered 127,600 kilometers (68,890 nautical miles). Captain Nares was no longer on board, having been recalled from Hong Kong to lead the British Arctic Expedition of 1875–76 into the Canadian Arctic. But while Nares was seeking the North Pole and

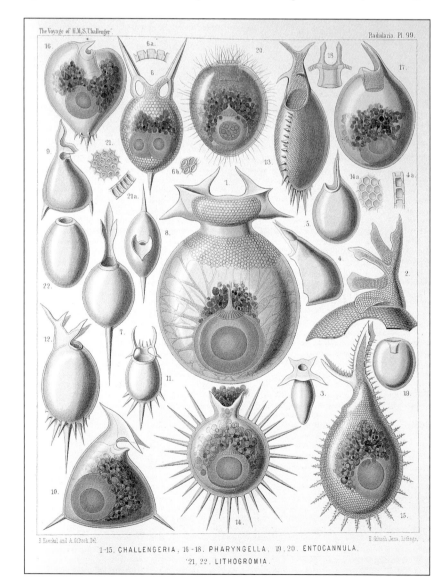

The Voyage of H.M.S. Challenger.　　　　Radiolaria. Pl. 99.

1–15. CHALLENGERIA, 16–18. PHARYNGELLA, 19, 20. ENTOCANNULA, 21, 22. LITHOGROMIA.

discovering the Challenger Mountains, *Challenger's* scientists and crew methodically continued their scientific work. They took daily magnetic observations while at sea. They collected flora, fauna, and geological samples on a range of sub-Antarctic islands, and acquired a wealth of information about ocean currents and temperatures, meteorology, and the seafloor of the basins of the Pacific, Atlantic, and Southern oceans. It had been a long, slow voyage because the scientific program required the ship to stop for most of a day every 320 kilometers (200 miles) to dredge and conduct soundings—but it was extremely productive, and provided information that is of value to this day.

▼ UNFAMILIAR MARINE CREATURE
The giant sea star was one of about 400 animal species brought to the surface by *Challenger's* dredges. At the time, such creatures had never before been seen by scientists.

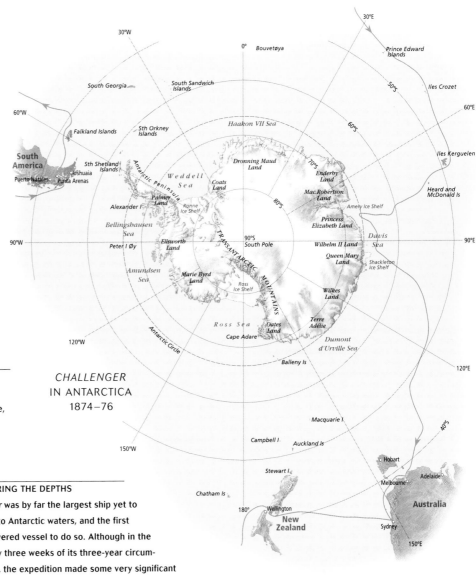

CHALLENGER
IN ANTARCTICA
1874–76

▼ EXPLORING THE DEPTHS
Challenger was by far the largest ship yet to venture into Antarctic waters, and the first steam-powered vessel to do so. Although in the ice for only three weeks of its three-year circumnavigation, the expedition made some very significant biological and geological discoveries, using dredging equipment never before employed in these latitudes.

A Commercial venture

1894–95

⌃ CAPTAIN OF COMMERCE
After several voyages to the Weddell Sea, Norwegian Carl Anton Larsen (1860–1924) went on to found South Georgia's lucrative whaling industry. Using explosive harpoons, Larsen became one of the most successful whaling captains of the early 1900s.

The powerful Australian interest in the Antarctic Continent can be traced back to Henryk Johan Bull, a Norwegian who had emigrated to Melbourne in 1885. In *The Cruise of the Antarctic*, Bull reported his disappointment when an attempt by Australian scientists to raise an Antarctic expedition failed, and his belief that "an expedition on commercial lines would possibly find supporters."

Whaling in Antarctic waters went back to the early nineteenth century, but at that time there were enough whales in the northern hemisphere to discourage the longer and more hazardous voyage south. Northern whale populations began to fall and alternative products such as mineral oils emerged, but the demand for baleen from right whales—mainly for the "whalebone" used in corsets—kept the quest alive. In 1873 a German expedition explored the Antarctic Peninsula but found only rorqual whales. Over the summer of 1892–93 a Scottish expedition explored the northwest Weddell Sea but found no right whales among the many it saw. A Norwegian expedition under Carl Anton Larsen worked the same area that summer, but with little success in terms of whaling. Larsen returned the next summer and had more luck with seals than whales, although his voyage was notable because he found petrified wood at Cape Seymour— the first fossils dicovered on Antarctica.

⌃ HAZARDS OF ICE AND STORM
The *Antarctic* dodging ice on its way to Cape Adare where Henryk Bull—with several others—made the first landing on East Antarctica. Bull was unsuccessful, however, in his primary task of locating baleen whales. On a later expedition in the Weddell Sea *Antarctic* was crushed by ice and sank.

Raising the money

Larsen named Foyn Land (now the Foyn Coast) in 1893 after Svend Foyn, the Norwegian inventor of the explosive harpoon, who went on to mount the new device on the steam-driven catchers that had replaced rowing boats. These innovations revolutionized the whaling industry and made Foyn rich. In 1893, when Bull failed to raise finance in Melbourne, he went to Norway to see Foyn. Within 15 minutes Foyn had promised him a ship—a 226-tonne steam whaler named *Kap Nor*, renamed *Antarctic* for the voyage. The ship ran aground on a preliminary whaling expedition to New Zealand's Campbell Island, but sailed from Melbourne Wharf on 26 September 1894, called at Hobart, and set off for Antarctica on 13 October. Her captain was Leonard Kristensen: Bull was on board in the loosely defined role of manager.

The expedition posed little threat to Antarctica's whale population, but it

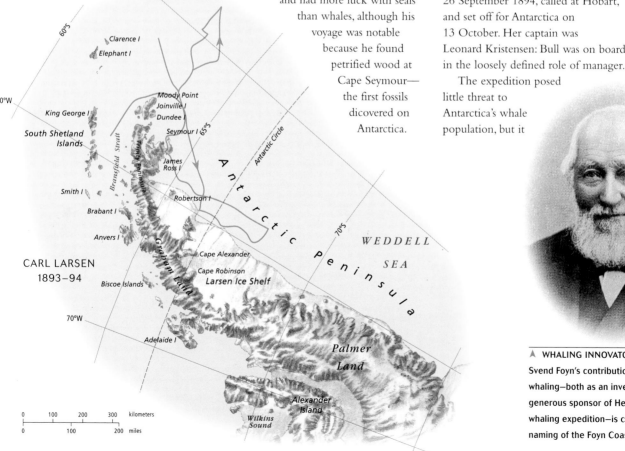

CARL LARSEN
1893–94

⌃ WHALING INNOVATOR AND SUPPORTER
Svend Foyn's contribution to commercial whaling—both as an inventor and later as a generous sponsor of Henryk Bull's exploratory whaling expedition—is commemorated in the naming of the Foyn Coast and Foyn Island.

added greatly to the controversy about who first set foot on the Antarctic Continent—if John Davis had not in fact landed on the Antarctic Peninsula in 1821, then Bull's party was the first recorded landing on the continent. The date was 24 January 1895; Bull wrote: "Cape Adare was made at midnight. The weather was now favorable for a landing, and at 1 am a party, including the Captain, second mate, Mr. Borchgrevink, and the writer, set off, landing on a pebbly beach of easy access after an hour's rowing through loose ice, negotiated without difficulty … The sensation of being the first men who had set foot on the real Antarctic mainland was both strange and pleasurable, although Mr Foyn would no doubt have preferred to exchange this pleasing sensation on our part for a right whale even of small dimensions."

Rival claims

Bull left the question of the first to step ashore open, but others were not so modest. Carsten Borchgrevink, who was born in Oslo and migrated to Australia when he was 24, was on board as a "generally useful hand." He claimed to have been a student at Christiania (Oslo) University, and Bull took him on because he could also fill the role of scientist. Bull acknowledged an ulterior

motive based on their common background as Norwegian immigrants to Australia: "His family was known to me … and the prospect of a real companion was so cheering that I promised to do my best for him." However, Borchgrevink's credentials were not rated highly; Bull added: "It is true we intended to build an extra cabin for … colonial men of science … but the qualifications and letters of introduction brought by Mr Borchgrevink were not such as to warrant … building extra cabins for him." Nevertheless, Borchgrevink was in the landing party, and subsequently (and repeatedly) claimed that he was first ashore. In one account he stated: "I do not know

> **EXPERIENCED HAND**
Carsten Borchgrevink, one of the claimants to the honor of being first to "set foot in South Victoria Land," had been employed by Bull as a seal shooter and "useful hand." In 1901 he returned to Antarctica with his own expedition ship, *Southern Cross* (shown here).

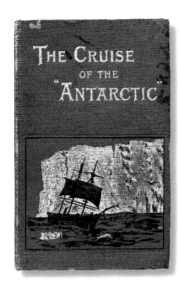

◄ ROSS SEA VOYAGE
Henryk Bull's *The Cruise of the Antarctic* fired enthusiasm for commercial whaling in the Ross Sea region and also sparked off the Heroic Age of Antarctic exploration.

► POLAR VETERAN
The whaler *Jason* took Carl Anton Larsen to the Arctic in 1881, and later on the Antarctic voyages that motivated him to set up a whaling industry on South Georgia.

▼ COLD-CLIMATE PLANTS
When the crew of *Antarctic* landed on the Possession Islands, they were greatly surprised to find lichen there. Until then, Antarctica was thought to be devoid of plant life. This photograph, taken at Fortuna Bay, South Georgia, shows that other plants can also survive there.

whether it was the desire to catch the jellyfish (seen in the shallows), or from a strong desire to be the first man to put foot on this *terra incognita*, but as soon as the order was given to stop pulling the oars, I jumped over the side of the boat."

His claim was disputed by at least two others. In his journal Kristensen wrote: "I was sitting foremost in the boat, and jumped ashore as the boat struck, saying 'I have then the honor of being the first man who has ever put foot on South Victoria Land.'" The final claimant has considerable merit: in view of the water temperature and shipboard protocol, there is reason to believe New Zealand crewman A. H. F. von Tunzelman, who maintained that he was in the bow and jumped out first to steady it so the captain could disembark.

This rather childish dispute aside, the *Antarctic* expedition was an important development in the exploration of Antarctica because the voyage opened the way for the land explorations of the Heroic Age. In

The Cruise of the Antarctic, Bull wrote: "We have proved that landing on Antarctica proper is not so difficult as it was hitherto considered, and that a wintering party have every chance of spending a safe and pleasant twelve-month at Cape Adare, with a fair chance of penetrating to, or nearly to, the magnetic pole by the aid of sledges and Norwegian skis."

The immediate commercial returns from the voyage were few: it had found no baleen whales, and while *Antarctic* had been exploring its namesake, its owner, Sven Foyn, had died, so that soon after it returned to Melbourne on 12 March 1895 the ship was instructed to return to Norway. But the expedition is justly remembered in scientific circles, and there is now a Bull Island in the Possession Islands group. This was where the crew of *Antarctic* landed and found lichen— the first vegetation ever discovered in Antarctica, an environment previously believed to be too bleak to allow any plant life to survive.

Adrien de Gerlache

1897–1899

At the Sixth International Geographical Congress in London, July 1895, English geographer Clements Markham said: "The exploration of the Antarctic regions is the greatest piece of geographical exploration yet to be undertaken." The Congress recommended that "this work should be undertaken before the close of the century."

Meanwhile the Brussels Geographical Society was assembling an Antarctic scientific expedition initiated and led by a young naval lieutenant, Adrien Victor Joseph, Baron de Gerlache de Gomery. The venture had an international flavor. In Norway, de Gerlache bought *Patric*, a 250-tonne whaler, and renamed it *Belgica*. His second-in-command was another Belgian, Lieutenant George Lecointe, and his crew included a Rumanian chief scientist, Emile Racovitza, Henryk Arçtowski, a Pole, as geologist, and a 25-year-old Norwegian, Roald Amundsen (then in training for his captain's certificate), as first mate. The ship's doctor, 32-year-old American Frederick Cook, had been in Greenland with Arctic explorer Robert Peary, who was later to claim to reach the North Pole.

Belgica left Antwerp on 16 August 1897 with new cabins and a new laboratory—additions that overloaded the vessel and made it painfully slow. The plan was to sail along the eastern side of the Antarctic Peninsula, to winter in Melbourne, and to visit Victoria Land the following summer. They reached Punta Arenas early in December, but de Gerlache decided to explore Tierra del Fuego, so they did not arrive off the Antarctic Peninsula until 20 January—almost the end of summer.

Ice-bound

Over the next few weeks they explored the islands that line what is now Gerlache Strait. They took some of the first photographs of Antarctica, and Cook recorded one occasion as "perfectly dazzling ... a photographic day." They crossed the Antarctic Circle on 15 February and encountered pack ice, but pushed on until it became clear that they were trapped in ice for the winter. De Gerlache was suspected of planning their late southward

passage from the outset, and there were recriminations. Amundsen wrote that they "faced the prospect of a winter in the Antarctic with no winter clothing … without adequate provisions … and even without lamps enough … It was a truly dreadful prospect."

Pack ice can move as fast as 16 kilometers (10 miles) a day, and it carried *Belgica* westward. De Gerlache had sailed well southwest of the Antarctic Peninsula, and *Belgica* became trapped at 71°30′S 85°16′W on 2 March 1898; she would not be free until 14 March 1899, at 70°30′S 103°W. It was a harrowing time. The sun sank below the horizon on 17 May, not to reappear for 70 days. Each day became a paler mirror of the one before. In *Through the First Antarctic Night* Cook outlined the daily routine: "Rise at 7.30 am; coffee at 8; 9 to 10, open air exercise; 10 to 12, scientific work … for the officers, and for the marines, bringing in the snow, melting snow for water, replenishing the ship's stores, repairing the ship, building new quarters, making new instruments, and doing anything which pertains to the regular work of the expedition; 12 to 2 pm dinner, and rest or recreation, 2 to 4, official work … 6 to 7, supper; 7 to 10, card-playing, music, mending, and on

> ⌃ **NEW CHALLENGES**
> Whether by accident or design, the scientific expedition led by young Belgian naval lieutenant Adrien de Gerlache, was the first to face the rigors of an Antarctic winter.

> ◂ **DE FACTO LEADER**
> When *Belgica* became icebound, ship's doctor Frederick Cook devised a regime of diet and activities to safeguard the physical and mental health of the crew. He also led the sawing of a channel through the ice to open water and freedom.

> ▶ **FIRST PHOTOGRAPHIC RECORDS**
> Watched by a curious Emperor penguin, *Belgica* is shown beset by ice in the Bellingshausen Sea in early March 1898. At first a novelty to the crew, the icy scene became monotonously familiar over the ensuing 377 days until the ship was freed.

ADRIEN DE GERLACHE 1898–99

▶ **FROZEN MEMORIAL**
Arriving on the west coast of the Antarctic Peninsula in late summer, the expedition explored and mapped the islands and coast bordering the strait (right) that is now named for de Gerlache. The mountainous coast commemorates Emile Danco, the expedition's geophysicist, who died later on the voyage, and Wiencke Island is named after a young seaman who fell overboard and drowned.

moonlight nights, excursions. At 10 o'clock we went to sleep." Cook proved a tireless source of strength, boosting morale and providing medical help when symptoms of distress, and even of insanity, surfaced. Amundsen wrote: "He … was the one man of unfaltering courage, unfailing hope, endless cheerfulness, and unwearied kindness."

Cold storage

Cook and Amundsen knew the importance of fresh meat in preventing scurvy, and had stored seal and penguin carcasses in the ice before all wildlife fled the advancing winter. But, as Amundsen related, "The commander … developed an aversion to the flesh of both … He was not content only to refuse to eat it himself, but he forbade any of the ship's company to indulge in it." De Gerlache and Lecointe took to their beds with scurvy, and the two Arctic veterans effectively took over. Cook provided a graphic description of penguin meat; it tasted like "a piece of beef, odiferous cod fish and a canvas-backed duck roasted together in a pot, with blood and cod-liver oil for sauce." But he persuaded de Gerlache to take it as a medicine, and soon they all started to recover.

Cook described the momentous occasion when the bright top of the sun reappeared: "For several minutes my companions did not speak … we could not … have found words with which to express the buoyant feeling of relief, and the emotion of the new life which was sent coursing through our arteries." His optimism was premature: summer progressed, but no way out of the ice presented itself and they faced the prospect of another winter trapped in the ice with little hope of survival. Antarctica's history contains some of the most harrowing tales of adventure imaginable, and the *Belgica*'s began on New Year's Eve when they saw open water 640 meters (2,100 ft) away. Cook suggested cutting a passage to freedom, although they had only three four-

foot saws. It was a desperate measure, particularly as it was only a small basin of water they were making for, and they were relying on Cook's faith that this was where any break to open water would appear. They worked day and night (indiscernible during this season of perpetual sunlight) for a month, cutting back toward the ship from the basin. By the end of January they had hacked through over 600 meters (2,000 ft) of ice. Then, as Amundsen wrote: "Imagine our horror on awakening to discover that the pressure from the surrounding ice pack had driven the banks of our channel together, and we were locked in as fast as ever."

Escape

Not surprisingly, depression set in—but that, too, was to prove unjustified. On 15 February the ice relented and the passage they had cut reopened and extended right to the ship. They took *Belgica* to the basin under power, but they were still 11 kilometers (7 miles) from freedom and winter was approaching. Amundsen wrote: "Other weary weeks passed … Then the miracle happened—

➤ TRAPPED FOR THE WINTER

As the long, dark winter deepened, so did the snow drifts around *Belgica*. Although some pressure ridges developed, the vessel was never seriously threatened. On moonlit nights the crew went on skiing excursions to relieve the cramped monotony of life on board, while the ship, its rigging glistening with ice, gleamed like a ghost.

exactly what Cook had predicted. The ice opened and the lane to the sea ran directly through our basin! Joy restored our energy, and with all speed we made our way to the open sea and safety." It was 14 March 1899, and *Belgica* returned to Antwerp on 5 November 1899.

Despite its problems, the expedition gathered much valuable scientific data. Arçtowski believed correctly that his bathymetric measurements suggested that the Antarctic Peninsula was an extension of the Andes—not directly across the Drake Passage but through the Scotia Arc. And, although the feat was achieved unwillingly, this was the first expedition to collect meteorological readings below the Antarctic Circle through a winter. The lowest temperature recorded was −43°C (−45°F).

The participants went on to varied futures. Amundsen became the first to reach the South Geographic Pole and de Gerlache completed several Arctic journeys. Cook won lavish praise for his leadership skills, but proved a flawed man: he subsequently claimed to have been the first to climb Mount McKinley, North America's highest mountain, and to be first to the North Geographic Pole, but in both cases he falsified his records. He was imprisoned for fraudulently using the United States mail in 1923, released in 1929, received a presidential pardon in 1940, and died the same year.

Tragedy—and light relief

Parts of the voyage of *Belgica* make sad reading. Two days after she reached the Antarctic Peninsula, a young sailor, Auguste-Karl Wiencke, fell overboard; he held onto a line for some time, but sank before he could be saved. Later, during the long Antarctic night of 1898, paranoia and fierce arguments developed. Men took to their bunks for days at a time, and 30-year-old Emile Danco, the popular geophysicist, died of cold and was buried at sea through a hole cut in the ice. There were lighter moments: on Easter Sunday, to keep up morale, a "Grand Concourse of Beautiful Women" was held, and won by "Princess de Chimay" and "Cléo de Mérode." But even the ebullient Cook wrote: "It is so long since we have seen a girl that I doubt our ability to pass judgment."

▼ BEAUTY PAGEANT

A "Grand Concourse of Beautiful Women," held on the Belgian King Leopold's birthday, was one of the entertainments devised by Cook to stave off boredom. It featured 464 Parisian pin-ups "representing all kinds of poses of dress and undress," which were judged by the men on a 21-point scale.

ANTARCTIC NIGHT

ANNOUNCEMENT BY

THE MINISTER OF ARTS, FEMININE BEAUTY, AND PUBLIC WORKS

GRAND CONCOURSE OF BEAUTIFUL WOMEN

ORGANISED IN THE COLD ANTARCTIC, HELD UNDER THE AUSPICES OF

S. M. ARTOCHO I.—KING OF THE POLAR ZONE

AND

S. A. ROALD, PRINCE OF THE KYODBOLLA

FIRST PART.—TOTAL OF VOTES FOR THE MOST BEAUTIFUL WOMEN.

DESCRIPTION FOR BALLOTING.	FIRST PRIZE.	SECOND PRIZE.	THIRD PRIZE.
I. Poses plastiques	252	217	218
II. Disposition (dreamy, fond of flattery)	183	326	339
III. Appearance, common	391	323	260
IV. Rosy complexion	306	245	264
V. Irreproachable character	94	88	210
VI. Grace, personified	209	230	319
VII. Elegant appearance (sweet disposition)	47	463	101
VIII. Underclothing	134	180	—
IX. Most artistic poses	274	404	391
X. Sporty girls	208	397	405
XI. Most graceful dancers	288	291	290

PART SECOND.—TOTAL OF VOTES ON THE EXCELLENCE OF SPECIAL PARTS

DESCRIPTION FOR BALLOTING.	FIRST PRIZE.	SECOND PRIZE.	THIRD PRIZE.
I. The most beautiful face	94	479	480
II. Luxuriant hair	308	320	282
III. Flashing eyes	312	88	—
IV. Mouth (Cupid's bow)	309	88	—
V. Shapely hands (tapering fingers)	311	217	191
VI. Arms	212	—	—
VII. Sloping, alabaster shoulders	212	—	—
VIII. Supple waist	218	—	—
IX. Les jambes	209	217	—
X. Feet	211	—	—

251

◄ POLISH PERSEVERANCE

Henryk Arçtowski, the expedition's Polish geologist, maintained his program of observations throughout the voyage and, despite poor conditions, made some significant discoveries. He was the first to propose that the Antarctic Peninsula was a continuation of the Andes chain through the islands of the Scotia Arc. This became an important piece in assembling the jigsaw puzzle of plate tectonics.

◄ FUNERAL OF DANCO

Emile Danco, the expedition's geophysicist, died of cold and exposure during the winter when temperatures fell as low as −43°C (−45°F). His body was lowered gently from the ice-shrouded ship and taken by sledge to a hole cut in the ice. This had to be kept open constantly while de Gerlache read the burial service, and then Danco was lowered into his frigid grave.

Carsten Borchgrevink

1898–1900

In 1894–95 Borchgrevink had accompanied Henryk Bull on the *Antarctic*, and when he returned he claimed that most of the achievements of that voyage were his and his alone. It was not an endearing performance but, as Hugh Robert Mill noted: "No one liked Borchgrevink very much but he had a dynamic quality and a set purpose to get out again to the unknown South that struck some of us as boding well for exploration."

British publisher Sir George Newnes had made his fortune publishing penny magazines such as *Tit-Bits*, *Country Life*, and *The Strand Magazine* for the newly educated masses. Newnes put up 40,000 pounds, with which Borchgrevink's British Antarctic Expedition bought the 277-tonne whaler *Pollux*, renamed it *Southern Cross*, and sailed from London on 22 August 1898. The aim was to spend a winter at Cape Adare and undertake some land exploration. This was the first land-based Antarctic expedition, and they took more than 70 dogs from Siberia and Greenland, handled by two Laplanders, to haul sledges. The ship's master was Bernhard Jensen, who had been second mate on *Antarctic*. The scientists were William Colbeck, of the Royal Naval Reserve, as cartographer, navigator, and magnetic observer; Nikolai Hanson, a Norwegian zoologist, assisted by Hugh Blackwall Evans; and Louis Bernacchi, a physicist from Tasmania.

They sailed from Hobart on 17 December 1898 and spotted their first ice on 30 December at 61°56′S. It took 43 days to get through the pack ice—fortunately *Southern Cross* had been ice-strengthened. They sighted the Antarctic Continent on 15 February 1899 and landed at a large black triangle of flat gravel on the western side of Cape Adare two days later. In ferocious storms it took 12 days to land the wintering party's supplies. At one point six men were stranded ashore, and survived only because the Laplanders had brought their tent and many of the dogs pushed into it and kept the men warm with their bodies. Soon afterwards they erected two prefabricated huts of Norwegian pine on the shore; this was Ridley Camp, named after Borchgrevink's mother and son.

Stir-crazy

Southern Cross departed for New Zealand on 1 March, leaving 10 men behind to brave the Antarctic winter. On 15 May the sun set, not to rise again until 27 July—72 days later. The small party retreated to the huts and cut outdoor work to a minimum. But confinement indoors had its hazards. Relations between Borchgrevink and the others worsened until he produced a petulant document stating that "the following things would be considered mutiny: to oppose C.E.B. [Borchgrevink] or induce others to do so, to speak ill of C.E.B., to ridicule Mr. C.E.B. or his work, to try and force C.E.B. to alter contracts." And there were more tangible problems: on 24 July the huts nearly burned down when Colbeck left a candle burning and set fire to his bunk, and on 1 September three of the party were almost asphyxiated when a wind shift filled the cabin with coal fumes. Hanson had been sick on the voyage down and through much of the winter, and on 14 October he died, probably from beri-beri, and never saw the thousands of Adélie penguins that crossed the ice toward the camp and their breeding grounds only two days later. Hanson was buried at the summit of Cape Adare on 20 October.

▲ FIRST WINTER EXPEDITION
After landing at Cape Adare with Henryk Bull in 1895, Carsten Borchgrevink returned there in February 1899 as leader of his own British Antarctic Expedition, determined to winter on the continent. Although lacking leadership skills, he held his team together sufficiently to prove this was possible, a breakthrough that paved the way to the major explorations of the Heroic Age.

CARSTEN BORCHGREVINK 1900

▲ TIRELESS RESEARCHER
Physicist Louis Bernacchi
carried out a full program of
work, even when ice-bound.
He returned to Antarctica with
Scott's *Discovery* expedition.

During the year at Cape Adare the land party made several survey trips deep into Robertson Bay, penetrated a short way into the interior, and carried out a program of scientific measurements. Much of our knowledge of the expedition comes from Louis Bernacchi's diary: for 1 January 1900 he noted "A new year and the last of the nineteenth century! A year spent entirely within the Antarctic Circle, and the ice and snow visible each day, for 365 days." The *Southern Cross* was expected back in January, and anxiety mounted as the days passed and the ship failed to appear. However, the entry for 28 January runs: "At about 8 o'clock in the morning when all were asleep in their bunks, a voice in the room calling 'Post!' broke upon our slumbers. It was the voice of Capt Jensen. With what hysterical joy we tumbled out of our bunks to welcome him!"

They left Cape Adare on 2 February. Bernacchi wrote: "We are not sorry to leave this gelid, desolate spot, our place of abode for so many dreary months!" Over the next few weeks they sailed into the Ross Sea, landing at Possession, Coulman, and Ross islands, and traveling southward across the Ross Ice Shelf to an estimated 78°50′S—the furthest south yet. They turned north on 22 February, crossed the Antarctic Circle on 28 February, and reached New Zealand's Stewart Island on 31 March.

Belated recognition

The news that it was possible to survive a winter ashore in Antarctica was noteworthy, but their return attracted little interest. The Boer War had intervened and Robert Falcon Scott's National Antarctic Expedition was preparing for departure amid great excitement. The public was interested to learn that there were no large

land animals in Antarctica—expeditions would not now need guns against Polar bears! Borchgrevink was made a Fellow of the Royal Geographical Society and was knighted by the king of Norway, but did not receive the recognition he considered his due. His account, *First on the Antarctic Continent*, was badly written and did not further his cause, but in 1930 the Royal Geographical Society belatedly awarded him the Patron's Medal, stating: "When the *Southern Cross* returned, this Society was engaged in fitting out Captain Scott to the same region … and the magnitude of the difficulties overcome by Mr Borchgrevink were underestimated … It was only after … that we were able to realize the improbability that any explorer could do more in the Cape Adare district than Mr. Borchgrevink had accomplished." He died in Norway on 21 April 1934.

Despite the eagerness with which they fled Antarctica after that grim winter, Bernacchi and Colbeck were soon heading south again—both returned to Cape Adare, Bernacchi with Scott on *Discovery* and Colbeck on *Morning* to rescue Scott's expedition.   DM

▲ CLOSE QUARTERS
Borchgrevink's huts of prefabricated Norwegian pine have withstood more than a century of gales and blizzards due to the ingenious system for interlocking the logs into a robust and rigid structure. The nearby huts of Scott's Northern party, 10 years younger, have virtually disintegrated. Though cramped and dark inside, the Norwegian huts provided adequate shelter, and were a model for later expeditions that wintered further south.

◄ BLEAK RAMPARTS
Cape Adare looms over desolate Ridley Beach, site of Borchgrevink's base. Although a good spot for the camp itself, the inaccessible terrain behind prevented the expedition from making any significant inland journeys of exploration. The body of zoologist Nicolai Hanson, who died during the long winter, was carried by his friends up the steep cliffs to a grave on the ridge above.

The Heroic Age

1901–1917

Robert Falcon Scott

British National Antarctic Expedition: 1901–04

Previous pages

◀ **SUPREME SACRIFICE**
J.C. Dollman's painting, simply titled *Captain Oates*, shows an ailing Lawrence "Titus" Oates stumbling into the polar night to die so that his companions might have a better chance of survival. There is no more poignant image of the selfless courage so often shown during the Heroic Age of Antarctic exploration.

▶ **THEIR NAMES LIVE ON**
From left to right, the officers and scientific staff on the stern of the *Discovery* in 1901 are Wilson, Shackleton, Armitage, Barne, Koettlitz, Skelton, Scott, Royds, Bernacchi, Ferrar, and Hodgson. Many of their names are now immortalized as place names in Antarctica.

It was fitting that Scott named his son, Peter Markham Scott, after Sir Clements Markham, president of the Royal Geographical Society, who took an unknown Royal Navy officer and transformed him into the legendary Scott of the Antarctic. Recent analysis has shown that Scott was neither a fortunate leader nor a good planner—but he achieved much for Antarctic science, and documented some of the definitive events of the Heroic Age of Antarctic exploration.

Markham presided over the Sixth International Geographical Congress, held in London in 1895, and wrote: "The exploration of the Antarctic regions is the greatest piece of geographical exploration yet to be undertaken." Three important scientific expeditions that flowed from that congress neatly divided the frozen continent: the German expedition led by Drygalski went to the region south of the Indian Ocean; Nordenskjöld traveled to the Antarctic Peninsula; and Britain focused on the Ross Sea.

The British National Antarctic Expedition

Markham recognized Scott's potential when he first met the 18-year-old midshipman in 1887, and when Scott volunteered to lead the British National Antarctic Expedition in June 1899, Markham accepted his offer. The expedition combined the might of the Royal Navy, the Royal Geographical Society, and the Royal Society, and its aims combined exploration and science. In keeping with these goals, the expeditioners were a mix of Royal Navy and Merchant Navy personnel, as well as an array of scientists.

Their vessel was *Discovery*—the first ship designed and built in Britain for scientific exploration. It was constructed in Dundee of oak strengthened with internal beams and clad in a steel bow. King Edward VII came to the ship at the Isle of Wight to wish them well, and *Discovery* sailed on 6 August 1901. The king's blessing did not prevent the slow but sturdy ship from pitching and rolling at sea, nor did it stop a persistent leak. The expedition called at Macquarie Island and Auckland Island on its way to Lyttelton, New Zealand. It had

◀ **BIRD'S-EYE VIEW**
Scott was the first Antarctic balloonist, ascending to 244 meters (800 ft) in the basket below a hydrogen balloon. Shackleton, the only other one in the party to go up, took aerial photographs. Their height of ascent was limited by the weight of the heavy tethering rope.

an enthusiastic send-off from there when it sailed on 21 December, although one unfortunate seaman fell to his death during the celebrations.

Discovery crossed the Antarctic Circle on 3 January 1902 and reached Cape Adare on 9 January. Louis Bernacchi was the only member of the expedition who had previously visited Antarctica—two years earlier as a member of Carsten Borchgrevink's expedition. Continuing into the Ross Sea, *Discovery* followed the edge of the Ross Ice Shelf eastward looking for the land reported by Ross, and eventually sighted the mountains of what is now Edward VII Land. On 4 February, at Balloon Bight (now the Bay of Whales), Scott rose in a tethered army-supplied hydrogen balloon and became the first person to fly over Antarctica. Sub-lieutenant Ernest Shackleton followed him up to become Antarctica's first aerial photographer.

The expedition established its base at Hut Point on Ross Island and began to make some excursions. On a badly managed trip to Cape Crozier, seaman George Vince died when he slipped over a cliff, and Frank Wild displayed Antarctic ingenuity by hammering nails into the soles of his boots to provide extra traction on the ice. The Adélie penguins around the base were a source of endless fascination. Scott noted that Emperor penguins began to head south at the start of winter—the first recorded observation of their remarkable breeding pattern.

On 23 April the sun disappeared below the horizon, not to return until 22 August. The officers and men continued to live on *Discovery* and remained busy throughout the winter. For diversion there was a series of theatrical performances and the monthly *South Polar Times*, edited by Ernest Shackleton. The officers and men messed apart but ate the same food. There were regular inspections and morning prayers were said on the mess deck. Of the inspections, Frank Wild wrote: "Somewhat unnecessary in the circumstances, but as one of the sailors remarked 'Oh well, it pleases him and it doesn't worry us.'"

In early October Charles Royds led a party to Cape Crozier where, on 12 October, Reginald Skelton found Emperor penguins with well-developed chicks—a clear indication that they had hatched during the bitter Antarctic winter.

▲ STARK REMINDER
Looking back from the tip of Hut Point across the Discovery hut to McMurdo Station, the cross in the foreground is a memorial to George Vince who died near this spot on 11 March 1902. On the first sledging expedition, Vince slid off an ice cliff into the sea, and his body was never found.

▲ PLACE OF REFUGE
The interior of the Discovery hut has been restored to look as it did when some of Shackleton's Ross Sea Party relied on it for their survival. The interior of the hut is stained black by smoke from the blubber that was their only fuel for heating, cooking, and lighting.

THE
SOUTH POLAR TIMES
1902 — 1908

THE ARMS OF THE 'DISCOVERY'

◄ WHILING AWAY WINTER
During winter, a monthly journal was produced within the hut. Shackleton was elected editor of the South Polar Times. The magazine contained scientific reports, humor, and fiction, all contributed anonymously. This coat of arms was drawn by Edward Wilson, principal artist for both the magazine and the expedition.

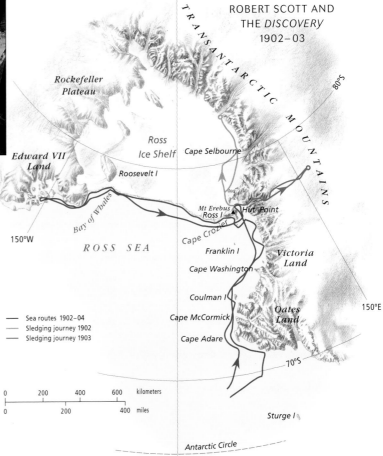

ROBERT SCOTT AND THE DISCOVERY 1902–03

Rockefeller Plateau

TRANSANTARCTIC MOUNTAINS

Ross Ice Shelf

Cape Selbourne

Edward VII Land

Roosevelt I

80°S

Bay of Whales

Mt Erebus
Ross I   Hut Point

Cape Crozier

150°W

ROSS SEA

Franklin I

Victoria Land

Cape Washington

Coulman I

Oates Land

150°E

— Sea routes 1902–04
— Sledging journey 1902
— Sledging journey 1903

Cape McCormick

Cape Adare

70°S

| 0 | 200 | 400 | 600 | kilometers |
| 0 | | 200 | | 400 | miles |

Sturge I

Antarctic Circle

180°

▲ MID-WINTER FESTIVAL

On 23 June 1902 Scott wrote: "The mess-deck was gaily decorated with designs in colored papers and festooned with chains and ropes of the same material, the tables loaded with plum pudding, mince pies, and cakes ... we left the men to enjoy their Christmas fare with an extra tot of grog."

A sledging expedition

The first-ever attempt to sledge to the South Pole left *Discovery* on 2 November 1902. The three-man party comprised Scott himself, Edward Wilson, and Ernest Shackleton, with 19 dogs. Twelve men had left earlier to deposit supply caches for the polar party on their return journey. The groups soon met up and traveled together until 15 November, when the polar party pressed on alone. Scott was confident at first, but progress was slow; the dogs were underfed and over-loaded, and the men mostly had to haul the sledges themselves. Scott and Shackleton began to annoy each other. Years later, the amiable Wilson recalled an event on the ice cap. After breakfast, while he and Shackleton were loading the sledges, Scott shouted: "Come here you bloody fools." When Wilson asked if Scott was addressing him, Scott replied "No." "Then it must have been me," said Shackleton, and when Scott was silent, continued: "Right, you're the worst bloody fool of the lot, and every time you dare to speak to me like that, you'll get it back." It was an odd exchange at the bottom of the world, where each relied on the other for survival.

Scott's attitude to dog sledging reveals what was to be a fatal flaw in his Antarctic philosophy. "In my mind," he wrote, "no journey ever made with dogs can approach the height of that fine conception which is

◄ ANTARCTICA'S BIG BIRD: WANDERING ALBATROSS

Scott wrote: "various devices were resorted to in our endeavors to capture birds for our collection, and sooner or later most of the species were brought on board. The larger albatrosses were caught by towing a small metal triangle, well baited ..."

◄ SUNLIT GLACIER
The sun stays low in the sky as it illuminates the Barne Glacier and the frozen sea of North Bay that lies at its base. The glacier, which descends from the western slopes of Mt Erebus, was discovered and named during Scott's expedition.

▼ INHOSPITABLE SHORE
In *The Voyages of the Morning* Doorly wrote of the "remarkable cone-shaped islet close off [Scott Island]" now named Haggits Pillar. He related how a party "after some difficulty" landed, claimed Scott Island, and left a record of its visit.

realized when a party of men go forth to face hardships, dangers, and difficulties with their own unaided efforts … Surely in this case the conquest is more nobly and splendidly won."

But they did not win. In fact, although they set a new furthest south of 82°16′ on 30 December 1902, they did not travel beyond the Ross Ice Shelf. By then all were suffering from scurvy, and the last few dogs died on the return journey. Shackleton, the largest man, suffered most from scurvy and lack of food, and even had to ride briefly on the sledge at the end of January. By 3 February 1903 they were back at Hut Point.

The second winter

A relief ship, *Morning*, reached Ross Island on 23 January under the command of William Colbeck, who had discovered what is now Scott Island on the voyage down the Ross Sea. However, Colbeck could not sail within 13 kilometers (8 miles) of *Discovery* because the sea ice had not fully melted over summer. It never did melt that summer, so *Discovery* remained set in ice for the coming winter, but at least the expedition had fresh supplies. Scott decided to send back "one or two undesirables" and, with the approval of surgeon Reginald Koettlitz (but not of Albert Armitage, his second-in-command), included Shackleton on the list, claiming that he was not fit enough to continue. The real reason may have been that Shackleton was popular and charismatic while Scott had to rely on rank to keep control. Reportedly Shackleton wept as he sailed away.

There was a series of excursions over the winter of 1903 and into the following summer. Bernacchi took over the editorship of the *South Polar Times*. Armitage had led a sledging trip into the mountains of South Victoria Land on the other side of McMurdo Sound in November 1902, and that group became the first to stand on the polar ice cap. Scott took a party into the same area on 26 October 1903, and journeyed well beyond the point Armitage had reached. When they returned on Christmas Day, after being away for 59 days and covering 1,170 kilometers (725 miles), Scott had formed a strong bond with the two seamen he had traveled with: Edgar Evans and William Lashly. Both were to accompany him on his next voyage.

Homeward bound

Next summer the British government sent *Morning* and *Terra Nova* back to Ross Island to evacuate the party. They arrived on 5 January 1904, with instructions that all the expeditioners were to leave by the end of summer, even if it meant abandoning *Discovery*. Luckily, the ice broke up in February (with the help of dynamite) and the relief ships were able to reach *Discovery* on 14 February—two days later *Discovery* herself was free. The relief vessels immediately sailed north, while Scott took *Discovery* on a surveying voyage along the coast of Victoria Land. They were back in New Zealand by the beginning of April, and arrived off Spithead, near Portsmouth, on 10 September 1904. Scott was promoted to the rank of captain on that day. His two-volume account of the expedition, *The Voyage of the Discovery*, has become a classic of Antarctic literature.   DM

▲ LOW IN THE WATER
The *Morning* was the relief ship for the *Discovery* expedition. Crewman Doorly wrote: "There are few oceans so tempestuous as that globe-encircling expanse to the southward … usually known as the Southern Ocean. Being very heavily laden, the great seas broke continually over the ship, and one night during a gale one of the quarter boats was dragged out of its tackle and swept away."

Erich von Drygalski

German Antarctic Expedition: 1901–03

▲ PIONEER GLACIOLOGIST
Erich Dagobert von Drygalski (1865–1949) was born in Prussia in what is now Kaliningrad. His interest in glaciology began in Greenland and took him to Antarctica and later to Spitsbergen. He spent many years meticulously preparing his expedition research for publication.

At the London International Geographical Congress of 1895, Sir Clements Markham, President of the British Royal Geographical Society, made a passionate plea to geographers and other scientists to take up the challenge of discovery in Antarctica, the most remote and least known region of the planet. His call met with an enthusiastic response, especially in Europe, and in 1898 the German South Polar Commission was formed to mount the first German Antarctic Expedition.

Erich von Drygalski, Professor of Geography and Geophysics at the University of Berlin, was appointed leader of the expedition. He had already led two expeditions to Greenland, and his scientific qualifications were impeccable. Generous state funding enabled him to commission a purpose-built ship, *Gauss*, and recruit a crew of 32, including five scientists. Scientific research and geographic discovery were the primary goals of the expedition, so Drygalski selected an unexplored area of the coast of East Antarctica as his field of operations.

Icebound

Gauss left Kiel in August 1901 and reached Iles Kerguelen five months later, then headed south at the end of January 1902. Ice conditions were difficult, but on 21 February, at about 90°E, Drygalski recorded the appearance of "coherent, uniform white contours … ending abruptly at the water's edge forming cliffs 40 to 50 meters high … the area behind rose gently to about 3,000 meters." Drygalski named this new land for Kaiser Wilhelm II. Attempting to approach more closely, *Gauss* became trapped between two ice ridges, and despite desperate attempts to free the ship, by early March it was clear that there was no escape.

Drygalski had ensured that the ship was provisioned and the crew prepared to winter in the ice. Life on board quickly developed a pleasant routine: "Sundays were beer nights, Wednesdays were lecture nights, Saturday nights were best of all … with a glass of grog, united in games or conversation." There were also card clubs, a smoking club, and a band. But scientific work was not neglected, and the expeditioners maintained continuous meteorological records and wildlife surveys. From mid-March Drygalski organized dog-sledging expeditions to the coast, and about 90 kilometers (56 miles) from the ship he discovered an extinct volcano, which he called Gaussberg. The explorers collected geological samples and made magnetic observations, but winter put an end to further excursions when one sledging party nearly perished in a prolonged snowstorm at the end of April.

A hot air balloon, the first to be used in Antarctica, took Drygalski to a height of nearly 500 meters (2,440 ft), from which, as he reported by telephone to the ship's deck, "the sight was grandiose. I could see the newly discovered Gaussberg … it was the only ice-free landmark in the surrounding area." On the basis of this aerial reconnaissance, Drygalski planned an inland journey to 72°S—much further south than Gaussberg (66°40′S). However, when the longer days and better weather of spring required for this journey arrived, Drygalski decided that his priority was to free the ship from the ice, so the idea was regretfully abandoned.

Making their escape

The crew set to with saws, pickaxes, and explosives to try to dig, cut, and blast their way out toward open water. But it was slow going until one day Drygalski noticed that, where soot from the funnel had stained the ice, melting was quite rapid. So all hands were called to lay a trail of garbage and any other dark, disposable matter to the ice edge. "Within a month we had a long water channel almost two meters deep. Although there were still four to five meters of ice underneath, the channel widened constantly." Even so, their patience was sorely tried; it was not until early February 1903, almost 11 months since the ship was trapped, that they "felt

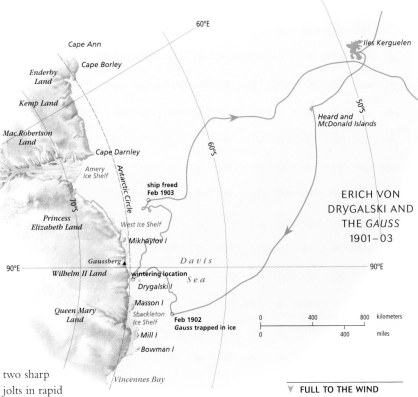

ERICH VON DRYGALSKI AND THE *GAUSS* 1901–03

two sharp jolts in rapid succession … it was like a revelation," as the ice finally broke up. Heading west, the expedition explored the coast for another two months in dangerous ice conditions that finally forced Drygalski to abandon the attempt and head north at the end of March, finally returning to Kiel in November 1903. Although the expedition discovered little new territory, it produced 20 volumes of scientific reports under Drygalski's expert editorship. LC

▼ FULL TO THE WIND
Drygalski took this photograph of *Gauss* under full sail. The ship was purpose-built and the specifications were sent to six shipyards for tender. The *Gauss* was a three-masted barquentine (with an auxiliary engine) of similar design to the *Fram*.

◄ HIS NAME SHALL NOT BE FORGOTTEN
The Drygalski Ice Tongue, 20 kilometers (12 miles) at its widest and 70 kilometers (43 miles) long, extends far off the coast of Victoria Land into the Ross Sea. Drygalski Island, near where *Gauss* wintered over, is another feature named after the renowned German explorer and scientist.

Nils Otto Gustaf Nordenskjöld

Swedish South Polar Expedition: 1901–04

A fictional account of Nordenskjöld's Swedish expedition would probably be dismissed as fanciful and overladen with coincidence. The plan was simple: *Antarctic*, Henryk Bull's old sealing ship, commanded by Carl Anton Larsen, would leave Nordenskjöld and a party of scientists to explore the eastern side of the Antarctic Peninsula for a year while others on the ship conducted scientific work elsewhere; the two parties would be reunited the following summer. The expedition leader, 32-year-old geologist Otto Nordenskjöld was the nephew of Adolf Nordenskjöld, who had discovered the Northeast Passage across the top of Siberia.

Parting company

Antarctic left Gothenburg on 16 October 1901 and arrived in the South Shetland Islands in January 1902. The expedition established that Trinity Peninsula and Danco Land were part of a single peninsula, and the ship sailed between the tip of the Antarctic Peninsula and Joinville Island through a passage that Nordenskjöld named Antarctic Strait (now Antarctic Sound), after his vessel. In February 1902, Nordenskjöld's scientific party of six landed at Snow Hill Island, where they erected a prefabricated hut measuring 4.1 by 6.4 meters (13½ by 21 ft), unaware that it would be their home for the next two years. In early October Nordenskjöld and two others made a dog-sledge journey of 600 kilometers (373 miles) south to the coast of King Oscar II Land. In December, Nordenskjöld discovered the fossil remains of a giant penguin—the largest ever found—on nearby Seymour Island. However, ice conditions remained severe all through the summer, and when the sea froze on 18 February 1903 they knew they were trapped for at least another winter. They killed 400 penguins, 30 seals, and some skuas as supplementary rations, and even throughout their second winter kept detailed meteorological records.

Antarctic wintered in South Georgia, Tierra del Fuego, and the Falkland Islands, and turned south on 5 November 1902 to collect the Snow Hill Island party. After surveying part of the west coast of the Antarctic Peninsula, the ship headed for the east coast but found the way blocked by heavy sea ice. Gunnar Andersson, Toralf Grunden, and Samuel Duse were put ashore at Hope Bay on 29 December to establish a depot and try to reach the scientists at Snow Hill Island over land.

Andersson and Larsen had prepared a contingency plan. Ideally, the ship and land parties would meet at Snow Hill Island, but if the ship arrived first and the land party had not arrived by 25 January, it would be

▲ LEADER FROM ACADEMIA
Otto Nordenskjöld (1869–1928) was a university lecturer in geology and mineralogy at Sweden's Uppsala University. Before going to Antarctica, he had already led a successful expedition to research glacial geology in Patagonia.

▲ MAPPING THE UNKNOWN
This map produced by Nordenskjöld and cartographer Duse shows the comprehensive survey conducted by the expedition. A decade later, Shackleton relied on Nordenskjöld's research when he planned to reach Paulet Island after the *Endurance* sank. Jason Land is now the Larsen Ice Shelf, named after the captain of the *Antarctic*.

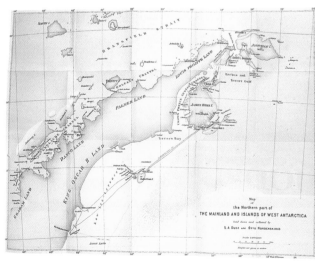

assumed that the overland attempt had failed, and the ship would return to collect them at Hope Bay between 25 February and 10 March. Alternatively, if *Antarctic* had not reached Snow Hill Island by 10 February, the scientists and overlanders would return to Hope Bay and wait for the ship.

Andersson, Grunden, and Duse set off southward. Their only map was an inaccurate chart by James Clark Ross in 1843, so they found a channel crowded with icebergs where they expected to find land. They crossed the ice to Vega Island, but open water prevented them from reaching Snow Hill Island, so they returned to Hope Bay by 13 January and waited for *Antarctic* to arrive.

▲▼ FOSSIL HUNTING BASE

While erecting (left) and living in the completed hut (right), the six men of the Snow Hill Island party little knew that their intended year here would stretch into two. The site was selected for the wealth of local fossils and is still being investigated today; Nordenskjöld's work remained unsurpassed for the next 50 years.

▼ PASSAGE OF DRIFTING ICEBERGS

The sound between Joinville Island and the tip of what is now the Trinity Peninsula was seen by Dumont d'Urville, but it was left to Nordenskjöld's company to sail through and name it Antarctic Sound after their ship. Vessels passing through often find the sound blocked with icebergs pushed out from the Weddell Sea.

▷ ISLAND HAVEN
Determined to retrieve Nordenskjöld and his men, Captain Larsen pushed the *Antarctic* hard into the ice. After the ship was trapped and damaged, constant pumping contained the leaks for several weeks. At last accepting that the ship was sinking, Larsen tried to sail to Paulet Island, and when the *Antarctic* finally sank, the men began a 14-day crossing of ice and water to reach the island.

▷ COOKING WITH KERO
The soot-free kerosene stove was invented in Sweden by Frans W. Lindqvist in 1882. This Primus stove was used by the Swedish South Polar Expedition and was recovered from Snow Hill Island. It is part of a small display of artefacts at Argentina's Antarctic Office on the dock at Ushuaia.

Shipwreck

Meanwhile, *Antarctic* had pushed deep into the ice of Erebus and Terror Gulf, and on 10 January 1903 the ship underwent the first of several violent squeezes from the ice pack. The damaged and sinking vessel (holding a large collection of scientific samples) was abandoned on 12 February, and the 20 men—and the ship's cat—crossed 40 kilometers (25 miles) of rough sea ice and patches of open water to Paulet Island. Arriving there on 28 February, they built a stone hut 7 by 10 meters (23 by 33 ft); its remains are still there. The flat basaltic rocks of the island were ideal building materials, and the hut even had two small windows. They had salvaged a ton of food from the ship's stores, and they killed 1,100 penguins as a winter food supply—they intended to kill more, but the last of Paulet's huge Adélie penguin colony quickly completed their molt and fled. Over the winter there was little to do, so meals and taking temperature readings became highlights of each day. On 7 June, seaman Ole Christian Wennersgaard died of a heart condition, and they had to store his body in a snowdrift until the ground thawed in spring and he could be buried.

The men at Snow Hill Island made several excursions and recorded hourly meteorological observations (which differed significantly from those of the first winter) but the three at Hope Bay had little but survival to occupy them. Realizing that the *Antarctic* would not return, they built a primitive stone hut with a tarpaulin roof, which could be warmed with a blubber stove to a relatively balmy few degrees below freezing. They took turns at cooking, and developed an elaborate ritual of formally thanking the cook after each meal. Andersson wrote of "the strong power that warm and honest friendship has, to proudly subdue the dark might of isolation and of extreme distress." Assuming that by then *Antarctic* was lost and their whereabouts unknown, they started off on 29 September 1903 toward Snow Hill Island and the hope of rescue.

That same day, Nordenskjöld and Ole Jonassen set out on an excursion; they soon had to turn back, but began again on 4 October. By 12 October they had explored the whole west coast of James Ross Island and were traveling along the north coast of Vega Island. Far out on the ice they saw what they first took for penguins—and then realized were men. It was the three from Hope Bay, and the two parties met with "delirious eagerness" at a point they renamed Cape Well-met. They all made their way back to the Snow Hill Island hut to enjoy a celebratory banquet.

Meanwhile, several search and rescue expeditions were being prepared, including one on an Argentinian corvette, *Uruguay*, under the command of Lieutenant Julian Irizar, Argentina's naval attaché to Britain. Arriving at Seymour Island on 7 November 1903, Irizar's crew found a boathook that Nordenskjöld had left on a signal cairn less than two weeks earlier. Investigating further, they spotted the nearby camp of two of Nordenskjöld's men, fortuitously on hand to lead Irizar and one of his officers, Lieutenant Yalour, to the Snow Hill hut, arriving on 8 November. The two

◁ RUINED REFUGE
Today, the remains of the hut on Paulet Island, built by the shipwrecked crew of *Antarctic,* is home to Adélie penguins. It was a remarkable building with double walls for insulation. The main living space was dominated by two wide stone beds along the walls accommodating 10 men in sleeping bags on each side.

◄ UNKEMPT BUT FIT AND WELL

Toralf Grunden, Gunnar Andersson, and Samuel
Duse (left to right), trapped at Hope Bay for nine
months, supplemented their meager supplies
with any penguin, seal, or fish they could catch
and cook. Their unshaven faces, blackened by
blubber and soot, show the state they were in
when reunited with the others at Snow Hill Island.

marooned land parties were saved but it seemed that
Antarctic and her crew were lost.

At about 10.30 that night the dogs started barking
at six men approaching the hut. Miraculously, it was
Captain Larsen and five others who reported that all
but one of *Antarctic*'s crew had survived and were safely
living on Paulet Island. Nordenskjöld wrote: "No pen
can describe the joy of this first moment … when I
saw amongst us these men, on whom I had only a few
minutes before been thinking with feelings of the greatest despondency … We conducted the newcomers in
triumph to the building."

Larsen explained that they had come the long way
round to Snow Hill Island. When the ice broke up in
October, they had rowed to Hope Bay in a whaleboat,
only to find that the three men stranded there had left
for Snow Hill Island five weeks earlier. With a tent pole
for a mast and the tarpaulin from Hope Bay as shelter,
they crossed treacherous Erebus and Terror Gulf in their
tiny craft, but had to walk the last 24 kilometers
(15 miles) to the island across solid sea ice.

Reunion

The presence of the *Uruguay* solved Larsen's last
problem: how to rescue the 14 men he had left on
Paulet Island. This group had been preparing for what
appeared to be another inevitable year on the island,
and had stockpiled 6,000 Adélie penguin eggs. One of
the party, botanist Carl Skottsberg, wrote of the events
of the evening of 11 November 1903—when he heard
a sound in the night, he thought: "I must have been
dreaming. The sound is repeated. It must be so … The
boat is here! … I thump at the sleepers beside me:
can't you hear it is the boat … The shouts are so
deafening that the penguins awake and join in the
cries; the cat, quite out of her wits, runs round and
round the walls of the room; everyone tries to be the
first out of doors."

The Swedish expeditioners were taken back to
Buenos Aires, and from there to Malmö, Sweden. On
the way they called into Hope Bay to pick up the rock
samples and paleobotanical fossils that Andersson had
collected. This was typical of an expedition that had carried out a full scientific program throughout a remarkable, harrowing adventure.   DM

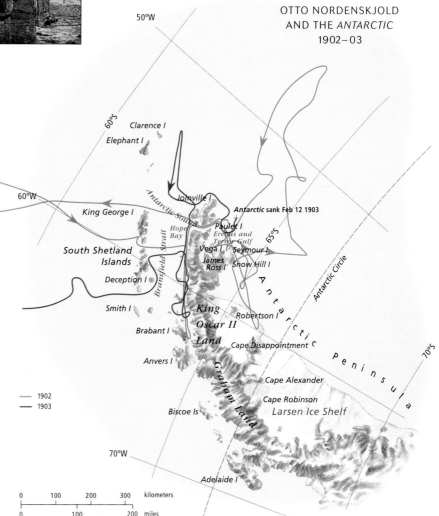

OTTO NORDENSKJOLD
AND THE *ANTARCTIC*
1902–03

50°W

60°S

Clarence I

Elephant I

60°W

Antarctic Sound

Joinville I

Antarctic sank Feb 12 1903

King George I

Hope
Bay

Paulet I

*Erebus and
Terror Gulf*

65°S

South Shetland
Islands

Bransfield Strait

Vega I   Seymour I

James
Ross I   Snow Hill I

Deception I

Antarctic Circle

Smith I

*King
Oscar II
Land*

Robertson I

A

n

t

a

r

c

t

i

c

Brabant I

Cape Disappointment C

P

e

n

i

n

s

u

l

a

Anvers I

Cape Alexander

70°S

Cape Robinson

Graham Land

Larsen Ice Shelf

Biscoe Is

70°W

Adelaide I

— 1902
— 1903

0   100   200   300   kilometers

0   100   200   miles

► HALTED BY WAR

When the *Uruguay* sailed into
Buenos Aires with the Swedish
expedition on board, the harbor was lined with festive craft
and celebrating spectators.
The far-thinking Nordenskjöld
later proposed a Swedish–
British expedition that might
have hastened the establishment of a scientific base in
Antarctica by some 25 years,
but his plan was forestalled by
the outbreak of World War I.

William Spiers Bruce

Scottish National Expedition: 1902–04

William Bruce, the leader of the Scottish National Expedition, first visited Antarctica at the age of 25 on a whaling expedition from Dundee to the Falkland Islands and beyond, in the summer of 1892–93. During this voyage, he and a fellow crewman also carried out oceanographic research, and this experience fired Bruce with the ambition to return with his own scientific expedition.

This ambition led both to his refusal of Scott's invitation to join the *Discovery* expedition as its naturalist, and to his eventual success in persuading the Coats bothers, prosperous Glasgow industrialists, to finance a private Scottish expedition after the British government had refused to help. Bruce invited the commander of his earlier whaling expedition, Thomas Robertson, to be captain of his ship, a Norwegian whaler renamed *Scotia*; and as well as a crew of 25, he recruited seven scientists, including a bagpipe-playing laboratory assistant. This exceptionally large contingent demonstrated that science, not pole-seeking, was the expedition's goal, with the focus on a hydrographic survey of the Weddell Sea and a wildlife survey of the South Orkney islands.

Working through the winter

After a two-month voyage from Scotland, the expedition reached the Falklands in January 1903, and the South Orkneys a month later. Encountering heavy ice, *Scotia* pushed on to the South Sandwich Islands but became beset not much further south, at 70°25′S. She broke free a week later, but further progress south was impossible, and it was only with great difficulty that Robertson worked the ship back toward the South Orkneys and found a winter anchorage in a sheltered bay on the south coast of Laurie Island. *Scotia* was soon frozen in, and Bruce established a shore base, with a stone hut as living quarters and other structures for the scientific work, including magnetic and meteorological observatories. Fieldwork also continued all winter, including detailed surveys of the coastline with soundings taken through holes cut in the sea ice, and botanical studies of the stunted vegetation of the island. In the spring, returning penguins were serenaded on the bagpipes by the laboratory assistant before being killed to provide much needed fresh meat; their skins were preserved by the expedition's zoologist.

It was not until late November 1903 that the *Scotia* was freed from the ice, by the combined efforts of the crew with ice saws and explosives and a timely northeast gale. Bruce left six men ashore, including the biologists to study the penguin colonies, and sailed north to refit the ship in Buenos Aires. He persuaded the Argentine government to continue the meteorological observations at the base on Laurie Island and brought three Argentine scientists back with him to take over, with the assistance of the expedition's meteorologist. The Scottish meteorologist and the cook were left behind when *Scotia* headed south again in late February 1904.

An important discovery

This time the ice was less trouble-some, and the ship reached 72°18′S quite quickly before being forced to a halt. A sounding showed that they were in comparatively shallow water, and on climbing to the crow's nest the captain was delighted to see an ice plateau to the southeast. Breaking out of the ice, the ship moved along a fairly clear lead parallel with this new coast, which Bruce named Coats Land in honor of his patrons, and followed it for 150 miles toward the southwest, reaching as far as 74°S. Bruce deduced that this long coastline must be an extension of Enderby Land, and therefore part of the continent rather than just an

▼ **THROUGH FOG AND ICE**
Laurie Island narrows to an isthmus at the modern Orcadas station where Ormond House was built previously. This view from the shore of Scotia Bay looks westward across the base to Jessie Bay behind, solidly packed with ice.

◄ SCIENTIFIC KNOW-HOW
Science was the motivation for the *Scotia* expedition and the staff on board had expertise in many disciplines. Dr Pirie, the ship's medical officer was also a geologist. Bruce was a keen oceanographer and he had with him a meteorologist, botanist, zoologist, and taxidermist.

➤ MARINE SAMPLE
This impressively large cuttle-fish taken in Scotia Bay at Signy Island was one of many marine invertebrate specimens that *Scotia* brought back from Antarctica. It was a collection that remained unequaled for several decades.

◄ CAPTIVE LISTENER
It may appear that this Emperor penguin admires piping—until one notices the string tying it to Kerr, the laboratory assistant. The expedition doctor noted that "these lethargic, phlegmatic birds" responded with complete indifference to a wide range of bagpiped tunes.

▲ INVESTIGATING THE WEDDELL SEA
Under Thomas Robertson's able command *Scotia* fulfilled an important role in oceanography—effectively supplementing the work of *Challenger* many years before. Measurements taken by *Scotia* showed that deep water in the south of the Weddell Sea was warmer than the water farther north.

island—a discovery as significant as Ross's of McMurdo Sound on the other side of the continent. They were trapped in the ice by a northeast gale and fearful of being frozen in for the winter (as happened to Shackleton's *Endurance* 12 years later in almost the same area), but a southwest gale luckily opened the pack again and *Scotia* fled gratefully northward, eventually reaching port in Northern Ireland in July 1904.

The four scientists and the cook left behind on Laurie Island continued their work throughout the winter, and were relieved the following summer by the Argentine ship *Uruguay*, which had rescued Nordenskjöld's expedition the previous year. Another Argentine party took over for 1905, initiating an unbroken series of expeditions that endures to the present, and makes Laurie Island the oldest continually operating base in Antarctica. LC

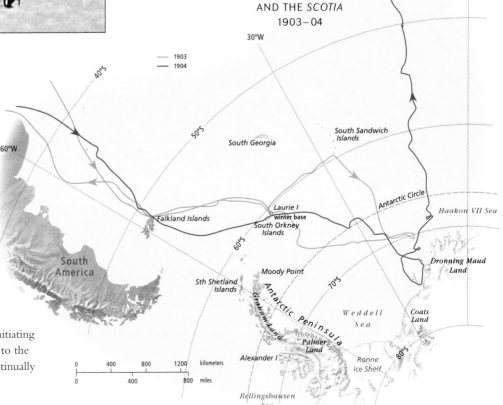

WILLIAM BRUCE
AND THE *SCOTIA*
1903–04

— 1903
— 1904

Jean-Baptiste-Etienne-Auguste Charcot

French Antarctic Expeditions: 1903–05; 1908–10

Polar history is littered with the exploits of people with famous relatives (Nordenskjöld and Ross, to mention only two), but very few had such an illustrious forebear as that of Jean-Baptiste Charcot of France. His father was Jean-Martin Charcot, the founder of modern neurology and an important influence on Sigmund Freud. Young Jean followed his father into medicine but was fascinated by the sea. When he inherited a fortune from his father, he invested much of it in the construction of *Français*, a purpose-built, three-masted polar exploration vessel of 245 tonnes displacement.

▲ ATTRACTED TO THE COLD
Jean-Baptiste Charcot (1867–1936) was born to wealth, but chose to spend much of his working life in harsh polar conditions. He always selflessly pointed out that Charcot Land and other features were named after his famous father, not himself.

The first expedition: *Français*

When news of the disappearance of Otto Nordenskjöld reached Europe, Charcot resolved to rush to his rescue. He wrote to his friend, industrialist Paul Pléneau, asking if he would like to participate in the expedition. Pléneau's famous reply was a telegram stating, with admirable economy of words: "Where you like. When you like. For as long as you like." Pléneau went along as the expedition's photographer. The *Français* expedition attracted strong official and public support, and finally left Le Havre on 27 August 1903 (the first departure, scheduled for 12 days earlier, was aborted when one of the crewmen was accidentally killed). The destination was the western side of the Antarctic Peninsula, where an ambitious program of surveying and science was planned as soon as the men of *Antarctic* had been rescued. However, their aid was not required—in Buenos Aires they met the recently rescued Nordenskjöld expeditioners, and so the French party was free to continue its own explorations. But Charcot was a rather diffident leader, and when the renowned Belgian explorer Adrien de Gerlache and two others decided to leave the ship in Buenos Aires (de Gerlache to return to his new wife), he required a personal reaffirmation of commitment to the expedition from all who remained before he would continue. *Français* left Ushuaia on 24 January 1904, and was in the ice and islands of the South Shetlands by the beginning of February.

Hampered by a temperamental ship's engine, the men of *Français* managed to discover and name the excellent harbor of Port Lockroy, and passed through the Lemaire Channel as far as the Biscoe Islands before retreating to the northern part of what is now Booth Island in the face of advancing winter ice. Here the crew moored the ship in a bay Charcot named Port Charcot after his father, and sat out the winter of 1904. Despite a large library, a supply of musical recordings, and expansive provisioning that extended to daily fresh bread and ample rations of wine and spirits, morale fell dramatically during the enforced inactivity of winter. Exploration continued the following summer, but on 15 January 1905 the ship hit a rock off Alexander Island and began to sink. The crew made a massive effort to stem the leaks and the vessel managed to limp back to Argentina, where Charcot sold it to the Argentinian government for use as a supply ship to its South Orkney Islands station. The expeditioners returned to France to be greeted as polar heroes, and the subsequent publication of 18 volumes of scientific reports amply justified that standing.

Map

60°W
King George I
South Shetland Islands
Deception I
Bransfield Strait
Joinville I
Dundee I
James Ross I
65°S
Robertson I
Brabant I
Anvers I
Booth I — wintering location
Biscoe Is
Graham Land
Cape Alexander
Cape Robinson
Larsen Ice Shelf
Antarctic Peninsula
70°S
70°W
Adelaide I
Marguerite Bay
Antarctic Circle
Palmer Land
Alexander I
Wilkins Sound
Charcot I
Latady I
Ronne Entrance

JEAN-BAPTISTE CHARCOT AND *FRANCAIS* 1904–05

0   100   200   300   kilometers
0   100   200   miles

▶ AVOIDING THE ICEBERGS
The narrow Lemaire Channel is tricky to navigate, even in modern craft. Charcot spoke of "An agonizing moment in which one's whole body contracts, as though trying to make the movement ... [of] the ship ... we glide along past this huge black wall, which fortunately for us, goes ... down to a tremendous depth."

▼ SHIP AGROUND

When *Français* struck a submerged rock near Cape Tuxen, some of the crew tried to stem the leak, while others hand-pumped for 23 hours a day. Even as the ship finally sailed to safety, the crew held to their original goal and completed their mapping of the Palmer Archipelago.

➤ ENDURING FRIENDSHIP

Gentoo penguins inhabit one of Pléneau Island's many rocky promontories at the southern end of the Lemaire Channel. The island was named after Paul Pléneau who first met Charcot only four months before they sailed but soon was a close friend. They stayed friends all their lives.

▲ FAVORITE NAME
The *Pourquoi-Pas?* pictured here was the third in the line. Charcot, at the age of four, painted *Pourquoi-Pas?* on a soapbox and set sail on a nearby lake; he was rescued by his nurse. The next *Pourquoi-Pas?* was a 15-tonne yacht that he bought as a young man.

The second expedition: *Pourquoi-Pas?*

As a young man Charcot had owned a yacht that he called *Pourquoi-Pas?* ("Why Not?"). When, buoyed by his earlier success and quite generous government funding, he started designing another polar vessel soon after his return, he gave it that name. The strong, well-equipped, and comfortable vessel weighed 800 tonnes, and had a new purpose-built 550 horse-power engine to avoid the engine problems that had plagued the first expedition. The ship sailed from Le Havre on 15 August 1908, bound for Antarctica with a complement of 30 officers, scientists, and crew—and Charcot's new wife, Marguerite, who was to leave the ship in Punta Arenas. *Pourquoi-Pas?* had an excellent wine cellar, and luxuries such as enough newspapers to dole out one a day through the winter gloom. Just before Christmas of 1908, she reached Deception Island, where Charcot learned that his chart of the Antarctic Peninsula was already in use by the whalers there in their constant quest for new areas to exploit.

On 1 January 1909, they found a sheltered small anchorage at Petermann Island, and named it Port Circumcision after the church feast day on which it was discovered. Only three days later, things started to go wrong, as first the expedition leader, then the ship, were nearly lost. On 4 January, Charcot (accompanied by Lieutenant E. Godfroy, who specialized in tides and

atmospherics, and E. Gourdon, geologist/glaciologist), took a boat south to Cape Tuxen and the Berthelot Islands. They planned to be away only for the day, so took no extra supplies, shelter, or clothing. It was a mistake that could have cost their lives. Soon after they finished their last food, the ice closed in and it began to snow. The three men were lucky to survive on the Berthelot Islands long enough to be rescued by *Pourquoi-Pas?* three days later. Then, mere hours after the rescue, the unfortunate history of the *Français* was repeated when *Pourquoi-Pas?* ran onto rocks near Cape Tuxen. Eventually, after considerable effort by everyone, they managed to shift enough cargo so that the ship could pull free of the rock at the top of the tide 24 hours later. They then sailed back to Petermann Island for makeshift repairs.

They spent the rest of the summer of 1909 charting south of the Antarctic Circle, sailing further south than Charcot (or Biscoe) had previously ventured. As winter approached, ice forced *Pourquoi-Pas?* back to Port Circumcision, where the crew securely moored the ship and ran steel wires across the mouth of the cove to keep icebergs out. Four huts built ashore for the scientific program were lit by electricity generated by the ship. Despite their precautions, a midwinter storm in June drove several icebergs into their bay, and one of these damaged the rudder of the ship so severely that it could be repaired only with difficulty. At the end of winter the men packed everything back on board the ship and sailed north to Deception Island in late November to pick up fresh provisions. There, a diver told Charcot in confidence that the ship's hull was gravely damaged, and that the vessel would sink if there was another incident. Nevertheless, Charcot resolved to turn south again to conduct his summer program. Beyond Marguerite Bay, which he had named for his wife the previous summer,

▼ WINTERING OVER
Looking north across the winter quarters on Petermann Island to (from left) Pléneau Island, Booth Island, and the Antarctic Peninsula behind, *Pourquoi-Pas?* lies at its winter mooring. It took a month to prepare for winter, and the greatest boon was electric light (even to the huts ashore).

▶ MODERN NAVIGATION
The small niche in the barren rocks of the foreground that is Port Circumcision remains the same today as when Charcot tied up there. However, he would be amazed by the ice-navigating capacities of the two Russian ships seen here between his anchorage and the Antarctic Peninsula.

> **SECURED FOR WINTER**
The *Pourquoi-Pas?* was a
much more comfortable place
to spend the winter months
than the *Français,* but Charcot
wrote that only work and
determination could save
them "from being completely
demoralized by the horrors of
this climate."

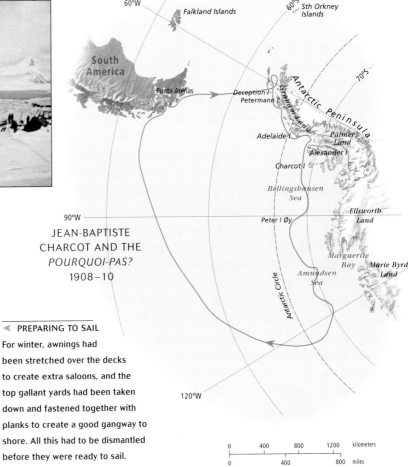

JEAN-BAPTISTE
CHARCOT AND THE
POURQUOI-PAS?
1908–10

< **PREPARING TO SAIL**
For winter, awnings had
been stretched over the decks
to create extra saloons, and the
top gallant yards had been taken
down and fastened together with
planks to create a good gangway to
shore. All this had to be dismantled
before they were ready to sail.

he discovered new land at 70°S, which was later named
Charcot Land (although it is now known to be an
island)—not after the modest explorer, but for his
famous father. Proceeding west along the edge of the
ice, *Pourquoi-Pas?* turned her bow to the north only on
22 January 1910, when she had reached 124°W. When
the crew saw Peter I Island in the distance, they were
the first to do so since Bellingshausen had passed by this
small volcanic landmass in 1821.

A victorious return

When *Pourquoi-Pas?* docked at Punta Arenas, Charcot
and his crew were greeted by enthusiastic local residents
and a flood of congratulatory telegrams. After the ship
had been substantially repaired in Montevideo, she
crossed the Atlantic to arrive in Rouen on 5 June 1910
to a tumultuous reception and an official escort. The fuss
was warranted. The expedition had charted 2,000 kilometers (1,240 miles) of coastline and had given form to
the whole western side of the Antarctic Peninsula, as
well as collecting so much scientific data that it would
take a decade and 28 volumes for it to be published.
Roald Amundsen, who described Charcot as "the
French savant and yachtsman," credited him with
"opening up a large extent of the unknown continent
… and the scientific results were extraordinarily rich …
The point that compels our special admiration in
Charcot's voyages is that he chose one of the most
difficult fields of the Antarctic zone to work in."

Charcot was showered with honors, and his courage
while serving with the British Royal Navy in World
War I was subsequently rewarded with a Distinguished
Service Cross. He never returned to Antarctica, but he
continued his research in the cold waters of the North
Atlantic. On 16 September 1936, *Pourquoi-Pas?* was
wrecked off Iceland; only one crew member survived.
Jean-Baptiste Charcot went down with his ship. DM

▲ **ROCKY MEMORIAL**
On Megalestris Hill behind
Port Circumcision stands a
cairn erected by Charcot's
expedition. Its original plaque
with the names of all 30 men
on the *Pourquoi-Pas?* is now in
Paris and the replica intended
to take its place was still
absent in 2001.

Ernest Henry Shackleton

British Antarctic Expedition: 1907–09

▲ LEADER OF MEN

▲ LEADER OF MEN
Ernest Shackleton (1874–1922), the most charismatic of all Antarctic explorers, once said: "The loyalty of your men is a sacred trust you carry. It is something which can never be betrayed, something you must live up to." He did.

When Ernest Shackleton arrived back in England on 12 June 1903, he found that Scott's 1901–04 expedition, from which he had been virtually sacked, was a controversial subject. Before departing, Scott had been told that the expedition was not to stay a second winter, and *Discovery* being icebound was the result of incompetence. Furthermore, Scott's expedition was broke, so the government would have to pay for the forthcoming relief voyage of *Morning* and *Terra Nova*. Shackleton, by contrast, was home and was a polar hero. He was soon declared fit for Antarctic service, and used his fame to raise his social standing. He also needed money. After a short period as a journalist he was appointed Secretary of the Royal Scottish Geographical Society, stood for Parliament in an election that he lost (along with his party), and set about organizing his own Antarctic expedition.

On 8 February 1907, William Beardmore, Shackleton's main sponsor, promised a large loan, and just three days later Shackleton was assembling his British Antarctic Expedition. However, he soon learned that Scott was planning to sail south again, and regarded not just the

Discovery base but the whole of Ross Island as rightfully his; he insisted Shackleton should stay away. Pressured by their mutual colleague Edward Wilson, Shackleton agreed to this outrageous constraint.

Establishing winter quarters

The expedition left England on 7 August 1907 aboard *Nimrod*. The ship also carried a motorcar: an Arrol-Johnston made in Scotland at a factory owned by Beardmore. Shackleton left later, and met the ship in Lyttelton; *Nimrod* could only make six knots, so catching up was not hard. The 300-ton *Nimrod* was 40 years old and rather battered, but Shackleton could not afford his first choice, *Bjørn*. To save coal, *Nimrod* was towed from New Zealand into the ice by the larger *Koonya*, owned by New Zealand's Union Steamship Company and captained by Frederick Evans. The ships left New Zealand on New Year's Day 1908, and *Koonya* turned north just past the Antarctic Circle on 15 January after two weeks of life-threatening storms. Shackleton noted that *Koonya* was the first steel vessel to cross the Antarctic Circle.

At 9.30 am on 23 January, they sighted the Ross Barrier (now the Ross Ice Shelf) and sailed along it, looking for the bay in the ice that Scott had named

▷ SLIPPERY ASCENT
On one of the few surviving recordings of Shackleton's voice, he recounts the first ascent of Mount Erebus. He took this photograph of the climbers setting out on the expedition's first sledging journey at about 9 am on 5 March 1908. He reported that they had trouble keeping on their feet in the soft snow.

▷ SMOKING MOUNT
The Mount Erebus ascent group included Edgeworth David, Douglas Mawson, and Alistair Mackay, plus a support group of three. The party reached the summit through savage blizzards to achieve the first ascent of Mount Erebus.

◀ LACK OF FORESIGHT
The car garaged at Cape
Royds was eventually returned
to England. The device (right)
is a maize crusher. Shipping a
car but not learning to ski was
typical of Shackleton—even on
the *Endurance* expedition, six
years later, he told Orde-Lees
that he "had no idea how
quickly it was possible for a
man on skis to get about."

Balloon Bight after he and Shackleton went aloft there. They passed the bay where Borchgrevink had landed in 1900, "but it had greatly changed." Then they came to a large harbor that Shackleton named the Bay of Whales because "it was a veritable playground for these monsters" before he fled the heavy ice that was driving into this bay. When Shackleton realized that this was where he had ballooned, but the configuration had changed dramatically because a large section of the ice shelf had broken away, he resolved to make his winter base on land, not ice. When pack ice prevented him from continuing east, he broke his promise to Scott and turned west toward Ross Island. But even there sea ice stopped *Nimrod* from approaching the old Discovery hut at Hut Point, so they erected their prefabricated hut at Cape Royds. *Nimrod* left a wintering party of 15 when she sailed for Lyttelton on 22 February.

Overland excursions

Their first major land excursion was the first ascent of Mount Erebus. Edgeworth David (the leader), Douglas Mawson, Alistair Mackay, and a support party of three left on 5 March, and struggled through savage blizzards to the summit (3,795 meters/12,450 ft) on 10 March; they were back two days later.

On 29 October, Shackleton, Frank Wild, Jameson Boyd Adams, and Eric Marshall set out for the South Geographic Pole. Shackleton planned to walk and use ponies, even though the advantages of skis and dogs for polar travel were by now widely recognized. The car, Antarctica's first, was virtually useless. By 26 November, Shackleton could again claim a record for the furthest south—further than he had penetrated with Scott—but he was worried that they were already rationing food.

▼ SCALING A VOLCANO
All six men in the climbing
party pushed on for the
summit, but Brocklehurst
suffered frostbite not far
from the top. Here, at the
summit, the men peer some
275 meters (900 ft) down into
the active crater—about 1 kilometer ($\frac{1}{2}$ mile) wide—on one
of the rare occasions when
the steam cloud was swept
aside by the wind.

◀ PONY POWER

Nansen, the greatest polar explorer of his time, personally advised Shackleton against the suitability of horses for drawing sledges in Antarctica. Even so, through the Hong Kong and Shanghai Bank, Shackleton ordered a dozen Manchurian ponies for his expedition.

▶ HOSTILE TERRAIN

The Taylor Glacier in the Dry Valleys was named by Douglas Mawson after geologist Griffith Taylor. While the long trudges across the Ice Shelf and over the polar plateau were feats of endurance, climbing the glaciers from one to the other was the biggest battle.

▼ RAISING THE FLAG FOR BRITAIN

Queen Alexandra gave the expedition a Union Jack, which the polar party erected at their furthest south—less than 160 kilometers (100 miles) from the Pole. Shackleton wrote: "While the Union Jack blew out stiffly in the icy gale that cut us to the bone, we looked south … but could see nothing but the dead white snow plain."

ERNEST
SHACKLETON
AND THE *NIMROD*
1907–09

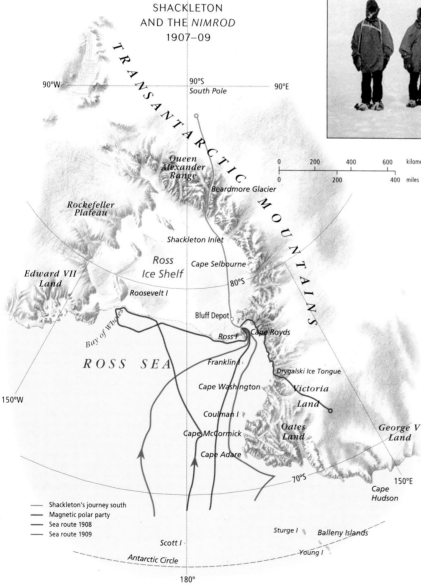

In early December they became the first men to reach the southern extent of the Ross Ice Shelf and to climb the vast Beardmore Glacier onto the polar ice cap. The climb was a nightmare, and on 7 December the last pony fell into a crevasse and died. Just after Christmas, the slope became easier and they were on the ice cap, but they had only one month's food to reach the Pole and return to their supply dump—a journey of almost 800 kilometers (500 miles). They were all suffering from altitude sickness, and Shackleton described the symptoms: "My head is very bad … as though the nerves were being twisted up with a corkscrew and then pulled out." He was torn between his ambition to reach the Pole and his sense of responsibility for his men. By 2 January he was writing: "I must … consider the lives of those who are with me. I feel that if we go on too far it will be impossible to get back … man can only do his best … time is going on and food is going also." On 9 January they left everything behind and made a one-day push to the south—to come within 160 kilometers (100 miles) of the Pole—and back. At 88°23′S they hoisted the Union Jack, took some photographs, and left a commemorative brass cylinder. Looking southward, Shackleton noted: "We feel sure that the goal we have failed to reach lies on this plain … we stayed only a few minutes …

Whatever regrets may be, we have done our best." By the time they returned to their camp that day, they had traveled 66 kilometers (41 miles).

On the return journey, they hoisted a sail on their sledge and quickly reached the top of the Beardmore Glacier, but descending to their food dump while weak and short of food was extremely draining: "I cannot describe adequately the mental and physical strain of the last 48 hours," wrote Shackleton of 26 and 27 January.

A narrow escape

They were near death, yet on 31 January Shackleton surreptitiously gave Frank Wild his breakfast biscuit. Wild wrote: "I do not suppose that anyone else in the world can thoroughly realize how much generosity and sympathy was shown by this; I do by God I shall never forget it." They all had acute dysentery and were "appallingly hungry," and by 21 February Shackleton was writing: "Our need is extreme and we must keep going … food lies ahead and death stalks us from behind." Two days later they arrived at Bluff Depot and food supplies, but Marshall was so ill that Shackleton and Wild left him with Adams and pressed on to arrange rescue. They found the building at Hut Point empty, but there was a note saying that the ship was nearby, so they set fire to a hut to attract attention. It worked; by

1 March both were on board the *Nimrod*. After "a good feed of bacon and fried bread," the indomitable Shackleton set off with three companions on the 18½ hour trek to rescue Adams and Marshall. Shackleton wrote: "We were all safe on board at 1 am on March 4." The polar party had walked 2,740 kilometers (1,700 miles) in 128 days.

Looking for the South Magnetic Pole

Meanwhile, the others had also had some harrowing experiences. Their most ambitious expedition was the Northern Sledging Party to the South Magnetic Pole, with a party of three: Edgeworth David, Douglas Mawson, and Alistair Mackay. David and Mawson were Australian geologists, and Mackay was a naval surgeon. David, the leader, turned 51 during the journey, Mawson was 26, and Mackay 30. They tried using the car to transport supplies across the sea ice, but soon abandoned it in favor of man-hauling—a challenging task, as one of the two sledges weighed 275 kilograms (606 lb) and they had some 1,600 kilometers (1,000 miles) to cover.

▲ FAILED HORSEPOWER
The motor car was adequate on hard ice but failed in the least depth of snow. On 22 September the car hauled sledges for eight miles on a depot-laying exercise before snow forced the men to take up the yokes themselves.

▲ MAGNIFICENT STAMINA
Tannatt William Edgeworth
David (1858–1934) turned 51
during the *Nimrod* expedition.
Nevertheless, he climbed to
the summit of Mount Erebus
and completed the difficult
2,028-kilometer (1,260-mile)
walk to the South Magnetic
Pole. He was knighted in 1920.

▼ SECOND PROUD MOMENT
The Magnetic Pole party (from
left: Mackay, David, Mawson)
claimed the area at 3.30 pm
on 16 January 1909. Mackay
and David set up the Union
Jack while Mawson positioned
the camera. David recalled
that he pulled the string to
trigger the camera: "Then we
gave three cheers for His
Majesty the King."

They left Cape Royds on 5 October, while supply depots were still being set up for the geographic polar bid. Shackleton was to cross the ice shelf while the sledging party walked northward on the sea ice along the edge of Victoria Land (which they claimed for Britain on 17 October), with Mawson as pathfinder. Heavy loads and soft snow slowed them down, so they cached some of their rations and relied on seal and penguin meat, cooked on an improvised stove. They could not abandon any sleeping gear; they already shared a three-man sleeping bag and kept each other awake. Their intended route up the Drygalski Ice Tongue proved impractical. David recalled that, by 20 December, "We had not yet climbed more than 100 feet [36 meters] or so above sea level … We knew that we had to travel at least 480 to 500 miles [770 to 800 kilometers] to the Magnetic Pole and back to our depot, and there remained only six weeks to accomplish this journey." They pushed on with a single sledge weighing 305 kilograms (670 lb), and climbed a route they named Backstairs Passage.

By 11 January 1909 they were on the plateau and more than 3,800 meters (7,000 ft) above sea level. Their compass was only 15 degrees off vertical, but that did not mean that they had only 15 nautical miles (equivalent to about 28 kilometers) more to travel, as they had to allow for the daily movement of the Magnetic Pole, and might even have to chase it. According to David, "Mawson considered that we were now practically at the Magnetic Pole, and that if we were to wait for twenty-four hours taking constant observations at this spot the Pole would, probably … come vertically beneath us." But rather than wait for the Pole to come to them, they decided to go to its approximate mean position. Like Shackleton, a week earlier and much further to the geographic south, they left everything and made a one-day rush for their goal on 16 January. They hoisted the Union Jack, claimed the area for the British Empire, gave three cheers for the King, and took a photograph. Then, "with a fervent 'Thank God' we all did a right about turn, and as quick a march as tired limbs would allow back in the direction of our little green tent in the wilderness of the snow."

Several years later, calculations from Mawson's observations showed that the party had been close to the area of polar oscillation but had not penetrated it. When Mawson learned of this in 1925, he amended his entry in *Who's Who in Australia* from "one of the discoverers of the South Magnetic Pole" to "Magnetic Pole journey 1908."

A fortunate rescue

To return to meet *Nimrod* at their depot on the Drygalski Ice Tongue, they had to average 27 kilometers (17 miles) for 15 days. They had full rations, but were short of tea and took to recycling tea bags that they had discarded on the outward journey and salvaged as they went back. They reached the depot on the morning of 3 February; only a few hours later they heard a rocket, and rushed from their tent to see *Nimrod* very close. A distinguished career was almost snuffed out when Mawson fell into a crevasse and had to be rescued, but then a journey of 2,030 kilometers (1,260 miles) over 122 days ended luxuriously with their first baths in four months.

Only later did they learn how lucky they were to be found. *Nimrod* had a daunting 320 kilometers (200 miles) of coast to search for their three small figures. John King Davis, First Officer, was on duty from 4 am to 8 am on 3 February. When Captain Evans came on deck before breakfast and confided that he thought that there was little chance of finding the land party, he asked Davis if his watch had examined the entire coast. Davis admitted that a small section had been obscured by icebergs. After balancing the extra fuel expenditure against the likelihood of the men being in just that location, they decided to steam back for four hours. Sailing into a narrow fault in the ice behind the obstructing icebergs, they saw that "upon the crest of a little knoll of ice was a green conical tent." When *Nimrod* had first passed this spot the expeditioners had been further up the glacier. If Davis had not missed it and decided to return, the Magnetic Pole party would have been left to make their own way back to Cape Royds. Much of the sea ice had broken up, and their chances of surviving would have been slight.

With David, Mackay, and Mawson on board, *Nimrod* returned to Cape Royds, and then to Hut Point where they found Shackleton and his party. On the way back to New Zealand, Shackleton sailed along the coast beyond Cape Adare to map the coast of Adélie Land as far west as possible, and reached 166°14′E—beyond any previous effort. They reached Lyttelton on 25 March and were back in England in June.

Shackleton was knighted for his achievements, but he thought he had failed, although he had paved the way to the Pole. Roald Amundsen was later to write of Shackleton: "Seldom has a man enjoyed a greater triumph; seldom has a man deserved it better." More specifically, he said: "I admire in the highest degree what [Shackleton] and his companions achieved with the equipment they had. Bravery, determination, strength they did not lack. A little more experience … would have crowned their success." Amundsen also wrote what could well be the defining judgment of Shackleton: "Sir Ernest Shackleton's name will for evermore be engraved with letters of fire in the history of Antarctic exploration. Courage and willpower can make miracles. I know of no better example than what that man has accomplished."   DM

➤ LIONIZED BY THE PRESS

After the *Nimrod* expedition, the *Daily Telegraph* gushed: "In his photographs Mr. Shackleton is nothing more than an intelligent but ordinary naval officer. In reality he radiates the fascination of an indefinable force ... If he has the face of a fighter, he has the look of a poet; one must be both a fighter and poet to accomplish what he has done."

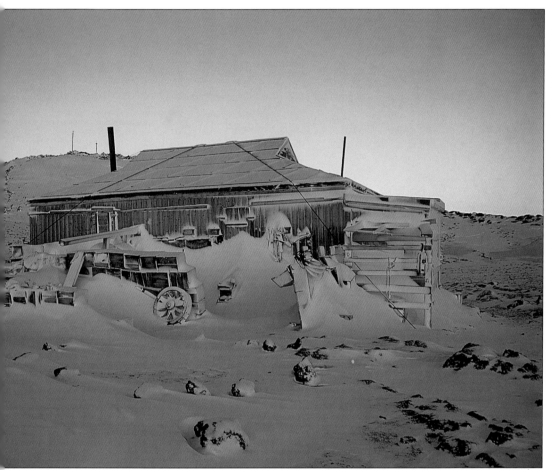

▲ LOOKING TO THE FUTURE

When Shackleton left Cape Royds in 1909, he noted: "I left ... stores sufficient to last fifteen men one year. The vicissitudes of life in the Antarctic are such that such a supply might prove of the greatest value to some future expedition."

▼ FOOTWEAR FOR POLAR CONDITIONS

Shackleton brought both ski boots and finnesko with him. Finnesko are boots of reindeer fur filled with sennegrass that both adds insulation and absorbs moisture so feet stay warm and dry.

▲ STANDING STILL

The hut at Cape Royds has changed little. A wheel for the motor car still lies near the collapsed garage. However, there has been considerable restoration— visitors in 1947 found boards off the roof of the hut and the garage had already collapsed. The roof was repaired and windows were re-glazed on later visits.

➤ TIME CAPSULE

The hut was restored by a New Zealand team in 1961 to look as it did in 1909. Shackleton wrote that the cooking range was designed "to burn anthracite coal continuously day and night and to heat a large superficial area of outer plate, so that there might be plenty of warmth given off in the hut."

Roald Engelbreth Gravning Amundsen

The Norwegian bid for the South Pole: 1909–11

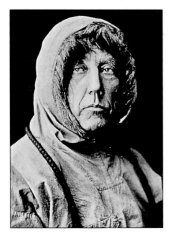

▲ HIGH ACHIEVER
It seems likely that Roald Amundsen (1872–1928) was not only the first to walk to the South Pole, but may also have reached the North Pole first (in an airship), as well as being the first to sail through the Northwest Passage.

▼ CUSTOM DESIGNED
Fram (meaning "Forward") was built with a rounded hull so that the pressure of ice would cause her to rise, rather than be crushed. The ship was launched in 1892, and had already performed well in the Arctic before Nansen lent her to Amundsen.

Amundsen has been described as a consummate professional in an age of amateurs. The story of his journey to the South Pole certainly reveals organizational skills that his competitors lacked, but there were also controversial elements, even errors, in the Norwegian polar bid.

Amundsen's chronicle of his expedition reveals a warmhearted, generous personality, far from the cunning manipulator that Scott's supporters painted him at the time. He rejected the title of adventurer: "For him [the explorer], adventure is only an unwanted interruption … He does not seek titillation, rather previously unknown facts … an adventure is simply an error in his calculations … Or is it a very unhappy proof that no one can take into consideration all eventualities? … Any explorer experiences adventures. They excite him and he looks back on them with pleasure, but he never seeks them out."

A change of direction

Amundsen had been the first to conquer the Northwest Passage, and had spent a winter in Antarctica with Adrien de Gerlache on *Belgica*. He was a meticulous planner, but in 1909, when he was aiming to be first to the North Pole, and had borrowed Fridtjof Nansen's *Fram* for the purpose, events forestalled him: in September of that year, both Robert Peary and Frederick Cook claimed the North Pole. Amundsen recalled: "Just as rapidly as the message had traveled over the cables I decided on my change of front—to turn to the right-about, and face to the South." With a secrecy that later attracted criticism, he told only his brother, and later Lieutenant Thorvald Nilsen, captain of *Fram*. The crew found out in Madeira; until then, they believed they were sailing around Cape Horn and up to the Bering

Strait to take advantage of the Arctic drift. They all agreed to stay for the new destination.

Fram left Christiania (now Oslo) on 7 June 1910, and sailed from Madeira on 9 September, with 19 men and 97 dogs—the number of dogs had risen to 116 by January 1911. Amundsen telegraphed Scott, stating: "Beg leave to inform you *Fram* proceeding Antarctic. Amundsen," and notified his sponsors and Nansen in the same way. Secrecy was not Amundsen's only odd trait: he also would not take a doctor on his expeditions, largely because he thought an expedition doctor, being well educated, could encourage dissent.

Having read reports of previous expeditions, Amundsen decided to establish his base on the ice at the Bay of Whales because it was closer to the Pole than Ross Island. *Fram* arrived there on 14 January, and the crew erected their prefabricated hut at "Framheim." They killed seals and penguins for food, and Amundsen wrote: "An Emperor penguin just came on a visit—soup-kettle." On 4 February, Scott's ship *Terra Nova* visited for a day, but Scott had already been left on Ross Island. Amundsen's men established supply depots as far as 82°S, and *Fram* departed northward before the Bay of Whales froze over.

Surprisingly, on an expedition so well planned, there was one key blunder: they had the Almanac for 1911, but had forgotten the Almanac for 1912, which contained essential information for fixing the position of the Pole. If Amundsen were to claim the Pole, he had to be there before the end of 1911.

◄ FOUR-LEGGED FRIENDS
On the voyage south, Rønne the sailmaker danced with one of the dogs—"in the absence of lady partners"—to a tune on the fiddle. Amundsen also reported that kenneling was adjusted to house dogs that were friendly with one another side by side, and they would be in the same teams.

When the sun appeared on 24 August after four months of darkness, the sledges were already packed for the polar run. Indeed, Amundsen set off for the South Pole in September—very early in spring—but had to return because conditions were too cold for survival. The dogs, after six months of leisure, did not "understand that a new era of toil had begun." On 19 October five men started for the Pole on skis and sledges, and by 7 November they were far enough across the Ross Ice Shelf to see mountains to the south. At the base of the climb to the polar ice cap, Amundsen outlined his plan: "The distance … from this spot to the Pole and back, was 683 miles [1,100 km]… we decided to take provisions and equipment for sixty days on the sledges, and to

▲ ONE-SHACK TOWN

"Framheim" in February 1911 was an orderly place. *Fram* was about to depart and the first of the depot laying parties was ready to set out. Some 900 cases had been unloaded and the hut was built. Its walls were tarred and the roof covered with tarred paper.

▼ STRATEGIC POSITIONING

Much of Amundsen's success came from his choice of sites. The Bay of Whales, seen here in early morning, was one degree closer to the Pole than McMurdo Sound. Shackleton had turned away from there fearing that the Barrier ice could calve and leave his whole camp adrift.

leave the remaining supplies … and outfit in depot … We now had forty-two dogs. Our plan was to take all forty-two to the plateau; there twenty-four of them were to be slaughtered, and the journey continued with three sledges and eighteen dogs. Of the last eighteen, it would be necessary, in our opinion, to slaughter six in order to bring the other twelve back to this point. As the number of dogs grew less, the sledges would become lighter and lighter, and when the time came for reducing their number to twelve, we should only have two sledges left."

His calculations were almost perfect: they took eight days less than the time allowed, and returned with 12 dogs. They climbed the Axel Heiberg Glacier from 17 to 21 November, and with some sadness, because a "trusty servant lost his life each time;" they shot the dogs at a camp they called the Butcher's Shop at the top of the glacier.

On 8 December they passed Shackleton's 88°23′S. It was a milestone for the 39-year-old Amundsen, who wrote: "No other moment of the whole trip affected me like this. The tears forced their way to my eyes …" They laid down their final depot here, and made a push for the South Pole.

▲ THE ART OF SKIING

Skis were heavy and a single pole was used. Amundsen had been skiing since a schoolboy in Norway; his first long trip was an 80-kilometer (50-mile) journey that took 20 hours.

▶ EXERCISING THE DOGS

This image of the Norwegian camp at 85°S first appeared in the *London Illustrated News*. Amundsen commented: "We had thought that a day's rest would be needed by the dogs for every degree of lati-tude but … instead of losing strength [they be-came] … more active every day."

Arriving at the South Pole

Of the events of 14 December 1911, Amundsen wrote:

"At three in the afternoon a simultaneous "Halt!" rang out from the drivers. They had carefully examined their sledge-meters and they all showed the full distance—our Pole by reckoning. The goal was reached, our journey ended. … I had better be honest and admit straight out that I have never known any man to be placed in such a diametrically opposed position to the goal of his desires as I was at that moment. The regions around the North Pole—well, yes, the North Pole itself—had attracted me from childhood, and here I was at the South Pole. Can anything more topsy-turvy be imagined?

"… Pride and affection shone in the five pairs of eyes that gazed upon the flag, as it unfurled itself with a sharp crack, and waved over the Pole. I had determined that the act of planting it—the historic event—should be equally divided among us all. It was not for one man to do this; it was for *all* who had staked their lives in the struggle, and held together through thick and thin.

"… Everyday life began again at once … Of course, there was a festivity in the tent that evening—not that champagne corks were popping and wine flowing—no, we contented ourselves with a little piece of seal meat each, and it tasted well and did us good."

With the exactness that characterized the whole expedition, they spent the next few days surveying around their camp until they were certain that they had reached the Pole by any calculations. On 17 December, at the point they concluded was the Pole, they erected a tent with a Norwegian flag on top and "inside the tent, in a little bag, I left a letter, addressed to HM the King, giving information of what we had accomplished. The way home was a long one, and so many things might happen …" They also left some clothes and a sextant, before lacing the tent and turning to the north.

The secret of success

By 6 January they were back on the ice shelf, and at 4 am on 25 January they reached Framheim, after a 99-day journey. They were probably fitter and healthier than when they left. The science of nutrition was in its infancy, but at Framheim the Norwegians ate a healthy mixture of wholemeal bread, berry preserves, and undercooked seal meat. Whereas Shackleton's party barely struggled back to rescue and Scott's party did not come back at all, because their rations resulted in starvation as well as nutritional deficiencies, Amundsen's

90°W   90°S South Pole 14 Dec 1911   90°E

TRANSANTARCTIC MOUNTAINS

ROALD AMUNDSEN
REACHES THE
SOUTH POLE 1911

Axel
Heiberg
Glacier

*Rockefeller
Plateau*

Ross
Ice Shelf

Cape Selbourne
80°S

*Edward VII
Land*

Roosevelt I

Bay of Whales

Mt Erebus
Ross I

150°E

150°W

ROSS   SEA

| 0 | 200 | 400 | 600 | kilometers |

| 0 | 200 | 400 | miles |

180°

men ate well, and had so much food on hand that he could write: "We are bringing the purveyors of our sledging samples of their goods that have made the journey to the South Pole and back in gratitude for the kind assistance they afforded us." Shackleton sacrificed a biscuit; Amundsen brought back food as souvenirs.

Five days after they arrived back at their coastal base, the men were sailing to Hobart on *Fram*. On 7 March 1912, Amundsen telegraphed news of his success to his brother, who announced it to the world. When Amundsen heard that Mawson needed dogs for his Australasian Antarctic Expedition; he donated 21, keeping only puppies and survivors from the conquest of the South Pole. DM

◀ **TRAVELER'S TALE**
Roald Amundsen's *The South Pole* reveals a charming man of self-deprecating humor.

▲ **THE SOUTH POLE**
After the "glacier's last farewell," Amundsen reported that "The surface at once became fine and even, with a splendid covering of snow everywhere, and we went rapidly on our way to the south with a feeling of security and safety."

▶ **MISSION ACCOMPLISHED**
Oscar Wisting, a naval gunner, seen here at the pole with his dog team, was not a volunteer for the expedition. He was highly recommended, and so Amundsen invited him. To ensure his polar claim could not be disputed, Amundsen surveyed the area thoroughly. Later calculations showed his survey was very accurate.

Robert Falcon Scott

The last voyage: 1910–12

▲ AUTHOR OF MERIT
Robert Falcon Scott
(1868–1912) entered the navy at the age of thirteen. He died on the ice aged forty-three. His legacy to the world was his writing, in which he captured the spirit of the Heroic Age of Antarctic exploration.

Scott arrived in Melbourne on 12 October 1910 on his way to Antarctica to unwelcome news: Roald Amundsen was also bound for the South Pole. The race had begun. As well as fretting about the unexpected personal competition, Scott was annoyed that the Norwegians were aiming for the Pole that he considered his by right. Science was important to his second polar expedition, but so was claiming the Pole for Britain. However, from the start Scott doubted that he would win the race; on 11 October, he wrote: "I don't know what to think of Amundsen's chances. If he gets to the Pole, it must be before we do, as he is bound to travel fast with dogs and pretty certain to start early."

Scott had announced his second Antarctic expedition in September 1909, but money was slow to come in, despite widespread endorsement. The *Terra Nova*, which had come to his support on the previous expedition, was to be the expedition ship. Edward "Teddy" Evans was to be his second-in-command. Lawrence "Titus" Oates and Apsley Cherry-Garrard each contributed a thousand pounds, and were recruited as an officer and assistant zoologist, respectively. Oates wrote to his mother: "Points in favor of going. It will help me professionally as in the army if they want a man to wash labels off bottles they would sooner employ a man who has been to the North Pole [*sic*] than one who has only got as far as the Mile End Road. The job is most suitable to my tastes. Scott is almost certain to get to the Pole and it is something to say you were with the first party. The climate is very healthy although inclined to be cold."

On the way to the Pole

Terra Nova left Britain at the beginning of June 1910, and Scott joined the ship in South Africa. The expedition photographer, Herbert Ponting, boarded in Lyttelton, New Zealand, with 19 Siberian ponies and 33 sledge dogs (and one collie bitch) shipped from Russia. After three weeks in the ice, the ship was off Ross Island on 4 January 1911. Hut Point and Cape Crozier were frozen in, so they erected their hut at Cape Evans. It had a wardroom for the 16 officers and scientists and a separate mess deck for the nine crewmen.

Before winter, a western party led by Griffith Taylor went to the Dry Valleys, and an eastern party sailed *Terra Nova* toward King Edward VII Land, and soon returned to tell of meeting Amundsen and the *Fram*, so the land party it carried had to move to Cape Adare. Scott was surprised to hear that the Norwegians were so close; he had assumed that they would start from the Weddell Sea side of the continent. He wrote: "There is no doubt that Amundsen's plan is a very serious threat to ours. He has a shorter distance to the Pole by 60 miles [97 kilometers]—I never thought he could have got so many dogs safely to the ice. His plan for running them seems excellent. But above and beyond all he can start his journey early in the season—an impossible condition with ponies."

Scott knew the limitations of his ponies. He had led the southern party intending to set up a supply depot at 80°S, but while the 26 dogs performed well, the eight ponies kept sinking in the soft snow. Even when the party switched to marching at night, when

◀ FAMILIAR VESSEL
Terra Nova was the 700-tonne Dundee whaler that came to Scott's rescue on the *Discovery* expedition. Cherry-Garrard reported that "there was not a square inch of the hold and between-decks which was not crammed almost to bursting ... Officers and men could hardly move in their living quarters."

▶ DOOMED TO DIE
Both "Titus" Oates and the ponies were setting out to their deaths. Oates, an expert horseman, had no part in selecting the Manchurian ponies. They had been bought in Harbin and shipped from Vladivostok to New Zealand where Oates first saw them and was not impressed with the quality of animals purchased.

⋏ A HOME AT CAPE EVANS

Beyond the Cape Evans hut stands the Greenpeace Antarctic base (now removed) and beyond that, Mount Erebus. The hut's interior appearance now owes as much to Shackleton's later Ross Sea Party, which was marooned here and lived off Scott's excess rations, as it does to the Scott expedition itself.

▷ APPRECIATING THE RETURN OF DAYLIGHT

One of the expeditioners looks across the hut and the stores to Mount Erebus. The return of the sun in spring was a major event. Scott wrote: "It changes the outlook on life of every individual, foul weather is robbed of its terrors; if it is stormy today it will be fine tomorrow or the next day, and each day's delay will mean a brighter outlook when the sky is clear."

◁ SCOTT'S SANCTUM AT CAPE EVANS

Scott, often uncomfortable in his dealings with people, was a skilled writer with a versatile mind. In the hut he appeared to be happiest reading and writing at his desk in his den, or presiding over academic discussions on a wide range of topics at the large oak table.

◄ CROWDED HOUSE
The hut had a big oak table for the officers and a smaller table for the men. The larger one (bearing a guest book) is still in the hut today. Herbert Ponting's darkroom is at the end of the room and the Tenements are on the left.

▽ NEGATIVE SPACE
Ponting's darkroom, at the hut's rear, today resembles a historical display of the early days of photography. Besides a tripod, there are chemicals and developing tanks and scientific equipment used in various experiments.

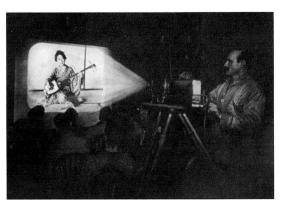

▲ CHEEK BY JOWL
The men lived in very close proximity in the Tenements. From left to right: Cherry-Garrard, Bowers, Oates, Meares, and Atkinson. Scott noted on 17 January 1911: "I saw Bowers making cubicles as I had arranged, but I soon saw these would not fit in, so instructed him to build a bulk-head of cases which shuts off the officer's space from the men's, I am quite sure to the satisfaction of both."

▲ PICTURE SHOW
On Monday 29 May 1911 Herbert Ponting gave a lecture on Japan using his own collection of illustrations. Scott noted: "He is happiest in his descriptions of the artistic side of the people, with which he is in fullest sympathy."

the snow was firmer, progress was slow. Finally, One Ton Camp was established 58 kilometers (equivalent to about 31½ nautical miles) short of their goal—a short-fall that proved fatal.

Scott insisted on a regular routine through the dark winter months. The most remarkable exploit was the bleak excursion recounted in Apsley Cherry-Garrard's *The Worst Journey in the World*. With Edward "Bill" Wilson and "Birdie" Bowers, he set out in the dark and bitter cold for Cape Crozier, 100 kilometers (62 miles) away to collect Emperor penguin eggs because Wilson believed they might provide a clue to the evolution of reptiles into birds. They returned five weeks later, with just three eggs, and ravaged faces. (The eggs were analyzed years later, with inconclusive results.) Cherry-Garrard's mas-terpiece of polar literature begins: "Polar exploration is at once the cleanest and most isolated way of having a bad time which has been devised," and ends: "if you have the desire for knowledge go out and explore … If you march your Winter Journeys you will have your reward, as long as all you want is a penguin's egg."

The assault on the Pole
On 24 October 1911 the motor sledges set off to lay depots, and on 1 November the main polar party, led by

Scott, left Cape Evans. The two groups comprised 16 men, 10 ponies, 33 dogs, and 15 sledges, including the two motorized ones. By 6 November the polar party had passed the abandoned motor sledges—their crew had continued south but were now man-hauling. On 21 November the two groups met, and on 9 December they reached the bottom of the Beardmore Glacier: Scott was literally following in Shackleton's footsteps. Here they shot the surviving ponies, and men and dogs carried on up the glacier. It was hard work, especially as they could not ski, and even Scott admitted that "it would be impossible to drag sledges on foot." On 11 December the dogs were sent back and 12 men continued, hauling three heavily laden sledges. On 14 December Amundsen reached the South Pole while Scott was still on the Beardmore Glacier, writing: "We got bogged again and again, and, do what we would, the sledge dragged like lead … Considering all things, we are getting better on

◄ EXPEDITION METEOROLOGIST
Scott described Dr George Simpson's well set up work area as "A mere glimpse of the intricate arrangements of a first-class meteorological station—the one and only station of that order that has been established in the polar regions."

ROBERT SCOTT:
THE LAST VOYAGE
1911–12

skis." By 21 December Amundsen was on his way home and Scott's party was at Upper Glacier Depot. One of the supporting sledge teams turned back here, and eight men continued with two sledges.

On 3 January Scott made a fatal decision. He told his last support team to turn back the next day, but that one of them, "Birdie" Bowers, would come with him to the Pole. The sledge was already fully laden with supplies for four and Bowers had no skis, so the team would be limited to his pace. Even so, on Thursday 4 January Tom Crean, "Teddy" Evans, and William Lashly headed back, while Scott, Wilson, Oates, Bowers, and Petty Officer Edgar Evans carried on—never to be seen alive again.

At first things went well. On 9 January they passed Shackleton's furthest south (88°23'S), and Scott wrote: "All is new ahead." But by 11 January he was expressing doubts: "We ought to do the trick, but oh! for a better surface." On 16 January they saw a cairn of Amundsen's; Scott's frustration is clear: "The worst has happened … sledge tracks and ski tracks going and coming and the clear trace of dog's paws … The Norwegians have forestalled us and are first at the Pole … All the day dreams must go; it will be a wearisome return. Certainly we are descending in altitude—certainly also the Norwegians found an easy way up."

They reached what they thought was the Pole the next day, and found the Norwegians' tent the day after. Scott wrote despairingly: "The Pole. Yes, but under very different circumstances from those expected. We have had a horrible day … Great God! this is an awful place and terrible enough for us to have labored to it without the reward of priority … Now for the run home and a desperate struggle."

▼ SOON ABANDONED
The Wolseley four-cylinder air-cooled motor sledge, operating with caterpillar track, was an idea ahead of its time, and a precursor to the oversnow vehicles of today. However, it was poorly made and came with inadequate spare parts.

▲ PRECISION INSTRUMENT
Lieutenant "Teddy" Evans surveying with the four-inch theodolite that was used to locate the South Pole. Evans had been considering leading his own expedition to Antarctica to explore what is now the Edward VII Peninsula, but Scott adopted the plan as his own and offered Evans the role of his second-in-command.

▷ EXHAUSTED AND WEATHERWORN
Ponting took this photograph soon after "Birdie" Bowers, Edward "Bill" Wilson, and Apsley Cherry-Garrard (left to right) returned from "the worst journey in the world." Cherry-Garrard was the only one of the trio who returned to England, where the three eggs they had gathered were finally analyzed.

The way back

The dispirited party started back, but by 23 January Scott reported that Evans, the strongest of them, "is a good deal run down," and that "Oates gets cold feet." They reached Upper Glacier Depot on 7 February. Although they had mislaid or misjudged a day's supply of biscuits, and although Scott had noticed that Evans was "going steadily downhill," they spent two days adding 16 kilograms (35 pounds) of mineral specimens collected at Mount Darwin to their sledging load. Scott noted: "The morraine was obviously so interesting that … I decided to camp and spend the rest of the day geologizing. It has been extremely interesting." The fossils that the rocks contained later proved valuable in assessing the geological history of Antarctica, but the time spent collecting them and the effort of carrying them contributed to their deaths.

On Saturday 17 February, near the foot of the Beardmore Glacier, Edgar Evans lost his life. He had had several bad falls, and was straggling behind, and then disappeared and was found partly undressed and delirious. He sank into a coma and died that night.

By early March they were all in poor condition, particularly Oates, who had severe frostbite in his feet. On 10 March Scott wrote: "Things steadily downhill. Oates' foot worse. He has rare pluck and must know that he can never get through. He asked Wilson if he had a chance this morning and of course Bill had to say he didn't know. In point of fact he has none. Apart from him, if he went under now, I doubt whether we could get through. With great care we might have a dog's chance, but no more."

A week later Scott wrote: "Friday March 16 or Saturday 17—Lost track of dates but think the last correct … At lunch the day before yesterday, poor Titus Oates said he couldn't go on: he proposed we should leave him in his sleeping bag. That we could not do." They struggled on, but the same entry records: "He slept through the night before last, hoping not to wake; but he woke in the morning—yesterday. It was blowing a blizzard. He said 'I am just going outside and may be some time.' He went out into the blizzard and we have not seen him since … I take this opportunity of saying that we have stuck to our sick companions to the last."

Scott's other expeditioners

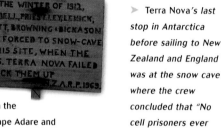

Often overlooked in accounts of Scott's expedition is the Northern Party, renamed the Eastern Party when, after failing to find a landing place on the eastern side of the Ross Sea, it moved to Cape Adare and erected a hut near the Borchgrevink expedition's base. Six men, led by Victor Campbell, landed from *Terra Nova* on 18 February 1911. Bad snow conditions greatly limited their explorations before they were picked up by *Terra Nova* almost a year later, on 3 January 1912. They decided to explore further along the coast of Victoria Land, starting at Terra Nova Bay on 8 January. They were due to be picked up around 18 February, and had ample supplies for their intended stay of about six weeks. However, the ship could not get back to them, so they had to dig a small snow cave on Inexpressible Island and spend the winter there. In poor health after a winter diet of penguins and seals, they left their cave at the end of September 1912 and traveled along the sea ice to Ross Island, surviving only because they found some food caches along the way. They reached Hut Point on 6 November, and learned of the fate of Scott's party from a note reporting that the polar party had not returned by the start of winter. They were recuperating at Cape Evans on 25 November when the search party returned with the news that they had found the bodies of Scott, Wilson, and Bowers.

➤ *Terra Nova's last stop in Antarctica before sailing to New Zealand and England was at the snow cave where the crew concluded that "No cell prisoners ever lived through such discomfort."*

➤ *The Northern Party first named the 12-kilometer (7-mile) long island "Southern Foothills" but later called it Inexpressible Island after living in the tiny snow cave.*

> ➤ PIPPED TO THE POLE
> When they finally reached the South Pole, Scott's party discovered that the Norwegians had been there before them. Bitter disappointment is etched on their faces after performing the empty formalities that they had expected to be a rite of victory. From left to right: Wilson, Evans, Scott, Oates, and Bowers.

On March 19 a blizzard held them down in their tent just 11 miles from One Ton Depot, and their fuel ran out three days later. Scott's diary entry for 29 March records his last words: "Outside the door of the tent it remains a scene of whirling drift. I do not think we can hope for any better things now. We shall stick it out to the end, but we are getting weaker… and the end cannot be far.

"It seems a pity but I do not think I can write more—

"R. Scott

"Last entry

"For God's sake look after our people."

Aftermath

Eight months later, on 12 November 1912, expedition survivors found their tent and their bodies. The tent was closed and they erected a cairn of ice above it, with skis forming a rough cross on top. They marched 20 miles south to look for the body of Oates, but found the area covered in snow so built a cairn in his memory. Finally, they erected a huge wooden cross on the summit of Ross Island's Observation Hill, with a dedication that concluded: "to strive, to seek, to find, and not to yield."

Judgment of the British Antarctic explorers of the Heroic Age has shifted in prevailing attitudes. Scott was cast as a hero for most of the twentieth century, but recent critics have suggested that his only achievements were to write well and to die well. Shackleton has even been called unpatriotic because in 1909 he had enough provisions to reach the Pole but turned back. He and his companions could have claimed the Pole for Britain and died on their way back. By not sacrificing himself and his men, Shackleton is held to have left it to a foreigner to be first to the Pole. Scott was unlikely to have made the same decision. On the contrary, his message to the public as he was dying ran, in part: "The causes of the disaster are not due to faulty organization but to misfortune in all risks which had to be undertaken … I do not regret this journey, which has shown that Englishmen can endure hardships, help one another and meet death with as great a fortitude as ever in the past." Scott's memorial service in St Paul's Cathedral in February 1913 was attended by King George V, and Scott's widow, Kathleen, received his knighthood. DM

▲ SCOTT'S FINAL WORDS
The last entry in Scott's diary was taken as a testament to British spirit. Huntford, Scott's biographer, states that the Admiralty even announced that Scott and his men were considered as having been "killed in action" and a naval sermon was given on "the glory of self-sacrifice, the blessing of failure."

▲ TRIBUTE TO HEROIC FAILURE
A 2.7-meter (9-ft) memorial cross, made from tough Australian jarrah, stands on the summit of Observation Hill. It lists the men's names, and is further inscribed: "Who died on their return from the Pole. March 1912. To strive, to seek, to find, and not to yield." The words by Lord Tennyson were Cherry-Garrard's suggestion.

Wilhelm Filchner

Second German Antarctic Expedition: 1911–12

Before setting out for Antarctica in 1911, Filchner had made expeditions to Russia, Central Asia, and Tibet. Swiss by birth, he was a Prussian military officer by profession, and prepared for Antarctica with military efficiency. He bought a Norwegian ship that had been purpose-built for polar work, renamed it *Deutschland*, and took six members of his team to Spitsbergen for training. His original goal was to cross the Antarctic Continent, launching two expeditions from the Ross and Weddell seas respectively, but limited funds curtailed his plans, and only *Deutschland* sailed for the Weddell Sea in May 1911, leaving South Georgia in December. Fighting its way south, at the end of January 1912 the expedition sighted an ice shelf 40 meters (130 ft) high (now named after Filchner), and steamed south along it to a bay at 77°45′S, which they named after the ship's captain, Richard Vahsel.

Here they started to build their base on ice, and it was almost finished when, on 18 February, tonnes of ice broke away from the ice shelf and took their base with it. *Deutschland* set off in hot pursuit, and, in an extremely

hazardous exercise, Filchner managed to salvage most of the materials and animals, including dogs and horses, and ferry them to the ship by lifeboat. Undeterred, he established two depots at—he hoped—more stable points on the ice shelf, before retreating to South Georgia for the winter.

A sighting disproved

Within days the ship was icebound. The crew settled into winter quarters, setting up tents for scientific observations and exercising the dogs and horses whenever the weather permitted. In June Filchner led a dangerous expedition to try to locate land sighted in 1823 by an American sealer, Benjamin Morell, in a position only 70 kilometers (44 miles) east of the ship. At 60 kilometers (37 miles) they saw nothing, and found no bottom at 1,500 meters (4,920 ft), thus disproving the existence of Morell Land. Over six months, *Deutschland* drifted northward through 10 degrees of latitude. The ship finally broke out of the ice in October 1912, and reached South Georgia two months later. LC

▼ SPECTACULAR CLEFT
Drygalski Fiord lies on the southeast coast of South Georgia. It was charted by the Filchner expedition of 1911–12 and named for the leader of the earlier German polar expedition of 1901–03.

Nobu Shirase

Japanese South Polar Expedition: 1910–12

The first Japanese Antarctic venture originated with the ambition and determination of an obscure lieutenant in the Japanese army. Nobu Shirase battled popular ridicule and government snubs, and eventually sought support from a former Premier of Japan, Count Shigenobu Okuma, whose prestige attracted sufficient public subscriptions to fund the expedition.

Shirase sailed from Tokyo in December 1910 in an antiquated whaling vessel, *Kainan Maru*. His expedition was dogged by vile weather and hostile—and outright racist—publicity. When *Kainan Maru* put in at Wellington in February 1911, the New Zealand press ridiculed its "crew of gorillas." The polar regions (they claimed) were no place for "such beasts of the forests." Rough seas and thick ice slowed the expedition's progress south to the Ross Sea, and although they sighted the coast of Victoria Land they could not land, and finally turned back at Coulman Island. When they reached Sydney in May 1911, the crew were refused accommodation ashore, and lived like beggars until Edgeworth David, a member of Shackleton's 1907 expedition, intervened on their behalf.

Raising the Japanese flag

Shirase's first objective had been the South Pole, but it was now clear that both Amundsen and Scott would be way ahead of him, so on his second foray south he concentrated on the exploration of Edward VII Land. In much better conditions than before, the expedition reached the Ross Ice Shelf in January 1912 and landed at Kainan Bay at the eastern end. Heading west, they were startled to see another ship ahead— it was *Fram*, waiting for Amundsen's party to return from the Pole.

Eventually, they reached the ice shelf by climbing a cliff 100 meters (330 ft) high, and Shirase and four companions set off on a "Dash Patrol" to the south with dog sledges. Although inexperienced and hampered by frequent blizzards, they covered nearly 300 kilometers (186 miles) before raising their flag at 80°05′S, and burying a cylinder containing a record of their achievement. The patrol returned to the Bay of Whales and *Kainan Maru*, and in June 1912 the expedition returned to Yokohama, where they were acclaimed as heroes. LC

> PATIENT PROGRESS
Members of the underfunded Japanese expedition showed the same resolve as many better known expeditions. The Japanese spent 60 hours cutting a path to the top of the Ross Ice Shelf before beginning a journey inland.

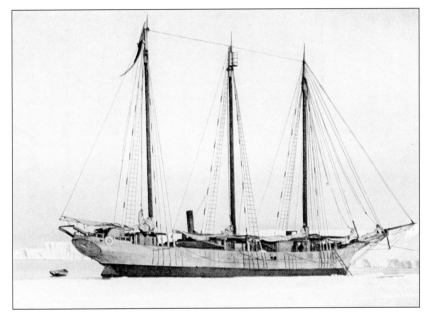

⌃ ADAPTED FOR ANTARCTICA
Kainan Maru (meaning "Southern Pioneer") was a three-masted wooden fishing boat from Shibaura, Tokyo. Its displacement was 199 tonnes, and an 18-horsepower auxiliary steam engine was added. The ship was clad in 6-millimeter (¼ in) sheet steel.

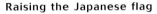

> FIRST VOYAGE LIMIT
Coulman Island marked the southern limit of the *Kainan Maru*'s first season in Antarctica. The Japanese reached the island on 14 March 1911, but here at 73°28′S the ice was too thick for them to proceed further; so, with winter approaching, they turned north.

Douglas Mawson

Australasian Antarctic Expedition: 1911–14

▲ COMPLETE COMMITMENT
Douglas Mawson (1882–1958) was an explorer, survivor, and geologist. His work in outback Australia earned him a Ph.D., but his experience in Antarctica brought him world renown. New Zealander Eric Webb described him as "an intellectual leader with utter motivation and selfless dedication to his objective."

After Shackleton's *Nimrod* expedition, Edgeworth David said: "Mawson was the real leader and was the soul of our party to the magnetic pole. We really have in him an Australian Nansen, of infinite resource, splendid physique, astonishing indifference to frost." Both Shackleton and Scott asked the Australian explorer and geologist to join their next expeditions. Mawson turned Scott down when Scott rejected his proposal to explore the coast west of Cape Adare. Shackleton and Mawson developed an extensive Antarctic exploration proposal, but it came to nothing. So Mawson promulgated his own Australasian Antarctic Expedition. Unlike other Antarctic explorers, he was not aiming for the South Pole; he was most interested in the part of Antarctica lying directly below Australia, in what he called the Australian quadrant.

He initially envisaged three land parties, but modified this to two: the main party of 18, under Mawson, was based in Adélie Land, and the western party of eight, led by the experienced Antarctic expeditioner Frank Wild, was based in Queen Mary Land.

Mawson's ship was *Aurora*, an old Dundee whaler of 612 tonnes, captained by John King Davis, with a complement of four officers and 19 crew. Mawson also bought an aircraft; it crashed in October 1911 during a fund-raising display in Adelaide, but he took the wingless fuselage to Antarctica to use as a motor sledge. The expedition left Hobart on 2 December 1911, established a five-man radio station at Macquarie Island, and continued southward. The main party landed on the Antarctic mainland on 8 January 1912, and *Aurora* sailed west with Wild's party on 19 January.

"The home of the blizzard"

The main party had landed in what seemed an ideal natural harbor, which Mawson named Commonwealth Bay. He called the point where they erected their hut Cape Denison. Too late, he found that they were at the base of a funnel for winds that came down from the polar ice cap; he wrote: "The climate [was] little more than one continuous blizzard …" It was a trying winter, but the expeditioners kept meteorological observations, and erected aerials to send Morse messages to Macquarie Island, the first use of radio in Antarctica—though it was only toward the end that they received replies.

In August, they excavated and provisioned a comfortable cave in the ice of a slope five miles south of the hut. They named it Aladdin's Cave, and it was to prove a lifesaver. Because of spring blizzards, it was November before the five planned sledging expeditions set out. All left between 8 and 10 November, knowing that they must be back to meet *Aurora* by 15 January. One party was aiming for the South Magnetic Pole, and got to within 80 kilometers (50 miles) of their goal, at an

◄ BROADCASTING STATION
Ham Shack near the summit of Wireless Hill (112 meters/370 ft) marks the site that Mawson chose for his wireless station—the living hut for the five-man party was located where the base is today. Building materials and supplies for Ham Shack were carried up the steep hillside by an aerial cable, 243 meters (800 ft) long.

▲ THE LIE OF THE LAND
This photograph looking south from Wireless Hill that was taken during Mawson's expedition gives a good overview of the northern end of Macquarie Island. The main difference between now and then is the development of the year-round Australian base that is the only human habitation on this sub-Antarctic island.

◁ *AURORA* 1914

Mawson described *Aurora* as "by no means young [but] still in good condition and capable of buffeting with the pack for many a year ... The hull was made of stout oak planks ... The bow ... was a mass of solid wood, armored with steel plates."

▷ "THE HOME OF THE BLIZZARD"

Landing stores and equipment at the head of the boat harbor, Cape Denison, were the first steps toward forming the main base station. In the distance men are sledging materials to where the hut is to be built.

altitude of 1,935 meters (5,900 ft), on 21 December before lack of time and food forced them to turn back. They returned to the hut on 11 January.

A second party planned to use the crippled aircraft to explore the hinterland of Adélie Land; once they even managed to drive it up a steep slope, but on the second day several pistons snapped, and they abandoned it. On 5 December they found a tiny black meteorite—the first ever discovered in Antarctica. They turned back on 26 December and after a difficult journey reached the base on 17 January.

The third and fourth parties traveled together for 75 kilometers (46 miles), and then the near-east group turned back to map the coast between the glacier named after the expedition's Swiss ski expert and

◁ ALADDIN'S CAVE

For Mawson the cave was "a truly magical world of glassy facets and scintillating crystals ... The purest ice for cooking could be immediately hacked from its walls ... Finally one neatly disposed of spare clothes by moistening the corner of each garment and pressing it against the wall for a few seconds, where it would freeze on and remain hanging until required."

◁ CHANGE OF PURPOSE

Long before this station was built, sealers had their huts at the southern end of the peninsula at the foot of Razorback Hill. According to Dr A.L. McLean, "everywhere [was] covered with the bones, bleached skeletons and putrid carcasses of sea elephants." Today, the isthmus is a peaceful place of scientific enquiry.

◄ EDGE OF A GLACIER
The Mertz Glacier is some 45 miles (72 kilometers) long and 20 miles (32 kilometers) wide, and where it reaches Cape Hurley, it extends into the water as the Mertz Ice Tongue seen here. These icy features were named by Mawson for Xavier Mertz who died on 7 January 1913 on the far-east sledge journey.

◄ FIELD WORK
At New Year the eastern coastal party set up camp at Penguin Point where Cecil Madigan described "a rock wall three hundred feet high formed the sea-face, jutting out from beneath the ice slopes of the buried land." Besides seals and Adélie penguins they found some new bird's eggs and mites living in the moss.

mountaineer, Xavier Mertz, and Cape Denison. In one small stretch they found an archipelago of 154 little islands. Meanwhile, the eastern coastal party continued eastward, making its difficult way across the rough Mertz Glacial Tongue, then sea ice, then the Ninnis Glacial Tongue, and then sea ice again as far as a towering coastal promontory of columnar lava that they named Horn Bluff—an expanse of spectacular "organ pipes," with outcrops of sandstone and coal. They started back on 21 December, but struck bad weather and soft snow, and passed several days with virtually nothing to eat before reaching a food depot. They were back in the comparative safety and warmth of the hut on 17 January.

▼ CLAIMED BY A CREVASSE
Belgrave Ninnis, a lieutenant of the Royal Fusiliers, was just 23 years old when he plunged to his death. Mawson wrote: "It was difficult to realize that Ninnis, who was a young giant in build, so jovial and so real but a few minutes before, should thus have vanished without even a sound."

◄ KILLED BY THE COLD
Dr Xavier Mertz was a 28-year-old lawyer from Berne, a ski jumping champion and an excellent mountaineer. His outer trousers and helmet were lost on Ninnis's sledge and becoming wet and sick, he died soon afterwards. Mawson wrote: "No one could have done better ... he had been a general favorite."

A beleaguered party

The fifth party was overdue. Mawson had left on 10 November, with Mertz and the ex-Royal Fusilier Belgrave Ninnis. With three sledges and 16 dogs, they made good time at first, but disaster struck on 14 December, not long after they had abandoned one of the sledges. Mertz was leading on skis, Mawson was behind with the first sledge, and Ninnis was in the rear with the other. Mawson wrote: "My sledge crossed a crevasse obliquely and I called back to Ninnis … to watch it, then went on, not thinking to look back again …" But when he did look back, Ninnis had disappeared through a hole in the ice bridge that spanned the crevasse. Mawson and Mertz peered over the edge and saw two dogs and a tent and food bag caught on a ledge; there was no sign of Ninnis. Lacking a rope long enough to climb down, they called fruitlessly for four hours, read the burial service, and took stock of their own situation.

And their situation was desperate indeed, for Ninnis, in the supposedly safer rear position and with the best dog team, had been carrying most of the food, the main tent, and other vital supplies. "May God help us," wrote Mawson, as he contemplated their return journey. They started back on a more southerly route than their outward one to avoid the most dangerous crevasses, surviving by killing and eating the six remaining dogs; the paws took the longest to boil into an edible stew. It was 28 December when they dispatched Ginger, the last of the dogs.

◄ RESTORED HUT AT CAPE DENISON
In the opinion of J. Gordon Hayes, Mawson's undertaking "judged by the magnitude both of its scale and of its achievements, was the greatest and most consummate expedition that ever sailed for Antarctica."

▼ CALLING ANTARCTICA HOME
When the main base hut was completed in February 1912, the Union Jack was hoisted as the men cheered. Mawson noted that soon after the hut was finished it "became so buried in packed snow that ever afterwards little beyond the roof was to be seen."

By 3 January 1913, Mertz was ravaged by dysentery and severe frostbite. Mawson told how Mertz tried to maintain his own courage at this time: "To convince himself he bit a considerable piece of the fleshy part off the end of one of [his fingers]." By 6 January Mertz was riding on the sledge; the following day he had several fits, and struggled to get out of his sleeping bag. Mawson woke a couple of hours later to find his companion "stiff in death." There have been insinuations that Mawson practiced cannibalism in these desperate straits, but he had enough food, and his actions in burying his comrade under blocks of snow and marking the position— and above all, his high principles—argue against this. He cut his sledge in half with a penknife to lighten his load, sewed a makeshift sail, and carried on.

The last lap

Mawson had recorded that the skin had peeled away from their groins, and on 11 January wrote: "My feet felt curiously lumpy and sore." The soles of his feet had come off, and had to be strapped on with bandages. But sudden death was more immediately threatening than frostbite. On 17 January Mawson fell through an ice bridge; suspended on a thin rope, and precariously anchored by

▲ MARKING TIME WITH MUSIC
Winter evenings passed slowly in the hut but books and a gramophone helped to entertain the expeditioners. Standing up: Mawson, Madigan, Ninnis, and Correll. Sitting round the table from left to right: Stillwell, Close, McLean, Hunter, Hannam, Hodgeman, Murphy, Laseron, Mertz, and Bage.

Roving camera

Frank Hurley, a Sydney suburban post-card photographer, had been accepted into the expedition after he inveigled his way into Mawson's overnight railway carriage, and went on to make a remarkable film of the expedition. He returned from Commonwealth Bay with most of the expeditioners, then sailed south again with *Aurora* to pick up Mawson, who said: "Hallo! Back again?" Two weeks after he returned to Australia with Mawson, he was filming a motoring trip into outback Australia when a black tracker arrived with a cabled invitation to join Shackleton's *Endurance* expedition. Two months later he sailed to Buenos Aires to meet the ship, and was soon trapped with it in the ice.

six hours earlier by three other members of the expedition, and learned that Amundsen had reached the South Pole, and that the rest of his own men were safe.

Mawson's problems were not over. The snow along his route turned to slippery ice, and he had discarded his crampons. He fell frequently, "until I expected to see my bones burst through my clothes," and made crampons from his wooden theodolite case and from nails and screws. He arrived at Aladdin's Cave on 1 February, only to be pinned down by a blizzard for a week. On 8 February he made his way to the Cape Denison hut, where he saw *Aurora* sailing into the distance, and "my hopes went down." Then three men waved from the harbor and ran to meet him: six men had stayed in case any of Mawson's party returned, and they wept when they heard of the deaths of Ninnis and Mertz.

A return in triumph

Aurora was immediately recalled by radio, but weather conditions made landing impossible, so she resumed her course to pick up Frank Wild's western party and returned to Hobart on 15 March 1913. Mawson believed that that storm may have saved his life, as he was gently nursed through the winter rather than having to withstand the rough Southern Ocean crossing. Sidney Jeffryes was the new radio operator, and his skill was such that they managed to communicate with Australia via Macquarie Island in the early part of the winter. However, Jeffryes suffered a mental breakdown and was able to work only intermittently in the second half of the year. *Aurora* returned to Commonwealth Bay on 13 December to collect Mawson's party, carried out marine research along the Antarctic coast for two months, and docked in Adelaide on 26 February 1914.

Mawson's scientific achievements had preceded him, and he was treated as a hero when he arrived home. On 31 March 1914 he married Paquita Delprat in Melbourne (J.K. Davis was best man); on 29 June 1914 (the day that Archduke Franz Ferdinand was assassinated in Sarejevo, triggering World War I) he was knighted at St James's Palace in London.

his half-sledge, Mawson thought: "So this is it ... then I thought of the uneaten food on the sledge ... A great effort brought a knot in the rope within my grasp, and after a moment's rest, I was able to draw myself up and reach another, and, at length, hauled my body on to the overhanging snow lip. Then when all appeared to be well ... a further section of the lip gave way, precipitating me once more to the full length of the rope."

He contemplated giving up by releasing himself from the rope, but hauled himself to the surface, where he "cooked and ate dog meat enough to give me a regular orgy," and wove a rope ladder that he tied to himself and the sledge—and subsequently used on several occasions. By 29 January he was close to the Cape Denison hut. That day he found a cache of food and a message left

DOUGLAS MAWSON AND THE *AURORA* 1911–14

George V Land

140°E

150°E

Terre Adélie

70°S

Mertz Glacier

Ninnis Gl.

Mertz dies 7 Jan 1913

Aladdin's Cave

food depot

main base
Cape Denison

Ninnis dies 14 Dec 1912

Dixson I.

Mt Murchison

Horn Bluff

Fisher Bay

Watt Bay

Mawson Pen.

Cape Hudson

Lauritzen Bay

Cook Ice Shelf

Cape Freshfield

Ninnis Glacier Tongue

Mertz Glacier Tongue

Commonwealth Bay

140°E

Antarctic Circle

150°E

———— Far eastern party
———— Eastern coastal party
———— Near eastern party

0   200   kilometers
0   100   miles

▶ AT THE WHIM OF THE WIND
Katabatic winds regularly whip Commonwealth Bay into a lather. Gusts there can also be very localized. Mawson, observing the behavior of the wind, noted: "Laseron one day was skinning at one end of a seal and remained in pefect calm, while McLean, at the other extremity, was on the edge of a furious vortex."

The Western Party

While Mawson's party endured their various ordeals, *Aurora* had taken Frank Wild's party to the Shackleton Ice Shelf of Queen Mary Land, 2,400 kilometers (1,500 miles) to the west. Vast quantities of supplies had to be dragged by aerial cable to the summit of the ice shelf before the ship sailed away. Their hut on the ice was habitable by 28 February, but the radio masts they erected were blown down by the first blizzard. On clear days they could see the mainland, 27 kilometers (17 miles) to the south; they visited it in mid-March and left a cache of supplies.

Wild wrote: "The most bigoted teetotaler could not call us an intemperate party. On each Saturday night, one drink per man was served out, the popular toast being 'Sweethearts and Wives.' The only other convivial meetings of our small symposium were on the birthdays of each member, Midwinter's Day and King's Birthday."

In August, just before they planned on starting their spring skiing program, the hut nearly burned down while the acetylene generators were being charged. The first journey was to the east; six men and three dogs left on 22 August and returned on 15 September. In summer there were two expeditions: one to the eastern coast under Wild, and one to the western coast under Evan Jones. One man was left at the hut.

Wild's group found the going very heavy, particularly across the crevasses of the Denman Glacier. They climbed to the peak of Mount Barr-Smith on 22 December, at the end of their outward journey, and were back at the hut on 6 January 1913. Jones's party found vast colonies of penguins and other nesting birds on their way. They spent Christmas at a small, bare, extinct volcano that Drygalski had named Gaussberg when the Germans sledged there 10 years before; the German cairns were still visible on the summit. They were 346 kilometers (215 miles) from their hut, but were back there on 21 January.

When *Aurora* had not returned by the end of January, the party began to stockpile seal meat and blubber in case they were marooned for another winter. Saturday 23 February was the anniversary of the departure of *Aurora*, and in a howling blizzard they faced the possibility that the ship would not return. But *Aurora* arrived the following day, with most of the other expeditioners, and they were quickly boarded and taken back to Australia.   DM

▲ VETERAN ANTARCTICAN Mawson thought Frank Wild "An excellent Petty Officer... and in some respects more than that ... He could not be excelled in intrepidity and had a full quota of sound horse sense. A very likeable fellow."

John King Davis

Veteran of Antarctica: 1907–30

▲ SKIPPER OF CONFIDENCE
John King Davis (1884–1967)
was sometimes irascible. On
the Australasian Antarctic
Expedition, Hurley wrote:
"Davis led on sea and Mawson
on land. Both possessed an
inflexible spirit in danger and
a rare ability to inspire and
encourage. They were equally
human and generous and
thought of their men first and
themselves afterwards."

The ablest and most experienced ship's captain of the Heroic Age of Antarctic exploration, Davis made no fewer than seven voyages to the ice, unerringly transporting and retrieving four expeditions led by Shackleton and Mawson. En route, he carried out pioneering oceanographic research, for which he was awarded the Murchison Medal of the Royal Geographical Society.

Running away to sea

Davis was a seaman first and last. At the age of 16, he ran away to sea to join the full rigged ship *Celtic Chief* as an apprentice. Her skipper, Captain John Jones, inspired Davis's subsequent career, providing a role model of "a shipmaster of great integrity and skill, vested with supreme authority over the lives of men, charged with the safety of the ship, her cargo and those who served in her, doomed to that loneliness and austerity inseparable from command." The arena in which Davis's career was played out was determined by a chance meeting in 1907 with Ernest Shackleton. This led to Davis's appointment as Chief Officer of *Nimrod*, the ship of the 1907–09 British Antarctic Expedition, on which Shackleton reached to within 160 kilometers (100 miles) of the South Pole. Also on the expedition was a young Australian geologist, Douglas Mawson, who, with two companions, was first to the South Magnetic Pole in Victoria Land. It was due to Davis's vigilance in searching for them that this party was picked up and returned safely.

His efficiency so impressed Mawson that he appointed Davis skipper of *Aurora*, the ship chosen for his 1911–14 Australasian Antarctic Expedition. Mawson's objective was to chart new territory west of Victoria Land, which gave enormous responsibility to his ship's captain. Davis was equal to it: three times he brought *Aurora* unscathed through the hazards of the unexplored Adélie Land pack ice and the almost impossible conditions at Mawson's main base at Commonwealth Bay.

It was for these qualities that Shackleton sought his services as skipper of *Endurance* on the Imperial Transantarctic Expedition of 1914–17. Davis refused, and thus escaped being trapped in the ice of the Weddell Sea for two years, but the tangled web of fate eventually saw Davis and Shackleton sailing together in Davis's old ship, the *Aurora,* to rescue the marooned Ross Sea Party of the Imperial Transantarctic Expedition.

An uneasy partnership

Davis's final call to Antarctica came in 1929, when he was appointed Captain of *Discovery*, Scott's old ship, and second-in-command to Mawson, of the British, Australian and New Zealand Research Expedition (BANZARE). Although undoubtedly the most experienced Antarctic navigator, Davis was 45, and had not held a seagoing command for more than 10 years. From the beginning there were disagreements between Mawson and Davis about their route, which stemmed from Davis's concern about *Discovery*'s limited coal capacity and poor sailing qualities. Davis also believed that his extensive experience of oceanographic work in *Aurora* should give him authority over the scientific program. Mawson saw this as his responsibility as expedition leader, and a series of conflicts led to Davis's resignation from the second BANZARE expedition, on the grounds that "it is a very thankless task [which] must always be the case in a marine expedition under the leadership of a layman."

The conflict was as much a clash of culture as of personality. In Davis's view, it was largely due to Mawson's inability to understand the perspective of a professional seaman, whose overriding responsibility is the safety of the ship and her complement. In addition, the enormous

➤ FORWARD PLANNING
When this photograph of
Davis, Shackleton, and
Mawson (left to right) was
taken in 1911, the planning
for Mawson's Australasian
Antarctic Expedition was
already well under way. Davis
was chosen to serve as both
captain of Aurora (he selected
the vessel) and second-in-
command of the expedition.

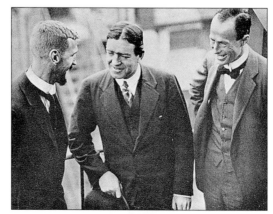

▼ ICY IMPEDIMENT
Heavy sea ice near Cape
Adare makes navigation hard.
In a tribute to the Antarctic
explorers of the nineteenth
century, Davis stated: "We in
the *Aurora* with our auxiliary
steam power and specially
strengthened hull, had an
immense advantage over our
gallant predecessors."

▶ EXERCISE AT SEA
Captain John King Davis paces
the deck of *Aurora* with
another Antarctic traveler,
Captain James Davis, a
whaling expert from Hobart.
J.K. Davis described *Aurora* as
"a considerably larger vessel
than the little *Nimrod* though
by present-day standards she
was small enough."

technical difficulties of navigating an underpowered,
overloaded ship in uncharted waters in extreme ice
and weather conditions impose intense physical and
psychological stresses on the captain. Mawson, for
his part, thought that Davis did not understand the
scientific and exploratory goals of the expedition, was
timid in pursuing them, and even wilfully obstructed
them by insisting that some oceanographic operations
endangered the ship, unless carried out his way. Their
diaries reveal that these disagreements led both men to
acts of petty spite and expressions of virulent personal
criticism that only exacerbated their antagonism.

   This animosity is one example—although possibly
the most extreme—of the stresses that are perhaps an
inevitable outcome of the different roles and imperatives
of ship's captain and expedition leader on any Antarctic
expedition, past, present, or future. Happily, it did not
lead to a lasting rift between the two men.   LC

▲ PENETRATING PACK ICE
While trying to get Wild's
western party ashore, Davis
tried to push through thick ice
to reach Knox Land. He later
recalled: "Had I set about the
task of demolishing the cabin
bulkhead with my own head I
should probably have had as
much success as, and hardly
less of a headache than we
achieved in attempting to
work our way through more
than fifty miles of pack ice in
the *Aurora*."

Ernest Shackleton

Imperial Transantarctic Expedition: 1914–17

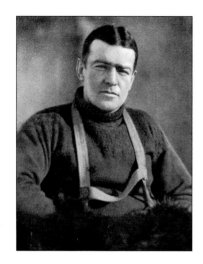

▲ A PLACE IN HISTORY
Ernest Shackleton
(1874–1922) faced his
greatest challenge on the
Endurance expedition.
Biographer Margot Morrell
stated "Shackleton never
planted a flag at the South
Pole, he never made any of his
goals, and he never earned all
the money he wanted. Yet he
was doing what he wanted to
do, and he did it well enough
to earn a place in history."

When the news broke that Amundsen had reached the Pole, Shackleton wrote: "The discovery of the South Pole will not be an end to Antarctic exploration. The next work is a transcontinental journey from sea to sea, crossing the Pole." He soon turned his ambitions towards achieving that goal himself. The project was not without critics—including Clements Markham, then 84 years old but still very influential. He also faced the embarrassment of his brother Frank being jailed for fraud (and possibly being involved in the theft of the Irish crown jewels), and his own finances were always uncertain. But, with determination and Irish charm, Shackleton pulled it all together.

Shackleton's first plan was to take one ship, disembark at Vahsel Bay in the Weddell Sea, and travel 2,900 kilometers (1,800 miles) overland while the ship

sailed around to meet them. If he was late and the ship had to depart, it would leave a small whaler behind and he could have a chance at another of his ambitions: an open boat journey across Antarctic waters. It was a foolhardy plan, and he was persuaded instead to use two ships, with a party on the Ross Sea establishing food depots for him on the Ice Shelf. Shackleton was now 40, but he optimistically predicted that he could complete the journey in 100 days, a rate faster than Amundsen's, yet over completely unknown terrain. It was a brave step into the unknown.

The ship for the Weddell Sea was the Norwegian-built *Polaris*, captained by New Zealander Frank Worsley and renamed *Endurance*, from Shackleton's family motto: "By endurance we conquer." For the Ross Sea Party, he bought Mawson's *Aurora* quite cheaply. *Endurance* sailed from London on 1 August 1914, after a good luck visit from Queen Alexandra. On 5 August England declared war on Germany and Shackleton offered his ship and

▶ CUTTING THE ICE
The sight of a clean wake
behind *Endurance,* as it
pushed deep into the ice of
the Weddell Sea was rare. On
14 December Shackleton
wrote: "The young ice did not
present difficulties to the
Endurance, which was able to
smash a way through, but the
lumps of older ice were more
formidable obstacles."

ERNEST SHACKLETON AND THE *ENDURANCE* 1914-16

30°W

0 400 800 1200 kilometers
0 400 800 miles

South Sandwich Islands

South Georgia
leaves Grytviken 5 Dec 1914
Shackleton reaches Stromness whaling station 20 May 1916

50°S

0°

Antarctic Circle

Haakon VII Sea

James Caird's route

60°W

Falkland Islands

60°S

Sth Orkney Islands

Shackleton departs 24 April 1916

Elephant I

boats launched 9 April 1916

drifting on ice floes

Dronning Maud Land

Sth Shetland Islands

70°S

Weddell Sea

Endurance trapped in ice 19 Jan 1915

Endurance sinks 21 Nov 1915

Graham Land

Antarctic Peninsula

Coats Land

Vahsel Bay

Palmer Land

Alexander I

Ronne Ice Shelf

80°S

Bellingshausen Sea

90°W

Ellsworth Land

90°S
South Pole

90°E

men to the war effort, but the Admiralty, led by Winston Churchill, then First Lord of the Admiralty, told him to continue south. *Endurance* sailed from Plymouth on 8 August. Three days out from Buenos Aires, a 19-year-old stowaway, Perce Blackborrow, was found. Shackleton castigated him—and then took him on as a member of the crew, bringing the numbers to 28 expeditioners and crew.

Shackleton made for the whaling settlement of Grytviken, about halfway along the north coast of South Georgia. He arrived in November 1914 to hear of the worst ice conditions ever reported in the Weddell Sea, but for Shackleton it was now or never. They sailed south on 5 December 1914 and were in solid pack ice by 11 December, with 1,600 kilometers (1,000 miles) to cover to Vahsel Bay.

Shackleton had selected his polar party: surgeon Alexander Macklin; Australian photographer Frank Hurley; second-in-command Frank Wild; polar veteran Tom Crean; and George Marston, the artist from the *Nimrod* expedition.

The ice closes in

They saw land on 10 January 1915, and by 18 January they were only 130 kilometers (80 miles) from their goal. But next day the ice closed around *Endurance*; she was frozen in "like an almond in toffee," as one man wrote in his diary, and began to drift gradually north with the pack. So began 10 months of entombment in the drifting ice. They could not even take the base equipment across the sea ice to the coast they could see, because that would mean lugging a hut and vast quantities of supplies and equipment across broken ice.

It was now that Shackleton's leadership qualities became apparent. The whole purpose of his expedition had been foiled within site of its goal. They might break free eventually—or the ship might be crushed in the ice. Meanwhile the other half of the expedition was laying supply depots on the Ross Ice Shelf that would never be used. Yet Shackleton betrayed no mental anguish; Macklin noted: "We could see our base, maddening, tantalizing, Shackleton at this time showed one of his sparks of real greatness. He did not rage at all, or show outwardly the slightest sign of disappointment; he told us simply and calmly that we must winter in the Pack, explained its dangers and possibilities; never lost his optimism, and prepared for winter."

They built small igloos—"dogloos"—on the ice for the huskies, and settled into a strict routine of shipboard life. Some parts of the ship were better insulated than others, and Shackleton arranged for warmer areas to be created by repositioning cargo. Scientific experiments, amateur theatricals, and entertainments passed the time, led by Shackleton, whose boundless good spirits kept up their optimism. He took part in high jinks such as mid-winter head-shaving and singalongs—tunelessly, it is reported.

Slowly they drifted northward. About the time the sun reappeared, there were moments of terror as

▷ **AT PATIENCE CAMP**
The strange device between Hurley (left) and Shackleton is a blubber-fuel stove that Hurley, who trained as a metal worker in his youth, made out of the *Endurance* ash chute and old oil drums. While resourceful, Hurley could be difficult, so the canny Shackleton picked him as a tent-mate to diffuse tension.

the pack ice formed pressure ridges and squeezed *Endurance* until her timbers groaned and bent. The pattern continued for three months, and on 18 October 1915 tilted the ship at an angle of 30 degrees before releasing it, now with substantial leaks. Worsley suggested that they might "have to get out and walk," and Shackleton decided that they must be ready to abandon ship at any time. The following day they retreated to five tents on the ice, amid "the groaning and cracking of splintering timbers." After their first night on the ice, Shackleton announced: "Ship and stores have gone—so now we'll go home." His matter-of-fact optimism was an inspiration.

It was another month before *Endurance* was finally crushed and sunk on 21 November. They had all been forbidden to go back on board, but Hurley wanted to salvage his film and photographic plates, now under mushy ice but in sealed tins in an ice chest. He stripped to the waist, hacked through the ice chest's walls, and saved his film. The glass plate negatives were too heavy to carry, so in what has been called one of history's great editing tasks, Hurley selected 120 plates and handed the rejected 400 to Shackleton, who smashed them on the ice so Hurley could not change his mind. He kept an album of prints, his movie film, and a box Brownie and three rolls of film, with which he continued to photograph the plight of the expedition.

▲ ICE PRESS
On October 18, Frank Hurley noted in his diary: "Shortly after 5 pm, we began to rise from the ice, much after the manner of a gigantic pip squeezed between the fingers. In the short space of seven seconds, we were ... thrown over to port, with a list of 30 degrees."

▽ BEGINNING OF THE END
Shackleton recalled on Sunday 24 October 1915: "The *Endurance* groaned and quivered as her starboard quarter was forced against the floe, twisting the stern-post and starting the heads and ends of planking. The ice had lateral as well as forward movement, and the ship was twisted and actually bent by the stresses."

◀ THE END

It took several weeks for the *Endurance* to sink. At various times, Shackleton noted: "The ship presented a painful spectacle of chaos and wreck" and "to a sailor [a ship] is more than a floating home … in the *Endurance* I had centered ambitions, hopes, and desires. Now … she is slowly giving up her sentient life."

▲ TOPPLING MASTS

Hurley was filming when the masts came down. However, by 29 October when Frank Wild took a last look at the ship Shackleton commented: "The ship is still afloat, with the spurs of the pack driven through her and holding her up … The wreckage lies around in dismal confusion." At 5 pm on 21 November, the Boss called "She's going boys" and they watched her go down "bows first, her stern raised in the air."

▲ SLEEPING QUARTERS

All 22 men slept under the two remaining boats upturned on rock walls 1 meter (3½ ft) high. Hurley wrote: "The 'Snuggery' grows more grimy day by day: everything is an oleaginous sooty black ... Oil mixed with reindeer hair, bits of meat, senegrass, and penguin feathers form a conglomeration which cement the stones together."

Above left

▼ WAVING FAREWELL, NOT JOYFUL HELLOS

In their books, Shackleton and Hurley both said this photograph depicted the rescue from Elephant Island. However, in *The Endurance*, Caroline Alexander pointed out that Worsley's book said it showed *James Caird* leaving for South Georgia. Hurley modified the negative to provide a fitting final image for the expedition.

▲ THOSE WHO ENDURED

On 10 May 1916 Hurley "Took photo of group—the most motley and unkempt assembly that ever was projected on a plate." Hurley and the seriously ill Blackborrow are absent. From left, back row: Greenstreet, McIlroy, Marston, Wordie, James, Holness, Hudson, Stephenson, McLeod, Clark, Lees, Kerr, and Macklin. Bottom row: Green, Wild, How, Cheetham, Rickinson (with Hussey behind), and Bakewell.

Camping on the ice

After *Endurance* sank, they could only wait for the ice where they were camped to drift north and break up. Towards Christmas they moved 16 kilometers (10 miles) from Ocean Camp to Patience Camp, planning to take to the three boats they had salvaged from *Endurance* in an attempt to reach land. Shackleton's leadership was critical: he had a knack of spotting imminent psychological collapse and bolstering the sufferer's self-esteem.

The men lived on the ice for five months, surviving toward the end largely on fresh seal and penguin meat. They shot most of the dogs in January, and the remaining two teams at the end of March. Then the ice began to break up, so on 9 April 1916 they took to the boats and the numbingly cold, stormy seas. They had drifted with the ice for 3,220 kilometers (2,000 miles), and had now passed the tip of the Antarctic Peninsula. Their only course to steer was for Elephant Island, a desolate lump of rock; at least it was not an ice floe.

They were in the boats for seven days and nights, sometimes camping on ice floes, before landing at Elephant Island's Cape Valentine. It was just in time. Shackleton elected the youngest member of the expedition, the stowaway Perce Blackborrow, to have the honor of being first ashore. He half pushed him off the boat, but the young man sat in the surf, unable to walk because his feet were frostbitten, while Hurley

bounded about recording the scene with his pocket camera. Shackleton recorded that some of the men were reeling about the beach. Others were so grateful to be on land after almost 500 days that they lay down on the shingles and poured stones over themselves. Another immediately killed 10 seals with an axe.

They could not stay where they were, as they could be washed away from this exposed shore. They found a more sheltered spot nine miles down the north coast, with a gravelly beach; it was still desolate, but there was plenty of wildlife, which they would need. The night they moved there and named it Cape Wild, a gale shredded their remaining tents. Morale was at its lowest: "Dejected men were dragged from their sleeping bags and set to work," were the words recorded in one diary.

Shackleton announced that he and five of the fittest men would sail to South Georgia for help in the largest of the boats, *James Caird*—more than 1,290 kilometers (800 miles) in an open boat, through some of the wildest seas in the world. Harry McNeish, the ship's carpenter, raised the gunwales of the two-masted 22-foot whale boat and covered the decking with canvas to keep out some of the water. They had to launch the boat before ballasting it with a ton of rocks. They set out on 24 April, with Worsley as captain and navigator; Shackleton, Crean, and McCarthy went too, with

McNeish and John Vincent, who may have been included to remove potential troublemakers from the Elephant Island party.

Unable to take more than four sun sightings over a fortnight, and with the setting of his chronometer uncertain, Worsley steered by dead reckoning and gut instinct, knowing that if he was wrong they would sail past South Georgia into the South Atlantic, and the men on Elephant Island would never be found. Of the 17 days of their journey, 10 were in full gales. One night they awoke while hove to in a gale to find the boat foundering and ice up to 38 centimeters (15 inches) thick encasing every sodden inch of wood and canvas. They had to crawl out on the glassy decking and hack the ice away with an axe. Their feet, which were constantly wet, were white and swollen, and had lost all surface feeling.

When they sighted South Georgia on 8 May, after 15 days at sea, they were on the eastern, windward side of the island. They landed at King Haakon Bay on the evening of 10 May after several near-disasters along the coast. The Norwegian whaling stations were on the other side of the island, over a wall of high mountains, and the island had never been crossed before. Only three were fit to attempt the crossing: Shackleton, Worsley, and Crean. Guided only by common sense, they made three attempts to find a pass through the mountains before the fourth took them over just as daylight was failing.

Shackleton knew that if they stayed at high elevation overnight they would freeze to death. Beneath them, disappearing into the mist and darkness, lay a long, steep precipice of snow. "It's a devil of a risk but we've got to take it. We'll slide," he said. Coiling their climbing rope beneath them, they sat down one behind the other, each locking arms and legs around the man in front, and pushed off into the darkness.

"We seemed to shoot into space," wrote Worsley. "For a moment my hair fairly stood on end. Then quite suddenly I felt a glow, and knew that I was grinning! I was actually enjoying it … I yelled with excitement and found that Shackleton and Crean were yelling too." They descended 460 meters (1,500 feet) in a matter of seconds. They stumbled on, falling and slipping, and barely conscious. At dawn Shackleton recognized a rock formation at Stromness Bay, and at 6.30 am heard the welcome sound of the steam whistle blast to wake up the Stromness workers. They were saved.

They still had to clamber down a glacier and struggle through a waterfall, and it was 3.00 pm, after 36 hours without rest, when they walked into the outskirts of Stromness whaling station. Filthy, their faces blackened by blubber smoke, with salt-matted hair and long beards, they were a fearsome sight; two small children—their first human contact—ran from them in fright. Eventually they met the station foreman, who led them to the manager's house. Two years earlier, Thoralf Sørlle had been invited on board *Endurance*.

"Well?" said Sørlle. "Don't you know me?" asked Shackleton. "I know your voice," he replied doubtfully. "You're the mate of the *Daisy*." "My name is Shackleton."

The amazed Norwegian whalers received the castaways with great warmth. A ship was sent around to the east coast to collect the other three crew of *James Caird* and the little boat itself; Worsley wrote: "Every man on the place claimed the honor of helping to haul her up to the wharf." But it was to be another five months before ice conditions made rescue of the men marooned on Elephant Island possible.

Shackleton made three increasingly desperate rescue bids before he finally managed to get through the pack ice in *Yelcho*, a little steel-hulled tug the Chilean government had lent him under the command of Luis Pardo. It took less than an hour to take all the men and their meager belongings from the camp to the ship. The rescue was just in time for Perce Blackborrow; his frostbitten feet had been operated on twice by the two surgeons, but bone infection had set in and he was gravely ill. Astonishingly, however, Shackleton had not lost a single man during the two-year ordeal.

Shackleton's task was not over: the Ross Sea Party had to be rescued, and money raised for the *Aurora* to go south again under the command of John King Davis. Shackleton sailed directly from South America to New Zealand to join the expedition. Davis found Shackleton "somewhat changed," and allowed him to sign on only as a supernumerary, but recorded that at sea Shackleton's "old greatness of spirit shone out."   DM

▲ SOUTH GEORGIA AHOY! George Marston painted *James Caird* arriving at the rugged south coast of South Georgia from material supplied by those who were there. Shackleton recorded that the first view "was a glad moment. Thirst-ridden, chilled and weak as we were, happiness irradiated us. The job was nearly done."

▼ NO WAY TO CROSS The night before he left to cross the island with Crean and Worsley, Shackleton was unable to sleep: "My mind was busy with the task of the following day … No man had ever penetrated a mile from the coast of South Georgia at any point, and the whalers I knew regarded the country as inaccessible."

Ernest Shackleton

The Ross Sea Party: 1915–17

The other part of the expedition, the Ross Sea Party on *Aurora*, experienced as many perils as those on *Endurance*; indeed, while Shackleton lost not a man under his direct command, three died on the other side of Antarctica.

Aurora was to moor off Ross Island for the winter as a base for a sledging party to lay food depots across the Ross Ice Shelf and onto the Beardmore Glacier for Shackleton's homeward journey. The ship left Hobart on Christmas Day 1914 and arrived at Ross Island on 9 January 1915, captained by Aeneas Mackintosh, who had lost an eye on Shackleton's *Nimrod* expedition. Also on board was an experienced polar veteran, Ernest Joyce, but Shackleton was vague about which of the two was in command, which led to friction and disaster.

Mackintosh decided to use Scott's Cape Evans hut, which was in excellent condition. On 24 January the first depot-laying expedition set out. Mackintosh was eager to get moving in case Shackleton arrived at the end of this first summer, and—against Joyce's advice—he pushed the dogs too hard; all but five died, and much of the task remained uncompleted.

Aurora vanishes

Aurora was to be a supply and accommodation base until spring, when the laying of supply depots would begin, and by mid-March she was secured at Cape Evans by her two anchors and seven steel hawsers. The 10 men who would lay the depots were ashore with minimal equipment. But on the night of 6 May there was a violent storm, and by 3.00 am, the ship had vanished—its mooring of ice had blown away, leaving only bent anchors and snapped hawsers. It had happened before,

and they expected *Aurora* to return, but the vessel was as firmly gripped in ice as *Endurance* on the other side of the continent; almost a year would pass before *Aurora* broke free and limped back to New Zealand.

After another blizzard, they knew that they were marooned for the winter. Mackintosh's first depot-laying expedition was still struggling back, and the four at Cape Evans frantically searched to see what Scott had left behind. They found tins of jam, flour, and oatmeal, and some pemmican, but no soap, tobacco, or medical supplies, and their only clothing was what they were wearing and some spare underwear. They killed seals for meat and fuel—and kept taking scientific observations.

By 2 June, the sea ice was firm enough for Mackintosh's party to get back to Cape Evans, where they learned of the loss of *Aurora*. Rescue might be two years away, but they still had to lay supply depots for Shackleton in the spring or he would certainly die on the ice shelf. Supplies left by Scott, including kerosene and two Primus stoves, were a life-saving discovery, and they also found cake, chocolate, sleeping bags, socks, underwear, and a tent that they cut up and sewed into sledging clothes.

After taking an inventory, they planned to carry 1,800 kilograms (4,000 lb) of supplies onto the Ice Shelf and establish the furthest depot at Mount Hope, at the foot of the Beardmore Glacier, at 83°40′S. Joyce was deferring to Mackintosh, but relations were strained. In mid-August Mackintosh and chief scientist Fred Stevens reached Shackleton's hut at Cape Royds and found cigars, tobacco, food, and soap.

During September they hauled supplies across to Hut Point, the starting point for sledging south. Joyce wrote: "Most of us wore the canvas trousers made from Scott's old tent, and they froze on us like boards." On 26 October they found a sledge left as a marker by Cherry-Garrard, a note for Scott, and six boxes of dog biscuits impregnated with cod liver oil; Joyce wrote: "At last we have struck gold in the Antarctic."

▲ DEEP FREEZE
The desiccated remains of one of the Ross Sea Party's dogs still lies in its collar by the Cape Evans hut. The work of the expedition was made infinitely harder by Captain Mackintosh pushing the dogs too hard too soon, so that most died before the main task of laying supplies.

◄ UNCERTAIN FATE
When the survivors were reunited at Cape Evans, those who crossed from Hut Point were sad but not surprised that Mackintosh and Hayward had not made it. With considerable under-statement, Dick Richards later recalled that "From July 1916 until January 1917, when the rescue ship arrived, it was a bit of a struggle to get by."

The first death

On 3 January 1916 one Primus was burning its own metal, and the party that was depending on it had to turn back. The men were failing, too: by 82°S Mackintosh was weakening and Arnold Patrick Spencer-Smith, the chaplain, had scurvy. They left him in a tent and pushed on to establish the southernmost supply depot at Mount Hope on 26 January. When they returned, Spencer-Smith could not walk; they carried him on the sledge until he died on 8 March and then they buried him in the ice.

By now they were all suffering from scurvy, and when a blizzard stopped them close to where Scott and his companions had perished, they seemed set for the same fate, but they reached Hut Point on 11 March. They had been away for more than six months, "without change of clothing or a bath." The four who had turned back earlier had reached Cape Evans, and now the survivors at Hut Point could eat seal meat and recover their strength. Joyce calculated that they must wait four months for the surface to be stable enough to cross to Cape Evans, but by 8 May after a blizzard had cleared much of the sea ice, Mackintosh announced that he and Hayward would walk across the 19 kilometers (12 miles) that remained. It was a suicidal decision: a blizzard arose, and Mackintosh and Hayward were never seen again. On the night of 15 July, the remaining three, Dick Richards, Ernest Joyce, and Ernest Wild, crossed to Cape Evans and rejoined the other four. The seven men then had to endure a long wait for rescue.

Meanwhile, Aurora and her crew had drifted 1,130 kilometers (700 miles) in the ice. North of the Antarctic Circle, as summer approached, they hoped that the ice would open up and release the ship, but the looser ice did even more damage; when Aurora was finally freed in March 1916, she was leaking and relied on a makeshift rudder to cover the 5,220 kilometers (2,000 miles) to New Zealand. On 3 April 1916 she was taken under tow to Port Chalmers, near Dunedin.

Shackleton returns

Rescue came to the Ross Sea on 10 January 1917. Dick Richards was outside looking for seals when he saw a ship's smoke. He walked calmly back into the hut and casually said: "There's a ship out there." When they saw that the ship was Aurora, the men shouted for joy. The ultimate miracle was when they realized that one of the men approaching them was Shackleton—coming from the north, not lost in the south as they had thought. John King Davis, who was commanding Aurora, wrote of the encounter: "They were just about the wildest looking gang of men I have ever seen ... But the marks of their great physical and mental privations went deeper than their appearance ... factors had combined to render these hapless individuals

as unlike ordinary human beings as any I have ever met ... They were dazed by it; we were excited."

Aurora docked in Wellington on 9 February 1917, and the Ross Sea Party's ordeal was over. Joyce's claim that their expedition was "without parallel in the annals of polar sledging" was justified. The Discovery sledging party of Scott, Shackleton, and Wilson was out for 93 days; Shackleton's "Farthest South" took 120 days of sledging; and Scott's tragic final venture lasted 150 days. The Ross Sea Party was on the ice for 199 days, without adequate clothing or support, and burdened with food for the failed polar party—an incredible achievement. DM

▲ **A SIGHT FOR SORE EYES**
When Ross Sea Party member Dick Richards stepped outside in search of seals on 10 January 1917, he saw a ship's smoke. He calmly reported to the others what he had seen, but once they realized it was the Aurora and that Shackleton was on board there was no containing their elation.

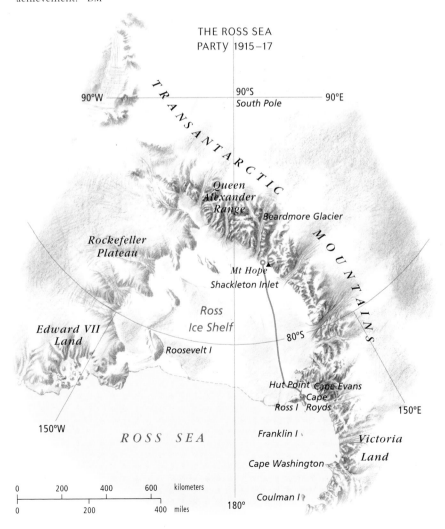

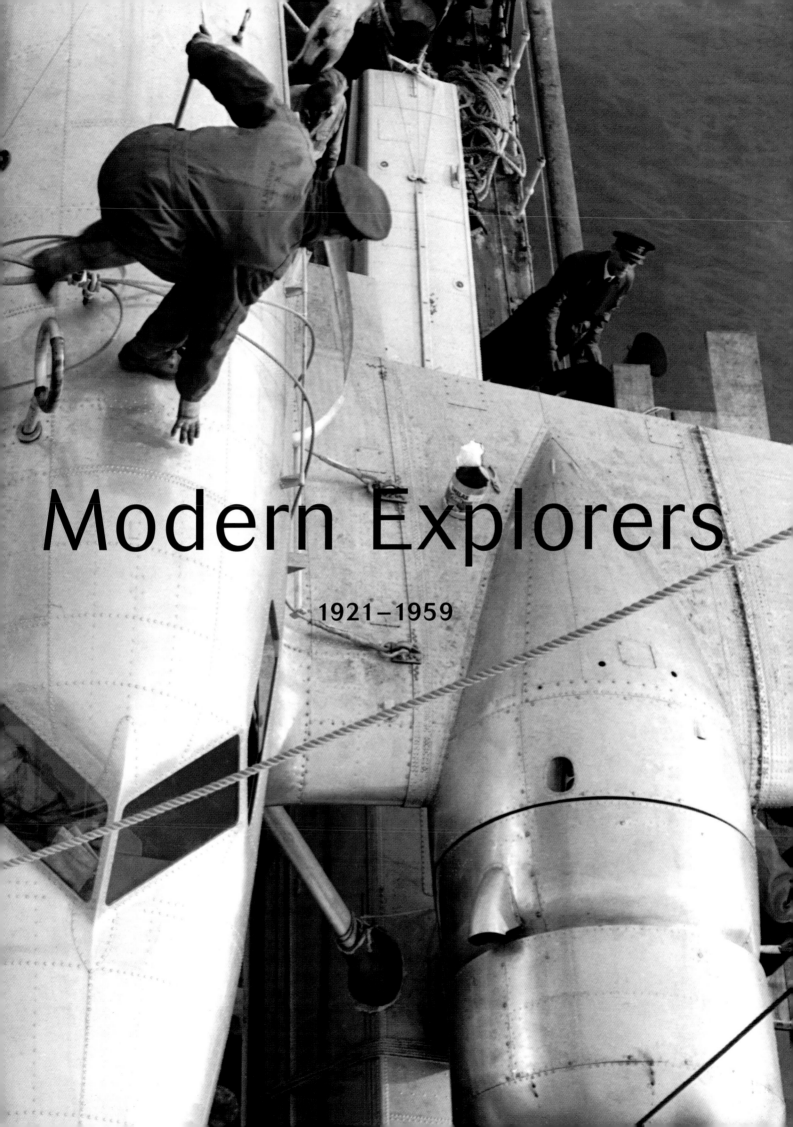

Modern Explorers

1921–1959

The Quest

1920–1922

▲ FINAL RESTING PLACE
Shackleton died in South
Georgia on 5 January 1922.
He was buried in the whalers'
cemetery at Grytviken in
South Georgia. When *Quest*
returned in April, his seven
comrades from *Endurance*
paid their respects at the
graveside. Frank Wild, his
trusted lieutenant, is second
from the left.

"You are always wanting me
to give up something," replied
Shackleton. "What do you
want me to give up now?" He
then suffered a massive heart
attack, and died at 2.50 am on
5 January. He was 47 years old.

After holding a memorial
service and erecting a cairn to
his memory, *Quest* continued
its voyage, but achieved little.
Shackleton's body was taken to
Montevideo, but Lady Emily
requested that he be buried at
South Georgia, and he was
interred there on 5 March.
Symbolically, his was the only
grave with its head facing to
the south, rather than the east.
Quest returned to England on
16 September 1922.

B y 1920 Ernest Shackleton was planning to revisit
the Antarctic region. A friend, John Rowett, put up
money for the new venture, and a dilapidated Norwegian
sealer was renamed *Quest*, at Emily Shackleton's suggestion. Frank Worsley was captain again, Frank Wild was
Shackleton's deputy, and Alexander Macklin was one of
three doctors. The aims of the expedition were vague—
oceanography; searching for sub-Antarctic islands;
mapping uncharted coast; looking for Captain Kidd's
treasure in Trinidad, and for
pearls in the Pacific.

Quest sailed from London
on 17 September 1921, and
then left Rio de Janeiro on
17 December. Shackleton
was obviously unwell, but on
2 January 1922 he wrote:
"At 1 pm we passed our first
berg. The old familiar sight
aroused in me memories …
Ah me! the years that have
gone since in the pride of
young manhood I first went
forth to the fight. I grow
old and tired but must always
lead on." On 4 January they
sighted South Georgia, and
Frank Wild recalled: "I recognized once more the old
buoyant, optimistic Boss."
After a cheerful evening in
Grytviken they returned to
the ship, where Macklin told
Shackleton to take it easy.

A companion's tribute

Worsley wrote: "I knew that I should never look upon
his like again. He was not only a great explorer: he was
a great man … by his genius for leadership he had kept
us all in good health… by sheer force of his personality
he had kept our spirits up… by his magnificent example,
he had enabled us to win through …

"His was a proud and dauntless spirit … We would
have gone anywhere without question just on his order.
What more glowing tribute
could any man wish for?" DM

Previous pages

◄ FLYING SOUTH
New technology—especially
aircraft—became the most
potent force in the modern
era of Antarctic exploration.
Richard Byrd, the pioneer of
polar aviation, added a sophisticated twin-engined seaplane,
the Barley-Grow T8P1, to his
fleet for his third expedition
in 1839–41.

◄ A MEMORIAL CAIRN
This simple cairn was built by
Shackleton's comrades from *Quest*
at Hope Point, on the northern
shore of King Edward Cove. The
grave itself, on the southern shore,
has a granite headstone bearing a
quotation from one of his favorite
poets, Robert Browning: "I hold
that a man should strive to the
uttermost for his life's prize."

British Imperial Expedition

1920–1922

Despite its grandiloquent name, there had never been a smaller expedition to winter in Antarctica than the British Imperial Expedition. Essentially, it consisted of just two men: 19-year-old geologist Thomas Bagshawe, and 22-year-old Royal Naval Reserve Lieutenant Michael Lester, who had some surveying experience. The two spent 10 months, from March 1921 to January 1922, at Waterboat Point on the northwest coast of Graham Land.

The expedition originally had two other members: its official leader, John Cope, who had been surgeon and biologist for the Ross Sea Party of Shackleton's 1914–17 expedition, and aviator Hubert Wilkins, who had spent four years in the Arctic with Vilhjalmur Stefánsson's 1913–18 expedition. Cope wanted to fly 12 war surplus planes to Edward VII Land and thence to the South Pole but could not secure financial backing, so he turned his sights to the Antarctic Peninsula, where whaling ships operated and transport was easier to arrange. He gave up the idea of flying, proposing instead to extend the work of Nordenskjöld's 1901–03 expedition with a dog-sledge journey southward from Snow Hill Island along the western and southern shores of the Weddell Sea.

The four expedition members met at Deception Island on Christmas Eve 1920, hoping to find a whaler to take them to Hope Bay on the tip of the Antarctic Peninsula, and then to make their way southward along the eastern coast of Graham Land. But Hope Bay was icebound, and the whaler deposited them at Waterboat Point on the west coast instead. They still hoped to cross the Peninsula, but there was no way over its mountainous spine. Cope, his credibility in tatters, and Wilkins, his patience exhausted, gave up and found a whaler to take them home, but Bagshawe and Lester "volunteered to stay for the winter and carry on … the scientific work …" Incredibly, they managed to do so, keeping a two-hourly meteorological log, an ice log, a natural history log, and, for a month, an hourly tidal log. Frank Debenham, first Director of the Scott Polar Research Institute, commented: "Never were there such devoted scientists … these two young men possessed that quality so annoying to the great Napoleon, of not having the sense to know when they were defeated."   LC

Messing about in a boat

Bagshawe and Lester made their home in the stranded boat, 9 meters long by 3 meters wide (30 by 10 ft), after which Waterboat Point was named, extending it with packing cases to provide a tiny "lounge" and kitchen, and sewing canvas food bags together to serve as windows. In these cramped quarters, which leaked in the frequent icy rain, they cooked on a stove made from an old oil drum, and lived almost entirely on seals and penguins. For Bagshawe, however, "visiting Antarctica was … sheer delight," and both he and Lester "felt miserable as we watched the hut disappear from sight … it was to us a haven of comfort."

▼ SPLENDID ISOLATION
Waterboat Point is today the site of the Chilean summer station Presidente Gabriel González Videla, though the ruins of Lester and Bagshawe's hut remain. The cliffs and soaring mountains that frustrated the expedition's efforts to cross to the Weddell Sea instead provided a spectacular setting for their stay.

Discovery Investigations

1921–51

The modern whaling industry of Antarctica began in 1904 in the waters around the Falkland Islands, South Georgia, and the South Shetland Islands, an area of acknowledged British sovereignty. Catches—and profits—soared, and in 1908 the British government established the Falkland Islands Dependencies (FID), with the aim of regulating the species and numbers of whales taken, and of licensing the shore flensing plants, where whales were processed. Part of each license fee was set aside for research, and in 1913 a Whaling Committee was established to consider whether whales needed protection from exploitation, and, if so, how this could be achieved. Its first report, published in 1915, concluded that "on its present scale, and with its present wasteful and indiscriminate methods, whaling is an industry which, by destroying its own resources, must soon expire." The wholesale slaughter of whales was exacerbated by the huge wartime demand for whale oil, which was a valuable source of glycerine for explosives; it reached a peak in 1915–16, when nearly 12,000 whales were taken.

Saving the whales

The committee reconvened after the war, and its 1920 report emphasized that the depletion of whale stocks could be avoided only by control of the industry. However, effective control was impossible because not enough was known about the habits and ecology of whales. Thus the Discovery Investigations were born; named after Robert Scott's ship *Discovery*, which became its first research vessel, the Investigations concentrated on "conducting research into the economic resources of the Antarctic with the particular object of providing a scientific foundation for the whaling industry." The Discovery Investigations continued until 1951, ranging in scope from "the greatest animals this world has ever seen down to the microscopic life of the sea." The committee's final report was published in 1980.

Stanley Wells Kemp, an experienced marine biologist, was appointed Director of Research responsible for "planning, organizing and carrying out one of the most comprehensive schemes of oceanographical research ever undertaken by any country in the world." Kemp led the first *Discovery* expedition, which left England in September 1925 and reached South Georgia in February 1926. The captain was John Stenhouse, an experienced Antarctic navigator and master of square-rigged sailing ships; he had taken command of *Aurora* in the Ross Sea during Shackleton's 1914–17 Transantarctic Expedition. Also on board were eight biologists and oceanographers, including Alistair Hardy, who spent hours "fishing for knowledge in this deep water world we cannot see" with nets, dredges, and water bottles at a series of "stations" across the Southern Ocean.

Scientists at work

A shore laboratory had already been set up in Cumberland Bay on South Georgia, near the whaling station at Grytviken, and the scientists continued their collecting, often working "on the platform among whales in all stages of dismemberment. Little figures busy with long handled knives like hockey sticks, look like flies as they work upon the huge carcasses, whose girth is not far short of a railway carriage …" From the data collected, they gradually determined the breeding times, gestation periods, rates of growth, and ages at maturity for various species.

The following season, in 1927–28, *Discovery* was joined by the smaller *William Scoresby*, which was equipped as a trawler and whale catcher as well as a research ship. Between them, the two ships covered a wide area between the Falklands and the South Shetlands, marking whales in order to track their migration routes, collecting krill and plankton, and taking water samples. In 1929 *Discovery* was replaced by *Discovery II*, a purpose-built 1,250 horse-power research ship 70 meters (230 ft) long. Between 1929 and 1951 *Discovery II* made six two-year voyages, including two

GRYTVIKEN, EARLY DAYS
When this was a working whaling station, whales were hauled onto the open space between the jetties and stripped of their blubber—flensed. The blubber was fed into huge pressure cookers in the building with chimneys (center). The oil was pumped into the tanks behind. Meat, bones, and guts were turned into fertilizer and animal feed.

▼ *DISCOVERY* AT SEA
In 1925 Scott's ship *Discovery* was recommissioned as a whaling research ship. After her escape from the ice of McMurdo Sound in 1903, her checkered career included six years in the Canadian Arctic, five years as a wartime supply ship in the Russian Arctic, and another four years laid up on the Thames.

▲ GRYTVIKEN TODAY
Comparing the 1920s picture (left), and today (above) shows how the whaling station grew and then decayed. The main buildings can still be seen, as can two wrecked whale catchers, *Albatross* (nearest jetty) and *Diaz*. The pink-roofed house, once the manager's residence, is now a museum. Shackleton's grave stands in the small graveyard on the further shore.

circumnavigations of Antarctica, while *William Scoresby* made eight voyages in all.

On its fourth commission, in 1935–37, *Discovery II* was diverted to rescue Lincoln Ellsworth and Hollick Kenyon from the Bay of Whales when their plane crashed nearby. Later the same season, a survey party at King George Island was nearly lost when its launch broke down and was wrecked. Happily, most Discovery Investigations were less eventful, but their work laid the foundation for scientific whale conservation, and led to the establishment of the International Whaling Convention in 1936, and, eventually, to the effective regulation of the whaling industry.   LC

▲ A STEAM WHALER
Scientific research on whales and whaling was first proposed in 1913, and this led to the launch of the Discovery Investigations in 1924—a program devoted to examining Antarctica's economic resources, in particular, whaling. However, by the time the *Discovery* voyages ceased in 1951, factory ships and steam whalers had drastically reduced the world's whale populations.

Sir George Hubert Wilkins

1928–29

▲ PIONEER AVIATOR

Sir Hubert Wilkins was the first person to fly an airplane in Antarctica. He was also the first to successfully fly across the Arctic Ocean, for which he was knighted in 1928. After his Antarctic exploits he was invited to join the airship *Graf Zeppelin* on the first round the world flight in 1929.

> MEMORIES AND RELICS

This hangar and derelict Beaver are relics of the British Antarctic Survey base that operated on Deception Island from 1943 until 1967, when a volcanic eruption covered the base in cinders, forcing a hurried evacuation. The black sand beach was used as an airstrip by both the BAS Beavers and Wilkins's Lockheed Vegas in 1928.

Possibly the greatest adventurer and explorer that Australia has ever produced, Hubert Wilkins was knighted in 1928 in recognition of his pioneering flight over polar ice from Alaska to Spitsbergen, in Norway. He had a distinguished and hair-raising photographic and flying career during World War I, and thenceforth regarded the airplane as the natural means of exploring the polar regions. He was involved with the ill-managed British Imperial Expedition to Graham Land in 1920–22 until he lost patience with the venture's cheerful amateurism—its aircraft were never even purchased—but the expedition did take him to Deception Island. With American aviator Carl Eilson, he then flew with great distinction in the Arctic from 1925 to 1928. The fame that this attracted secured financial backing from American newspaper magnate William Randolf Hearst (who is believed to be the model for Orson Welles's reclusive Citizen Kane). In return for the news and radio rights, Hearst raised the money for an Antarctic expedition in the southern summer of 1928, after the Australian government declined to finance Wilkins's dream of exploring and photographing the south polar regions and perhaps reaching the South Pole.

The first flight

Early in November 1928, Wilkins's small expedition of just five men, with Eilson as chief pilot, reached its proposed base in the relative shelter of Deception Island, the doughnut-shaped volcanic caldera that lies some 80 kilometers (50 miles) north-west of the Antarctic Peninsula. *Hektoria*, the Norwegian whaling ship that took them there, carried two single-engined Lockheed Vega high-winged monoplanes, one of which Wilkins and Eilson had used to good effect in the Arctic. Twelve days later, on 16 November, the two aviators made the first powered flight in Antarctic skies. They had originally planned to use skis for take-off and landing, but inadequate ice in Deception's harbor forced them to use conventional wheels, albeit on a somewhat unconventional runway. They took off from a very rough and less-than-straight airstrip scraped out of the black sand beach. The first flight over

Antarctica lasted a mere 20 minutes; after a few brief circuits of the island, and with the weather deteriorating, the plane was back on the ground. But the exploration of Antarctica by air had begun.

Ten days later both aircraft flew, again locally over the island. On landing, however, one of them skidded from the ice into the water, where it lay suspended only by its wings, to be salvaged with considerable difficulty. Finally, on 20 December, after persistent frustrations with the weather, Eilson and Wilkins were able to make the historic flight of over 2,100 kilometers (1,300 miles) that was to prove the feasibility of exploring the south polar regions by air, and to achieve many of the expedition's aims. They flew for 11 hours from Deception and back, across to the mainland of the Antarctic Peninsula, and then south to the Antarctic Circle and beyond, over much territory that had never before been seen, let alone explored, Wilkins recording their journey lavishly in photographs and sketches. Some of his observations and deductions were inaccurate—for example, he erroneously concluded that some parts of the Antarctic Peninsula were part of an archipelago, rather than an extension of the mainland—but the value of the airplane in Antarctic exploration had been demonstrated beyond doubt.

The transcontinental dream

In November of the following year, Wilkins returned to Deception Island to collect the two aircraft, which had been stored there for the winter. Again, he was financed largely by Hearst, even though the American Eilson was not with him this time. Remembering the difficulties of taking off and landing on Deception's rugged volcanic lava, he sailed as far south as the Antarctic Circle in search of easier flying bases. He found none; nevertheless, he made a number of flights with the aircraft rigged on floats, although none matched his epic journey of 20 December 1928.

For several years Wilkins continued with his dream of flying across the Southern Continent from the Weddell Sea to the Ross Sea, but his involvement with Antarctic aviation continued only at sea level. In a non-flying capacity, he helped and advised the millionaire Lincoln Ellsworth in his frequently frustrated attempts to make a major transcontinental flight in 1934–35, and he also played the same role during Ellsworth's final expedition, in 1938–39. **PL**

⋏ STEPPING OUT
In 1929 Wilkins married an Australian actress, Suzanne Bennett, whom he had met at a reception in New York in honor of his successful trans-Arctic flight of the previous year. They were married for 29 years. They had no children.

⋏ ON BOARD SHACKLETON'S *QUEST*
In 1921, Wilkins joined Shackleton's last expedition on *Quest* as a naturalist. His work so impressed the British Museum of Natural History staff that he was asked to lead an expedition to the Torres Strait islands. Seen here with other expedition members, Wilkins is on the far right, with Frank Wild in the center.

➤ WALKABOUT ROCKS
Wilkins claimed the Vestfold Hills area (now Davis station) for Australia in the summer of 1938–39. This cairn is known as Walkabout Rocks because he protected his handwritten proclamation by wrapping it in a copy of the Australian magazine *Walkabout*.

Richard Byrd

1928–41

American flying ace Richard Byrd claimed to have been the first to fly over the North Pole, on 9 May 1926, and in 1927 he was pipped at the post by Charles Lindbergh in the famous race to fly solo and nonstop across the Atlantic from New York to Paris. By the late 1920s, Byrd had one overriding ambition: he wanted to be the first to fly over the South Pole. It was clearly a more difficult proposition than the North Pole, but Byrd was backed by such great American financial names as Edsel Ford, who provided the Ford Trimotor aircraft that was to be the backbone of Byrd's operation, and the millionaire philanthropist John D. Rockefeller, who had extensive media interests. Rather than seek flying bases in milder conditions further north, Byrd established a camp, which he called Little America, on the Ross Ice Shelf, a little way inland from the Bay of Whales, at the end of December 1928. The expedition of more than 80 men carried sophisticated radio equipment for maintaining contact with the outside world—mainly America—to recompense his media backers.

First over the South Pole: 1928–30

Byrd's first Antarctic flight took place on 15 January 1929, followed two days later, on 17 January, by a significant flight during which he discovered a mountain range that he promptly named after Rockefeller. For the next month, the expedition concentrated on the hitherto unexplored lands east of Little America, using the Ford Trimotor and the other two expedition aircraft, a single-engined Fokker Universal, and a Fairchild. By March, however, flying conditions had become virtually impossible; the Fokker was destroyed on the ground in a blizzard, and the remaining two aircraft were deliberately buried in large pits to protect them for the winter. About half the crew wintered underground, in the most sophisticated Antarctic camp constructed up to that time, and it was not until early October that a land-based geological party was able to embark on an extended expedition to the Queen Maud Mountains, and final plans for the polar flight could be finalized. Byrd was doubtless spurred on by the news that Wilkins intended to return to Antarctica: the thought of being thwarted by the Australian in his quest to be the first to fly to the South Pole was a driving force.

The Ford Trimotor lacked the range to fly to the Pole and back nonstop, so on 19 November, mounted on skis, it first flew to the bottom of the Axel Heiberg Glacier to establish a fuel dump and thereby reduce fuel dependency for the polar flight by 700 kilometers (435 miles). Possibly too much was deposited; on the way back the aircraft ran out of fuel and had to make an emergency landing 160 kilometers (100 miles) from Little America. The plane came down safely, but had to wait three days for more fuel to be flown in.

The attempt on the Pole began on the afternoon of 28 November with a crew of four, including Bernt Balchen as pilot and Byrd as navigator. Encouragingly, the geological party at Queen Maud Mountains had reported clear weather there, but no one knew what the interior had in store. The principal barrier was the inland plateau, and the greatest drama of the flight unfolded as Balchen tried to fly up the Liv glacier into the skies above the plateau. Severe turbulence hammered the aircraft, and it was unable to climb. They faced an unenviable choice: turn back, or jettison weight in the form of either vitally needed fuel or food that would be essential in the case of a forced landing. It was the food that went, and the aircraft finally gained enough height—just—to clear the top of the glacier.

The rest of the flight was largely uneventful, and they flew over the South Pole at 1.14 am on 29 November, nine hours and 56 minutes after departure. Little was visible of a fairly featureless landscape; later, Byrd commented: "One gets there and that is all there is for the telling. It is the effort to get there that counts." With fuel so important a consideration, the aircraft turned around after just nine minutes and headed back. Refueling en route went smoothly, and the aircraft landed back at Little America shortly after 10.00 am.

Back to the Pole: 1933–35

Byrd returned to the acclamation of a proud and grateful nation. The fame and fortune that he craved was his for the asking, in the form of honors, adulation—and further financial backing, so that when he turned his sights southward again, some four years later, he was able to enlist the support of many of the industrial giants of the United States. Edsel Ford was again a backer, but the mainstay of Byrd's aerial fleet this time was a large, new Curtiss Condor biplane, supported by a Fokker, a Fairchild, and an autogiro. Like the previous venture, this expedition was large, and its base, reached in December 1933, was the site of Little America, which had been left to the elements four years earlier. They built a new camp, restoring parts of the original and adding to it.

Rather than making record-breaking flights—if any were left to be made—the aim of this expedition was to explore and map the great uncharted *terra nullius* that extended far to the east. Scientific research was another major goal, particularly study of the meteor-ology and geology of Antarctica. From January 1934 until February 1935 (except for the very significant over-wintering time, when no flying was possible), well over one million square kilometers (386,000 sq. miles) were photographed and surveyed from the air, including vast tracts of Antarctic lands previously unexplored, indeed unseen. In the process, the autogiro was wrecked and the Fokker crashed in flying accidents, but no lives were lost. Ground parties also carried out extensive studies, although over an understandably smaller area.

▼ POLAR COMPETITOR
Richard Byrd was already an American hero for his success in flying over the North Pole in 1926, when he turned his sights to the South Pole. With the advice of Roald Amundsen, the leader of the only party to return from the Pole, and substantial corporate backing, he was sure of his ability to succeed.

Byrd's third expedition set up two bases, one on the Antarctic Peninsula, the other at the Bay of Whales. In charge of this West Base was Dr Paul Siple, a veteran of Byrd's two previous expeditions, and destined to follow him as leader of the post-war US Antarctic Program. He is seen here celebrating the Christmas of 1941 with other expedition members.

▲ FLY LIKE A BYRD
The fuselage of the Ford Trimotor is unloaded onto the Ross Ice Shelf from the *City of New York* in December 1928. Byrd made his first flight in January 1929, two months after Wilkins's inaugural Antarctic flight. After spending the winter in an ice coffin, the Trimotor also made a successful polar flight the following November.

Government takes the reins: 1939–41

Byrd's 1933–35 undertaking was among the last of the large, privately funded and operated Antarctic expeditions: thereafter governments became involved. German interest in Antarctica worried the United States government, and in late 1939 Byrd went south again, this time as head of a government-funded expedition. Once more, aircraft played a leading role; the expedition included a fleet of two Curtiss Condors, a Beechcraft monoplane, and a twin-engined, all-metal Barley-Grow floatplane. By now, Antarctic ventures had become almost routine, and the showmanship of earlier expeditions had faded away. Byrd's third expedition conducted extensive survey work and scientific research and evaluation from two bases until March 1941, when the expedition came to an end. PL

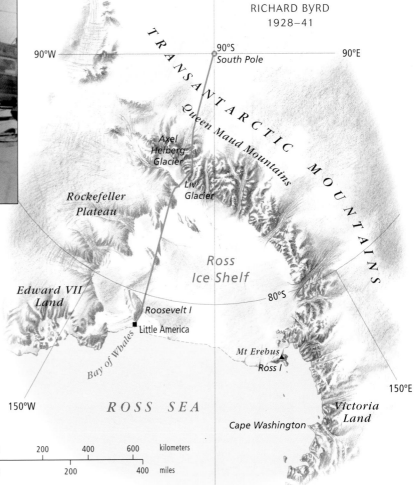

RICHARD BYRD
1928–41

90°W   90°S
South Pole   90°E

TRANSANTARCTIC MOUNTAINS

Queen Maud Mountains

Axel Heiberg Glacier

Liv Glacier

Rockefeller Plateau

Ross Ice Shelf

Edward VII Land   80°S

Roosevelt I

Bay of Whales   Little America

Mt Erebus
Ross I

150°W   150°E

ROSS SEA

Victoria Land

Cape Washington

0   200   400   600   kilometers

0   200   400   miles

180°

BANZARE

1929–31

The 1920s saw a surge of interest in Antarctica by several nations. Norway was eager to extend its whaling grounds from the South Atlantic to the Ross Sea and East Antarctica. France wanted to protect its interests in Adélie Land, based on d'Urville's voyage of 1840, as well as in Iles Kerguelen and other sub-Antarctic islands. In response, in 1923 Britain established a territorial claim to the Ross Dependency, both to control and to profit from the expansion of whaling, and at the 1926 Imperial Conference articulated a policy to "paint Antarctica red"—a reference to the practice at the time of mapping the British Empire in red. In particular, Britain had set its sights on the so-called Australian sector, which had been explored by Douglas Mawson's 1911–14 Australasian Antarctic Expedition (AAE). Mawson himself was keen to extend his discoveries, emphasizing that "over half the circumference of the globe remains to be charted in high southern latitudes," and that, in particular, "the great Antarctic region lying to the southwards … is a heritage for Australians … nearer to the Commonwealth than the distance between the east and west coasts of our own continent."

A hard-working ship

Mawson's persistence led to an agreement between the British and Australian governments to send an expedition to chart the coast of Antarctica from 160°E to 85°E. This was BANZARE—the British, Australian and New Zealand Antarctic Research Expedition. The Falkland Islands agreed to lend Scott's old ship, *Discovery*, which was being used for the Discovery Investigations, for two seasons, from 1929 to 1931. Mawson was expedition leader, and the captain for the first voyage was J. K. Davis, Mawson's skipper from the AAE. The expedition left London in August 1929.

As well as survey work, the plans included a full program of biological and hydrographic research, so there were 12 scientists on board, as well as the ship's crew of 17, plus a Gipsy Moth biplane for aerial reconnaissance and mapping. The ship was not just over-crowded; Davis was convinced that "a more unsuitable ship for oceanographical work than *Discovery* could not have been built." She was very seaworthy, but she rolled so badly that it was like "a sea going rodeo," and was slow, under-powered, and lacking in coal-carrying capacity.

BANZARE sailed from Cape Town in October, first to Iles Kerguelen and to Heard Island, then southward to Enderby Land—which had not yet been claimed—to forestall the Norwegian expedition, under Riiser Larsen, that was already in the area. Mawson sighted land several times and made three flights over it, but he was able to land only once, on Possession Island (53°37'E) on 13 January 1930. Next day, at 49°E heading west, he met Larsen in *Norvegia*, heading east; by mutual agreement, and in accord with their governments' respective policies, the two expeditions turned around, marking 45°E as the boundary between Norway's claim to the west and Britain's claim to the east. In mid-January, with coal running out, *Discovery* turned homeward, arriving in Adelaide two months later.

Staking a national claim

Discovery sailed south again from Hobart in November, once more led again by Mawson, but with K. N. Mackenzie as captain and a similar complement of scientists and surveyors. The route led south to Macquarie Island, where they found the derelict AAE huts and wireless masts, and then to the Ross Sea, where *Discovery* refueled from the huge whaling factory ship *Sir James Clark Ross*—a striking contrast between old and new. Following the coast westwards, they encountered heavy pack

▲ **LOST OPPORTUNITY**
Sir Douglas Mawson, who was knighted after his Australasian Antarctic Expedition of 1911-14, was 47 when he returned as leader of BANZARE. Although committed to its program of scientific studies, the expedition offered Mawson little of the overland exploration at which he excelled, and he was disappointed with its results.

◀ **TEAM SPIRIT**
The expedition carried a scientific staff off twelve, seen here with members of the crew. The photographer Frank Hurley, who had served with both Mawson on the AAE and Shackleton on the *Endurance* expedition, is in the front row (with scarf). This image of Mawson in a balaclava (behind Hurley) featured on Australia's first $100 note.

▶ **HOME SWEET HOME**
The scientists went ashore for 10 days on Heard Island where they used a small hexagonal hut at Atlas Cove, built by Norwegians as a refuge for shipwrecked whalers. After barricading the door against marauding elephant seals, Frank Hurley praised his temporary home as "warm, dry, rent free, and no taxes."

▼ MAWSON AT RED DOME
The expedition spent 10 days at Iles Kerguelen and made extensive surveys along the many fiords that pierce the coast. The scientists noted the contrast between the stunted vegetation of the mainland, devastated by rabbits, and the luxuriant growth on the isolated smaller islands

▲ FOOD FOR THOUGHT
After leaving Cape Town the expedition called first at Iles Crozet, where the scientists enjoyed a picnic on the beach of Possession Island—a lighter interlude in an unpleasant visit, for they found the crew of a South African sealing ship slaughtering elephant seals. First officer McKenzie called "the murder of a whole beach, bulls cows and pups ... an absolute outrage."

> THE GIPSY MOTH

A new De Havilland DH60 Gipsy Moth was purchased in England for the BANZARE voyages. It was first flown—for ice reconnaissance—on 31 December 1929 by Stuart Campbell and Eric Douglas. Mawson's first flight was one hour—he flew over what he named Mac.Robertson Land on January 5 1920.

ice and could not land until they reached Commonwealth Bay, the site of the AAE's main base in 1911–14, where Mawson found his old hut filled with ice. A Proclamation claiming the region for Britain was read and the expedition pressed on, sailing over "land" reported by Charles Wilkes in the 1840s; Mawson reasoned that it must have been an ice tongue. Further west, they landed and raised the flag at Scullin Monolith and Cape Bruce; they also sighted the Casey, David, and Masson ranges of Mac.Robertson Land, and, with the help of the Gipsy Moth, discovered the large indentation of Prydz/Mackenzie Bay.

Bad weather and heavy ice hampered both of the BANZARE voyages, and Mawson was disappointed with the results; he was unable to undertake as much overland exploration as he would have liked, and had to rely on aerial reconnaissance. Even so, the expedition gathered a wealth of information about a previously unknown part of Antarctica, and resulted in the 1933 affirmation of Britain's claim to the Southern Continent between 160°E and 45°E; later, in 1936, this was to become the Australian Antarctic Territory. **LC**

◄ FLYING THE FLAG

At noon on 5 January 1931 at his old base at Cape Denison the Union Jack was hoisted, the National Anthem sung, and Sir Douglas Mawson read a Proclamation in the name of the reigning monarch taking possession of all the area around Commonwealth Bay as King George V Land.

BANZARE VOYAGES
1929–31

Prince Edward Islands
Iles Crozet
Iles Kerguelen
Heard and McDonald Is
30°E
40°S
50°S
60°E
60°S
70°S
80°S
Haakon VII Sea
Antarctic Circle
Dronning Maud Land
Enderby Land
Mac. Robertson Land
Amery Ice Shelf
Prydz Bay
Princess Elizabeth Land
Davis Sea
90°E
Sth Shetland Islands
Weddell Sea
Coats Land
Palmer Land
Ronne Ice Shelf
Alexander I
Antarctic Peninsula
Bellingshausen Sea
Peter I Øy
Ellsworth Land
Wilhelm II Land
Queen Mary Land
Shackleton Ice Shelf
90°S
South Pole
TRANSANTARCTIC MOUNTAINS
Amundsen Sea
Marie Byrd Land
Ross Ice Shelf
Wilkes Land
120°E
Ross Sea
Oates Land
George V Land
Terre Adélie
Dumont d'Urville Sea
Possession I
Cape Adare
Commonwealth Bay
Antarctic Circle
Balleny Is
150°W
Macquarie I
150°E
Campbell I.
Auckland Is
Hobart
Adelaide
Melbourne
Australia
Stewart I
Chatham Is
30°S
Wellington
New Zealand
Sydney
180°
150°E

0   500   1000   1500   kilometers
0   500   1000   miles

—— 1929-30
—— 1930-31

◄ EBBS AND FLOES

On BANZARE's second voyage, easier ice conditions allowed *Discovery* to sail closer to the shore and into the great indentation of Prydz Bay, amongst the tabular icebergs spawned by the Amery Ice Shelf. Many whalers also took advantage of the good season, several operating as far west as Prydz Bay.

British Graham Land Expedition

1934–37

The British Graham Land Expedition (BGLE) represents the watershed between the heroic and the modern phases of Antarctic exploration, delimited by the outbreak of World War II. It was the last primarily private expedition for several decades, with the last mainly sail-powered ship, and it was the first to combine the traditional use of dogs and skis with modern tractors, motor boats, and aircraft. Its leader was John Rymill, a South Australian farmer and grazier with a lifelong passion for polar regions. He was a member of the first British Arctic Air Route Expedition to Greenland under Gino Watkins in 1930–31, and he led a second expedition (1932–33) after Watkins's death in a kayaking accident in 1932. Undeterred, Rymill resolved to realize their common ambition of organizing an Antarctic expedition. He managed to raise £20,000 from the Royal Geographical Society and the Colonial Office, and used it to buy a 34-meter (112-ft) three-masted topsail schooner (called *Penola* after his family farm), a single-engined Gipsy Moth biplane, boats, dogs, sledges, a tractor, a hut, a hangar, and all the other equipment necessary for two years in Antarctica.

▲ PENOLA TO THE RESCUE
The expedition ship *Penola* was a 30-year-old schooner, only capable of four knots under sail, and with an unreliable engine. But under her determined Navy captain "Red" Ryder, she made a major contribution to the success of the expedition. After wintering in the Argentine Islands, the ship was sent to the Falklands for a refit before returning to retrieve the expeditioners.

An economical venture

The expedition ran on a shoestring, but Rymill had no trouble recruiting four veteran companions of his Arctic expeditions and three young scientists, as well as a naval crew to sail the ship. They set off in September 1934 for the west coast of Graham Land, intent on determining whether it was a peninsula or an archipelago. The question had been raised by Hubert Wilkins's pioneering flights south from Deception Island in 1929, on which he claimed to have identified ice-filled fiords cutting through from the Bellingshausen Sea on the west to the Weddell Sea on the east.

Penola's engines broke down just south of the Falklands, but they sailed on and coaxed the ship into an anchorage in the Argentine Islands with the plane and the motor boat. There Rymill established his first base, and used this novel air–sea reconnaissance technique to lay depots further south for winter sledging journeys. During a flight in February 1935, he sighted Marguerite Bay, south of Adelaide Island, and this became the expedition's next target, after a winter exploring the islands and coast by tractor, dog sledges, and skis. Strong currents and katabatic winds sweeping down from the mainland glaciers made for treacherous sea ice; John

Bechervaise, in his biography of Rymill, *The Will and the Way of John Riddoch Rymill*, wrote that the conditions tested to the full Rymill's "almost uncanny judgement in respect of sea ice. He seemed to know by instinct the degree of risk, over sea ice and crevasses, which could sensibly be taken." Bechervaise also recorded a fellow expeditioner's comment that Rymill's "strong sense of responsibility not just to see the expedition through but for the party he had selected … made a remarkably happy expedition under a leader whom all of us liked and greatly respected."

Peninsula or archipelago?

In January 1936 *Penola* moved the expedition south to the Debenham Islands in Marguerite Bay, returning to South Georgia for a refit during the second winter while the land-based party was making its most significant discoveries. Reconnaissance flights in March revealed the complex archipelago between Adelaide Island and the mainland to the north, but cast doubt on Wilkins's postulated sea-level links with the Weddell Sea. Once the sea ice firmed, the ski-equipped plane could range more widely, and in August a narrow sound running south between Alexander I Land and the mainland was revealed.

In early September a major dog sledge expedition set off to explore the sound, and penetrated more than 960 kilometers (600 miles) along it. Here the sound opened out to the southwest, suggesting that Alexander I Land was an island (as it later proved to be), but they found no sea-level channels to the east. A second sledging trip of 375 kilometers (233 miles) inland across the plateau also failed to uncover any evidence of links to the Weddell Sea.

It was with great satisfaction, tinged with sadness at having to shoot the dogs who had made the achievement possible, that Rymill headed north when *Penola* returned to collect the land party in March 1937. Having followed the coast from 64°30'S to 72°30'S, the expedition had demonstrated that Graham Land was not an archipelago but a peninsula. This discovery, and the extensive scientific data collected, made the 16 men of the British Graham Land Expedition some of the most productive Antarctic explorers ever—and some of the most cost-effective. LC

➤ VERSATILE TRANSPORT
The single-engined De Havilland Fox Moth aircraft could fly with either skis or floats, and was small enough to be towed by *Stella,* the expedition's motor launch, as seen here. This versatility was enhanced by the use of dog and motor sledges.

> STRAIT AND NARROW
Penola Strait, named after the expedition ship, separates the Argentine Islands from the mainland of the Antarctic Peninsula. The site of the expedition's first base on Winter Island became Base F of the post-war British network of stations set up to counter-act Chilean and Argentine claims to the area. Base F later became Faraday station and is now the Ukrainian Vernadsky station.

> A NEW SOUTHERN BASE
The Moth was invaluable in reconnoitering a route that *Penola* could follow from the Argentine Islands to a more southerly base for the second year's exploration. A suitable site was found at Marguerite Bay in the Debenham Islands where dogs, stores, and equipment were landed.

◁ RYMILL'S RETURN
Rymill (on the right) and the expedition's doctor Edward Bingham photographed on their return from crossing Graham Land east to the Weddell Sea. They spent a month exploring valleys and glaciers on the Weddell Sea coast, but found no evidence of Wilkins's channels cutting through the peninsula.

Lincoln Ellsworth

1934–39

With the South Pole "conquered" by the mid-1930s, the next challenge was trans-Antarctic flights. The ebullient American millionaire Lincoln Ellsworth had flown in Arctic regions, and had financed an unsuccessful attempt by Hubert Wilkins (who was perhaps in search of a change of altitude) to reach the North Pole in 1931 by submarine. The two teamed up again when Ellsworth decided to organize an expedition to the south, Ellsworth to fly and Wilkins to provide the managerial expertise from sea level. The plan was to fly from the Ross Sea to the Weddell Sea and back. Although hardly a crossing of the continent at its widest point, this would still be a return flight of about 5,500 kilometers (3,420 miles), and would cover substantially unexplored territory over much of the area between the flights made by Wilkins in 1928 and Byrd in 1929.

▲ HIGH HOPES DASHED
Ellsworth (left) and his co-pilot Norwegian Bernt Balchen in the cockpit of *Polar Star*; in January 1934 they planned to fly across Antarctica from the Ross Sea to the Weddell Sea and back. But the plane was badly damaged when the ice shelf on which it had been unloaded from the ship broke up, and the flight was delayed for a year.

A chapter of accidents

Ellsworth and Wilkins acquired a ship, *Wyatt Earp*; recruited Bernt Balchen, who had been Byrd's pilot to the South Pole in 1929; and took delivery of a newly designed and constructed single-engined two-seater Northrop Gamma metal monoplane. But thereafter, success did not come readily. The pair reached the Bay of Whales in January 1934, but after a brief test flight the Northrop Gamma was damaged beyond any hope of on-site repair when the ice shelf broke apart beneath it. They retreated northward to rebuild the aircraft and returned to Deception Island in November with a modified plan to fly one way, in the opposite direction, and finish at Little America. This time, a smashed engine rod on the very first engine run—and no spare—forced *Wyatt Earp* to go back to South America for a replacement. When she returned, the ice at Deception had melted (the aircraft was to use skis), and they were forced to relocate 160 kilometers (100 miles) further south. Here the weather delayed them for more than a month; they tried to take off on the record-setting flight on 3 January 1935, but poor weather forced Balchen to turn back after an hour. Ellsworth abandoned the project for the summer, but it was still some time before the ship was able to sail north through the pack ice.

◄ TORTUOUS TRIP
Ellsworth (right) with Herbert Hollick-Kenyon, co-pilot of *Polar Star* on the successful flight from Dundee Island to (nearly) Byrd's Little America base. Anticipating a 14-hour flight, they finally reached Little America on foot 22 days later.

A dream achieved

For his third attempt, in November 1935, Ellsworth replaced Balchen with English-born Herbert Hollick-Kenyon. A northern base was established on Dundee Island, just off the northeastern tip of the Antarctic Peninsula, but on 20 November disaster struck again. A mere couple of hours into the proposed flight of 3,700 kilometers (2,300 miles)—the longest period that Ellsworth had achieved in Antarctic skies—a fuel gauge threatened to disintegrate, and they had to abandon the attempt. The next day, after some five hours in the air, bad weather forced Hollick-Kenyon to again turn back. Two days later, on 23 November, they took off again, and finally achieved Ellsworth's dream—although hardly in the shortest possible time. Part way through the flight, as they crossed lands never before seen from the air, their radio failed and communication with *Wyatt Earp* was cut—but this time they did not turn back.

Some 14 hours into the flight—roughly two thirds of the way—they landed, stopping for about 19 hours for Ellsworth to pursue territorial claims and stretch his legs. The weather deteriorated once they were back in the air, and they had to make two forced landings. The first delayed them for three days, the next for a blizzard-racked eight, before they at last took off for the final 800 kilometers (500 miles) to the Bay of Whales—only to run out of fuel less than half an hour from their destination. Eventually, on 15 December, after a long and tortuous 10-day trek, they reached the Ross Sea on foot and found the remains of Little America II, abandoned in February of that year. Ellsworth had left a detailed emergency plan with Wilkins on the ship, which had included provision for a possible rescue, so the party settled down to wait for *Wyatt Earp* to collect them.

Help did arrive, but from an unexpected source. When *Wyatt Earp* had lost radio contact with Ellsworth back on 23 November, Wilkins had raised the alarm (he could not have delayed because of the ship's contractual arrangements with United States media interests). The Australian government, prompted by Douglas Mawson, immediately asked the British to provide a suitable rescue ship, and *Discovery II* left Hobart for the Bay of Whales. On 14 January 1936, an Australian Airforce DH60 Gipsy Moth from the ship flew over the stranded party and directed them to *Discovery II*. Only five days later, *Wyatt Earp* arrived at the Bay of Whales, delayed by heavy ice and the need to investigate various collection points that Ellsworth had pre-arranged in his emergency plan.

Another try

Ellsworth made one more expedition, to the other side of the Southern Continent, in the summer of 1938–39. Again he was accompanied by Wilkins, but in a bizarre twist each was now making, or reinforcing existing, territorial claims to the same area on behalf of their

▲ **THE RESCUE PARTY**
Ellsworth and Hollick-Kenyon's trans-Antarctic flight ended at Byrd's Little America, where they waited for their ship to arrive. On losing radio contact, however, their ship had raised the alarm and Australia sent *Discovery II* to rescue them. Ellsworth reported that "I saw through the fog, which magnifies frightfully in those regions, what appeared to be a whole army marching towards me; in reality there were six men."

A slightly unwelcome "rescue"

Despite the major and minor disasters that pursued him, Ellsworth's epic journey of 1935 was the longest transcontinental flight thus far completed. When *Discovery II* and *Wyatt Earp* reached the Bay of Whales, the Northrop Gamma was salvaged and loaded onto *Wyatt Earp*, and Ellsworth headed north to Australia on *Discovery II*. He was prolific in his thanks for Australia's assistance, but always maintained that his plan-ning had been so meticulous that help had never really been necessary. Possibly both Ellsworth and Wilkins long regretted that heavy ice and investigative land-falls had delayed *Wyatt Earp* during her southward search, and that she had not beaten *Discovery II* to the Bay of Whales.

▲ **HISTORIC AIRCRAFT**
Ellsworth's Northrop Gamma monoplane, *Polar Star*, finally fulfilled its destiny to make the first trans-Antarctic flight. Despite running out of fuel only 40 kilometers (25 miles) short of its destination, it was recovered by the *Wyatt Earp* and is now on display at the National Air and Space Museum in Washington DC.

respective governments, Ellsworth from the air and Wilkins from the ground. Ellsworth may also have been stung by suggestions that his epic flight of 1935 could not truly be described as trans-Antarctic. At the back of his mind, therefore, was the possibility of a flight across the Southern Continent to the Pole, and then on to the Bay of Whales on the other side.

Dreadful weather and heavy ice marred and prolonged the journey south from Cape Town, and it took about two months to reach the Antarctic mainland. *Wyatt Earp* spent much of the first two weeks of 1939 frantically maneuvering to avoid icebergs. Three brief flights were made in an Aeronca aircraft, which had narrowly avoided being destroyed by an engine fire; then, on 11 January, a flight of greater interest took place in a single-engined Lockheed Delta. An inadequate ski runway meant that it could not carry enough fuel for a trans-Antarctic flight; instead, in three hours Ellsworth and his pilot, Canadian Jack Lymeburner, flew inland and laid claim from the air to an area previously claimed by Australia. (In the three preceding days, Wilkins had landed three times to reinforce Australia's claim!) Just a few days later, a serious injury to a member of the ship's crew forced them to turn back to Australia, making another tempestuous crossing of the Southern Ocean. Ellsworth continued to dream of returning to Antarctica—specifically to the South Pole, for a protracted expedition, including an overwintering—but these dreams eventually came to nothing. PL

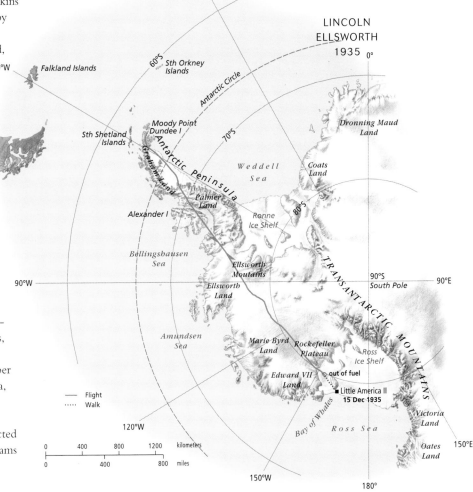

LINCOLN ELLSWORTH 1935

Flight
Walk

The Schwabenland Expedition

December 1938–January 1939

One of the more bizarre byproducts of Germany's Third Reich, the Schwabenland Expedition was the shortest Antarctic expedition ever undertaken, involving only three weeks south of the Antarctic Circle. Mounted by Air Chief Marshall Hermann Göring in a bid to claim Antarctic territory for Germany, it was led by Captain Alfred Ritscher of the German Navy, and was the first attempt by any nation at large-scale aerial mapping of Antarctica. It employed two 10-tonne

January 1939. Over just seven days, its two aircraft made 16 flights and surveyed 600,000 square kilometers (232,000 sq. miles), about half of which the crew recorded in 11,000 photographs. The expeditioners ventured 600 kilometers (370 miles) inland, and discovered several mountain ranges, whose peaks geographers later named after famous German explorers—among them Erich von Drygalski, Alexander von Humboldt, and Wilhelm Filchner.

The expedition's aerial surveys ranged from 4°50′W to 16°30′E, but although the supporting ground parties landed three times on the ice edge, poor ground control limited the value of the expedition's photographs for accurate mapping; furthermore, many of the photographs were lost during World War II. The aircraft never landed on the Antarctic Continent, but at short intervals along their flight path aluminum darts topped with swastikas were dropped in an attempt to establish the German claim to "Neu-Schwabenland."

Territorial squabbles

Germany's claim to Antarctic land was quickly disputed by Norway—the same area had been explored more thoroughly, and several landings had been made, by Norwegian Riiser Larsen's two expeditions of 1927–30. The Royal Decree proclaiming Norwegian sovereignty over the area between 20°W and 45°E had been issued on 14 January 1939, just five days before the Schwabenland expedition arrived, and was

ABANDONED CLAIMS
Germany's pre-war Antarctica forays of 1938–39 were short-lived; with World War II, Allied strategists (above) were soon more interested in captured German maps of Europe than in German attempts to map and claim Queen Maud land.

Dornier Wal flying boats, launched by catapult from their mother ship, *Schwabenland*. On return, the aircraft landed near the ship, and had to be hoisted on board for the next flight. Both the ship and the aircraft were borrowed from the German airline Lufthansa, which normally used them on the Atlantic run, the ship being stationed in mid-ocean as a refueling point.

The expedition left Hamburg in December 1938, and reached the coast of Queen Maud Land in late

recognized by all other claimant nations of the time: Australia, Britain, France, and New Zealand. But the territorial dispute over Antarctica was soon overshadowed by far more significant conflicts in Europe—the outbreak of World War II. After the war, Ritscher published the surviving aerial photographs, providing valuable new information about a previously unknown region of Antarctica, but Germany's claim to the area was never pursued. LC

Operation Highjump

1946–47

An entirely new concept in Antarctic exploration, Operation Highjump was the first primarily military Antarctic operation. Run by the United States Navy, it was the biggest Antarctic venture ever, with 13 ships, 23 aircraft, and over 4,700 personnel. The operation originated in the massive demobilization of the United States armed forces after World War II, the escalating cold war between the the United States and the Union of Soviet Socialist Republics, and the realization that Antarctica was one theater in which this competition could be played out. So, in August 1946, the United States Navy initiated the Antarctic Development Project to meet this challenge. Its main purpose was to build up personnel, material, and logistics capacity in polar regions, but it was also to consolidate United States interest in areas previously explored, find suitable base sites, and determine the feasibility of building and maintaining permanent bases, including air fields. Scientific research was only a secondary concern.

The immediate objective was to establish a base on the Ross Ice Shelf, near Richard Byrd's 1940-41 Little America III, and to photograph as wide an area as possible from the air. There was also an ambitious program of aerial mapping of the entire coastline.

▷ NOSTALGIC LUNCH
Admiral Richard Byrd (right), Officer in Charge of Operation Highjump, and his Scientific Advisor Dr Paul Siple, on their fourth expedition. Siple had led the Little America III venture in 1940–41 and he found his old base more than 3 kilometers (2 miles) north-west of its original position, and 1.5 meters (5 feet) under ice. He dug his way in for a lunch with comrades from the previous expedition.

▽ GROUND CONTROL SITE RE-USED
First mapped from the air, Peterson Island in the Windmill Islands was used by Operation Windmill in 1948, which gave its name to the group of about 50 islands. A plaque commemorating an Operation Highjump landing is sited close to the fiberglass Apple hut now used by Australian scientists from nearby Casey station.

▲ ICE SCULPTURE

A spectacular windscour in the
coastal ice cliffs surrounding
O'Brien Bay, near Casey
station in Wilkes Land. The
bay is named after US Navy
Lieutenant Clement O'Brien
who served as a radio
communications officer with
Operation Windmill.

A three-part armada

The central group, comprising an icebreaker, two supply
ships, a submarine, and a command ship under Admiral
Richard Cruzen, was to penetrate the Ross Sea pack ice
and establish the base. In one of the worst ice years on
record, this was a difficult undertaking. The submarine
proved to be a dangerous liability, and had to be
escorted out of the ice by the icebreaker *Northwind*,
leaving the thin-skinned supply ships to fend for themselves; one of them tried to blast its way through the
pack ice with a cannon to escape a menacing iceberg.
Eventually, however, *Northwind* brought the convoy
safely to the Ross Ice Shelf.

Little America IV, a 200-tent camp with an ice runway, was set up to receive the six R4D aircraft, which
were brought to the edge of the pack ice by aircraft
carrier. The ship's bridge was in the middle of the flight
deck and the aircraft were too wide to get past it, so
they could use only half the length of the flight deck

and had to be launched by JATO (Jet Assisted Take Off),
but with Byrd on board they landed safely on the ice on
29–30 January 1947. From there, Byrd made his second
flight over the Pole, and 160 kilometers (100 miles)
beyond. In over 200 hours of flying, sophisticated aerial
mapping cameras recorded vast areas of new territory,
defining the shape of the Ross Ice Shelf, revealing the
nature of the Dry Valleys, and confirming the existence
of the unbroken chain of the Transantarctic Mountains.
At the end of February, Little America IV camp was
evacuated; the tents, aircraft, and equipment of this
most expensive and technologically advanced Antarctic
operation were abandoned, never to be used again.

The eastern group, comprising three Martin Mariner
PBM seaplanes and a seaplane tender, and accompanied
by a tanker and a destroyer under Admiral George
Dufek, was assigned the area from the Amundsen Sea to
the Weddell Sea. This was the worst sector for weather
and sea conditions, and there was an early setback when

they lost a seaplane and three crew, and had to mount a protracted rescue of the other six. They did some useful mapping along the coast of Marie Byrd Land and south Graham Land, but none could be done in the Weddell Sea, and the overall results were disappointing.

The western group, with similar resources under Captain Charles Bond, worked westward from the Balleny Islands. They encountered better weather, and made the most significant discovery of the entire operation on 1 February 1947, when Commander David Bunger sighted the ice-free oasis in Queen Mary Land that now bears his name: Bunger Hills. This group mapped over a third of the coastline, from 160°E to 40°E, and in places penetrated 800 kilometers (500 miles) inland.

A limited success

For all its massive scale, and for all the claims made for it, Operation Highjump photo-mapped a mere 25 percent of its initial objective. More than half the photographs that were taken were useless, due to lack of ground control, although this was partially remedied by the use of helicopters in Operation Windmill the following year, and only 35 percent were incorporated into published maps. Nonetheless, this was a huge advance on existing maps of Antarctica—and Operation Highjump certainly succeeded in making the United States Navy a world leader in the militarization of Antarctic research and exploration. LC

▲ SHORT-LIVED ENCAMPMENT
Supplies are unloaded onto the sea ice in the Bay of Whales (USS *Yancey* in foreground). All equipment had to be taken by tractor-drawn 10-ton sledges to the site of Little America IV two kilometers (over 1 mile) away. After the three-month operation the base was abandoned, never to be used—or even seen—again.

◄ ICE-FREE OASES
Photo-mapping during Operation Highjump revealed that the Dry Valleys were far more extensive than previously thought. The operation also discovered new ice-free oases, most notably the Bunger Hills in Queen Mary Land.

ANARE

1947–present

Δuring the "cold war" that followed World War II, both the United States and the Union of Soviet Socialist Republics looked to Antarctica as a possible sphere of economic advantage and influence. This put pressure on Australia to safeguard its claim to 42 percent of the Southern Continent. Douglas Mawson was instrumental in persuading the Australian government to act decisively, and he joined other polar veterans, including J. K. Davis, on the Executive Planning Committee charged with implementing a policy of "effective occupation" of Australia's Antarctic sector. As a result of his efforts, the Australian National Antarctic Research Expeditions (ANARE) were launched in 1947. Stuart Campbell, another Antarctic veteran, was chosen to lead the first ANARE, which set up bases on Heard Island in December 1947 and Macquarie Island in March 1948, each with a staff of fourteen. The only ship available was a war-surplus HMAS LST 3501 (landing ship, tanks), which "quivered like a springboard … the decks bend and ripple like a caterpillar in progress" in the swells of the Southern Ocean. A simultaneous attempt to reach the Antarctic Continent in *Wyatt Earp*,

▼ DISTINGUISHED SERVICE

Nella Dan was the longest serving, and best loved, of the *Dan* ships that provided access to the continent for ANARE for over 30 years. Her stranding and subsequent scuttling at Macquarie Island in 1986 marked the end of an era. In 1990 she was replaced by a purpose-built research and resupply icebreaker *Aurora Australis*.

► CELESTIAL DISPLAY

The Antarctic atmosphere produces spectacular effects like this solar pillar caused by refraction of light by ice crystals, photographed near Davis station. Davis, in the ice-free oasis of the Vestfold Hills, is known as the 'Riviera of the South' for its warm sunny summers and comparatively mild winters.

▼ MAWSON'S GROWTH

The green-roofed hut (right) is one of the station's original (1954) buildings; others were added over time: aluminum workshops and sleeping quarters (center) (1960-70s), the red and green buildings (1980s). The large geodesic dome which houses the satellite communication system was built in the 1990s.

the old wooden sealing ship that Lincoln Ellsworth had used in 1934–35, failed due to heavy ice and mechanical problems. It was not until 1954, with the charter of the Danish ice-strengthened cargo vessel *Kista Dan*—the first in a long line of *Dan* ships that serviced ANARE until 1986, when *Nella Dan* was wrecked at Macquarie Island—that access to the Southern Continent could be guaranteed and a permanent base established.

Horseshoe Harbor, in Mac.Robertson Land, was selected as an ideal location, and there Mawson station opened on 13 February 1954. The Heard Island station was closed in 1955, and much equipment was transferred to the new base, including dog teams and sledges. These became the backbone of an extensive program of exploration and mapping, which reached south to the Prince Charles Mountains and east and west along the coast of Enderby Land and Mac.Robertson Land. Dog sledges were supplemented by tractors, and from 1956 by RAAF Beaver and Auster aircraft. The Lambert Glacier—the world's largest—was discovered, and enormous tracts of land, including 2,000 kilometers (1,240 miles) of coastline were photographed.

Australia's Antarctic role

Australia has had a continuing commitment to the scientific study of Antarctica, and this was reflected in her major participation in the International Geophysical Year (IGY) of 1957–58, and in her later role in negotiating the Antarctic Treaty and as founding signatory in 1959. During the IGY a second Australian station, Davis, was built in the Vestfold Hills, and the program of Antarctic meteorological, magnetic, and upper atmosphere and auroral observations was stepped up. The Vestfold Hills are one of the largest of the ice-free Antarctic oases, and Davis station became a focus for studying the unique ecology of its lakes, as well as contributing to the geophysical program.

▼ OLD AND NEW AT CASEY STATION

Casey station's unique caterpillar-on-stilts design was used to avoid drifts building up by allowing snow to blow harmlessly beneath. It worked well, but its position so close to the shore exposed it to salt spray that weakened the supporting structure, so the new Casey was built higher up the hill.

Both the United States and the Union of Soviet Socialist Republics built stations in the Australian Antarctic Territory for the IGY; the American station, Wilkes, was on the coast, and the five Russian stations included Vostok, the highest and coldest of the Antarctic bases. In 1959 ANARE took over Wilkes station, and in 1962 six men traveled by tractor from Wilkes to Vostok and back, a distance of 3,000 kilometers (1,860 miles)—

the most ambitious overland Antarctic journey attempted to that time. They took ice core samples and fired seismic shots to determine the thickness of the ice cap, and also made magnetic and meteorological observations in an area never traversed before—or since. By 1965 Wilkes station was almost buried in snow, so Davis station closed down for four years to release funds to build a new Australian station, named Casey, to replace Wilkes.   LC

Phillip Law

Born in northern Victoria and educated in Melbourne, Phillip Law was appointed senior scientific officer of the newly conceived Australian National Antarctic Research Expeditions (ANARE) in 1947; two years later he became its director, a position he held until 1966. Over that period he made 28 voyages to Antarctica or the sub-Antarctic and developed a technique of "hit and run" exploration that accurately charted over 5,000 kilometers (3,100 miles) of the coast of the Australian Antarctic Territory (AAT). When Law retired after 17 years as Director of ANARE, the main geographic features of the AAT were known and he had established a strong focus on scientific research, both on land and at sea. He had also defined the ANARE style—inventive, resourceful, egalitarian yet highly professional—that is still its distinguishing characteristic.

▲ LAW FIELD BASE IN THE LARSEMANN HILLS

As well as permanent stations, ANARE operates a number of summer bases in areas of particular scientific significance. Scientists live in traditional polar pyramid tents (right) or the unique Australian-designed red Apple huts that are easily transported slung beneath a helicopter.

The International Geophysical Year

1957–58

The International Geophysical Year (IGY) focused attention on Antarctic science, and was a stepping stone to the Antarctic Treaty of 1959. The idea originated in a discussion among American and British physicists in 1950, and took its theme from the International Polar Years of 1882–83 and 1932–33. The second had been a period of minimum sunspot activity; 1957–58 would be a period of maximum activity, so findings would provide valuable comparisons. The proposal was endorsed by the International Council of Scientific Unions, which set up a committee in 1952 to organize what became the IGY. Its scientific program focused on two areas, Antarctica and outer space, and ran for 18 months from 1 July 1957.

Twelve nations—Argentina, Australia, Belgium, Britain, Chile, France, Japan, New Zealand, Norway, South Africa, the United States, and the Union of Soviet Socialist Republics—established 50 stations in Antarctica, with a total of more than 5,000 personnel, to conduct a research program focusing on glaciology, meteorology, geology, geomagnetism, and upper atmosphere physics. The largest contributors were the Union of Soviet Socialist Republics, with seven Antarctic stations, including one at the Pole of Inaccessibility (the point on the Southern Continent furthest from all its coasts), and America, also with seven stations, including Amundsen-Scott station at the South Pole. (Britain

had a total of 13 stations, but 11 were on the Antarctic Peninsula and adjacent sub-Antarctic islands and most pre-dated the IGY.)

Operation Deepfreeze

This was the four-year program through which the United States met its IGY objectives, under the overall command of Admiral George Dufek. Dufek's task was to coordinate the resources of the United States Navy, Army, Air Force, Marines, and Coast Guard with those of other nations, particularly New Zealand, and to have all seven bases operational, equipped, and staffed with civilian scientists, by July 1957—a formidable challenge, even though staged over several seasons.

Deepfreeze I (1955–56) established McMurdo station on Ross Island and Little America V on the Ross Ice Shelf. Eight Air Force Globemaster aircraft flew from Christchurch to an ice runway at McMurdo, and seven ships "converged on Antarctica and constructed Little America and McMurdo stations on schedule." In March 1956 the ships and aircraft departed, leaving 93 men at McMurdo and 73 at Little America for the winter.

From this bridgehead, Deepfreeze II (1956–57) brought 12 ships, 13 aircraft, and 4,000 men to undertake the far more difficult task of establishing Amundsen–Scott base at the South Pole, and Byrd base, 1,000 kilometers (650 miles) inland from Little America. Army engineers dynamited through a crevasse field 13 kilometers (8 miles) wide to build an "all weather

◄ **HUB OF THE ANTARCTIC**
McMurdo station on Ross Island was first established in 1955-56 by Operation Deepfreeze I. It was the advance staging post for building Amundsen Scott base at the South Pole in 1957, and has since become by far the largest base. Housing up to 2,000 people in summer, it develops the ambience of a mining town, with its dusty gravel roads.

➤ **OLD COMRADES**
Although nominally in command of Operation Deepfreeze, Admiral Byrd (left) was 68 when he visited Antarctica for the last time in 1955, leaving management to Admiral George Dufek. Paul Siple (right) again visited Little America with his old boss before taking charge of Amundsen-Scott base for IGY.

highway" for the main supply route to Byrd, where CBs—members of the Construction Battalion—constructed the base in the eight days between Christmas and New Year.

Dufek was only the eleventh person to stand at the South Pole when the first ski-equipped Navy Skytrain landed there on 30 October 1956 after a four-hour flight from McMurdo. It was accompanied by a wheeled Air Force Globemaster, which circled overhead ready to drop survival equipment if the Skytrain crashed on landing or take-off. In fact the Skytrain's skis froze to the ice during the 50-minute stopover, and 15 JATO (Jet Assisted Take Off) bottles were required to get it airborne. The advance CB team landed in late November to begin construction, and over the next month the Globemasters parachuted 750 tonnes of equipment onto the site. The base was finished by March 1957, and 17 men were left to winter over, led by Paul Siple, a veteran of Richard Byrd's early expeditions and of Operation Highjump. Meanwhile, a joint United States/New Zealand base was established at Cape Hallett on the western shore of McMurdo Sound; Wilkes base in Knox land was built in 17 days; and two icebreakers battled through the Weddell Sea to set up Ellsworth base on its southwestern shore.

During Deepfreeze III (1957–58), 350 men wintered at the seven United States bases to conduct research. Little America V was Weather Control for the entire Antarctic meteorological network, and an international staff coordinated daily observations. Seismic soundings at the South Pole revealed that the ice cap at the Pole was nearly 2,750 meters (9,000 ft) thick, and glaciological traverses helped to delineate the contours of the bedrock far below.

▼ **DOOMED STATION**
Wilkes station, built in 1957, was run for the IGY by the US Navy. In 1959 it was handed over to ANARE, but it soon became clear that the site—in a hollow—had been chosen in haste and was prone to snow drift. It was largely abandoned by ANARE in 1965. It is now under ice and its removal presents a major challenge.

◄ SUCCESS DESPITE SETBACKS
Sir Edmund Hillary, New Zealand conqueror of
Everest (left) and Dr Vivian Fuchs, Director of the
British Antarctic Survey, were a formidable team to
lead the first successful crossing of the Antarctic
Continent. They shared the honors—Fuchs complet-
ing the whole crossing and Hillary becoming the
first to reach the South Pole overland since Scott.

The Commonwealth Transantarctic Expedition

The IGY also saw the fulfilment of Ernest Shackleton's dream of 40 years earlier when a joint Commonwealth team made the greatest traverse of all—crossing the Antarctic Continent from coast to coast. As Shackleton had planned for his Imperial Transantarctic Expedition,

the Commonwealth Transantarctic Expedition required a two-pronged attack.

The main expedition, led by Vivian Fuchs, started from the Weddell Sea and was preceded by a two-year effort to establish a base camp, first in 1955 at Vahsel Bay, where eight men spent the winter of 1956. The following summer an advance base was established by air 440 kilometers (275 miles) inland at South Ice, and three men wintered there. Thirteen others, including Fuchs, spent the winter at Vahsel Bay, preparing the vehicles and aircraft for the crossing the following summer. Bad weather delayed the departure of the route-finding party to South Ice until early October; even then, and with the help of dog teams scouting ahead, the journey took until mid-November. The main party reached South Ice in late December, and started for the Pole on Christmas Day. It consisted of four Snocats, three Weasels, and a Muskeg tractor—the most advanced over-snow vehicles available—with two dog teams pioneering the route. Even so, the steep ice ridges with frequent vertical drops, and numerous crevasse fields, took their toll. Three vehicles had to be abandoned, but the remainder finally reached the Pole on 19 January 1957.

Fuchs's party was forestalled, however, by the irrepressible Sir Edmund Hillary, leader of the Ross Sea party. Hillary's designated role was to support the crossing party by laying depots from the Ross Sea toward the Pole. He first established Scott base, a little south of McMurdo, on Ross Island in January 1957, and spent the rest of the summer finding a route over the Skelton Glacier onto the plateau, and stocking advance depots from the air. After wintering at Scott Base with a party of 22, Hillary and three others set out with two dog teams, three Ferguson tractors fitted with rubber wheel tracks, towing a caravan for living in, and sledges loaded with fuel and cargo. With air support they established two more depots, the last only 800 kilometers (500 miles) from the Pole. All their vehicles were in good shape and they had plenty of fuel, so Hillary decided to press on rather than wait for Fuchs, who was already behind schedule. The last lap, over the rough terrain of the polar plateau at an altitude of 3,000 meters (10,000 ft), lowered fuel efficiency, and they arrived at the Pole on 4 January 1958 with less than 100 liters (20 gallons) of fuel left. Hillary joked that "our tractor train was a bit of a laugh;" nonetheless, this was the first expedition to reach the Pole overland since Robert Falcon Scott's, nearly 46 years earlier.

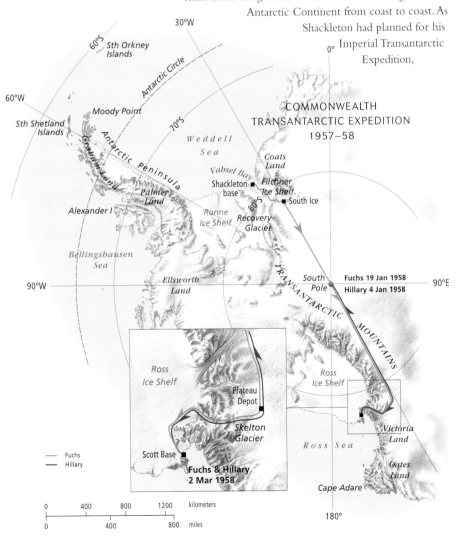

COMMONWEALTH
TRANSANTARCTIC EXPEDITION
1957–58

Fuchs's party completed the crossing of the continent, joined by Hillary's group for the final stretch down the Skelton Glacier. It was plagued by frequent breakdowns, the Snocats opened crevasses that the lighter Fergusons had negotiated easily, and navigation was tricky due to the proximity of the south magnetic pole, but they finally reached Scott Base on 2 March 1958 after traveling 3,460 kilometers (2,160 miles) in 99 days. The journey was a superb achievement in itself, but the seismic survey and other scientific investigations provided unique data on a completely unknown area of the continent, and added substantially to the scientific value of the IGY. So much data was collected that three world centers were established to evaluate it, further extending the spirit of international cooperation.

"A continent for peace and science"

Perhaps the IGY's greatest success was that it focused attention on the need for a permanent international regime to manage Antarctica, and prevent it from falling prey to cold war conflict between the superpowers. After 18 months of negotiations, the Antarctic Treaty was signed in December 1959 by the 12 nations that had participated in the IGY. The Treaty banned military activity, guaranteed free access for scientific research, and solved the problem of territorial claims by stating that the Treaty did not "endorse, support or deny any territorial claim," and prohibiting any further claims. Under this regime, Antarctica has become a unique area. **LC**

▲ DELAYED WELCOME

Below Mount Erebus on Ross Island, the crew of one of the three surviving Snocats of Fuchs's triumphant crossing party are greeted as they reach Scott Base after a 99-day marathon.

▼ SMALL BEGINNINGS

Scott Base, established by Sir Edmund Hillary on the southern tip of Ross Island, is about one-tenth the size of its mighty US neighbor, McMurdo. The hut in which Hillary and his companion spent the winter of 1957 can be seen on the far left in front of the main buildings. It is now a museum.

Arctic Exploration

Arctic exploration

Previous pages

◄ **AN ILL-FATED VENTURE**
Edward William Cooke's
painting shows *Erebus* and
Terror in icy seas during
Sir John Franklin's Arctic
expedition of 1845. Franklin
was 59 years old, but could
not be dissuaded from the
trip. He and all his crew
perished, but it took 10 years
to confirm their fate.

For many years, Antarctica was little but a vague
notion of a Great South Land; by contrast, the Arctic
was a known populated area, and the far north was
regularly perceived as a potential fast route to the riches
of the East. Many guesses about the nature of the Arctic
were wildly wrong, and some expeditions were laugh-
able, others tragic—but at least navigators did not have
to travel far to set out on an Arctic adventure.

Aristotle believed that both poles were too cold to
be habitable, and his assumption held as long as there
was no definite empirical information. However, there
were more fanciful suggestions. One early map showed a
magnetic rock at the North Pole, ringed by land divided
into four by streams, so that it resembled a donut sliced
like a pie. Some reports talked of whales and Polar bears,
others of sea monsters, a bottomless pit, and tribes who
lived in comfort at the Pole itself.

Early sightings

The history of Antarctic exploration is largely precise
and defined, based on ships' logs and personal accounts.
In the Arctic, on the other hand, the distinctions
between settlement, exploration, invasion, and exploitation are much less clear. Indigenous peoples had been
living north of the Arctic Circle for thousands of years,
but exploration from southern regions probably began

> MAKING THE BEST OF THINGS
In 1596, when Dutch explorer Willem Barents made his third attempt to find the Northeast Passage—the fastest route from Europe to the East—his ship broke up in the ice. Undaunted, his men built longboats from the fragments, but Barents died on the voyage home.

in the fourth century BC, when the Greek explorer Pytheas may have seen the pack ice off Greenland. It is likely that Irish monks reached Iceland in cowskin boats in the eighth century AD, and in the following century the Vikings certainly sailed north of the Arctic Circle, but settled south of it, in Greenland and Iceland.

The riches of the East

The dream of accessing the riches of Cathay across the top of Russia had been mooted in the mid-fifteenth century, but it was only after the Treaty of Tordesillas in 1494 granted all the southern routes to Spain (the west) and Portugal (the east) that the prospect was seriously investigated. In 1553, an expedition financed by London merchants and under the command of Sir Hugh Willoughby set out from London for the Northeast Passage. The ships soon became separated in a storm. The *Edward Bonaventure*, under Richard Chancellor, reached the top of Russia and made contacts in Moscow that led to the formation of the Muscovy Company; Willoughby and the other two ships may have sailed as far as the islands of Novaya Zemlya, but the crew froze to death over the winter. Their bodies were found by Russian fishermen the following summer.

In 1594 Willem Barents entered the scene. Barents was one of the great Arctic expeditioners, noted for the extent of his explorations and the accuracy of his reporting. On his first expedition he was the first person to successfully reach the northern coast of Novaya Zemlya. The next year he intended to sail further east, but left it too late and was stopped by early winter ice. In 1596 he set off again and discovered

◄ AN IMAGINARY CONTINENT
"Polus Arcticus," drawn in 1602, was based on a 1569 projection by Flemish geographer Gerardus Mercator. It shows recognizable outlines of some north polar regions—modern Greenland, Iceland, Northern Scandinavia, the Davis Strait, and fragments of Svalbard and Novaya Zemlya. The then unexplored North Pole, however, appears as a fantasy continent quartered by four rivers.

▼ A LIFE AT SEA
Yorkshire-born Sir Martin Frobisher went to sea as a boy. In 1576, fired by England's expansionist ambitions, he sailed *Gabriel* and *Michael* to find a Northwest Passage to Cathay and the East. He lost *Michael*, but reached Labrador, discovered Frobisher Bay, off Canada's Baffin Island, and brought an Inuit prisoner back to England.

Spitsbergen; proceeding back to Novaya Zemlya, he rounded the top of the islands, but then was beset by ice. The crew wintered in a hut they built on the island and set out in open boats the following June. Barents died during the boat journey, but 12 of the crew survived. The first time the name "Barents Sea" appeared on a map to signify this area was in 1853. Eighteen years later, in 1871, a Norwegian ship found the remains of Barents's winter hut just as the crew had left it 274 years earlier. Some of the articles recovered are in Amsterdam's Rijksmuseum.

During this period a number of expeditions had attempted to reach the Far East by traveling west. Many were British, and names on the map of Canada today commemorate the achievements of Martin Frobisher, John Davis, Henry Hudson, and William Baffin between 1576 and 1616. Hudson's report of the whales he saw off Spitsbergen triggered the Arctic whaling industry. Hudson himself made a last voyage to Hudson Bay in 1610. The ship became trapped in the ice, quarrels broke

▲ "WHEREIN WE WINTERED"
Trapped in the north polar ice, Willem Barents and his crew used parts of their shattered ship to build a hut, modeling its interior on Dutch domestic architecture. They spent the winter of 1596–97 in these quarters, which were discovered intact by a Norwegian fisherman in 1871. In 1875, another expedition found part of Barents's journal.

▲ A MAN OF PARTS
Lieutenant Frederick William
Beechey, second-in-command
of William Edward Parry's 1819
expedition in search of the
North-west Passage, was a man
of many talents. A determined
explorer and a fine seaman,
he created this dramatic wash
drawing of the Davis Strait,
and also directed the amateur
theatricals that Parry organized to combat the trip's most
insidious problem—boredom.

▶ A FAMILY TRADITION
Sir John Ross was at the fore-
front of a new phase of polar
exploration, that of wintering
over. With two specially reinforced whaling ships, *Isabella*
and *Alexander*, he sailed north
on 30 August 1818 to look for
the Northwest Passage; a
reward of £5,000 had been
offered. Among his crew was
his nephew, James Clark Ross,
later to win Antarctic fame.

out, and when the ship was finally free to sail home the
following summer, mutineers put Hudson, his son, and
seven others into an open boat on Hudson Bay on
22 June 1611. They were never seen again.

Obviously a northern route was useful only if there
was a passage to the south at the end of it. In 1724 Peter
the Great, Tsar of Russia, sent Vitus Jonassen Bering to
map the Russian coast north of Kamchatka. It took
Bering three and a half years to get his expedition
across Russia and to build the ship. He sailed from the
Kamchatka River on 13 July 1728, but proved a timid
explorer; he managed to sail through what is now the
Bering Strait without seeing much of the coast of Asia
to the west or any of America to the east, and he
retreated before he reached the pack ice to the north.
Fifty years later, James Cook charted the strait and
generously named it after Bering.

The next attempted voyage to the North Pole was
somewhat bizarre. The Honorable Daines Barrington
was an English aristocrat who was not very bright
but had become a judge through family influence.
Barrington became obsessed with a theory that became
known as the Open Polar Sea. He believed that the
24-hour summer sunlight would melt the polar
ice; some peripheral ice might remain, but the
Pole itself would be ice-free, and a ship could
then simply sail across. In 1773 he presented
proposals for a polar expedition to the Royal
Society, which forwarded them to the Admiralty.
The expedition sailed with two vessels,
HMS *Carcass* and HMS *Racehorse*, under
the command of Captain Constantine
John Phipps, on 26 May 1773. In a significant setback for Barrington's theory,
they were in ice by 5 July and were
stuck fast by the beginning of August.
Fortunately, the ice parted 10 days later
and the ships escaped and returned home.

Historically, the most interesting episode of the journey occurred when a 14-year-old midshipman set off
across the ice in pursuit of a bear. If the captain had not

fired a gun to scare off the bear, young Horatio Nelson
might not have lived to fulfill his destiny. In any case, the
failure of the expedition cooled British enthusiasm for
finding a route across the Pole. Interest reverted to the
Northwest Passage, and in 1776 James Cook was sent to
look for a route from the Pacific side. Waiting in Hawaii
for summer to melt the polar sea ice, Cook was killed
when a boat was stolen and a fight broke out.

In search of the Northwest Passage

When the Napoleonic wars ended in 1815, Britain
was left with a large, idle navy and the threat of Russian
expansion in the north Pacific. In 1816, John Barrow,
Second Secretary to the Admiralty—and the man who
would dominate Arctic exploration for half a century—
wrote: "To what purpose could … our naval force be …
more usefully employed than in completing those details
of geographical … science of which the grand outlines
have been boldly and broadly sketched by … our
countrymen?"

So began the great quest for the Northwest Passage.
In 1818 John Ross sailed to Baffin Bay, but got no further
than Lancaster Sound. William Edward Parry made
better progress in 1819–20 and wintered at
Melville Island. At the same time, John Franklin
attempted to travel overland to Canada's
northern coast and to rendezvous with Parry.
This 1819–22 expedition was a harbinger
of Franklin's last expedition: most of his party
starved to death. In 1821–23, Parry took *Fury*
and *Hecla* to find a route south of Baffin
Island, but failed. Undeterred, he tried
again in 1822–24, but *Fury* was
wrecked, and Parry abandoned the
ship's stores in Prince Regent Inlet.
Franklin returned to northern
Canada in 1825–27, and mapped more
than 1,600 kilometers (1,000 miles) of coastline.
In 1827 Parry was less successful in his bid to reach the
North Pole, traveling no further than 82°45´N. At least
his voyage was relatively brief: John Ross, on the other

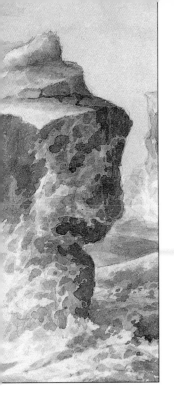

◄ "WHITE BEAR"
The gifted Lieutenant Frederick William Beechey produced this watercolor of a polar bear during Edward Parry's north polar expedition of 1819. The expedition reported and hunted a wealth of Arctic wildlife: reindeer, musk oxen, wolves, Arctic foxes, geese, and ptarmigan. They also found traces of Inuit encampments.

▼ DRESSED FOR THE WEATHER
An artist on the 1829 Ross expedition painted this watercolor of Ervick, an Inuit from Prince Regent Inlet, west of Baffin Bay. The Inuit had much to teach about surviving polar conditions; later, Roald Amundsen copied their clothing on his successful South Pole assault.

► GOLDEN OPINIONS
William Edward Parry was 29 when he led the north polar expedition of 1819. Sir Clements Markham, President of the Royal Geographical Society, described him as "unsurpassed as a leader of men ... knowing when to take risks and when to avoid them ... the ideal of an Arctic officer."

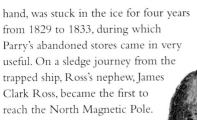

hand, was stuck in the ice for four years from 1829 to 1833, during which Parry's abandoned stores came in very useful. On a sledge journey from the trapped ship, Ross's nephew, James Clark Ross, became the first to reach the North Magnetic Pole.

The Franklin expedition

In 1845, two years after James Clark Ross returned from Antarctica with *Terror* and *Erebus*, John Franklin took the two ships on another voyage to try to find a Northwest Passage. Ten years passed before Franklin's fate was finally known. During this time, the search for his ill-fated expedition, promoted by his devoted wife involved some 30 ventures—and eventually disclosed the long-sought Northwest Passage. James Clark Ross went looking for Franklin in 1848, but returned unsuccessful in 1849. An overland search from 1848 to 1851, led by John Richardson assisted by John Rae, was equally fruitless. In 1850–51, vessels sponsored by Lady Franklin renewed the search, together with others commanded by Horatio Austin and James Clark Ross, and some United States ships under the command of Elisha Kent Kane.

In 1850 Robert McClure in *Investigator* tried to find a passage from the western side. When his ship became icebound, he walked eastward and met Edward Belcher's expedition in 1854. Despite the loss of four of its five vessels, Belcher finally found the Northwest Passage. Meanwhile, Lady Franklin's second search for her husband, in 1851–52, had no more success than her first attempt. Finally in 1854, John Rae reported that he had discovered the fate of the Franklin expedition from Inuits who had sold him some objects recovered from the bodies. During his voyage of 1857–59 Leopold McClintock

► A DOOMED VOYAGE
In 1845, Sir John Franklin headed north in *Erebus* and *Terror*, looking for the Northwest Passage. His whole party perished. Franklin's wife, Lady Jane, used all her influence to promote a search for her husband, but it took 10 years for Inuit reports to confirm his fate.

► LOOKING FOR FRANKLIN
Commanded by Robert McLure, and lured by the British Admiralty's reward of £20,000, HMS *Investigator* set out in January 1850 to search for John Franklin's lost expedition, and for the elusive Northwest Passage. Shown here off the north coast of Baring Island, the ship was icebound for three years, and did not return to England until 1853.

◄ REACHING NORTH
HMS *Alert*, a 751-tonne sloop with 17 guns commanded by Sir Albert Hastings Markham, and the 668-tonne *Discovery*, under George Nares, sailed north from Portsmouth on 29 May 1875. On 11 May 1876, Markham reached the highest northern latitude so far attained—83°20′—his team of 16 men hauling their sledges over the tumultuous pack ice. In triumph they built their camp against a backdrop of towering icebergs.

▼ *JEANNETTE* IN ICE
The voyage of *Jeannette* in 1879–81, led by United States Navy Lieutenant De Long and financed by James Gordon Bennett of the *New York Herald*, ended in tragedy. The vessel, an old Admiralty gun-ship, was fast but not built for polar ice. She struggled for 22 months in rough polar seas then sank on 13 June 1881.

found three corpses and a note confirming the deaths of Franklin and most of his men.

Bids for the Pole

In 1871, an American expedition under Charles Francis Hall made a well prepared attempt to reach the Pole. However, Hall mysteriously died of arsenic poisoning, and his crew were lucky to survive; some had to winter on an ice floe, and the others were forced to abandon ship and take to the boats. Both groups were rescued by hunting vessels in the summer of 1873.

After a series of generally unsuccessful German polar expeditions between 1868 and 1871, the Austro-Hungarian North Pole Expedition aboard *Tegetthoff* set off from Bremerhaven in June 1872. The ship was soon trapped in ice that carried it to the north, where land was sighted in August 1873. This turned out to be a large archipelago, which they named Franz Josef Land after their emperor. It was to be another year in the ice before sledge parties could explore the new land. Subsequently, the crew took to the boats, and were rescued off the coast of Russia.

The British Arctic Expedition, under the command of George Strong Nares, set off in 1875 with two ships, *Alert* and *Discovery*, and high hopes of reaching the Pole. *Alert* reached 82°28′N off Ellesmere Island—the furthest

▼ NEAR THE NORTH POLE
Scottish explorer and naval commander Sir George Strong Nares returned from Antarctica in 1874, and immediately turned his attention to the North Pole. In 1875 he sailed *Alert* and *Discovery* to within 645 kilometers (400 miles) of the Pole before being driven back by heavy ice and scurvy.

north that any vessel had sailed. After the winter, two parties set off hauling sledges by hand, although most other nations relied on dog sledging; this set a pattern that was to continue at the other end of the globe. They came no closer than 644 kilometers (400 miles) to the Pole, and when they returned one newspaper reported that the expedition "went out like a rocket and came back like a stick."

The American expedition of 1879–81, under Commander George Washington De Long aboard *Jeannette*, was jointly organized by the United States Navy and the New York *Herald*. Hoping to repeat its circulation success after sending Henry Morton Stanley to Africa to "find" David Livingstone, the newspaper's proprietor, James Gordon Bennett, put two journalists on board and named the ship after his sister. The

➤ THE NORTHEAST
PASSAGE CONQUERED
In 1832, Finlander Baron Nils
Adolf Erik Nordenskjöld took
Swedish citizenship and began
a career of polar navigation. In
1878–79, commanding the
sail-and-steam ship *Vega*, he
successfully navigated the
Northeast Passage from the
Atlantic to the Pacific around
the northern coast of Asia,
across the tip of Siberia.
Icebound in September 1878,
Vega's scientists built a snow
observatory and studied
native survival techniques.

expedition was aiming for the Pole, but was first directed
to "rescue" Swedish explorer Adolf Erik Nordenskjöld.
In 1868, Nordenskjöld had set a new furthest north of
81°42′N and, when *Jeannette* sailed, was completing the
first successful transit of the Northeast Passage. De
Long's expedition, on the other hand, failed to reach
even 78°N, the ship sank, and 20 of the 33 men died.

Nansen and *Fram*

Norwegian Fridtjof Nansen was the most illustrious
Arctic explorer, combining intellectual rigor and reckless
adventure in a series of remarkable exploits. In 1888 he
led the first expedition to cross Greenland, but his most
extraordinary venture left Oslo in 1893. The rationale
was simple, the plan breathtakingly ambitious. In 1884
some artefacts from the *Jeannette* had been found on the
shore of southwest Greenland, having apparently been
carried in the ice from the New Siberian Islands.
Nansen's plan was to build a ship with a rounded hull
capable of withstanding winter ice;
he named the ship *Fram* (meaning
"forward"). He would sail into the
ice and conduct scientific experiments for the three to five years it
would take for the ice to carry the
ship across the Pole and release it
on the other side.

By September 1893, *Fram* was
set in the ice north of the New
Siberian Islands. Forced inactivity
did not suit Nansen, so on

➤ BALLOONING TO THE POLE
Swede Salomon Andrée believed that
a hydrogen balloon was the way to
reach the North Pole. The king of
Sweden and the philanthropist Alfred
Nobel lent support, and a balloon was
built in Paris and named *Ornen*—Swedish
for "eagle." After an unsuccessful launch
from Spitsbergen in 1896, Andrée set
off in July 1897. He never returned.

14 March 1895 he and Hjalmar Johansen left the
ship on their dog sledges, with enough food for
100 days, and set off for the Pole. They had just
under 666 kilometers (415 miles) to cover to the
Pole; they then planned to continue onward and
use their kayaks to reach Svalbard or Franz Josef
Land and safety. However, difficult ice
conditions and the southward drift of the
polar ice defeated them, and they were
forced to turn southward on 8 April, at
86°13′N. They reached Franz Josef Land
on 14 August, but had to spend the winter in a stone hut they erected. Strangely,
although they were close friends, it was
only after a year alone together that they
began to call each other by their first
names. On 17 June 1896, after a series
of perilously close shaves, they chanced
upon the only humans in the area, a British
scientific group led by Frederick Jackson.
The expedition's relief ship took them back to
Tromsø, where they arrived just one day after
Fram put in after completing her three-year drift
in the ice off Svalbard a week earlier.

Nansen combined daring and luck, but
Swede Salomon August Andrée's attempt to fly a
hydrogen balloon over the North Pole was sheer
foolhardiness, even though it was supported by King
Oscar II, Adolf Nordenskjöld, and Alfred Nobel. After
an abortive attempt in 1896, Andrée and two others
took off from Spitsbergen on 11 July 1897, but the
balloon came down on the ice after three days. More
than 30 years later, in August 1931, the men's last camp
was found on White Island. Their journals revealed that
they had survived until 17 October. Exposed film was
also found, and was later processed successfully.

▲ ICY DEDUCTION
When *Jeannette*'s wreckage
surfaced thousands of miles
from where she had sunk,
Fridtjof Nansen reasoned that
a sea that could carry wreckage could carry a ship. He
planned to drift across the
Pole in *Fram*, which had a
rounded hull to lift under ice.

⟩ THREE POLAR STARS

Formally clad—a far cry from their polar gear—three great heroes of polar exploration are immortalized in an official photograph. Sir Ernest Shackleton (left), discovered the South Magnetic Pole, Commander Robert Peary of the United States Navy (center), discovered the North Pole, and Roald Amundsen (right), was the first person to reach the South Pole.

▽ HUSKY COMPANIONS

Robert Peary poses with some of the dogs that made his eight Arctic explorations possible. Strong, agile, and intelligent, the dogs, which are members of the wolf family, work in teams to draw the heavy polar sledges rapidly across the ice. Their splayed feet are perfectly adapted for movement over snow and ice.

▲ THE RACE FOR THE POLE

A French cartoon reflects early twentieth century interest in polar exploration. Two Americans battle to be first to the North Pole—Frederick Cook, whose claim was disproved, and Robert Peary, who came close, but may have miscalculated his exact position.

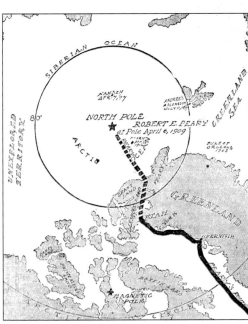

◀ THE ROUTE TO THE NORTH POLE

Peary's assault on the Pole was a highly nationalistic affair. His ship was named *Theodore Roosevelt*, and the American President himself turned out to bid him goodbye. Heading north from Ellesmere Island, Peary had provisions for 60 days, and traveled with five companions, 133 dogs and 19 sledges, as well as 17 Inuit, valued for their knowledge of polar survival.

American ambitions

Robert Edwin Peary from America was possibly the most driven of Arctic explorers. He once wrote to his mother "I must have fame"—and he worked hard for it. Like Nansen, he began his explorations in Greenland, and on his first attempt to reach the Pole in 1898–99 he stated: "a few toes aren't much to give to achieve the Pole." He made further bids in 1900 and 1902, and in 1905 had to turn back at 87°6′N after passing the 86°16′N reached by an Italian expedition of 1899–1900.

Peary was not the only American explorer seeking the Pole, but none posed such a threat to his ambitions as Frederick Cook, who had been to Greenland with Peary in 1892, and to Antarctica with de Gerlache's *Belgica* expedition in 1897–99. Cook was likable and widely admired as an explorer and expedition doctor, but his claim in 1906 to have scaled Mount McKinley, North America's highest mountain, was later disproved by his companions. Undaunted, in 1908, he set off from Greenland for the Pole with two Inuit companions. Later, in purple prose, he wrote of reaching the Pole on 21 April 1908: "We step over colored fields of sparkle, climbing walls of purple and gold—finally, under skies of crystal blue, with flaming clouds of glory, we touch the mark! … We are at the top of the world!" In fact he was probably nowhere near the Pole: his journal and astronomical observations were soon discredited; he claimed to have seen land that did not exist, and failed to record land that he should have seen; his two companions also stated they were never more than a couple of days from land. But Cook was likable and Peary was not, and many wanted the more obvious rogue to triumph. As one commentator succinctly put it: "Cook was a gentleman and a liar; Peary was neither."

Cook made his claim on 2 September 1909, when he returned from Greenland and telegraphed from the Shetland Islands. He was widely lauded until his story was subjected to intense scrutiny and cracks began to appear—especially as on 6 April 1909, only four days later, Peary announced from Canada that he had reached the Pole.

Peary had left New York on 6 July 1908 for his final attempt on the Pole. He headed north from the top of Ellesmere Island on 1 March 1909 and left the last of his support parties on 1 April, continuing with only Matt Henson, the black manservant who always accompanied him, and four Inuit. On 6 April he wrote, uncharacteristically, "Yet with the Pole actually in

> THE LAST FAREWELL
Sailors on the deck of the United States ship
Pontcharian wave to *Nautilus* on her way to
north polar regions. *Nautilus* was formerly a
United States Navy submarine torpedo boat but
was refitted for polar conditions by Australian
explorer Sir George Hubert Wilkins. He used
the submarine to explore the underside of the
polar ice cap, but failed to pass under the Pole.

sight I was too weary to take the last few
steps," but once he had reached the Pole,
he wrote in his diary: "The Pole at last!
The prize of three centuries. My dream
and goal for 20 years. Mine at last! I cannot bring myself to realize it. It seems all
so simple and commonplace." Naturally,
his calculations were examined closely
on his return. His patrons the New York
Times and the National Geographical
Society stood by him, but his claimed
average progress of 61 kilometers
(38 miles) a day during the last push was about twice
what was reasonable; possibly he turned back early and
falsified his figures. Polar veteran Wally Herbert has
suggested that Peary did not allow sufficiently for the
westward drift of the polar ice, and so may have missed
the Pole by 80 kilometers (50 miles), even if he did
cover the distance he claimed. It will never be known
whether either Peary or Cook reached the Pole, but the
current balance of opinion suggests that neither did so.

By air and under sea

In 1903–06 Norwegian Roald Amundsen was the first
to navigate the Northwest Passage, and in 1911 he was
the first to reach the South Pole. In 1925 he tried to
fly to the North Pole with Lincoln Ellsworth but was
forced to land (for three weeks), beyond 88°N but short
of the Pole. The following year, Richard Byrd and Floyd
Bennett claimed to have flown over the Pole at 9.02 am
on 9 May 1926. Their claim was widely accepted at the
time, but recent analysis suggests that they could not
have averaged the speed they claimed.

When Byrd and Bennett landed at Spitsbergen,
Amundsen and Ellsworth were there, about to try for
the Pole again in Italian aviator Umberto Nobile's airship *Norge*. They left on 11 May with a crew of 16, flew
low over the Pole and dropped national flags, then flew
on to Teller, Alaska (near Nome). Amundsen was accompanied by Oscar Wisting, who had also traveled with
him to the South Pole. Remarkably, it now seems likely
that these two were the first to reach both Poles. The
Norge expedition appeared to confirm the value of airships for polar exploration. Yet in 1928 Nobile returned
to the Arctic with a newer airship, *Italia*, and reached the
Pole (seven of the crew were the first ever to return to
a Pole), but crashed on the way back. Half the crew of
16 perished, and Amundsen disappeared during an aerial
search for survivors that left Tromsø on 18 June 1928.

In 1931 Hubert Wilkins of Australia, who had made
the first powered flight in Antarctica in 1928, tried to

reach the North Pole in a
submarine, *Nautilus*. The
expedition failed, but on
4 August 1958 the United
States nuclear submarine
Nautilus, commanded by
W. R. Anderson, passed
under the North Pole at
a depth of 122 meters
(400 ft). The following year
another United States submarine, *Skate*, surfaced at
the Pole and scattered the
ashes of Wilkins, who had
died in December 1958.

Standing at the Pole

If the claims of Peary and Cook are discounted, who
was the first to stand at the North Pole? In all probability,
the honor belongs to the 24 members of a Soviet aerial
scientific expedition that landed at the Pole on 23 April
1948 and remained on the ice for three days. The expedition was shrouded in secrecy for 50 years, but in 1999
the British publication *Polar Record* published a report
that included interviews with one of the party, Pavel
Senko. Just over 20 years later, on 29 May 1969, Wally
Herbert's British Trans-Arctic Expedition completed the
first land crossing of the north polar ice, from Point
Barrow to Svalbard.

Visiting the Pole today

During the late twentieth century, improvements in icebreaker technology made the Northwest Passage viable
and the Northeast Passage a popular trade route across
Russia, with freight and passenger ships following routes
once regarded as impossible. The first Russian nuclear
icebreaker to reach the Pole was *Arktika*, on 17 August
1977; today, the Russian icebreakers *Yamal* and *Sovietskiy
Soyuz* regularly take tourists to the North Pole. DM

▲ BREAKING THE ICE
Steel icebreakers, such as the
Sovietsky Soyuz, shown here,
have strong, thick plating and
a raked stem designed to ride
up on the ice and crush it with
their weight. This allows them
to forge through thick polar
ice. Devices such as bubbler
jets lubricate the ice, and multiple screws also provide extra
power and maneuverability.

Part V
LIFE AT THE POLES

Antarctica is a vast and unique laboratory for the study of life on earth—and of the earth itself. In a rare spirit of international cooperation, scientists from all over the world live and work in this most extreme of environments, in the hope of understanding the mysteries of the planet's past, and—more significantly—learning valuable lessons about what the future may hold. The world's last wild continent is a place where settlement has been determined by science, not conquest, and where both science and tourism remain strictly controlled.

Managing the Poles

Which nations claim what?

Previous pages

◄ PRESERVING THE POLES
The most bleak and hostile
environment in the world, the
polar regions nevertheless
have an austere beauty of
their own. Advances in ship-
building techniques and
communications mean that
humans are reaching the
poles in increasing numbers.

Uniquely among the continents of the world, Antarctica belongs to no country. This is the outcome of decades of negotiation and dispute, both on the ice itself and through remote diplomatic channels. By the 1920s, there were seven claimants to territory in Antarctica. Their claims were based on discovery and "a sufficient display of authority" to show that the land was occupied and controlled by the claimants.

Australia claimed the biggest slice—42 percent, largely based on the work of Sir Douglas Mawson. In 1924 France claimed a small slice (a 6° arc) within the Australian claim as Terre Adélie, because of the exploration and claim made by Dumont d'Urville in 1840. Since Britain handed it over in 1923, New Zealand claims the Ross Dependency—effectively the 50° sector below New Zealand that embraces the whole of the Ross Sea. Norway claimed Dronning Maud Land in 1939 through its whaling activity; however, while the other nations claim a slice from the South Geographic Pole to latitude 60°S, Norway does not define the outer limit of its claim but was really interested only in a 65° arc of the coastal strip. There is a section of West Antarctica that is not claimed by any nation.

The problem sector centers around the Antarctic Peninsula, where, like a fan of playing cards, the claims of Chile (37° arc), Argentina (49° arc), and Britain (60° arc) overlap. However, in geopolitical terms, two absentees from the list of claimants were just as important as the claimants in the development of the unique solution that is represented by the Antarctic Treaty. In 1924, the United States adopted the Hughes Doctrine (named for Charles Hughes, Secretary of State from 1921 to 1925) which states that mere discovery is not sufficient for territorial claim—actual settlement is required. The United States could have based a substantial claim on the explorations of Wilkes, Byrd, and Ellsworth, but it has not done so, nor has it recognized any other claims to Antarctica. When Norway claimed Dronning Maud Land in 1939, the then Soviet Union refused to recognize any claims to Antarctica, even though it could have based an all-encompassing claim on Bellingshausen's 1819–21 circumnavigation. Subsequently, Hitler ordered the annexation of the Norwegian claim by Germany, and sent Dornier flying boats to drop steel javelins capped with swastikas right across the region.

◄ A STATION OVERCOME BY ICE
Wilkes station was built by the United States in 1957, and in 1958 it was agreed to run it jointly with Australia. Within two years the United States withdrew, and eventually the ice took over.

▼ ANTARCTIC PRESENCE
Argentina's Cámara Base on Half Moon Island opened in 1953 and closed in 1997–98. Argentina has 14 Antarctic bases, more than any other nation, though some have been unoccupied in recent years.

> **THE ANTARCTIC "PIE"**

The Antarctica region is divided into segments shaped like slices of a pie, each slice being under the jurisdiction of a nation signatory to the Antarctic Treaty of 1 December 1959, which arose out of the International Geophysical Year of June 1957 to December 1958.

Warfare averted—twice

During and immediately after World War II, there was a flurry of base-building by Britain, Chile, and Argentina. Britain suggested taking disputed claims to the International Court of Justice, but Chile and Argentina rejected this solution and wanted an international conference to resolve all claims over the continent. In 1948 Argentina and Britain sent warships to the region, and armed conflict seemed likely.

The peace-making suggestion of the United States was that Antarctica should fall under the trusteeship of the United Nations, and this evolved into the concept of a condominium (joint sovereignty) by the seven claimant nations and the United States—and excluding the Soviet Union. Only Britain and New Zealand were in favor of this contentious idea. Meanwhile, a legal adviser to the Chilean government, Professor Julio Escudero, put forward the "modus vivendi" proposal that all national claims be effectively frozen so that scientific programs could continue. This intelligent compromise became the heart of the Antarctic Treaty 11 years later.

However, the mere proposal was not the end of the matter. In January 1949, the Soviet Union announced that it would not accept the legitimacy of any Antarctic agreement to which it was not a party. The closest that Antarctica came to warfare was in February 1952, when Argentinean troops fired over the heads of members of a British party who were rebuilding the Hope Bay station, which had been destroyed by fire in 1948. (South Georgia and the Falklands lie outside the Treaty area, so the 1982 conflict does not count.)

A continent for the world

The remarkable spirit of scientific cooperation (at the height of the Cold War) shown in the International Geophysical Year that ran for 18 months from June 1957 to December 1958 showed that nations could work together in Antarctica. The 12 nations that participated in the IGY operated 60 field stations across Antarctica and its islands, and produced huge amounts of research data. From the IGY the Antarctic Treaty was born, and on 1 December 1959, 12 nations signed the Treaty: Argentina, Australia, Belgium, Britain, Chile, France, Japan, New Zealand, Norway, South Africa, the United States, and the Soviet Union (now Russian Federation). There are now 44 signatories to the Antarctic Treaty. Strengthened by the Madrid Protocol of 1991, Antarctica belongs to no nation, and remains a nuclear-free continent for science and exploration. DM

ANTARCTIC CLAIMS

▼ **SOLD FOR A SONG**

The break up of the Soviet Union gave rise to some interesting complications in Antarctica. Russia kept all the bases that were previously Soviet, but of the other Soviet states only the Ukraine now has an Antarctic presence. The Ukrainian government purchased Vernadsky (shown here), in the Argentine Islands adjacent to the Antarctic Peninsula, for a nominal sum from Britain in 1996.

◄ **CARRYING ON**

An expedition based at Russia's Mirny station is continuing the national scientific program in Antarctica that began in 1956, the year that Mirny opened. In August 1992 President Boris Yeltsin authorized the renaming of the former Soviet Antarctic Expedition as the Russian Antarctic Expedition.

Antarctic law

> **FLYING THE FLAG**
The flags of the 12 original
Antarctic signatories fly at
the South Pole. The nations
represented are: Argentina,
Australia, Belgium, Chile,
France, Japan, New Zealand,
Norway, Russia (now Russian
Federation), South Africa,
the United Kingdom, and the
United States. The treaty was
made in 1959 and came into
effect in 1961.

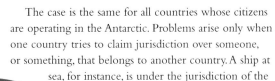

Antarctica in the third millennium is a unique continent—a nature reserve consecrated to world peace and to scientific research. It is also unique in international law, not only because of its isolation and its climatic extremes but because it has no permanent human populations, so that there is no natural sovereignty over land, and it also contains the only significant unclaimed territory on earth. This uncertain jurisdiction can lead to precarious circumstances: for example, the claims of Britain, Chile, and Argentina overlap across the Antarctic Peninsula, but some other nations do not regard these claims as valid.

Antarctica is underpinned by a unique legal system. A group of 26 "consultative countries," with another 18 looking on, uses the 1959 Antarctic Treaty as the foundation for a legal regime that applies to the whole area south of latitude 60°S, including, to some extent, the "high seas." The rules are formulated at the annual meetings of the 26 decision-making countries. Each country's legal system must endorse these decisions, and then they become binding on the citizens of each country, as well as on those of the 18 countries that do not participate directly but have agreed to abide by the rules.

World heritage

By popular consensus Antarctica belongs to the world, and countries and individuals working there must obey rules that have been formulated over half a century to protect this remarkable part of the earth. The members of the Antarctic Treaty are not owners of the Antarctic Continent but custodians and keepers of mutually sanctioned laws.

Participating countries have become "polar police," albeit armed only with diplomacy. They have the power to scrutinize each other's behavior, guided by the rules that they themselves have made, and using environmental impact assessments, mandatory inspections, and exchange of information to monitor each other's compliance. Ultimate authority, however, rests with independent nations and their power to invoke their laws against their citizens. Australia, for example, claims 42 percent of the Antarctic Continent as sovereign territory, and applies territorial law and ordinances to its Australian Antarctic Territory (AAT), but an Australian citizen anywhere in Antarctica is subject to Australian law. It is not necessary for Australian sovereignty to be recognized by any other country for the Commonwealth Government to make laws specifically for the AAT, or to apply other appropriate civil and criminal law generally. So if an Australian citizen breaks any Antarctic law (even outside the AAT), the full force of national law can be invoked.

The case is the same for all countries whose citizens are operating in the Antarctic. Problems arise only when one country tries to claim jurisdiction over someone, or something, that belongs to another country. A ship at sea, for instance, is under the jurisdiction of the country of its registration, but the owners, the master, the tour operators, the charterers—even the passengers—may have to contend with many different laws in the event of violation. Each circumstance requires an individual approach, depending on the nature of the violation and the applicable law. Fundamentally, however, on land south of latitude 60°S, all persons are theoretically subject to the laws of their country of citizenship.

Peaceful solutions

If a whole range of Antarctic Treaty countries were to be involved in a crime committed in Antarctica, which might lead to a dispute about jurisdiction, they could call upon the Treaty's dispute resolution procedure, which states quite clearly that the participating countries should try to resolve the argument among themselves in the first instance. Because all countries active in the

▲ **CLAIMING TERRITORY**
While this obelisk at Chile's
Presidente Eduardo Frei
Montalva base (generally
known simply as "Frei") sym-
bolizes the cooperative nature
of the Antarctic community,
the base itself supports
Chile's territorial claim over
this slice of Antarctica. An
obelisk in Patagonia marks the
mid-point of Chile's claimed
territory—from the Peruvian
border to the South Pole.

1959 ANTARCTIC TREATY

Original 12 signatories (all consultative parties)
32 other contracting parties (14 consultative parties)

Treaty meeting:
recommendations, measures,
decisions, resolutions

1972
Convention for the
Conservation of
Antarctic Seals

1982
Convention on the
Conservation of Antarctic
Marine Living Resources

1991
Protocol on Environmental Protection

➤ **TREATY ACTIVITIES**
A flow chart illustrates how, from the original Antarctic Treaty of 1959 with its 12 signatories, membership of the Antarctic community has grown and measures have been adopted to protect the region's marine life and its unique environment.

Antarctic are reasonably unanimous about its heritage values, disputes are most likely to be resolved peacefully and to the mutual satisfaction of all, and this has been the case so far. If the crime also involves a country outside the Treaty group, they can attempt to summon international sentiment against the outside party and try to reach a peaceful accommodation through diplomatic means, which is how the Treaty asks its members to behave. Because the 44 Treaty countries include almost all those with a capacity to operate in this environment, diplomacy has thus far proved an adequate weapon. JG

▼ **PRACTICAL COOPERATION**
Cooperation between nations in Antarctica is not merely an abstract discussion point at meetings, but a daily necessity. Here an Italian resupply cargo vessel and a United States fuel tanker use a path cut through the pack ice by the United States Coastguard icebreaker *Polar Sea*.

Antarctic Treaty parties

STATE	EFFECTIVE DATE	STATUS
Argentina	23.6.61	ATCP/OS/C
Australia	23.6.61	ATCP/OS/C
Austria	25.8.87	Acceding state
Belgium	26.7.60	ATCP/OS
Brazil	16.5.75	ATCP 12.9.83
Bulgaria	11.9.78	ATCP 25.5.98
Canada	04.5.88	Acceding state
Chile	23.6.61	ATCP/OS/C
China	08.6.83	ATCP 07.10.85
Colombia	31.1.89	Acceding state
Cuba	16.8.84	Acceding state
Czech Republic	14.6.62	Succeeding state [1]
DPR Korea	21.1.87	Acceding state
Denmark	20.5.65	Acceding state
Ecuador	15.9.87	ATCP 19.11.90
Finland	15.5.84	ATCP 09.10.89
France	16.9.60	ATCP/OS/C
Germany [2]	05.2.79	ATCP 03.3.81
Greece	08.1.87	Acceding state
Guatemala	31.7.91	Acceding state
Hungary	27.1.84	Acceding state
India	19.8.83	ATCP 12.9.83
Italy	18.3.81	ATCP 05.10.87
Japan	04.8.60	ATCP/OS
Netherlands	30.3.67	ATCP 19.11.90
New Zealand	01.11.60	ATCP/OS/C
Norway	24.8.60	ATCP/OS/C
Papua New Guinea	16.3.81	Acceding state
Peru	10.4.81	ATCP 09.10.89
Poland	08.6.61	ATCP 29.7.77
Republic of Korea	28.11.86	ATCP 09.10.89
Romania	15.9.71	Acceding state
Russian Federation	02.11.60	Succeeding state [3] ATCP/OS
Slovak Republic	14.6.62	Succeeding state [4]
South Africa	21.6.60	ATCP/OS
Spain	31.3.82	ATCP 21.9.88
Sweden	24.4.84	ATCP 21.9.88
Switzerland	15.11.90	Acceding state
Turkey	24.1.96	Acceding state
United Kingdom	31.5.60	ATCP/OS/C
United States of America	18.8.60	ATCP/OS
Uruguay	11.1.80	ATCP 07.10.85
Ukraine	28.10.92	Succeeding state [5] OS
Venezuela	24.3.99	Acceding state

KEY
ATCP — Antarctic Treaty consultative party
OS   — Original signatory (automatically ATCP)
C   — Claimant state (automatically ATCP)

[1] The Czech and Slovak Republics inherited Czechoslovakia's obligations as a "contracting party" with effect from 1 January 1993, the date of their succession to the Treaty.

[2] The German Democratic Republic was united with the Federal Republic of Germany on 2.10.90. GDR acceded to the Treaty on 19.11.74 and was recognized as an ATCP on 5.10.87.

[3] Following the dissolution of the USSR, Russia assumed the rights and obligations of being a party to the Treaty. The USSR had been an original signatory to the Treaty.

[4] See footnote 1 above.

[5] Ukraine has asserted that it succeeded to the Treaty following the dissolution of the USSR and should be entitled to ATCP status. Formal application has not yet been received.

Tourism in Antarctica

▼ SAFETY FIRST—ALWAYS
Arranging a private expedition
to Antarctica is a daunting
exercise. These expeditioners
crossing the Neumayer
Glacier are bound by their
nations' commitment to the
Antarctic Treaty. Their government will approve the venture
only with proof that all eventualities have been planned for.

The traveler who decides to go south for a holiday in a place as remote and hostile as Antarctica may be questioned with disbelief by colleagues, friends, and relatives, who may ask: what is there to see there but snow, ice, Eskimos, igloos, and maybe a few Polar bears? This, at once, necessitates passing on the basic information that there are no indigenous people in Antarctica, and that Polar bears are found only in the other polar region—the Arctic. So what is it about this place that is worth the thousands of dollars that the trip will cost?

The continent of Antarctica covers the bottom of the world, and adventurous tourists have visited it since the 1950s. Lars-Eric Lindblad, of Lindblad Travel, New York, started tourism on a regular basis in 1966 when he chartered an Argentine vessel, the *Lapataia*, to take a group of tourists from South America to the South Shetland Islands, 1,000 kilometers (621 miles) across the stormy Drake Passage. They, like the tourists who have followed them, traveled south to find the last wilderness on earth, the final frontier on the planet, and to feel like their heroes: Shackleton, Mawson, Amundsen, and Scott. They wanted to experience the cold, the isolation, and the mystique of this faraway place, with its raging seas, mountain scenery, towering icebergs, and never-ending days, and they longed to encounter its profusion of wildlife: penguins, flying seabirds, seals, and whales.

Ship-based tourism is still the most dominant form of tourism in Antarctica, allowing travelers to view the southern continent from a floating hotel. The alternative for the average tourist is to participate in overflights of the Ross Sea region that are organized by Croydon Travel in Melbourne and carried out by QANTAS. A few intrepid souls also participate in polar adventure tours organized by Adventure Network International (ANI) and climb Vinson Massif 4,897 meters (16,066 ft), or even ski to the Geographic South Pole.

Antarctica is a unique place—there are no hotels ashore, no local markets to visit, and shopping opportunities are limited to a few gift shops at the stations

of some of the Antarctic Treaty Party countries that administer Antarctic affairs through the complex web known as the Antarctic Treaty System. Most tourist landings are carried out using inflatable boats called Zodiacs, and more than 90 percent of all tourism activities take place in the Antarctic Peninsula region and on the offshore islands to the west of the Peninsula.

The season for visiting the south runs from the middle of November to the middle of March—for the remainder of the year, thick sea ice surrounds the continent, making tourist landings impossible.

Visitor numbers have been increasing steadily during the past decade. Enticed by the prospect of spending the turn of the millennium in a very special place, a record 14,623 ship-based tourists visited in the 1999/2000 season—roughly the same number as during the entire 1980s. Numerous expeditions also went south to the Geographic South Pole on skis, or even using parachutes.

The management of Antarctic tourism is an interesting issue in itself. When the Antarctic Treaty came into force in 1961,

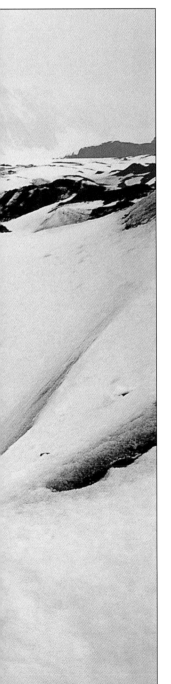

▲ DECEPTIVE STRENGTH
The sea ice and bergs along the coast of the Antarctic Peninsula dampen wave action—a welcome change after the often tempestuous Drake Passage. Ice arches contain hundreds of tonnes of apparently solid ice, but they are unstable and can collapse at any time.

▲ DECEPTIVE STRENGTH
The sea ice and bergs along the coast of the Antarctic Peninsula dampen wave action—a welcome change after the often tempestuous Drake Passage. Ice arches contain hundreds of tonnes of apparently solid ice, but they are unstable and can collapse at any time.

it did not specifically mention non-governmental activities, such as tourism. At that time, it was not apparent that significant numbers of tourists would one day want to visit a place that scientists on government-sponsored expeditions had claimed for themselves. As tourism increased, the Treaty Parties passed several recommendations that have affected the way tourism is conducted. However, it was not until 1998, when the Protocol on Environmental Protection to the Antarctic Treaty (also known as the Madrid Protocol) came into force, that the continent had a comprehensive and systematic management regime that specifically addressed tourism issues. Under the protocol, all human activities are subject to an environmental impact assessment process, and all activities with more than a transitory impact require a comprehensive environmental impact evaluation. The protocol is given effect through legislative measures in the various countries that are signatories to it. In the United States, for example, it is an offence to "harass" Antarctic wildlife (that includes approaching a penguin too closely to get a better photograph, or frightening a flying

▲ WILDLIFE WATCHERS
Antarctica's wildlife is so abundant and approachable that anyone can develop a tourist mentality. Standing on the sea ice, the crew of a United States icebreaker watch a Minke whale taking advantage of the channel their vessel has created.

▼ KEEPING A DISTANCE

One of the most intense of Antarctic experiences is landing
on Albatross Island in South Georgia's Bay of Whales to
observe nesting Wandering albatrosses and their chicks without
disturbing them. Recently a plan has been drafted to limit such
close encounters.

▼ CHEAP AT THE PRICE

A visit to South Georgia adds
a week to the length of an
Antarctic Peninsula cruise—
and several thousand dollars
to the cost. However, it is well
worth the extra time and
money for the chance to
observe vast throngs of the
most colorful of penguins,
the King, at sites such as
Salisbury Plain.

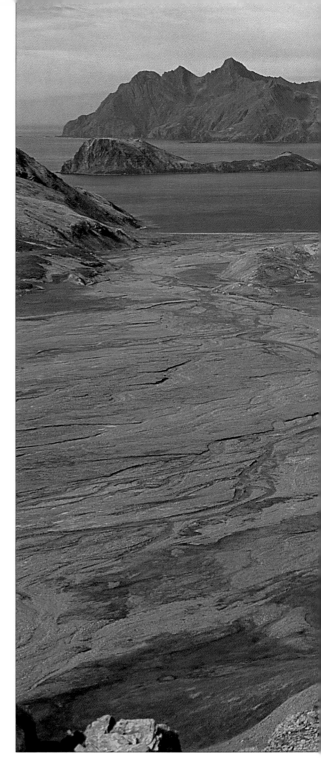

seabird so that it departs its nest). The penalties for
infringement can include jail terms and fines of up
to $US10,000.

To prevent tourists from having a negative impact on
the environment (and to self-regulate their own activities)
Antarctic tour operators cooperated as early as 1991,
when they established the International Association of
Antarctica Tour Operators (IAATO). IAATO developed
its own guidelines for the safe and responsible conduct
of tourism. It established a code of conduct for tourists
that was subsequently modified by the Treaty Parties and
formed the basis for Recommendation XVIII-I, which
provides the *Guidance for Visitors to the Antarctic* and the
*Guidance for Those Organizing and Conducting Tourism and
Non-Governmental Activities in the Antarctic.* Tourists
aboard all IAATO member vessels are required to attend
compulsory lectures to familiarize them with the code
of conduct for shore visits. They learn the minimum

distances that must be kept between people and wildlife; that there are no toilet facilities at landing sites; and that taking food or drink ashore is strictly prohibited, as is smoking during landings. The people who pay a lot of money to visit Antarctica expect to find a pristine place. This puts pressure on the tour operators and travelers to do the right thing and to keep Antarctica clean.

Antarctic tourism today is the best managed tourism in the world, and other destinations are encouraged to adopt the measures used in Antarctica to improve their own tourist practices. A visit to Antarctica is often the highlight of people's lives—they come back changed, having experienced nature at the most fundamental level. They have also developed a yardstick against which to measure environmental degradation at home.

What will the future bring for the world's southern-most tourist destination? Research suggests that the current pattern of ship-based tourism will continue, and that Russian-registered vessels will still play a significant role. There is, however, also the possibility that some day large passenger aircraft will make the short flight across the Drake Passage to an improved airstrip on King George Island, from where people could be transported to waiting cruise ships. This would save around four uncomfortable days of crossing and re-crossing the stormy Drake Passage, and would probably open up new and less adventurous segments of the market.

Should this scenario become a reality, tourism will certainly spread further south along the west coast of the Antarctic Peninsula and will encroach on previously unvisited sites, thus increasing the potential for negative impact. Strict regulation will be necessary (perhaps in the form of a quota system like those applied in other natural and cultural sites around the world) in order to safeguard the wildlife and scenic beauty of the most unspoiled wilderness on this planet. TB

▲ IN THE MASTER'S STEPS
Hikers on South Georgia are literally following in the footsteps of Ernest Shackleton as they descend into the ruins of Stromness whaling station, though the building is now too dangerous for them to follow his path through the station itself. Growing public interest in Shackleton and his men's amazing battle for survival during the *Endurance* expedition has resulted in some specialized tourist voyages.

Arctic tourism

From the farms and highways of Lapland to the ice-bound shores of Greenland, the Arctic has a great variety of landscapes. It also has diverse wildlife and a fascinating cultural heritage—from the ancient Viking settlements of western Greenland to the modern lives of the Inuit, Saami, and Chukchi.

There are zones of Arctic tourism that circle the ice with the North Pole as the bull's-eye. Those who simply wish to cross the Arctic Circle can drive there from anywhere in North America or Europe. Further north lie the northern coasts of Canada, Alaska, and Siberia, and beyond are the islands of the high Arctic, from whose frozen expanses the venturesome tourist can reach the floating ice of the Pole itself.

In 1913, the noted Antarctic explorer Adrien de Gerlache and a young Norwegian ship owner called Lars Christensen planned tourist voyages to Greenland and Spitsbergen. Their plan fell through and *Polaris*, a 10-cabin ship they had built, was sold to Ernest Shackleton and became *Endurance* of Antarctic legend. However, these failed entrepreneurs were not the first to consider the financial benefits of polar tourism. In 1892 a Captain Bade organized tours to Spitsbergen from Germany; he continued operating until 1907. In 1896 a rival company began to offer weekly departures. Regular tourist voyages to Spitsbergen continued—interrupted only by the two world wars—until 1975, when the airport at Longyearbyen opened.

Most tourism to the high Arctic is still ship-based, perhaps because the Arctic summer is too short to justify an extensive hotel infrastructure. In addition, as in the Antarctic, many more ships are available for cruising the Arctic since the Soviet Union collapsed in 1991 and the government agencies that owned the ships had to seek commercial revenue. For Western charterers the Arctic season perfectly offsets the Antarctic season.

▼ **WILDLIFE GALORE**
Dundas Harbor, on the south coast of Devon Island, has been the site of a Royal Canadian Mounted Police post from the 1930s. The island and its surrounding waters are home to a wide range of wildlife, from Polar bears, foxes, hares, and Musk oxen on land, to Narwhals, Bearded seals, and Beluga whales in the water.

➤ **BIRD HAVEN**

Kolyuchin Island, off the north coast of Siberia, has a small scientific base on the plateau above the landing beach. Captain Cook named it Burney's Island, but it is now known by its Russian name. The island is a haven for several bird species, so most cruise ships make the effort to call here and see the puffins, murres, cormorants, and guillemots that nest along the cliffs.

All polar operators take the educational side of their programs very seriously indeed. Lecturers are required to have a high degree of expertise and to be available to passengers at all times, especially while ashore. The open-bridge policy invites passengers to wander onto the ship's bridge whenever they wish, to check the charts and weather maps, view the radar and GPS, and chat to the officers on the watch. This ensures that real friendships develop between the Russian crew, the expedition staff, and the passengers.

The main center of Arctic operations is the North Atlantic, both from the Canadian and European sides, where the goals are Greenland, Spitsbergen, and around Baffin Island. A few voyages may run through the Bering Strait, and perhaps as far as Russia's Wrangel Island. In the northern summer most of these destinations are accessible to ice-strengthened ships.

There are two great traverses along the edges of the frozen Arctic Sea: the Northwest Passage and the Northeast Passage. The Northwest Passage skirts around the top of Canada; the Northeast Passage runs through the seas and islands at the top of Siberia. Orientation can be tricky toward the North Pole—watches have to be changed by an hour as the ship enters a new time zone every couple of days.

There is a distinction between icebreakers and ice-strengthened ships. While the latter can push sea ice aside, and is perfectly adequate for travel around the edges of the ice pack, an icebreaker cuts deep into the ice—a uniquely exciting experience. The *Kapitan Khlebnikov* and the *Kapitan Dranitsyn* are near-identical twin "regular" icebreakers. Each has six engines, which can produce

▲ **HELP FROM HELICOPTERS**

Icebreakers carry helicopters that fly ahead to find the thinnest ice. The helicopters may also take cruise passengers ashore or on scenic overflights. When the sea ice is too thick for Zodiacs, helicopters may be the only practical way to make a landing.

a total of 24,000 horsepower, and can cut through 2 meters (6 ft) of first year ice at 16 kilometers per hour (10 mph). While ice-strengthened ships may sometimes successfully pass through the Northwest Passage, only icebreakers can be sure of success.

Lars-Eric Lindblad was the pioneer of modern Arctic tourism, just as he was of Antarctic tourism. His purpose-built Lindblad *Explorer* was launched in December 1969, and in 1972 went further north than any other passenger ship: she reached 82°12′N. In 1984 she made it right through the Northwest Passage.

On 22 August 1994 there was a historic meeting of three icebreakers at the North Pole: *Yamal* of Russia, *Louis S. St-Laurent* of Canada, and *Polar Sea* of the United States. *Yamal* has returned to the North Pole many times since, carrying paying passengers. She, and her sister ship, *Sovetskiy Soyuz*, are nuclear icebreakers. Their power plants are governed down to 75,000 horsepower, and they can break up to 8 meters (26 ft) of first-year ice at 20 kilometers per hour (12 mph). Icebreakers carry helicopters for ice reconnaissance, and these are used on commercial voyages to ferry passengers to shore and for sightseeing overflights. DM

▲ **ICE IN SUMMER**

Ice conditions in the High Arctic can be so dense, even in summer, that only a full icebreaker can proceed. The *Kapitan Dranitsyn* is a Russian icebreaker owned by the Murmansk Shipping Company. It was built in 1981 by the Wartsila Company, Finland, and is 131 meters (400 ft) long, displaces 10,471 tonnes, and is propelled by six diesel-electric engines generating a total of 24,000 horsepower.

Keeping a balance

▲ HISTORY AND WILDLIFE
Tourists visiting Antarctica are
particularly drawn to historic
sites and to aggregations of
wildlife. Port Lockroy, with its
colony of Gentoo penguins,
historic research station, and
collection of whalebones left
by the whaling factories, is
among the most visited sites.
Despite many thousands of
visitors, the population of
Gentoo penguins appears to
be increasing.

Unlike the Arctic, which has indigenous populations and is surrounded by industrialized nations, Antarctica was long protected from the impact of humans by the barrier of the Southern Ocean. Now, however, scientists and tourists are crossing that barrier in increasing numbers, and its isolation will no longer protect the environment of Antarctica. Human activity affects Antarctica at three levels: locally, as scientific and tourist visits; regionally, as resource exploitation such as whaling and fishing; and globally, as industrially caused changes such as ozone depletion and climate change.

Local impacts

The types of environmental impacts caused by people in the Antarctic are much the same as elsewhere, with the important difference that extremely high standards of environmental stewardship apply in Antarctica. It has not always been so; in the early days of Antarctic exploration, few people realized that human activity could harm the environment. Attitudes gradually changed as science became the focus of Antarctic activities with the establishment of the Antarctic Treaty System (ATS) in 1959. Conservation was on the agenda of the first Antarctic Treaty Consultative Meeting in 1961, and since then the environment has become the treaty's major concern.

The Protocol on Environmental Protection to the Antarctic Treaty sets out to provide comprehensive protection of the Antarctic environment and its fauna and flora. The Protocol also protects the wilderness and esthetic values of Antarctica, and its value as an area for science.

The Protocol recognizes that people will visit Antarctica and that some impacts are inevitable. Any building changes the landscape, and can destroy plant and animal habitats; stations generate human waste and exhaust fumes; even walking damages the fragile soils. The Protocol requires that activities should be planned on the basis of information sufficient to allow prior assessment of impacts. This involves research to ensure the consequences of human activities are known.

◄ ALTERNATIVE TECHNOLOGIES
Until recently, Antarctic stations relied on
fossil fuels, but burning fuel contributes
to pollution and greenhouse gases, and
transport of fuel oil carries the risk of oil
spills. Alternative technologies, such as
this wind generator at Casey station and
solar power, are being tested in Antarctica
to reduce the dependence on fossil fuels.

Tourists and scientists alike want to experience the unique wildlife of Antarctica at close quarters, but approaching animals too closely can cause stress and lower their chances of breeding successfully. Research into the response of Antarctic animals to human activities is being used as the basis for guidelines to allow visitors to enjoy the region's wildlife without endangering it.

Other local environmental impacts are not inevitable, but they do constitute a definite risk. All human activity in Antarctica necessarily relies on fuel, but conditions in this harshest of regions make handling fuel containers difficult, and small fuel spills are a common occurrence. More seriously, shipping accidents in Antarctica have released large quantities of fuel, and there may well be

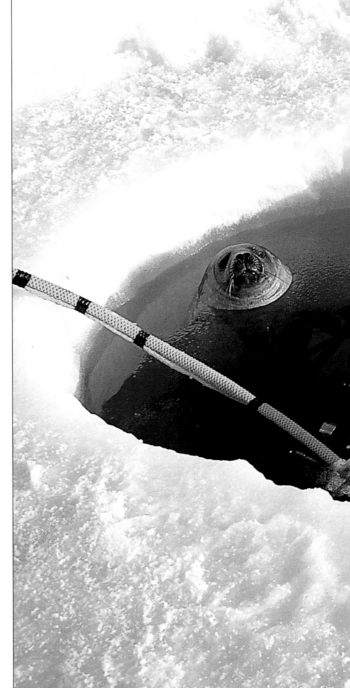

similar large-scale spills in future. The formulation of contingency plans to reduce the damage when these spills occur is a top priority.

One of the most environmentally damaging events anywhere in the world is the introduction of species, whether deliberate or inadvertent, because once established they are practically impossible to eradicate. The ecologies of almost all the sub-Antarctic islands have been significantly changed by introduced plants and anmals. Introduced pathogens could also cause havoc; there is no evidence that human activity has yet introduced disease to Antarctic wildlife, but it could happen. Measures to reduce the risk are currently being formulated within the Antarctic Treaty System.

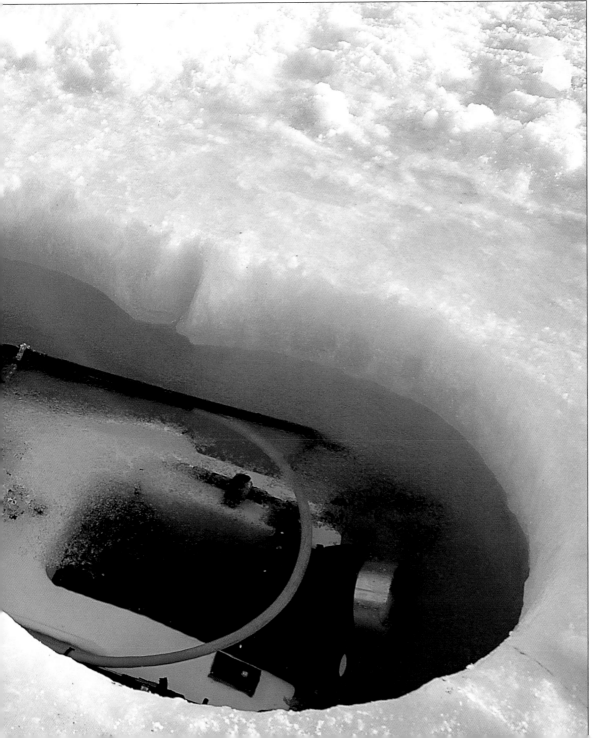

▲ NOT AT HOME HERE
These handsome reindeer seem perfectly at home at Stromness, on South Georgia, but reindeer do not belong in the southern hemisphere. Introduced species such as these can be the cause of serious harm to the delicate plant communities found on sub-Antarctic islands.

◄ ANIMAL WELFARE
Antarctica's unique wildlife is the focus of extensive scientific attention. Research protocols are carefully set up and scrutinized by independent animal ethics committees, with the aim of ensuring that intrusive techniques do not generate unnecessary pain, disturbance, or stress for the animals being studied.

◄ OPERATING SAFELY IN ICE
All national Antarctic programs rely on icebreakers and ice-strengthened ships such as the MV *Polar Bird* to re-supply their stations with fuel, food, and heavy equipment. These vessels operate in some of the most hazardous conditions in the world, and must conform to international standards of construction and operation to ensure the safety of personnel and the environment.

▲ PLASTIC PERIL
The Southern Ocean is the most remote of the world's seas; even here, however, discarded litter, particularly plastics, can be dangerous for animals. The neck of this fur seal is being slashed by fishing line that would shortly have killed it had it not been disentangled and removed.

Most visitors to the Antarctic are either members of one of the national Antarctic programs or tourists. Many more people visit Antarctica for recreation (14,000 in 1999–2000) than as part of a national program (4,000 for the same period), but the total number of person-days in Antarctica for national programs far exceeds the number for tourism. National programs also have a far greater infrastructure presence in Antarctica than tourist operations; there are as yet no permanent tourist facilities there. People visit Antarctica because of its special environment, whether as tourists to experience the unique wildlife and landscapes, or as scientists to study phenomena of global importance. It would be ironic if their visits were to threaten the very reasons for going there.

Hunting and fishing

The earliest human impact on the Antarctic was the completely uncontrolled exploitation of the region's wildlife. The first regular visitors to Antarctica and the Southern Ocean went to hunt its wildlife, and they very quickly had an enormous impact on its seal populations. As early as 1830—just 60 years after James Cook's first foray south of the Antarctic Circle—sealers had destroyed many of the once-teeming fur seal colonies of the sub-Antarctic islands. From seals they shifted their attention to penguins as a source of oil, and then to the great whales. The

sealing industry declined when it became economically unviable, but whaling, which began on a large scale in the Southern Ocean in the 1900s, has been regulated since 1949 by the International Whaling Commission in an attempt to maintain its viability. Nevertheless, many companies stopped whaling only because of falling catches and diminishing profits. Recently the commission has introduced more effective protective measures: by the 1960s Blue and Humpback whales were fully protected; Fin and Sei whales were protected by the 1970s; and in 1986 the commission suspended all commercial whaling. Whaling is now limited to "scientific whaling."

The effects of early wildlife exploitation are apparent today. Fur seal populations in many locations have exploded, probably as a result of ecological imbalances caused by hunting them and their competitors. King penguin populations in most localities are now growing enormously, however, the populations of long-lived great whales have not yet recovered.

Now that large-scale sealing and whaling have been prohibited, fishing is the only major resource extraction activity in the Antarctic Treaty area. Fishing can cause

The last bastion

Antarctica has a symbolic value: if the world cannot protect this most remote of lands, what hope is there for the rest of the planet? In the 1970s the Antarctic Treaty System (ATS) realized that Antarctica's mineral resources could well attract commercial interests, and that control would be necessary. Throughout the 1980s, the ATS worked toward the establishment of the Convention on the Regulation of Antarctic Mineral Resource Activities, based on the assumption that mineral extraction would occur, but with strong measures to ensure the least possible environmental impact. In 1989, just as the convention was ready for signing, it was set aside when Australia and France declined to sign and sought permanent prohibition of mining in Antarctica and comprehensive protection of the Antarctic environment. The Protocol on Environmental Protection to

the ATS was adopted in 1991 and came into force in 1998, committing treaty parties to comprehensive protection of the Antarctic environment, specifically prohibiting commercial mining, and designating Antarctica as a natural reserve devoted to peace and science.

However, when the protocol was negotiated the international community was not ready to lock up the mineral resources of Antarctica for ever. The protocol is open for review in 2048, 50 years from the date of its coming into force; at that time, and given the agreement of a majority of the nations engaged in Antarctic activities, it could be amended to permit commercial mining. But if the world decides to turn to Antarctica for its mineral resources, it will mean that no lessons have been learned from the wasteful and short-sighted use of mineral resources that characterized the twentieth century.

adverse impacts by taking too many fish, by reducing the food available for natural predators of the targeted fish, by destruction of habitat, and by killing non-target species, such as albatrosses caught on long lines. Fisheries in the Southern Ocean are regulated by the Convention for the Conservation of Antarctic Marine Living Resources, established in 1980, primarily to regulate the growing krill fishing industry, but with responsibility for management of all activities affecting living resources in the Southern Ocean. When established, the convention was unique because it recognized the importance of maintaining ecological relationships between harvested and dependent populations.

Global impacts

From the International Geophysical Year in 1957–58, Antarctica has been a laboratory for studying both surface and atmospheric processes on a global scale. Since then there has been a growing recognition of the importance of Antarctica for understanding global change caused by human activity against the background of natural variability. In 1985, the British Antarctic Survey announced the discovery of the springtime depletion of ozone over Antarctica; this was among the first indications that human activity is causing environmental impacts at a global scale. At about the same time, the scientific world became aware of growing evidence for climatic warming—the phenomenon now known as the greenhouse effect. The study of climate change and greenhouse gases is now perhaps the major focus of Antarctic science. These global changes might also directly harm the environment of Antarctica and its fauna and flora. Increasing ultraviolet radiation caused by ozone depletion could bring about changes in Antarctic phytoplankton that may create effects throughout the food web. Global warming may contribute to the breaking up of ice shelves and the loss of animal habitats. And global change may have effects in Antarctica that could have serious consequences elsewhere in the world—for example, changes in the amount of water frozen in the polar ice sheet would radically alter global sea levels.

Managing environmental change

During the Cold War that followed World War II, the international community, through the Antarctic Treaty System, acknowledged the value of Antarctica for bridging political chasms. In recent years, however, political preoccupations have given way to a growing awareness of the effects of industrialization on the global environment, and of Antarctica's unique role in analyzing and containing them. There is no quick or easy solution to the world's environmental problems—but the importance of Antarctica for understanding these problems can only increase.

▼ RECYLING RUBBISH
Waste generated at McMurdo Station is sorted for recycling before being shipped out of Antarctica. In the past it was standard practice simply to deposit the waste from Antarctic stations in open waste-disposal tips. Happily, this is no longer permitted, and most waste is now taken back to the country of origin.

▲ CLEANING UP ANTARCTICA
Rubbish like this at abandoned Wilkes station must now be taken away, provided the site has not been declared historic, or unless removal would do more damage than leaving it where it is.

Harvesting the polar regions

Antarctica was the last continent to be discovered, yet the exploitation of its biological resources proceeded so quickly that by the end of the twentieth century populations of many species had been seriously depleted. How did this occur?

Seals, whales, and penguins

The first wave of exploitation followed shortly after James Cook's Antarctic explorations and his description of fur seals on the beaches of South Georgia in 1773. The first Antarctic sealers reached South Georgia in 1778, and by 1791 more than 100 ships were harvesting fur and elephant seals from the island. Further south, sealing activity peaked in 1820–21, when a quarter of a million fur seals were taken, and by 1822 there were essentially no seal colonies on the Antarctic Peninsula. As early as 1788 a proposal to license sealers and regulate their catches came to nothing, and exploitation was

so intensive that fur seal numbers have only recently recovered. Since the 1950s, however, fur seal populations on some South Atlantic islands have revived so successfully that they may be causing environmental problems. Populations on southern Indian Ocean islands are still sparse, although numbers are increasing rapidly.

Elephant seals were harvested mainly on the sub-Antarctic islands of the South Atlantic Ocean and the southern Indian Ocean, but some were also taken in the

▷ **A PITEOUS SIGHT**
Elephant seals were hunted ruthlessly in Antarctic regions, for the yield and quality of their oil. Harvesting of these seals was stopped in 1964, and in some areas populations seem to be recovering.

◁ **A VERSATILE BOTANIST**
H. Hamilton, who studied the plant life of Macquarie Island during Douglas Mawson's 1911–14 Australasian Antarctic Expedition, systematically quarters an elephant seal for its blubber—an invaluable source of fuel in Antarctica.

South Shetlands. Elephant seal harvesting ceased in 1964, when South Georgia's sealing industry closed down. Elephant seal numbers in the South Atlantic appear to have recovered, but there is evidence that their abundance in the southern Indian Ocean has recently declined.

There were three attempts at commercial exploitation of pack ice seals—Weddell seals, Leopard seals, Ross seals, and Crabeater seals—between the 1960s and the 1980s. A few thousand seals were taken, but these enterprises seem to have been economically unprofitable.

▷ **A BRUTAL SPECTACLE**
The daily scene when South Georgia's whaling station at Grytviken was still fully operational was one of scarcely imaginable violence. Hundreds of whales were roped and chained together, and then slaughtered and stripped of every profit-making substance, and the remains left to rot or pollute the environment.

◁ **HUNTERS' PARADISE**
With northern regions all but hunted out, sealers were quick to see the potential of James Cook's sighting of fur seals off South Georgia. Grytviken, on the shores of the island's Cumberland Bay, was a commercial whaling station from 1904 to 1965, but sealers and whalers had arrived to plunder the region as early as 1778.

▲ PENGUIN PROVENDER
Teeming colonies of penguins such as these Kings were an invaluable source of protein and fats for the first Antarctic explorers: the land party of Otto Nordenskjöld's Swedish expedition of 1901–04 stockpiled 1,000 Adélies for their first Antarctic winter (they would have killed more, but the birds prudently fled!), and took 6,000 Adélie eggs when faced with a second winter on the ice.

A Norwegian company established the first whaling station in the Antarctic region in 1904, and a single catcher boat took 195 whales. By 1912 there were six land stations in the Antarctic and sub-Antarctic. By 1930–31, 40,000 whales were being killed each year, and this level of harvesting continued until the late 1960s. Humpbacks, abundant in the inshore waters near the whaling stations, were the initial target, but with the advent of factory ships, steam-powered catcher vessels, and explosive harpoons came the pursuit of oceanic species such as the Blue and the Fin. As these large species became depleted the focus shifted to Sperm, Sei, and the smaller Minke. In 1986 a moratorium on commercial whaling came into effect, although there has been an annual catch of Minkes by Japan under a "scientific whaling" exemption. At least 1.5 million whales were harvested from Antarctic waters, and the effects on the ecosystem are still being debated.

Birds have not been commercially important, though penguins have been taken for oil—150,000 Kings were taken on Macquarie Island from 1895 to 1919.

Fish

Commercial fishing in Antarctic waters began in 1905, when salt fish from South Georgia were exported to Buenos Aires. In the early 1950s commercial fishing at South Georgia recommenced, but full-scale commercial trawling did not begin until the late 1960s, again off South Georgia and the South Orkneys. Populations of Antarctic rock cod and icefish fell rapidly and have not

yet recovered, even after 30 years of protection. In the South Indian Ocean, commercial fishing began around Iles Kerguelen in the late 1960s and still goes on. Icefish and rock cod have recently given way to more profitable longline fishing for Patagonian toothfish. In the Pacific Ocean sector, little fishing took place until the search for new toothfish populations began in the late 1990s, partly due to a lack of known concentrations of commercial species and partly because of problems in fishing in this remote, icy region. In recent years pirate longline fishing for toothfish has depleted fish stocks and taken a huge accidental toll on albatrosses and other seabirds.

Krill

In 1901, Von Drygalski summarized problems that still plague the krill fishing industry when he reported that krill "tasted quite good, but were rather small and tiresome to peel." Early twentieth century studies indicated the huge size of the krill population, and during the 1960s interest was further spurred by the suggestion that there existed a "surplus" of krill—the vast quantities that the great whales would have consumed before their over-exploitation. The assumption was naive, but it did indicate that there was immense commercial potential for Antarctic krill. With the depletion of traditional fisheries in the late 1970s and the worldwide proliferation of Exclusive Economic Zones, krill appeared to offer unrivaled potential for fishing in international waters.

The Soviet Union began exploratory krill fishing in 1961, and Soviet vessels made sporadic small catches throughout the 1960s as technologists developed better gear and scientists found where the best concentrations lay. The Soviet Union set up a permanent Southern Ocean fishery in 1972, and by the mid-1970s full-scale Soviet and Japanese commercial operations were under way. Currently, vessels from Japan, Poland, Korea, Argentina, and the Ukraine are fishing for krill. Most krill is caught in the South Atlantic, fisheries moving south as the ice recedes in summer and north to the ice-free waters of South Georgia in winter. In the Indian Ocean, krill fishing is concentrated along the continental shelf break, but the fishing season is much shorter there because of the abundance of sea ice.

The krill catch peaked in 1982, when over half a million tonnes were taken—93 percent of this quantity by the Soviet Union. With the break up of the Soviet Union, catches of distant-water, low-value products such as krill began to decline. The low current krill catch—around 100,000 tonnes a year—reflects lack of demand rather than supply difficulties, but there has recently been renewed interest in krill fishing by India, Norway, the United States, Britain, Canada, and Australia.

The expansion of the fishery is likely to be driven by advances in food processing, pharmaceuticals, and aquaculture. Krill is a near-perfect food for farmed fish, and the krill fishery may become strongly oriented toward aquaculture in future. Antarctica could sustain an annual harvest of around 5 million tonnes of krill, ensuring that attention remains focused on the Antarctic as a potential source of marine protein, and it is probably only a matter of time before a much larger krill fishery develops.

▶ FISHING FOR THE FUTURE
Krill (from the Norwegian *kril,*
"young fish") are shrimp-like
crustaceans that swarm in
immense numbers in Antarctic
waters; the map shows the
main concentrations. A major
krill-fishing industry is a real
possibility for the future.

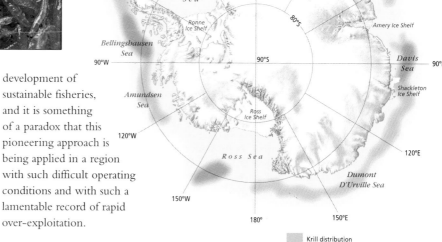

ANTARCTIC KRILL

Krill distribution
Areas of high concentration

Regulating resource exploitation

Initial attempts at regulation were difficult because of conflicting national claims over sections of Antarctica, and because of the high-seas nature of many fisheries. Gradually, however, a number of international treaties and conventions were devised to regulate harvesting in the Antarctic region. Commercial exploitation of all species is now covered by either the International Whaling Commission (IWC), the Convention for the Conservation of Antarctic Seals (CCAS), or the Convention for the Conservation of Antarctic Marine Living Resources (CCAMLR).

The early sealing industry exploited seals so rapidly that by the time it became obvious that there was a problem, there was no industry to regulate. Much later, in 1972, the Convention for the Conservation of Antarctic Seals was adopted, specifying total protection for fur, Elephant and Ross seals south of 60°S, and catch limits on Crabeater, Leopard, and Weddell seals.

Whales were exploited over a much longer period than seals, but despite early concern about the possible decline of whale stocks, international regulation of whaling did not begin until 1935, with the Convention for the Regulation of Whaling, in response to over-exploitation, an over-supply of whale oil, and a collapse of the market for whale products. The convention was largely ineffectual, and the International Convention for the Regulation of Whaling took over the management of whaling in 1946. This Convention established the International Whaling Commission (IWC), which is still responsible for managing whaling in all oceans. Since the formation of the IWC, the vast majority of whales harvested have been in Antarctic waters.

The Convention on the Conservation of Antarctic Marine Living Resources (CCAMLR) came into force in 1982, and regulates the remaining Southern Ocean fisheries. CCAMLR arose from concerns about the expanding krill fishery. Preemptive action was necessary to avoid the over-exploitation that had characterized Antarctic sealing, whaling, and fishing. The Convention was worded to ensure that krill fishing would not harm krill-dependent ecosystems, and in particular would not hinder the recovery of baleen whale populations.

At first CCAMLR developed protective measures for Antarctic fish; in later years it has introduced management measures for krill. Conservation measures are now in place for all harvested species within the CCAMLR area, and the Convention is now embarking on a much more difficult task: ecosystem management. Its efforts to protect ecosystems are right at the leading edge of the development of sustainable fisheries, and it is something of a paradox that this pioneering approach is being applied in a region with such difficult operating conditions and with such a lamentable record of rapid over-exploitation.

Exploitation of the Arctic

Arctic resources have been exploited for far longer than their southern counterparts. The Arctic differs from the Antarctic in having indigenous peoples who relied on subsistence hunting and fishing for thousands of years. Furthermore, because it is close to population centers in the northern hemisphere, Arctic resources were discovered and exploited on a large scale much earlier: Basque whalers were hunting Bowhead whales in the Arctic as early as the sixteenth century, and Europeans were capturing North American seals and Walruses in the seventeenth and eighteenth centuries.

Commercial whaling in the Arctic ended early in the twentieth century, although there has recently been a revival of Minke whaling by Norway, and subsistence hunting of some whale species is permitted. Sealing has continued in many Arctic areas, both at a subsistence level and on a commercial scale. Arctic waters have also been extensively fished; the Bering and Barents seas are the sites of some of the world's largest fisheries. However, many of these fishing operations take place in the territorial waters of Arctic rim nations, so regulation has been easier than it is for their Antarctic equivalents.   SN

▼ FISHING POLAR WATERS
Fishing vessels are much the
same the world over. This
fishing craft, pictured trawling
Arctic waters, differs little
from the small but efficient
fishing boats that can be
seen from any coastal settlement, from polar regions
through temperate and subtropical zones to the equator.

Conquering the Poles

Mapping Antarctica

Maps are the public record of Antarctic discovery, often incorporating thematic and human content as well as perceptions of topography. A hypothetical *Terra Australis* was represented in two early cartographic models, but the one that best prefigured later discoveries was that drawn by Roman philosopher Ambrosius Macrobius in the fifth century AD; he resurrected the Greek model of a zonal world map, postulating frozen polar landmasses separated by symmetrical temperate zones and continuous ocean systems.

A shrinking continent

More refined concepts of the geography of Antarctica followed Ferdinand Magellan's circumnavigation of the world in the early 1520s. In 1531 French geographer Orance Finé published a map that for the first time identified a circumscribed Antarctic landmass; he called it *Terra Australis*, and delineated its northern border along Magellan's route across the Pacific Ocean. Gerhard Mercator incorporated much of this map into his epic wall map of 1569, but over the next two centuries, as explorers traversed empty southern oceans, the hypothetical southern landmass shrank to within the Antarctic Circle.

A more honest—but equally incorrect—map of the southern hemisphere also appeared within 20 years of Magellan's voyage. In 1540 Sebastian Munster published a world map in his edition of Ptolemy's *Geographia*, in which he strove to modernize the traditional Ptolemaic format and to incorporate only the latest and most accurate geographic detail. The only "Antarctic" landmass in Munster's map was south of the Strait of Magellan (he did not realize that Tierra del Fuego was an island), and his was the first map to name the Pacific Ocean. It became a template for the classic Dutch maps drawn after Abel Tasman's circumnavigation of Australia in 1642, and world maps by the great Dutch cartographers of the seventeenth century, such as Nicolas Visscher, Joan Blaeu, and Pieter Goos, show empty oceans south of Australia.

Regional maps emphasize European interest in unknown Antarctic land from the late sixteenth century. Early regional maps of Tierra del Fuego—for example, the 1599 map of Cornelis Wytfliet—were followed by polar projections. The paring away of the hypothetical Antarctic landmass is illustrated in a series of polar projections based on the 1639 map *Polus Antarcticus*,

▼ THE SOUTH IN 1600

Chica sive Patagonica et Australis Terra—an almost exact copy, drawn in 1600 by Matthias Quad, of a 1599 map by Cornelis Wytfliet—is one of the first detailed representations of the southern tip of South America. It shows the Strait of Magellan and Tierra del Fuego, and the speculative *Terrae Australis* ("South Lands").

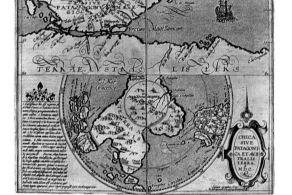

▼ THE LATE SIXTEENTH CENTURY

In 1595, Rumold Mercator, son of the influential Flemish cartographer Gerhard Mercator, produced a world map, *Orbis Terrae Compendiosa Descriptio* ("A complete map of the world"). He modeled it on a 1570 map by Ortelius, which was, in turn, a simplified version of the elder Mercator's iconic wall map of 1569.

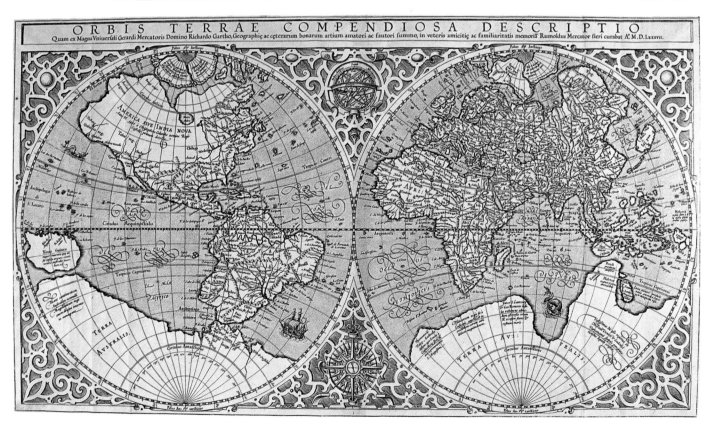

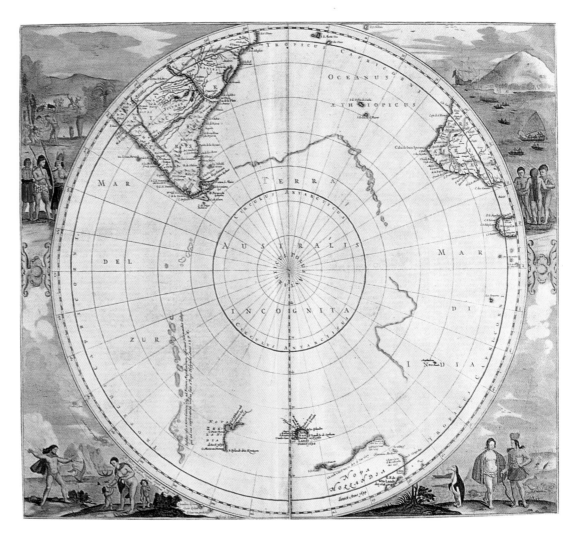

▶ THE MID-1600s
Framed with charming (albeit inaccurate) decorative motifs, Johannes Jansson's 1650 hemisphere map, *Polus Antarcticus*, was re-engraved from the plate of a 1639 map by Henricus Hondius. Jansson's edition was one of the first printed maps to incorporate some of the information brought back by Abel Tasman from his 1642 circumnavigation of Australia.

by Henricus Hondius. A French series of maps, *Hemisphere Meridional pour voir plus distinctement Les Terres Australis*, followed, emphasizing the French contribution to discovery in the southern hemisphere. The series was based on the 1714 map by Guillaume de l'Isle. It was completed in 1782 by Philippe Buache, who included Cook's 1772–75 circumnavigation of Antarctica, during which he confined any major landmass to within the Antarctic Circle and discovered the South Sandwich Islands and South Georgia—discoveries that led to sealers advancing down the Scotia Arc to the Antarctic Peninsula 50 years later.

The continent takes shape

Modern mapping of the Antarctic Continent began with the records of Cook's second voyage, in which limits of a landmass were set. Circumnavigations south of Cook's tracks by Thaddeus von Bellingshausen in 1819–21 and John Biscoe in 1830–32 further diminished the hypothetical Antarctic Continent. Biscoe, a sealer who worked for the Enderby Brothers, achieved the first sighting of West Antarctica under terrible conditions, and Enderby, presenting the log of Biscoe's ship, *Tula*, to the Royal Geographical Society in 1833, concluded that "the probability seems thus to be revived of the existence of a great Southern Land." This was important, as doubt had been raised by a map created by another Enderby employee, James Weddell, who had recorded penetrating the southern oceans to 74°S in 1822–24 without sight of land or significant ice.

The maps that accompanied the published journals of three national scientific expeditions in the middle of the nineteenth century represented significant progress in the mapping of Antarctica's coastline. These charts recorded the discovery of Terre Adélie by Jules

▶ FRANCE—THE 1800s
The journals of the French Antarctic scientific expedition of 1837–40, under the leadership of Jules Dumont d'Urville in *Astrolabe* and *Zélée*, were published in 1841 and included maps of his coastal sightings of Terre Adélie. This inset detail shows the expedition's plotting of the South Magnetic Pole.

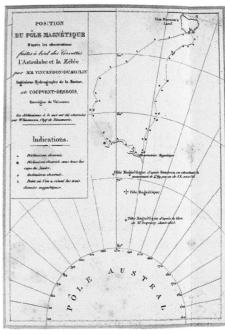

Dumont d'Urville; of about 1,600 kilometers (1,000 miles) of the coastline below Australia by Charles Wilkes; and of the Ross Ice Shelf and Victoria Land by James Ross. Two important maps compiled by the Hydrographic Office of the British Admiralty followed this haphazard collection of discoveries. The first was the *Chart of the South Polar Sea* (1845), the second the *Ice Chart of the Southern Hemisphere* (1874). Each contained all available information on discovery, including comments on ice conditions and wildlife observations, and the tracks of all previous exploration. A broad outline of the perimeter of

Antarctica was deceptive, as there was little validation of scattered sightings, significant discrepancies, still no landfall, and no proof that Antarctica was indeed a single landmass. However, European and American atlases competed to include the scanty detail that was available into regional maps of substance.

The most important maps made at the end of the nineteenth century were those that accompanied calls for a renewal of interest in Antarctic exploration. The loudest voices came from John Murray of the *Challenger* expedition, and from Sir Clements Markham (1898), president of the Royal Geographical Society. A detailed map published by Bartholomew, initially in 1894, accompanied various presentations, which collectively launched the Heroic Age of Antarctic exploration. Though national interest was stirred, the driving force behind exploration was private enterprise, and the maps that record these ventures either accompanied published journals of discovery, or were published by their sponsoring bodies, in particular the Royal Geographical Society. From its inception in 1832 until the Heroic

Age, this organization had been intricately involved in recording Antarctic discovery; it published the logs and charts of sealers employed by the Enderby Brothers, and its officers played a leading role in initiating and sponsoring expeditions. The society also published the charts of the British National Antarctic Expedition of 1901–04, led by Robert Scott. These charts of Victoria Land—the first new information since this area was discovered by James Ross in 1841—represent the beginning of detailed coastal and land mapping of the Antarctic Continent. By contrast, most of the maps accompanying the journals of discovery are simple overviews; for instance, the maps of Amundsen and Scott recording their dramatic race to the South Pole (1911 and 1912, respectively) are little more than sketches of direct passages from base to pole.

The emergence of nationalism

Cartography between the two world wars was strongly influenced by increasingly sophisticated technology and an awakening nationalism with respect to the world's only unexplored continent; the intense personal competitiveness of the Heroic Age gave way to national rivalries. The map of Antarctica was progressively replaced by accurate charts and blocks of color in the publications of government bodies managing national claims. Britain, in particular, knew the value of charts, considering "the whole Antarctica within the British Empire," with her own zone of influence extending from the Falkland Island Dependencies to the Ross Sea area. Land to the east and west of 160° was considered to be dependencies of Britain's dominions, New Zealand and Australia, respectively. To consolidate claims in this "Commonwealth Sector," Sir Douglas Mawson led the BANZARE expedition of 1929–31, mapping the coastline west of 160° to Enderby Land. Following this epic venture, and encouraged by Mawson, the Australian government produced a landmark map of Antarctica on a scale of 1:10,000,000, which was published in 1939, accompanied by a comprehensive and informative *Handbook and Index*. Even so, the overwhelming impression conveyed by this 1939 map is the sketchiness of the shape and content of Antarctica. Over the past 60 years the Australian government has produced a series of seven maps at 1:10,000,000 scale to record progressive charting and mapping of the Antarctic Continent.

A second important event for cartography in 1939 was the first expedition to be coordinated by a single government service. Rear

▼ THE UNITED STATES—1855
A mid-nineteenth century map of the Southern Regions by American cartographer Joseph Colton represents an increased interest in accurate mapping. It also reflects the burgeoning spirit of nationalism in the exploration and charting of Antarctica: the discoveries of the controversial American explorer Charles Wilkes are depicted in great detail in the lower left-hand quadrant, whereas other features are comparatively vague.

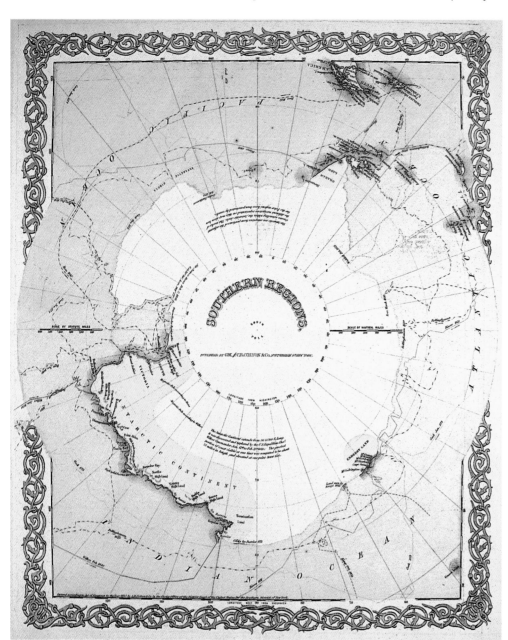

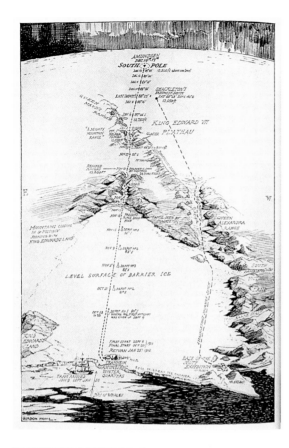

▲ THE EARLY TWENTIETH CENTURY
At the beginning of the twentieth century, overland forays into the Antarctic Continent itself began to make inland mapping possible. Nevertheless, early records were really just sketch maps of the journeys. Shown here are Roald Amundsen's 1911 route to the South Pole (left) and Robert Scott's 1912 route (right).

◄ THE 1920s AND THE 1930s
Twentieth century technology made the task of mapping Antarctica infinitely easier. In addition to its commercial interests, the whaling firm of Lars Christensen used sea-planes like this one to take thousands of aerial photos of the Antarctic coastline.

▲ THE MID-1930s
John Rymill's British Graham Land Expedition was the first Antarctic venture to use aircraft and motorized tractors and boats, as well as dogs and skis. But it was a dog-sledging expedition that enabled Rymill to prove that Graham Land was a peninsula, not a group of islands. Sadly, the animals had to be shot when Rymill returned north.

Admiral Richard Byrd of the United Stated Navy, on his third visit to Antarctica, to his base at Little America on the Ross Ice Shelf, planned to establish the first permanent base, but the attempt was cut short in 1941 by World War II. (On his first expedition, in 1928, Byrd had confirmed the potential for aerial navigation with an epic 19-hour flight to the South Pole and back. On his second trip, in 1934, he had surveyed over a million kilometers to the east by air, and had put paid to the idea that Antarctica might consist of two landmasses.) In the west, the whaling firm of Lars Christensen surveyed 6,440 kilometers (4,000 miles) of coastline, documented in 2,000 aerial photos, between 1927 and 1937. In 1934–37, John Rymill led a British expedition to Graham Land, and established a model for small expeditions by combining air, land, and sea activities to coordinate efficient survey. Thus mapping of the Antarctic between the world wars was an amalgam of the individualistic philosophy of the Heroic Age, the beginning of well funded national programs, and the development of team science with the tightly organized expedition led by Rymill.

Technology enters the picture

Detailed mapping of Antarctica has been a post World War II event, directly related to nationalism and the development of permanent bases, and to technology. In 1946, the United States Navy organized Operation Highjump to "consolidate and extend US potential sovereignty." Sailing east and west around Antarctica, in one month this vast expedition surveyed 60 percent of the coastline and took no fewer than 70,000 photos. When it was realized that there were no reference points, a supplementary program, Operation Windmill, identified 30 prominent landmarks, using helicopters operating from icebreakers. Since that time the technological revolution has continuously extended the role of mapping in Antarctica, providing topographical detail and thematic maps of physical features ranging from weather patterns through ice depth to animal migration routes. Modern maps of Antarctica based on satellite and digital technology provide a continuous, accurate, and changing record of the form of Antarctica. RC

◄ 1980 AND ONWARD
By the end of the twentieth century, artificial satellites were being routinely used for mapping. The information transmitted from Landsat satellites is turned into images that can be combined and colored to highlight various details and sent around the world. Here the information being conveyed is the weather pattern over the Weddell Sea.

Phantom islands

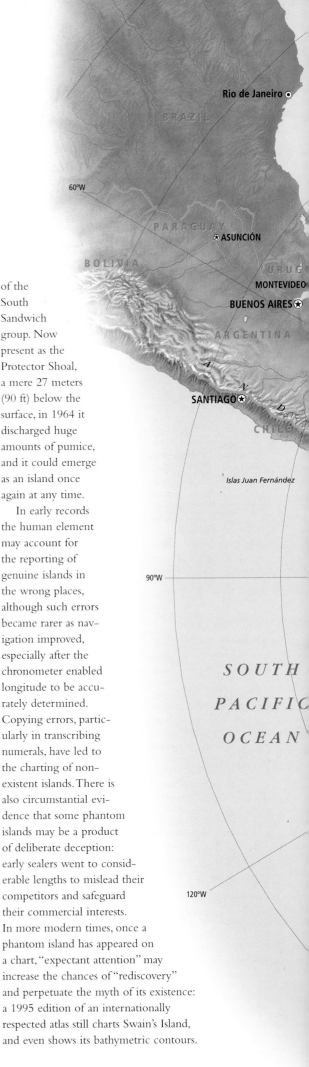

Islands that do not exist have been reported in all the world's oceans, but the waters around Antarctica have more than their share of these intriguing phenomena. "Phantom islands" are charted, and replicated on subsequent charts, because cartographers must err on the side of caution: if a suspect island or rocky outcrop might be a navigational hazard, it is far safer to chart it than to ignore it. On the downside, charting a non-existent island may mean loss of life if shipwreck survivors or damaged vessels seek refuge there, so a serious investigation is essential before an island can be "expunged from the charts." Until satellite surveillance became possible, the best way—indeed, the only way—to confirm the existence of an unapproachable island in poor conditions was to circumnavigate it and map the surrounding sea floor, but prudent ships' masters, especially in the days of sailing ships, were reluctant to take such risks.

Phantom sightings

In Antarctic waters, where there are long periods of fog in the summer and polar darkness in the winter, it is deceptively easy to mistake an iceberg for a small landform. On a gray day—and there are many gray days in the southern oceans—an iceberg can look very much like an island or a rocky outcrop, especially if it is covered with snow. A morainic iceberg contains rock fragments that make it look like land, and bird colonies often cover icebergs with pinkish-orange guano that makes them likely to resemble land. Floating debris is another possible source of mistaken sightings: shipping wreckage, bloated whale corpses, drifts of kelp, rafts of volcanic pumice, particularly when the flotsam is covered with ice and snow or when birds are converging upon it. Mirages and similar visual phenomena often occur in the clear, cold, highly refractive air of polar regions, and this, combined with the effects of cloud and other meteorological conditions, may account for some reports of phantom islands. Another natural phenomenon, volcanic activity, has created and destroyed many islands: one example is the potential twelfth island of the South Sandwich group. Now present as the Protector Shoal, a mere 27 meters (90 ft) below the surface, in 1964 it discharged huge amounts of pumice, and it could emerge as an island once again at any time.

In early records the human element may account for the reporting of genuine islands in the wrong places, although such errors became rarer as navigation improved, especially after the chronometer enabled longitude to be accurately determined. Copying errors, particularly in transcribing numerals, have led to the charting of non-existent islands. There is also circumstantial evidence that some phantom islands may be a product of deliberate deception: early sealers went to considerable lengths to mislead their competitors and safeguard their commercial interests. In more modern times, once a phantom island has appeared on a chart, "expectant attention" may increase the chances of "rediscovery" and perpetuate the myth of its existence: a 1995 edition of an internationally respected atlas still charts Swain's Island, and even shows its bathymetric contours.

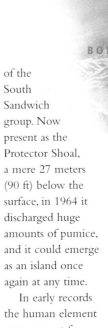

EASILY DECEIVED
The Southern Ocean environment with its wild seas, violent winds, and gray skies gave rise to reports of numerous "phantom islands." Early mariners played it safe and did not investigate such sightings, but simply assumed that they were permanent islands and marked them on their charts.

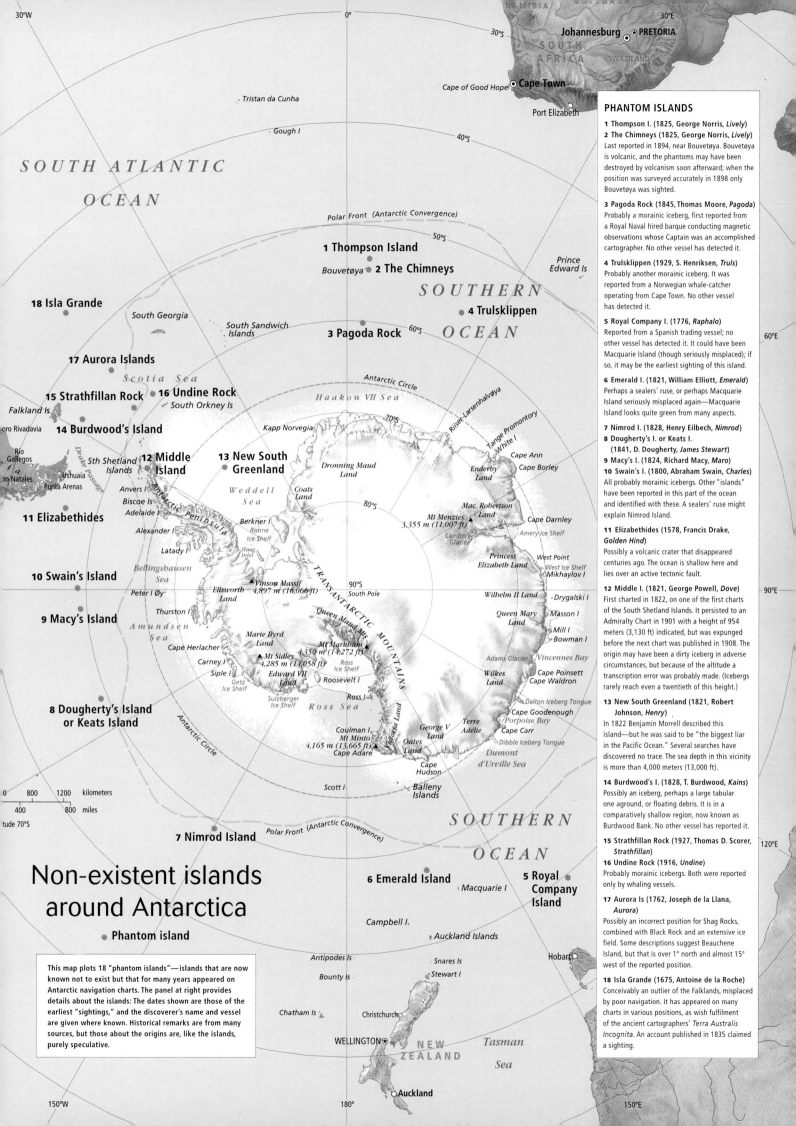

Non-existent islands around Antarctica

● Phantom island

This map plots 18 "phantom islands"—islands that are now known not to exist but that for many years appeared on Antarctic navigation charts. The panel at right provides details about the islands: The dates shown are those of the earliest "sightings," and the discoverer's name and vessel are given where known. Historical remarks are from many sources, but those about the origins are, like the islands, purely speculative.

PHANTOM ISLANDS

1 Thompson I. (1825, George Norris, *Lively*)
2 The Chimneys (1825, George Norris, *Lively*)
Last reported in 1894, near Bouvetøya. Bouvetøya is volcanic, and the phantoms may have been destroyed by volcanism soon afterward; when the position was surveyed accurately in 1898 only Bouvetøya was sighted.

3 Pagoda Rock (1845, Thomas Moore, *Pagoda*)
Probably a morainic iceberg, first reported from a Royal Naval hired barque conducting magnetic observations whose Captain was an accomplished cartographer. No other vessel has detected it.

4 Trulsklippen (1929, S. Henriksen, *Truls*)
Probably another morainic iceberg. It was reported from a Norwegian whale-catcher operating from Cape Town. No other vessel has detected it.

5 Royal Company I. (1776, *Raphalo*)
Reported from a Spanish trading vessel; no other vessel has detected it. It could have been Macquarie Island (though seriously misplaced); if so, it may be the earliest sighting of this island.

6 Emerald I. (1821, William Elliott, *Emerald*)
Perhaps a sealers' ruse, or perhaps Macquarie Island seriously misplaced again—Macquarie Island looks quite green from many aspects.

7 Nimrod I. (1828, Henry Eilbech, *Nimrod*)
8 Dougherty's I. or Keats I.
(1841, D. Dougherty, *James Stewart*)
9 Macy's I. (1824, Richard Macy, *Maro*)
10 Swain's I. (1800, Abraham Swain, *Charles*)
All probably morainic icebergs. Other "islands" have been reported in this part of the ocean and identified with these. A sealers' ruse might explain Nimrod Island.

11 Elizabethides (1578, Francis Drake, *Golden Hind*)
Possibly a volcanic crater that disappeared centuries ago. The ocean is shallow here and lies over an active tectonic fault.

12 Middle I. (1821, George Powell, *Dove*)
First charted in 1822, on one of the first charts of the South Shetland Islands. It persisted to an Admiralty Chart in 1901 with a height of 954 meters (3,130 ft) indicated, but was expunged before the next chart was published in 1908. The origin may have been a dirty iceberg in adverse circumstances, but because of the altitude a transcription error was probably made. (Icebergs rarely reach even a twentieth of this height.)

13 New South Greenland (1821, Robert Johnson, *Henry*)
In 1822 Benjamin Morrell described this island—but he was said to be "the biggest liar in the Pacific Ocean." Several searches have discovered no trace. The sea depth in this vicinity is more than 4,000 meters (13,000 ft).

14 Burdwood's I. (1828, T. Burdwood, *Kains*)
Possibly an iceberg, perhaps a large tabular one aground, or floating debris. It is in a comparatively shallow region, now known as Burdwood Bank. No other vessel has reported it.

15 Strathfillan Rock (1927, Thomas D. Scorer, *Strathfillan*)
16 Undine Rock (1916, *Undine*)
Probably morainic icebergs. Both were reported only by whaling vessels.

17 Aurora Is (1762, Joseph de la Llana, *Aurora*)
Possibly an incorrect position for Shag Rocks, combined with Black Rock and an extensive ice field. Some descriptions suggest Beauchene Island, but that is over 1° north and almost 15° west of the reported position.

18 Isla Grande (1675, Antoine de la Roche)
Conceivably an outlier of the Falklands, misplaced by poor navigation. It has appeared on many charts in various positions, as wish fulfilment of the ancient cartographers' *Terra Australis Incognita*. An account published in 1835 claimed a sighting.

Polar shipping

The earliest maritime explorers, in polar regions as elsewhere, were not motivated so much by the discovery of new lands as by the commercial advantages of pioneering new trade routes to the East or exploiting newfound resources, whether animal, vegetable, or mineral. In the sixteenth century, northern polar navigators such as Henry Hudson and Willem Barents at least knew that the Arctic Ocean and its ice occupied the high northern latitudes and were surrounded by more-or-less known landmasses, but no one knew whether the Antarctic Continent even existed, still less what it might be like. In the fourth century BC, the Greek navigator Pytheas speculated that it was there to balance the landmasses in the northern hemisphere; his speculation

had to allow for the incalculable times at sea, the ports to be served, the weather to be survived, and the cargoes to be carried. But there was almost no provision for sailing through pack ice, even though whalers and seekers of the Northwest and Northeast passages frequently encountered Arctic sea ice. Ships were sometimes fitted with hardwood sheathing at the waterline to withstand the abrasion of ice floes, and beams and frames might be stiffened to resist pressure, but no special attention was given to navigation in polar ice until late in the nineteenth century, when Norwegian explorer Fridtjof Nansen purpose-built his ship *Fram* for breasting the polar pack ice—and in the hitherto unimaginable interests of scientific investigation, rather than commerce.

Sailing the polar seas

Early sailing ships were not designed for ice navigation, for a very simple reason: it was so difficult to make way through ice under sail that the best strategy in polar regions was to avoid the ice in the first place. It could well have been an ice navigator on a sailing ship who first remarked: "The longest way around may be the shortest way home." Even in the twenty-first century, the most powerful icebreakers stay as clear of the ice as possible, venturing into it only when there is no obvious route through open water.

Amazingly, from 1818 virtually unmodified British Navy sailing warships penetrated deep into the Canadian Arctic in search of the Northwest Passage, whereas for almost two centuries ships of other nations had used the Northeast Passage (the Siberian Northern Sea route). Of course, all of these expeditions to both Canadian and Siberian waters were equipped to winter over for several seasons. It is true that the entire Northwest Passage expedition of 1848, under the command of John Franklin, was lost, but it is remarkable how many other ships survived polar ordeals: *Erebus* and *Terror*, both lost with Franklin in the Canadian Arctic, had explored Antarctica in 1839–43 under the command of James Clark Ross, and HMS *Resolute*, abandoned in 1854 during the fruitless search for Franklin, drifted into the open water of Davis Strait and was salvaged by an American whaler in August 1855.

The first serious scientific attempt to find the Antarctic Continent was in 1772, when Captain James Cook set out on his second voyage of discovery with two small wooden sailing ships, *Resolution* and *Adventure*. Cook crossed the Antarctic Circle on 17 January 1773, but was too far off to sight land and was confronted by pack ice. He prudently turned back into open waters, but continued to probe southeast for two months more, threading his way between masses of icebergs before giving up and making for New Zealand. His Antarctic experience probably influenced his retreat from the Arctic ice of the Chukchi Sea on his third voyage, in 1778; he must have realized that small ships were not equipped to battle polar pack ice.

▲ FORCING THROUGH ICE
An artist's impression of HMS *Erebus* battling through icebergs during James Clark Ross's Antarctic expedition of 1839–43 dramatically depicts the perils that Ross and his fleet faced. *Erebus* was eventually lost with the Franklin Expedition during the search for the Canadian Northwest Passage in 1845–48.

was right, but his reasoning was wrong. The ancients also guessed that the south polar region was as inhospitable as the Arctic, whether or not there was land to be found there, so it was not an attractive destination for adventurers out to make fortunes in trade monopolies.

As explorers penetrated the Arctic and whalers plugged the gaps in knowledge of the north polar seas, they exploited the region's wildlife almost to extinction, and then turned their attention southward in search of new stocks of whales and seals. Secrecy is the watchword in all commercial enterprises, and whaling captains and adventurers tried to keep their favorite hunting grounds and routes to themselves, even deceiving their rivals into searching elsewhere. Thus the records of some of the earliest voyages into the Southern Ocean went with their authors to the grave.

The first circumnavigations of the globe, which took place in the sixteenth century by Portuguese navigator Ferdinand Magellan and English explorer Francis Drake, spurred much more daring ocean voyages. Shipbuilders

◁ A PURPOSE-BUILT SHIP
For his journey to the South
Pole in December 1911,
Amundsen borrowed Fridtjof
Nansen's ship *Fram*, designed
to drift through the north
polar pack ice. Under sail,
because the steam engine
generated little power and
coal was in short supply, the
ship made agonizingly slow
progress through the drift ice.

The steam revolution

Cook's epic voyages contained lessons for
ship construction techniques, but seafarers
are notoriously conservative, and ships
continued to evolve at a snail's pace.
Moreover, the push was towards the
expansion of tropical trade empires, and
polar navigation was a low priority.

It was the harnessing of a power
source other than wind—the application
of steam-power, originally generated by
paddle-wheels—that began to make the
difference to ice navigation. But paddle-wheels could
not function in ice without risking serious damage from
floes, and the totally submerged screw propeller was the
answer to propulsion independent of the wind. Iron-
hulled vessels were built in greater numbers from about
1830, but they were not used in ice until the end of the
nineteenth century, probably because the cast iron plates
were rivetted and ice would spring the seams and cause
damage impossible to repair at sea, whereas wooden
ships could be repaired at sea, all the materials being to
hand to scarf in a new piece of timber and caulk the
seams with oakum and tar.

Canada claims to have been the first nation to
acquire a steel icebreaker, *Stanley*, built in Scotland in
1888 as a winter ferry between Prince Edward Island
and the Canadian mainland. She became the first escort
icebreaker, assisting shipping in the region on an opportunity basis. Russia made a determined effort to build

a steel icebreaker to escort ships along the Siberian
Northern Sea Route, and in 1899 Admiral Stepan
Makarov (1849–1904), "the father of Russian icebreaking," was instrumental in designing and operating the
first of several icebreakers to bear the name *Yermak*. But
years earlier, in 1892, Norwegian Fridtjof Nansen had
built the small wooden *Fram*, in which he planned to
drift across the North Pole in the Arctic ice for three
years. She had a rounded hull that would be lifted rather
than crushed by ice pressure, and internal strengthening
of beams, frames, and longitudinal members
to withstand pressure, but she was not made
for forcing through ice. Her steam power was
rudimentary; her main propulsion in open
water was a full set of sails. In 1896 she completed her Arctic mission, and was later used
by fellow Norwegian Roald Amundsen in his
assault on the South Pole in 1911, proving
that even a wooden vessel—if well
enough designed—could defeat the
ice of both polar regions.

In the early twentieth century
Heroic Age of Antarctic exploration,
most explorers set out in converted
whaling ships, some of them well past
their prime, and all built of wood. *Endurance*,
purpose-built for polar voyages in 1913 and used
for Ernest Shackleton's doomed Transantarctic
Expedition, still followed the traditional design of
the wooden sailing whalers, with only auxiliary
steam power. Robert Falcon Scott's *Terra Nova*
and Shackleton's *Nimrod* were leaky ex-whalers,
sail-driven with auxiliary steam power, and not
designed for forcing through polar ice. *Discovery*
was purpose-built in 1900 for Scott's first
Antarctic expedition, but still followed the old
wooden whaler design, albeit greatly reinforced.

Steam power was jealously conserved, since
coal was needed to warm living quarters ashore.
In March 1898 Adrien de Gerlache's expedition
to the Antarctic Peninsula became—unintention-
ally—the first expedition to winter in Antarctica.
De Gerlache's ship, *Belgica*, was a small sealer
of 250 tonnes displacement: she was only
30 meters (100 ft) long, with a beam of

▽ MEETING AN ICEBERG
Avoiding icebergs under sail is
difficult, and close encounters
are common. *Belgica*, the
Belgian Antarctic Expedition
ship of 1897–99, collided with
an iceberg bow-on. Luckily,
the damage was above the
waterline, and comparatively
easy to repair.

◁ IN PERIL ON THE SEA
Trapped in the ice of the
Weddell Sea in January 1915
Shackleton's *Endurance* was
still upright and under full
sail, giving hope that she
would eventually break free.
Ten months later, she sank
after drifting 920 kilometers
(575 miles) locked in the ice
that crushed her.

▲ HELICOPTER RECONNAISSANCE
A helicopter takes off from the flight deck of an icebreaker in the pack ice of the Ross Sea to map the ice conditions ahead. Aerial reconnaissance is an essential tool for ice navigators, who must seek the best route to follow, and must find alternative routes when beset by heavy ice.

▲ AT THE NORTH POLE
The nuclear-powered Russian icebreaker *Yamal* pauses at the North Pole for passengers and crew to celebrate their achievement in navigating the polar pack ice. *Yamal* is one of the largest and most powerful icebreakers in the world, but even she can reach the North Pole only in summer ice conditions.

6.5 meters (21 ft). Her steam engine developed a mere 35 horsepower, but still required more coal than they could afford when she was trapped in ice for almost a year. Mawson used the 40-year old *Aurora* for his 1911–14 Australasian Antarctic Expedition, and returned to Antarctica with Scott's *Discovery*, by then 30 years old, for his 1929–31 expedition

Steam, steel, and diesel

Interest in Antarctica waned once Amundsen had won the race to reach the Pole, and the next major effort at scientific discovery—as opposed to geographic exploration—came after World War II. By then, steel ships had replaced "wooden walls," steam and diesel propulsion had taken over from sail, and oil fuel had eliminated the need to carry vast quantities of coal. Icebreakers had been developed into a specialized class of ship, mainly by Russia, Finland, Canada, and the United States.

Design and technology advances were encouraged by the expansion of wintertime marine trade in the Baltic, in the eastern Canadian waters, on the Russian Northern Sea Route, and by the Antarctic supply operation for the rapidly expanding scientific community. The construction in the 1950s of the Distant Early Warning (DEW) Line of Arctic radar stations across North

America and Greenland was another incentive for icebreaker development and construction, with the Russians engaged in similar activity on their side of the Arctic Ocean. The result was to provide the world with a fleet of dedicated icebreakers, and, as a spin-off, with cargo vessels capable of navigating polar ice without icebreaker escort.

Why are icebreakers different?

In the first place, their hulls are much stronger than those of most ships, with everything, from frames, to plating, to stiffeners, made of especially strong steel with greater than standard dimensions: for example, the hull plating of the Russian nuclear-powered icebreaker *Yamal* is 4.5 centimeters ($1\frac{3}{4}$ in) thick—about double the plating thickness of a cargo ship. They have a sharply raked stem, which allows them to ride up onto the ice and crush it with their weight. They have no bilge keels or other external hull fittings that might be caught on ice, or ripped off by it. They are powerful for their size, which means that they can twist and turn to escape from thick pack ice. They have multiple screws—usually twin screws, but larger vessels have triple screws, partly for maneuverability, but also as a fallback if a propeller is damaged by old, hard ice.

Most modern icebreakers have a flight deck so that a helicopter can reconnoiter the way ahead, and a high wheelhouse to provide a clear view over the ice. To help them through heavy ice they have devices such as bubbler jets, which blow air along the hull beneath the waterline to bring water into the ice alongside, wetting it to make it more slippery, and low-friction hull paint, with non-stick properties to stop ice from adhering to the hull. Icebreaker hulls may be shaped into "reamers" at the forward shoulders to assist in breaking ice and in turning; they may also be fitted with internal ballast tanks that can quickly heel or trim the vessel by transferring water from side to side or from end to end, which can be very effective in freeing them when beset by heavy ice. Many icebreaking freighters have all the features of a full icebreaker, but the addition of cargo-carrying capability increases their size and reduces their maneuverability; even so, they seldom need icebreaker escort.

➤ ENCOUNTERING PACK ICE
Pack ice in the Antarctic Peninsula slows down the progress of the Russian polar research vessel *Professor Molchanov* as she heads towards the ice-edge, visible as a dark line on the horizon. The darker blue of the sky at the ice edge is a sign of open water in that direction, whereas "ice-blink" with a bright sky indicates the presence of ice.

Ships strengthened for ice navigation must be certified by national marine classification societies, individual Ice-Classes being rated by a ship's capacity to resist ice damage. But there are as yet no global standards, although the International Maritime Organization (IMO) has been working for many years on a global classification system for polar ice vessels.

The Organization is proposing seven classifications for Polar Class (PC) vessels, ranging from PC1 (Polar Icebreaker permitted to operate year-round in all polar waters without restriction) to PC7 (Icebreaker confined to summer/autumn operation in thin first-year ice—up to 70 cm/28 in thick—with old ice inclusions). Freighters, tankers, or passenger vessels may also be built to a PC specification, but will probably go no higher than PC4 for reasons of economy.

Non-polar class vessels may be strengthened to varying degrees to resist ice damage, but none are considered to be icebreakers and their movements in polar ice are strictly confined to the summer and autumn ice navigation seasons in both northern and southern polar regions. Again, there are as yet no international standards for such vessels. Strengthening for ice navigation does not make a vessel more capable of ice navigation; it merely ensures that she can resist ice damage from either impact or pressure—as long as she is handled properly. In ice, such vessels must proceed with utmost caution; they must never come into contact with old ice; they must not take any "aggressive" action, such as ramming deliberately into heavy ice; and in extremely difficult conditions they may need icebreaker escort.

▶ **LIVING IN AN ICEBREAKER**
In the Russian icebreaker *Kapitan Khlebnikov*, living quarters for crew and passengers are high above the hull, away from the noise of ice grinding along the shell-plates. The wheelhouse is perched well above the ice for the best possible surveillance of the route between the ice floes.

Tracking the ice

Whether aboard a yacht or a state-of-the-art icebreaker, a venturer into polar regions needs to know sea-ice conditions as soon as possible, in order to plan a route from departure point to destination. But once a voyage is under way, the primary need is for short-term information. Polar ice is unpredictable; it can become impassable in a matter of hours, or, conversely, an ice-bound ship might have to wait only a few days, or even hours, before conditions improve enough to carry on.

Before reliable remote surveillance systems were developed in the 1960s, ice navigators had to rely mostly on local knowledge, past experience, and guesswork, which was hardly conducive to a rapid or safe passage through dangerous ice. In the late 1980s satellite surveillance was introduced to gather ice data by radar and infra-red sensing, providing a global picture of the sea-ice cover, able to show ice floes as small as 10 meters (33 ft) in diameter. Reconnaissance aircraft use similar equipment to plot local ice conditions in greater detail from an altitude of 9,150 meters (30,000 ft), and low-flying aircraft carry ice observers—meteorologists who can visually distinguish old ice from less dangerous newer ice, and can estimate percentages of sea ice-cover. Whenever possible, remote surveillance data are verified by observers on land or on ships in the ice.

All these polar ice data are transmitted to central offices in countries with polar interests, which broadcast them almost daily during their respective navigation seasons: May to November in the Arctic, and October to March in the Antarctic. Centers for Arctic ice data are maintained by Canada, the United States, and Russia, while the United States and Australia are the chief sources of ice information for ships in Antarctic waters. Australia's Antarctic Meteorological Center at Casey station on the Antarctic Continent also collates data and broadcasts ice maps, whenever possible, for the coastal sector from Proclamation Island to the Ross Sea, where most Antarctic sea traffic is concentrated.

In Antarctic waters, ships of all nations draw on United States and Australian ice information, which may be broadcast on marine radio frequencies or downloaded from internet sites. Ice bulletins are also transmitted in text form by radiotelephone for vessels without internet or facsimile capability.

The Egg-code

It is in these American and Australian information centers that ice data from all sources are collated into detailed ice maps. The prevailing ice conditions in each section of the map are represented by the internationally recognized Egg-code—an ice code devised by the World Meteorological Organization and generally adopted in the 1970s. The code takes its name from egg-shaped capsules that identify each zone on the ice-map. Each capsule contains numerically coded ice information, and navigators can consult the code to find the concentration of ice cover in any given zone, and to know whether the ice is new and easy to break, or old and impossible to break. On the basis of this information they can decide whether a given zone is safe to enter or is best avoided. But an ice map is not infallible: conditions can deteriorate so rapidly that a ship entering a zone showing navigable ice may become ice-bound and unable to move in any direction. Normally the consequences are no more serious than a delay in arrival time at the destination port, but they could condemn a ship to pass a winter in the ice, as many nineteenth-century sailing ships were forced to do in the Arctic. In a worst-case scenario, a ship could be crushed in the ice and lost, as Shackleton's *Endurance* was in 1915.   PT

TAKING ADVANTAGE
The crew of *Aurora*, Mawson's Australian Scientific Antarctic Expedition ship of 1911–13, hoisted blocks of old drift ice aboard and melted them to replenish the ship's supply of fresh water. Sea ice more than three years old freshens because the salt leaches back into the sea in the form of brine during the first two or three melt seasons.

Cape Adare

PHOTOGRAPHING THE ANTARCTIC
NOAA (National Oceanic and Atmospheric Administration) polar orbiting satellites pass over the poles and transmit very high resolution data, which are received at Casey station. The area scanned by each path of the satellite is about 2,600 kilometers (1,600 miles) wide, and transmission of the data is 15 minutes in duration. Sea ice images that originate from NOAA satellites are enhanced to give an optimum view of sea ice conditions. These images are transmitted to the Australian Antarctic Division for downloading to vessels.

▲ CLOSE ESCORT

In heavy ice conditions in Siberia's Kara Sea, the Russian nuclear-powered icebreaker *Sovetskiy Soyuz* comes to the aid of the Russian diesel-electric icebreaker *Kapitan Dranitsyn*. The two ships are about to link up in close-coupled tow, their combined power and weight able to make steady progress as one unit; singly, neither ship could move without great difficulty.

Ice on summer seas

Icebergs tend to be ignored by the ice-reporting systems, because seafarers know the limits within which icebergs are usually found, and so ships within those limits take precautions as a matter of course. But there are thousands of icebergs, and they move comparatively rapidly, driven by winds and currents. It is very hard to keep track of an individual berg, unless it is one of the huge monsters that occasionally break away from the Antarctic ice shelves. These giant bergs start to calve smaller bergs as they melt, and the smaller the bergs, the more dangerous they are to shipping, because they are more difficult to detect.

▲ STEAMING THROUGH ICE

The Russian diesel-electric icebreaker *Kapitan Khlebnikov* makes its tortuous way through dense polar pack ice. There are no openings in the ice for the ship to exploit, and nowhere to push the displaced ice. This ice field could be in a polar region at either end of the earth, and in such ice it is very difficult to make significant progress.

Flying south

▼ NEVER AIRBORNE

The first heavier-than-air
flying machine reached
Antarctica in 1911 – with
Douglas Mawson, but without
wings. The flimsy Vickers fuselage and its engine were used
briefly to pull loads on the
ground, but its use was soon
discontinued. It would be
another 17 years before an
aircraft flew in Antarctic skies.

In the rush for fame—if not fortune—that was so much a part of Antarctic exploration, the first flights were at most incidental to a greater purpose. The aptly named Robert Falcon Scott became the first to take, somewhat shakily, to the air over the Antarctic Continent, albeit secured to the ground by a wire hawser. The place was the Bay of Whales, at the edge of the Ross Ice Shelf, on 4 February 1902. The vessel was a hydrogen observation balloon, from which was suspended a flimsy basket and a nervous passenger. The purpose was to make a reconnaissance flight for the *Discovery* expedition, and after some 60 minutes aloft, Scott was followed by Ernest Shackleton, armed with a camera. Neither appears to have enjoyed the experience, and the balloon was not used again. Nearly two months later, on the opposite side of the continent, the German explorer Erich von Drygalski, who headed the German Antarctic Expedition of 1901–03, also made use of a balloon, and managed to reach almost twice the altitude that Scott and Shackleton had done— nearly 500 meters (1,640 ft).

The fuselage of a Vickers aircraft arrived in Antarctica with Douglas Mawson in 1911 on the first Australasian expedition. Mawson had originally intended to use the plane to make the first powered flight in Antarctica, but a training accident en route, in Adelaide, had deprived it of a certain necessary feature—its wings. Making the best of things, they took the damaged craft with them and used the fuselage and engine on land as a tractor, but this alternative use was not very successful.

Aviation's Golden Age

It was not until the late 1920s that the use of heavier-than-air machines in the Antarctic region began to receive serious attention. By that time the Golden Age of Aviation had begun, and intrepid fliers were making record-breaking flights to more and more remote and hazardous parts of the world; it was little wonder that some of these daring aviators turned their attention and ambitions to the Frozen South. Aircraft engineering had made great strides; the capabilities, range, and strength of aircraft had improved significantly, and the late 1920s and early 1930s became a much publicized arena for private flights and exploration in the Antarctic area, in most cases combined with some scientific research.

In this arena the main players were bankrolled by American backers such as Randolph Hearst, Edsel Ford, and John D. Rockefeller, with their massive financial resources. Australian explorer Sir Hubert Wilkins and American pilot Carl Eilson were the first to venture into Antarctic skies in a powered machine—a Lockheed Vega—from Deception Island, north of the Peninsula mainland, on 16 November 1928: they made a 20-minute flight, which was curtailed by the inclement weather of the area. It was their subsequent reconnaissance flight further south in late December, which lasted for 11 hours, covered

◄ AIRCRAFT IN ANTARCTICA
By 1939, twin-engined aircraft
and metal construction were
laying the foundations for much
future Antarctic aviation. A
Barley Grow floatplane was
an integral part of American
Richard Byrd's third Antarctic
expedition; it is shown here
being secured on the deck of
USS *Bear* in the Bay of Whales.

➤ THE SPINOFFS OF WAR
The technical advances that came with and after World War II, and the building of semi-permanent ice runways to service certain Antarctic bases, revolutionized aviation in Antarctica. The four-engined Hercules became a workhorse, and operations in daylight months became possible, even in temperatures to −50°C (−58°F).

2,170 kilometers (1,350 miles), and overflew extensive areas of the Antarctic Peninsula, that justifiably received publicity all over the world.

Wilkins's achievement was a spur to American flying hero Richard Byrd, the first man to claim to have flown to the North Pole. In 1928, at much the same time as Wilkins, Byrd had set up camp on the Antarctic mainland with the avowed intent of also becoming the first man to fly to the South Pole, and he did not want to be forestalled by Wilkins. However, Byrd's anxiety was unfounded. While Wilkins did pursue a further flying program in the latter part of 1929, he did not penetrate much further south than the Antarctic Circle. Meantime, in a Ford Trimotor piloted by Bernt Balchen, Byrd became the first flier to reach the South Pole, early on 29 November 1929, after a flight of just under 10 hours.

After 1929 other aircraft were used in the Antarctic, some mainly for short-distance observation and exploration, largely coastal. BANZARE (the British, Australian and New Zealand Antarctic Expedition of 1929–31), which was much less generously funded than those of Byrd or Wilkins, made extensive use of a DH60 Moth; at around the same time, Norwegian whaling interests also started to carry aircraft on their ships.

The pouring of lavish American private finance into early Antarctic flights continued with a further Byrd expedition in 1933-35, in which the focus was more on exploration and scientific research. In 1934, self-funded American millionaire Lincoln Ellsworth also attempted—and finally achieved, in December 1935—the first significant transcontinental flight, of some 3,200 kilometers (1,990 miles). Over a period of almost a fortnight, Ellsworth and British pilot Hollick-Kenyon flew from Dundee Island to within a short distance of Byrd's base, Little America (by then disused), but lost radio contact with their ship before they could report their whereabouts. The resultant search, together with the involvement of the Australian, British, and New Zealand governments, and of private bodies in America, generated a rash of publicity around the world.

Governments step in

From the mid-1930s the focus changed, with private finance and personal ambition rapidly giving way to government involvement and territorial claims in Antarctica, many of them competing. At the beginning of 1939 the German Government mounted a brief but very efficient Dornier flying boat expedition, operating from a seaplane tender, taking photographs of 250,000 square kilometers (96,500 sq. miles) of Dronning Maud Land, and laying claim to sections of the Antarctic already claimed by Norway. At the end of 1939, the United States returned to the Antarctic in the shape of a government-funded expedition that lasted until early 1941.

The outbreak of World War II, and particularly the entry of the United States when Japan attacked Pearl Harbor in December 1941, put a temporary stop to most exploration and aviation in the Antarctic. However, it also led to massive advances in aircraft technology and communications, and after the war the importance of aircraft in Antarctic exploration increased dramatically. In 1946 the United States mounted Operation Highjump, under the leadership of Richard Byrd (now a Rear Admiral), from an ice and snow airstrip hacked out from the Ross Sea Ice Shelf. This huge operation employed United States Navy R4Ds, Martin Mariner flying boats, and ship-borne helicopters. Highjump was a combination of scientific research, exploration (they claimed to have photographed huge areas of the Antarctic mainland and coastlines—up to a third of the continent), and, not

▲ HAULING BY HELICOPTER
A helicopter over the Amery Ice Shelf prepares to raise two reels of hose in a sling net at a signal from the ice. The helicopter in Antarctica followed the first gyrocopter (the Byrd Expedition of 1933–35), and helicopters were used extensively in Operation Highjump (1946). In ensuing decades, increased load capacity made these craft invaluable for supply work in Antarctica.

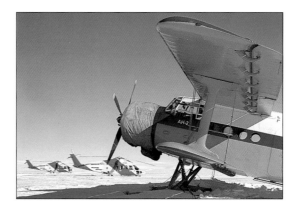

▷ DESERTS OF IMMENSITY
An aircraft at rest on a sea-ice runway is massive, but is still dwarfed by frozen surroundings that stretch as far as the eye can see from Arrival Heights at McMurdo Sound. The immensity of the Antarctic has been partially tamed by aviation, but it is to be hoped that its challenge will never be completely overcome.

△ AIRCRAFT TO THE AID OF SCIENCE
A blanket insulates the engine of a Russian AN-2 on the Amery Ice Shelf. Like the Australian twin-engined Sikorski helicopters in the background, such craft were used for long-range surveillance and survey along the coast and deep into the interior, opening up a whole new range of scientific research opportunities.

▷ A CRAFT WITHIN A CRAFT
The immense jaws of a US Galaxy (the second largest aircraft type in the world) open to regurgitate another, smaller craft—but one that would still have left the first Antarctic aviators gaping with amazement. Unloading takes place on a sea-ice runway 2 meters (6½ ft) thick at McMurdo Sound.

least, an assessment of operational capacities that might prove useful in a possible Arctic confrontation with the Union of Soviet Socialist Republics.

The 1950s saw the establishment of permanent bases in the Antarctic by a number of countries, followed by a proliferation of bases during the International Geophysical Year of 1957–58. (The IGY, and the Antarctic Treaty of 1959, saw a remarkable degree of peaceful cooperation among nations, and a shift in emphasis in the Antarctic towards scientific research and away from territorial claims, which were largely put "on

hold" by the Treaty.) Aircraft were increasingly used for local operations and for research by some of the bases; eventually a number of member countries used aircraft to partly supply their increasingly sophisticated bases for brief periods during the southern summer, and flying to Antarctica became almost routine. In 1957 the United States Navy started regular flights from Christchurch, New Zealand, to McMurdo Station on the Ross Sea Ice Shelf, with a one-way flying time of about 12 hours. Initially they used C-124 Globemasters, but these were replaced in the 1960s by that great flying workhorse, the

Lockheed Hercules. In due course, supply aircraft began to fly to the Antarctic from Chile and Argentina, and from Britain's Falkland Islands, to airstrips established on the Antarctic Peninsula by these countries, and aircraft are now integral to certain parts of the Antarctic scene.

Tourist flights over Antarctica

The Australian airline QANTAS and Air New Zealand began to schedule regular tourist flights over Antarctica in 1977. Two years later, an Air New Zealand DC10 crashed into Mount Erebus. At the time it was one of the worst air disasters in history, with total loss of life, and all tourist flights were suspended until QANTAS resumed in 1994, using Boeing 747s. Lan Chile operates sporadic tourist flights from Punta Arenas, and Adventure Network International runs private flights from Punta Arenas, mainly taking—at a price—adventurers and expeditioners to the Patriot Hills area, initially in Douglas DC4s, and more recently in C130 Hercules. Adventure Network International also has a DHC Twin Otter based at Patriot Hills to cater for tourist flights to the South Pole and other Antarctic destinations.

Working dogs

▼ PREPARING A SLEDGE
Members of Scott's *Terra Nova* expedition prepare a sledge for his last polar journey. Scott obstinately placed more faith in ponies than in dogs: "I withhold my opinion of the dogs ... but the ponies are going to be real good," he wrote.

For at least 2,000 years large dogs of strong build and dense fur have been used to haul sledges in the Arctic. Often called huskies, they are trained to accept harness and run in teams. Modern breeders recognize three or four strains: long-legged Malmutes of Alaska, shorter and stockier Samoyeds and Siberians of northern Asia, and powerful Eskimo huskies from Greenland and Labrador. However, sledge dogs everywhere show much variety of size, body shape, color, and disposition.

Dogs in Antarctica

For almost a century dogs were also used by Antarctic expeditions, sometimes providing the only form of transport, more often running in combination with manhauling, tractors, or aircraft. Antarctica's first sledge dogs were the 75 huskies, of mixed Siberian and Greenland origin, brought by Carsten Borchgrevink on his British Antarctic (*Southern Cross*) Ex-

pedition of 1898–1900. Cape Adare, their landing point, offered few opportunities for sledging, so the dogs had little chance to show their worth. Very much more effective were those of Otto Nordenskjöld's Swedish South Polar Expedition (1901–04). Though inexperienced drivers, Nordenskjöld and his companions ran small teams of five or six dogs, hauling light sleds on short survey journeys across sea ice and crevassed shelf ice. Their longest journey, which covered 610 kilometers (380 miles) in 38 days, showed that even beginners could use dogs effectively.

By contrast, the 23 huskies of Robert Scott's British National Antarctic Expedition (1901–04) proved sadly ineffective. Their amateur handlers showed little aptitude for looking after them: not surprisingly, the dogs' performance was disappointing, and ultimately disastrous. Of 19 that left Hut Point on 2 November 1902 to cross the Ross Ice Shelf, five died of malnutrition within the first six weeks. The rest died or were killed for food over the following five weeks, leaving the explorers to manhaul their way home. Ernest Shackleton, who had shared in the disaster, took only nine dogs for his own British Antarctic (*Nimrod*) Expedition of 1907–09, relying instead on Manchurian ponies and manpower. Scott in his last expedition (1910–12) did the same. The combination proved near-lethal for Shackleton, and fatal for Scott and his unfortunate companions.

▲ A PRIDE IN THEIR WORK
Douglas Mawson wrote that "the dogs enjoy their work," and vividly described how "['Cherub'] Ninnis and [Xavier] Mertz ran a tailoring service for the dogs, who were brought in one by one ... to be measured for harness ... the huskies look quite sharp."

▶ TRAINING RUN
An expeditioner takes a team of huskies for a daily training run across the slick surface of Horseshoe Harbor, Mawson. Runs like this—now supplanted by the "tin dogs" of the mechanical age—developed field and camping skills, and instilled a sense of camaraderie unsurpassed in other modes of polar transport.

A breed apart

First to use dogs to their full capacity in Antarctica was Roald Amundsen, leader of the Norwegian South Polar Expedition (1910–12). Years in the Arctic had convinced him that sledge dogs were essential to polar travel. To his station in the Bay of Whales Amundsen brought 100 of the finest Greenland huskies, and colleagues who knew how to handle them. That they would reach the Pole by an untrodden route was by no means certain, but careful planning and a thorough knowledge of dogs and sledging brought success. Amundsen left base with 55 dogs and drove in style to the South Pole and back, a distance of 2,995 kilometers (1,860 miles) in 89 days—although killing 41 dogs along the way to feed the 14 that brought him home implies a ruthlessness that few dog drivers find admirable. Meanwhile, unwilling rivals in an unequal contest, Scott and his party killed their wretched ponies and man-hauled to a dismal end.

Douglas Mawson's Australasian Antarctic Expedition (1911–14) used dogs to explore one of Antarctica's windiest and most uncomfortable corners. Richard Byrd's first and second expeditions (1928–30 and 1933–35) brought aircraft and aerial survey to the Antarctic, plus teams of huskies—Alaskan Malmutes in the first, a variety of heavier strains in the second—to provide ground control and collect scientific specimens. After World War II, huskies were fully and effectively employed in ground surveys by various long-term national expeditions. Bred from the best Greenland and Labrador stocks and spanning 10 or more generations in Antarctica, they were chosen for their intelligence, good humor, and willingness, and were trained and run by drivers with the skills to appreciate their qualities.

The end of an era

Sadly, dogs' days in Antarctica were numbered. Tractors, skidoos, and other motorized vehicles began to replace sledge dogs at Antarctic bases. Dog-driving became recreational rather than an essential Antarctic skill, although the enthusiasm of the dogs themselves, and their therapeutic value at the various bases, remained undiminished. The end came in 1994, when signatories to the Antarctic Treaty agreed to remove all dogs from the Treaty area, on the general principle of removing all non-indigenous species from the region. Some people contended that existing dogs should be allowed to live out their lives where they had been born and bred, but their argument was not heard. By the deadline date of 1 April, all had gone, many to premature death, and an era in Antarctic exploration was over. BS

▲ A FAITHFUL COMPANION
The evening sunlight shines golden on the contemplative face of Plato, a young husky pup. Until recently the rearing of huskies in Antarctica and their daily welcome were a constant source of delight in the often sterile environs of a winter station. Ever alert and playful, pups were given the run of the station once the summer wildlife had departed.

▶ MACHINES TAKE OVER
Dogs were a primary means of transport for many early Antarctic expeditions. They were widely used until the mid-1970s, when snowmobiles like this one, and larger over-snow mechanical transport, largely usurped their role.

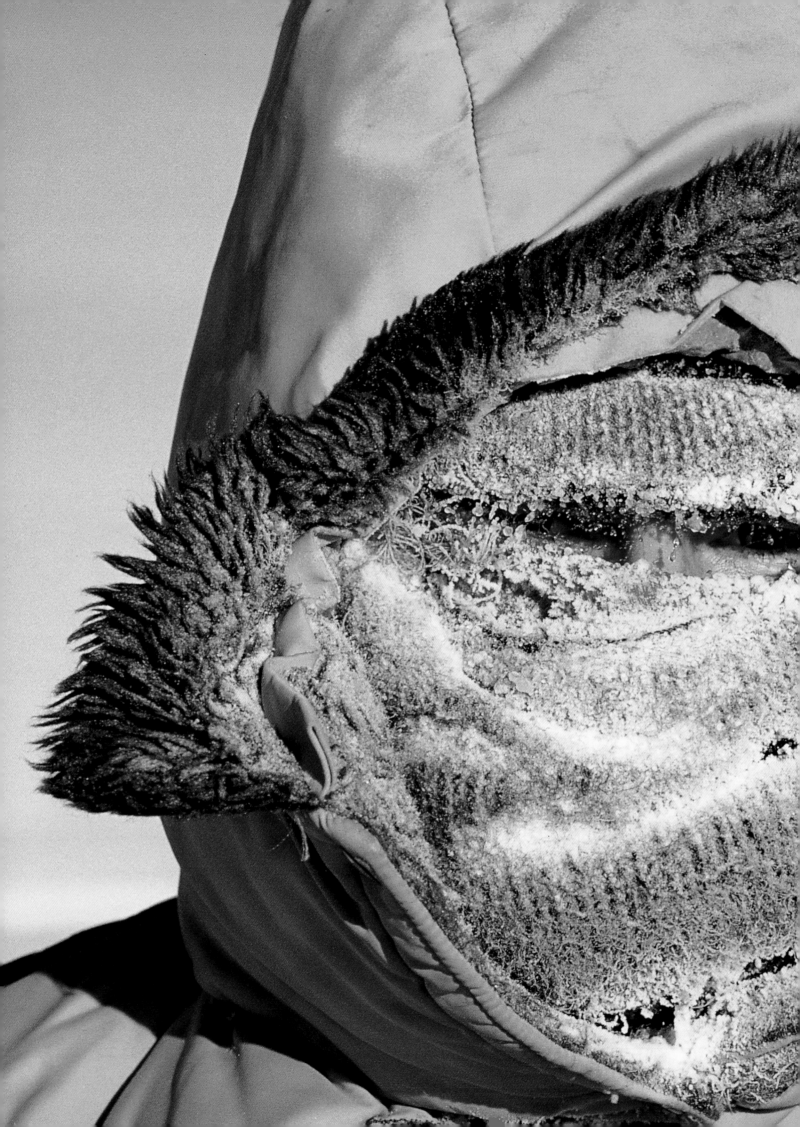

Living and Working in the Cold

Wintering over

Previous pages
◀ **RUGGING UP**
It is all too easy to be cold in Antarctica, even with the best of modern outdoor clothing! Choosing what to wear is a balance between retaining heat and practicality—extra layers of clothing increase warmth but decrease mobility.

▼ **FAREWELL WITH FLARES**
Left behind for the winter, a small group of expeditioners gives the last ship of the season out of Horseshoe Harbor a traditional send-off. Not until January or February of the following year will another ship reach Mawson to resupply the station.

Why would any sane person want to spend a year in Antarctica? It is a demanding, even life-threatening, environment. Far from family and friends, you must endure long periods of darkness in temperatures plummeting far below freezing point. Your companions are not of your choosing, but private life is not an option.

For me, as for most who fall under Antarctica's spell, there was a combination of factors. There was the job challenge: I was to be Station Leader at Australia's Mawson Station, the fourth woman in this role, and I was "to create and maintain a productive and harmonious community" of up to 70 people during the summer and 22 during the winter. But the main attraction was Antarctica itself, with its austere beauty, its enchanting wildlife, the challenge of mastering the skills to live confidently in a hostile environment, and the joy of forging the trust and interdependence that are essential to a long field expedition.

The Antarctic year begins in late spring, when the sea ice has retreated enough for ships to reach the coast and helicopters to reach the research stations. The days are long and the weather benign, and the station swarms with construction and maintenance teams, and with

▶ **FIERY INFERNO**
Folklore and legend has it that the aurora was the flickering of distant fires far over the horizon. Even Galileo attributed it to sunlight reflecting off the atmosphere. Blazing red displays like this at Mawson are, in fact, due to low energy electrons crashing into oxygen atoms in the rarefied upper atmosphere above the station.

scientists seizing the brief opportunity to study this most extreme of environments. Biologists studied a colony of Adélie penguins on a nearby island, geologists made forays into the mountains, and glaciologists trekked deep into the ice cap to measure its depth and ground contours. A field expedition camped out in the Prince Charles Mountains, studying its geology and collecting mosses and lichens from its sheltered slopes.

By the end of March the last ship had borne away most of the scientists and tradesmen, delivered a year's supplies, and removed more than a year's rubbish, leaving 22 of us behind. But we were not idle. First we took the bulldozers and sleds back to the Prince Charles Mountains to leave fuel for the following summer, and to retrieve rubbish and equipment needing repair. We mapped our multi-million-dollar traverse by radar, following a route marked by bamboo canes topped with empty beer cans. I drove one of the bulldozers back, covering 400 kilometers (250 miles) of featureless ice and snow at a painfully slow 6 kilometers an hour (less than 4 mph).

In the month we had been away, the sea had frozen over. We visited the nearby islands on foot, on skis, on four-wheeled motor bikes, and in snow vehicles. At the Auster Emperor penguin colony in the grounded icebergs to the east, the males huddled together, incubating the eggs balanced on their feet under a thick fold of skin. For two months of the harshest winter weather, they fast until the chicks hatch and the females return to feed them—a severe vigil, profoundly moving to witness.

Midwinter blues

The weeks passed and the sun swung lower in the sky until it disappeared beneath the northern horizon. For a month we worked in a few hours of twilight each day. But there were auroras and flaming sunsets: once we watched the southern sky paling to delicate pastels while a huge full moon floated above the translucent turquoise ice cap speared by the jagged purple peaks of the inland mountains. In June we celebrated the

◀ **PENGUIN LABORATORY**
Bechervaise Island, just across Kista Strait from Mawson Station, is the site of a long-running program studying the breeding behavior of Adélie penguins. The weight of penguins entering and leaving the nearby rookery is recorded so the total food/energy input into the chicks in the rookery can be calculated.

Apart from humans, only two warm-blooded animals stay in Antarctica throughout winter—Weddell seals and Emperor penguins. A flurry of wind-borne snow has caused these chicks to burrow in under their parents' bellies—though they have almost outgrown this cosy spot.

winter solstice, the turning point of the Antarctic year, with a banquet and a performance of *Cinderella* in drag, followed by an evening of original entertainment—and some monumental hangovers.

July and August are the "dog days" in Antarctica: the darkness seems interminable, morale falls, and tempers get short. Canine company was our salvation. Until 1994, when the Madrid Protocol mandated their removal, sledge dogs were a feature of life on Mawson station. In September we took two dog teams to count Emperor penguins at a rookery 350 kilometers (220 miles) away, carrying everything on sledges and

camping each night on the ice. This is a technology so simple, and so perfect for its purpose, that it has remained virtually unchanged since it took Amundsen to the South Pole.

Lengthening days heralded the Antarctic spring and the first ship of the season. Ice had become our only reality, and as we went to collect the supplies, I seemed to be falling off the edge of the known world into another medium—water! The ship carried new expeditioners too; our year in Antarctica was over—a year of a myriad vivid impressions but mostly of deep gratitude for the privilege of such a rare experience. LC

▷ MEANINGFUL COLORS
Australian National Antarctic Research Expeditions (ANARE) research and supply ship, *Aurora Australis*, brings fuel, food, and equipment to Mawson Station. Like all of Australia's stations, Mawson's buildings are color coded—red for accommodation, green for storage, and blue for the station's powerhouse.

Keeping out the cold

Hypothermia, frostbite, ice burns … keeping warm in the most ferocious climate on the planet has been the greatest challenge facing humans since they first began visiting Antarctica.

Polar clothing must insulate effectively against the cold and also allow excess heat and sweat to escape. Activity causes sweating, which, if it cannot escape, turns to ice once the body cools. This cycle is not only uncomfortable, it can be lethal, as can the wind chill factor. A 10-knot breeze causes an air temperature of –5°C (23°F) to plunge to –24°C (–11°F), dramatically accelerating body heat loss.

The first explorer to design clothing specifically for Antarctic conditions was James Cook. Sailing further south than anyone before him, Cook was concerned for his men: "I caused the sleeves of their jackets (which were short as to expose their arms) to be lengthened with baize; and had a cap made for each man of the same stuff, together with canvas." His men called them "Magellan jackets," and they were very popular.

Nineteenth-century explorers generally outfitted their expeditioners in heavy woolens, hats, and coats— standard cold-weather gear for European conditions. The first to recognize the correlation between clothing and expedition success was Roald Amundsen, whose conquest of the South Pole depended partly on minimizing the exposure of his team to heat and cold stress.

▼ SKIN SIDE INSIDE

SKIN SIDE INSIDE
Roald Amundsen, dressed here as he went to the Pole, learnt from the Arctic peoples. He wrote "no one had yet wintered on the Barrier so we had to be prepared for anything … with the richest assortment of reindeer-skin clothing … after the pattern of the Netchelli Eskimo."

> **WIND-CHILL FACTOR**
> With good clothing, working in temperatures of –30°C (–22°F) or even –50°C (–58°F) is no great problem, until the wind springs up. When blizzards rage, humans simply cannot survive without shelter, no matter what they are wearing.

Lessons from the Arctic

Amundsen's Arctic explorations and research among the Netchelli Esquimaux had demonstrated the value of dressing in skins and furs. He designed most of his Antarctic expedition gear himself, cladding his crew in easily adjustable layers: woolen underclothes, suits made from felted ex-Navy blankets, thick jumpers and stockings, sealskin suits, and oilskins. His shore party were clothed in hooded anoraks and trousers of reindeer skin, which weighed only a third of their woolen counterparts, and formed a loose, easily adjustable outer shell over several inner layers. He also provided wind-proof gaberdine oversuits with attached peaked hoods to protect the face, and yellow-tinted goggles to prevent snow-blindness.

Footwear posed particular problems. Tight-fitting boots restrict movement of the toes, when sweat turns to ice, frostbite sets in. Amundsen equipped his men with roomy finnesko boots, made from the most durable part of reindeer hides. Inner socks were padded with dried sennegrass, which absorbed sweat and could be shaken to remove ice crystals.

Except for wolfskin mitts and finnesko boots, Scott and Shackleton eschewed furs, preferring tight, ill-fitting clothes with complicated clasps and separate headwear that let in cold air around the neck. Choice of clothing may well have contributed to the failure of their expeditions; it certainly facilitated Amundsen's success.

Lessons from nature

Clothing design for polar conditions has changed little since Amundsen's day. Time continues to prove the validity of the layer theory: many thin layers of clothing trap pockets of air, just as our hairs stand up and trap air next to the body when it is cold. Thermal underwear has evolved from natural fibers to modern fabrics such as chlorofiber and polypropylene, which "wick" mois-ture away from the skin, keeping the body dry. A second layer—a wool shirt, polar fleece, or down vest—captures a warm layer of insulating air, and a windproof outer

➤ FUR SIDE OUTSIDE
Skin clothes, worn here by Olaf Bjaaland, had disadvantages. On 9 February 1911 Amundsen noted: "they proved to be too warm ... In low temperatures these reindeer clothes are beyond comparison the best, but here in the South we did not as a rule have low temperatures on our sledge journeys ..."

shell completes the outfit. Plastic zips and webbing, velcro, and snap buckles are easier to use and seal out the cold far more efficiently than the buttons, belts, and buckles of earlier times.

Like Amundsen's team, modern expeditioners wear inner gloves or mitts, generally made of silk, wool, or synthetic thermal fabric, that allow finger movement. These are covered with outer windproof mitts. Modern jackets are anorak style with attached hoods. Boots and mukluks are generally of leather (for summer use) or synthetic with several linings of warm inner soles, which

can be taken out and shaken to remove the ice—probably less efficient than the Eskimo finneskos, but more readily obtainable. In the early 1970s Soviet expeditioners experimented with electrically heated suits to survive the bitter conditions at Bellingshausen station. More recently, new, "breathable" waterproof fabrics such as Gore-Tex have been heralded as "the biggest breakthrough since the rucksack." But, even so, the challenge still remains to develop ever more lightweight, efficient protection against the most savage environment on earth. AR

▲ LAYERS FOR WARMTH
Quality materials and layers are combined for maximum warmth—several thinner layers instead of one or two thick ones. Once this man pulls his goggles over his eyes to ride on an unprotected motorized toboggan, every bit of skin will be covered by two or more layers of insulating material.

Polar medicine

In March 1879 an advertisement appeared on the noticeboard at Marischal College, Aberdeen: "SEAL FISHING—wanted immediately a surgeon for *SS Mazinthien* sailing from Peterhead on Thursday or Friday next (24th or 25th). A junior student not objected to, must be a good shot."

This advertisement epitomizes the challenge faced by polar medicine over the past couple of centuries—and its traditional solution. Whalers and sealers from Peterhead operated in waters off Greenland. Vessels sailed in late spring, and some penetrated further than 81°N, hunting whales in melting ice, and seals on the ice itself. British government regulations obliged these vessels to carry a qualified doctor, but medical students were often used. The surgeon was comparatively well paid, and it was an excellent way to meet medical school fees. The medical experience gained was minimal because crews were young, and were healthier than the general population. Occasional wound suturing and bone-setting apart, shooting seals and keeping the log filled the day.

The most famous Arctic whaler surgeon was Sir Arthur Conan Doyle, the creator of Sherlock Holmes, who worked for seven months on the *Hope* of Peterhead when he was a medical student in 1880. He made such a favorable impression that the captain offered him double pay as harpooner/surgeon for another voyage. He declined the offer, but there is no doubt that his experiences on the *Hope* formed the basis of his short story *The Captain of the Polestar*.

Doubling up

Clearly, the traditional solution to the problem of providing health care in this remote and hostile environment was to hire young (and therefore cheap) medical attendants. The main problem—boredom—was averted, and value for money obtained, by apportioning additional tasks to them. Fortunately for science, nineteenth-century medical courses included much natural history, particularly comparative anatomy and botany, so that polar surgeons

often doubled as naturalists. A classic example was the James Clark Ross Antarctic expedition of 1839–43; all four surgeons were also hired as zoologists or botanists, including Joseph Dalton Hooker (later Director of Kew Gardens) and Robert McCormick, who had been with Darwin on the *Beagle*. Edward Wilson, who perished on Robert Scott's doomed quest for the South Pole of 1910, took the role of surgeon/naturalist to the extreme: although he was medically qualified, his title was "Chief of the Scientific Staff, and Zoologist." Wilson's inability to relieve the medical problems of Edgar Evans or Lawrence Oates on the return from the Pole contrasts with the remarkable surgery carried out by the surgeons on Ernest Shackleton's *Endurance*, who aseptically amputated an expeditioner's gangrenous foot under chloroform on Elephant Island in June 1916. But the brutal and immediate challenges faced by these brave doctors were typical of the Heroic Age of Antarctic exploration. Frostbite, snow blindness, and inadequate diets leading to vitamin deficiency and starvation were common on the Antarctic expeditions of that time, and the presence of doctors did not prevent them. The key preventive developments—advances in nutritional science and improved equipment and clothing—were yet to come.

Polar medicine today

Antarctic expedition members of today are young and are screened for pre-existing medical problems, so that mortality and morbidity rates in the Antarctic are very low. Most serious health risks in modern Antarctica relate not to environmental extremes, which are now well understood and well guarded against, but to hazardous activities such as light engineering and the use of vehicles and aircraft, and Antarctic medical practice in the twenty-first century is mainly concerned with the treatment of occupational injuries.

The British Antarctic Survey (BAS) is a good example of current medical practice in Antarctica. Its medical unit combines the traditional approach—doctors resident on Antarctic bases—with modern approaches such as paramedical training for all staff, electronic access to specialists

▲ **MASK OF ICE**
Cecil Madigan, meteorologist with Mawson, had to take readings outside. Mawson wrote: "One's face inside the funnel of the Burberry helmet became rapidly packed with snow, which, by the warmth of the skin and breath, was changed into a mask of ice. This adhered firmly to the rim of the helmet and to the beard and face."

◀ **THE DANGERS OF FROSTBITE**
Petty Officer Evans bound up Dr Atkinson's frostbitten hand after he was lost in a blizzard. His face was frostbitten, too. Scott noted: "Atkinson had a bad hand today ... blisters on every finger giving them the appearance of sausages ... this bit of experience has done more than all the talking I could ever have accomplished."

▶ **EMERGENCY SERVICE**
This ambulance from the United States McMurdo Station is maintained in top condition. It was photographed in late August when the sun makes its first appearance for four months. Any group staying in Antarctica over winter needs to be self-sufficient in every way possible. They are a long way from medical—or any other—help.

in the United Kingdom, and telemedicine. Until 1962, BAS was called the Falkland Islands Dependencies Survey, and its employees were affectionately referred to as "Fids." With classic Antarctic humor, the BAS medical handbook is called *Kurafid*. Written in the 1960s to meet the needs of sledging parties, it has expanded into a comprehensive first-aid manual. Like their nineteenth-century counterparts, BAS doctors need to spend only a fraction of their time on medical duties, so that they can maintain nearly two hundred years of Antarctic tradition by conducting high quality research as well. HP

◄ STANDING IN FOR THE BEES
Pollinating vegetable flowers in the hydroponics unit at Scott Base must be done by hand. During the Antarctic winter, fresh vegetables are a welcome addition to meals. Food is one of the most important morale boosting factors for any isolated group.

▼ PROVISIONS FOR A QUARTER
This member of Mawson's expedition is preparing sledging rations—enough for three men for three months. Many early expeditions faced two crucial problems: not enough was known about nutrition and not enough food was carried when they set off inland.

Supporting science in Antarctica

The search for the Magnetic Pole, the race to be first to the South Pole—all the Antarctic ventures that explorers and scientists have undertaken in quest of their Antarctic goals—have succeeded or failed in direct relation to the logistics that supported their endeavors.

The Antarctic provides unique opportunities for research, and logistics managers must be able to respond to a wide range of requests from scientists. Researchers may be required to work on projects on glaciers, in the ocean, on mountains, on or under the sea ice, in and above the atmosphere, in every type of weather, at any time of the year, in the highest, driest, coldest, windiest, and loneliest continent on earth.

Summer and winter

Summer is the most productive time for research scientists, and logistics organizations work at a frenetic pace throughout this season. At this time it is relatively easy to reach Antarctica. The sun shines, the sea ice recedes, animal and bird life around the coast is abundant, and the weather is comparatively clement. Even so, traveling to and from Antarctica is not simple; despite all the sophisticated systems in place, it is not unusual for ships to be beset by ice, for aircraft to be unable to fly, and for planned transit times to be doubled. The distances are huge, and the isolation almost complete. Stores, hospitals, industrial centers, fire brigades,

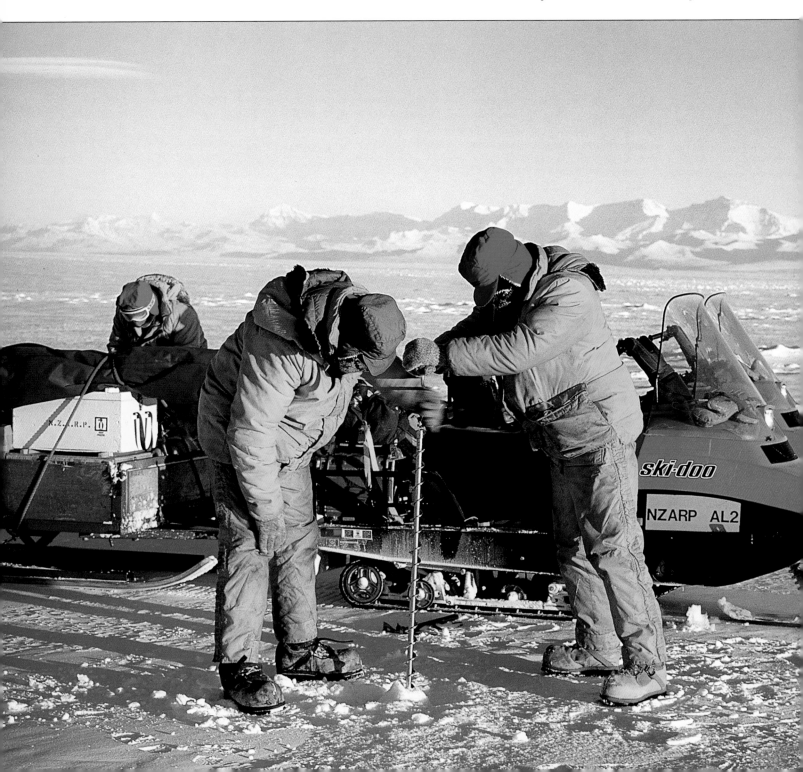

recreation facilities, repair yards—not to mention family and friends—are usually thousands of kilometers away, so the need for self-sufficiency is high.

But there are also excellent research opportunities in winter, and many nations operate their bases year-round. At most Antarctic bases, winter isolation is absolute; the self-sufficiency required in summer pales into insignificance with that required for wintering over. For most wintering parties, there is no possibility of support from anywhere other than their own resources. Perhaps for this reason, NASA is drawing on the experiences of researchers in Antarctica as a model for those that might occur on extended space flights.

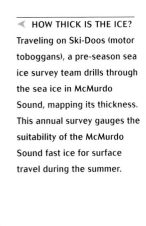

◄ **HOW THICK IS THE ICE?** Traveling on Ski-Doos (motor toboggans), a pre-season sea ice survey team drills through the sea ice in McMurdo Sound, mapping its thickness. This annual survey gauges the suitability of the McMurdo Sound fast ice for surface travel during the summer.

Bases, camps, and transport

It is no simple task to conduct research in the Antarctic, and many nations have built bases and research stations to provide the infrastructure to support their programs. All have the primary aim of providing a hub from which to conduct research programs. Some are the size of small towns, with hundreds of people; others are little more than an collection of tents and custom-modified shipping containers that provide rudimentary living quarters for four to twenty people.

For research at a distance from a base or station, field camps provide shelter, basic laboratory equipment, and cooking facilities. Travel between field camps and their parent stations can be hazardous and time-consuming;

▲ **FORMIDABLE WIND** Frank Hurley's photograph shows men "getting ice for domestic purposes." Hurley told how he built a shelter to photograph "other members of the party as they struggled about bent—very literally bent—on their duties. On one occasion ... I was lifted bodily, carried some fifteen yards [about 14 meters] ... and dumped on the rocks."

◄ **ROAD TOLL** Accidents can and do happen in Antarctica. This passenger carrier (empty at the time) slid off an icy road and came to rest at the bottom of a long slope. Fortunately, the driver jumped to safety unhurt.

▶ **BREAKING THE ICE** Before the annual resupply of McMurdo Station, an ice-breaker needs to open a channel through 20–59 kilometers (10–30 miles) of fast ice—sea ice still anchored to the shore. Here the United States Coastguard icebreaker *Polar Sea* (far right) continues to widen the channel it has broken through McMurdo Sound. Also in the channel are an Italian cargo vessel and an United States fuel tanker.

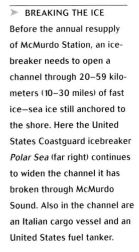

◄ RESUPPLY 2001
When conditions are good it is all hands on deck for the resupply of Macquarie Island station. Incoming supplies are sorted and stored, outgoing crates are marked "RTA" (Return to Australia). Often the success of a resupply voyage means the success or otherwise of a particular research program for that season.

▼ DEFYING MURPHY'S LAW
"CHECK: reliable vehicle, fuel, tents, sleeping bags and mats, food, cooking equipment, radio, scientific equipment, survival gear, first-aid kit …" every field trip is planned with care. Murphy's Law (If anything can go wrong—it will) is well known to those who live and work in Antarctica.

aircraft, helicopters, and all-terrain vehicles of many shapes and sizes are used to make these traverses, and each machine has special capabilities that makes it uniquely important for supporting Antarctic science. Most programs use all types of vehicles at some stage in their work.

New technologies such as telecommunications, advanced electronics, remote control and monitoring systems, and alternative energy programs are making the logistics of working in Antarctica more flexible and have the potential to greatly enhance research programs. Many data sets are now being collected and transmitted in real time from sites in Antarctica to scientists operating in laboratories many thousands of kilometers away. Remotely controlled and automated systems reduce the need for scientists to spend months traveling to and from Antarctica.

◄ FIT TO BE TIED!
Each summer colored flags mark routes around Hut Point Peninsula, and over the sea ice and the Ross Ice Shelf. McMurdo Station staff traditionally invite Scott Base staff to a party with free beer, a live band, food … and lots of red and green flags that need tying to cane poles.

Looking after Antarctica

The importance of Antarctica for studying global climate change cannot be overestimated, and this, coupled with the international commitment to protect the Antarctic, has led many nations to improve their support facilities to ensure that carrying out scientific research does not damage this most precious of environments. Older sites that have outlived their usefulness are being cleaned up, and when new facilities are built they are designed for easy dismantling when no longer required. Leading nations in this field are developing their facilities in such a way that on completion of their research, the station or base can be removed and the environment left unspoiled and in pristine condition.

Making things happen in Antarctica is an exciting and rewarding activity. Those entrusted with the task of supporting scientific research programs never lose sight of their responsibility to protect the Antarctic environment and ensure the safety of everyone involved.   KP

Breaking the silence

DATE: 21 December 1998
POSITION: 85°28.722'S; 176°26.676'W

I was standing at the top of the Shackleton Glacier, a thousand kilometers from Ross Island and the coast. The polar wind swirled down from the ice plateau, numbing my face and making it difficult to speak. Fingers clumsy with cold, I fumbled for my Iridium satellite phone, punched in a number, and spoke:

Peter Hillary calling TV One News.

Auckland City will be 28°C … And now, calling live from Antarctica, here's Peter Hillary.

Greetings from Antarctica! We've just climbed up onto the edge of the Polar Plateau at 2800 meters elevation, and the temperature is about the same as you have there—except with a minus in front of it! It's brutally cold …

The call ended, and I was back in the isolation and immensity of Antarctica, newly aware that the world is a big place.

Then—and now

When Roald Amundsen and Robert Scott raced for the South Pole in 1911–12, the outside world did not learn of their success, nor of Scott's death, until their respective parties returned to civilization. Without radio communications, Ernest Shackleton's Imperial Trans-antarctic Expedition of 1914–17 pursued its disastrous course completely unaware that most of the inhabited world was in the grip of a devastating war.

Radio communications changed the nature of Antarctic exploration, making the aerial ventures of the 1940s and 1950s much safer, and enabling ground-based expeditions to coordinate their plans, and even call for rescue. In 1957–58, the Commonwealth Transantarctic Expedition of Vivian Fuchs and Sir Edmund Hillary used radio regularly to communicate position reports between the two Pole-bound parties, to update their support crews, and to relay messages to New Zealand and Britain.

However, radio had its limitations: it was single-band voice communication, and was vulnerable to interference from sun flare activity. It was not until satellite technology was introduced that reliable and detailed communications became a reality in the Antarctic. In the twenty-first century, e-mail and telephone links mean that Antarctic bases can transmit complex verbal information and images to research and administration centers around the world—and to the public through the media. PH

► WIRELESS CONTACT
Mawson's expedition was the first to have radio contact with the outside world. Walter Hannam sent messages that were received at Macquarie Island but the replies did not come through. Only in February 1913, after Hannam had left, was the first two-way communication established, relaying through the radio base at Macquarie Island.

► FIELD PARTY TO BASE
Before the helicopter that delivered them departs, it is important for this field party to establish radio contact with base. Communications are kept between Antarctic field parties and their bases, typically at the same time each day. These are called "skeds" (from schedules).

◄ THE WIND WINS IN ADELIE LAND
Mawson's men struggle to raise the lower section of the northern wireless mast. Because of the winds, it took from April until September 1912 to erect the masts and, after a month of transmitting with no response, the masts blew down on 13 October.

▼ MODERN COMMUNICATIONS
While winter may leave Antarctic bases physically isolated, modern expeditioners can stay in touch with home and head office by phone, fax, and e-mail. Scott Base has a freely available pay phone and these communication lines at McMurdo Station carry both vital and chatty fax and e-mail messages.

Current scientific research

Antarctica is a vast scientific laboratory, with research initiatives in all the major fields. Since the signing of the Antarctic Treaty in 1961, an increasing number of nations have been establishing research programs to study Antarctica's atmosphere, its ice, its dry land, its freshwater lakes, and its seas.

All treaty signatories are entitled to send representatives to the Scientific Committee on Antarctic Research (SCAR)—the committee that coordinates international research programs. This committee takes advice from international working groups established to serve the scientific disciplines, with the role of articulating research questions in their fields and coordinating research activities. It also funds workshops and symposiums during the planning and operating phases of the research.

One of the newest advisory groups to be formed is coordinating research into lakes of water that lie below thousands of meters of Antarctic ice. These sub-glacial lakes are very puzzling. How did they form? Do they contain microscopic life forms? What can the sediment of their floors reveal about the history of Antarctica's climate? The focus of international attention is Lake Vostok, the largest of these lakes, which lies under about 4,000 meters (11,500 ft) of ice in central East Antarctica. A phased approach to its exploration is being developed, with the central requirement that no foreign material be introduced into the lake when its icy ceiling is breached. The task will require sophisticated remote sensing and robotic technologies, and the lake will not be accessed for many years to come—but when it is, the findings are likely to be amazing.

The main focus of biological research in Antarctica is the region's biodiversity, including how organisms are able to live, grow, and reproduce in the most hostile environment on Earth, and an assessment of the consequences of climate change. A newly established program, Regional Sensitivity to Climate Change, is using an internationally agreed methodology to monitor the effects of climate and environmental change on vegetation and invertebrate animals—a long-term project, since the phenomena studied change very slowly. Collaborators worldwide are pooling their data, in the spirit of openness and cooperation that is the hallmark of Antarctic science.

Studying the ice

Considerable effort is being devoted to studying the build-up, movement, and behavior of ice. Contemporary concern about global warming makes this research of great importance, and there are programs to monitor the frequency with which icebergs calve from floating ice sheets. There are also studies to determine whether ice loss at the edge is matched with new ice from precipitation, and it appears that (unlike at the Arctic,

where the ice is becoming thinner) the Antarctic ice cap is roughly in balance at present.

Sea ice is also important climatically. Vast amounts of heat are exchanged between the ice and the underlying water as the sea ice waxes and wanes annually, influencing the temperature of the Southern Ocean. These dynamic interactions have a major effect on world climate, and studies are providing valuable information for long-term climate predictions.

Ice cores drilled from the Antarctic ice cap provide an excellent record of the earth's past climate. The composition of minute gas bubbles trapped in the ice reveals much about the prevailing temperatures, and the amount of methane gives an indication of variations in biological productivity. Traces of sulfur in the ice tell the history of volcanic activity, and the ratio of isotopes of oxygen provides a measure of the temperature when the ice was formed. Ice cores from near Casey station show that the amount of carbon dioxide in the atmosphere has doubled since the Industrial Revolution. Comparison of ice cores from Greenland and Antarctica

▲ TEST CASES

A marine mammal specialist meets some of his "subjects"— Hooker's sea lions. Earlier, he had glued onto this nursing mother dive depth and duration recorders and location trans- mitters. The radio antenna helps him relocate his subject when she returns to land. This study yielded startling results: dives to more than 450 meters (1,470 ft), the deepest ever recorded for sea lions.

> FRIGID DIP

Glaciologists stand suspended in a cage over the frozen Southern Ocean from the research vessel *Aurora Australis* during a winter polynya cruise. Measurements of the rate of production and types of sea ice yield valuable information about the exchange of heat between atmosphere and ocean. Coastal polynyas are regions of perennial open water, which are often the site of prodigious sea ice formation in calm weather that is then blown out to sea by fierce katabatic winds.

⌃ PORTABLE POOL

Drilling ice cores has revealed a rich history of retreat and advance of massive northern hemisphere ice sheets of similar dimensions to modern-day Antarctica. Recent studies on ice shelves have adopted hot water techniques to melt access boreholes to the ocean cavity below. This camp on the Amery Ice Shelf shows the portable pool of melted snow, ready for drilling, in the foreground.

◄ MUD DIPPING

Scientists probe for a shallow sediment core through thin sea ice off Mitchell Peninsula. The continual rain of organic matter and the spent calcite shells of tiny marine organisms in the ocean slowly builds layers of muddy ooze on the sea floor, thereby preserving a textured history of local climatic conditions.

show many similarities but also many differences, suggesting that the northern and southern hemispheres have had somewhat different histories, at least since the break up of Gondwana created the Antarctic Circumpolar Current, when the mean temperature of the once luxuriant Antarctica fell.

Studying the seas

In the Southern Ocean, research focuses on the health of the marine ecosystem. Fishing is permitted in Antarctic waters, but a close watch is kept on how it affects penguin and seal populations. In 1999, the world's population of pack-ice seals was monitored in one of the largest wildlife surveys ever undertaken, establishing a baseline against which to measure future change. The relationship between penguin numbers and krill stocks has been monitored for more than 10 years, and consequently a precautionary limit on the krill fishery has been set. Satellite observations of ocean coloration measure the productivity of the Southern Ocean—the tiny plants that occupy the upper layer of the ocean are green and blue. The sea-ice zone is of particular biological interest, and it appears that variations in the extent of the winter sea ice partly drive the effectiveness of reproduction of plankton and krill. Changes in the composition of plankton fauna—the tiny, shrimp-like creatures that feed on the microscopic plants—are monitored by regular surveys carried out with automatic plankton sampling devices towed behind supply vessels.

▲ SEAL TRACKING
Scientists regularly fit satellite dive recorders to Crabeater seals to track their seasonal movement and foraging behavior. The recorders are attached to an animal's fur with adhesive, preferably to post-molt individuals, when the device is much more likely to remain in place and be active for several months.

Studying the land

SCAR is coordinating several large-scale international studies on dry land. The age and evolution of Antarctica is the subject of a study currently being undertaken by 15 nations. Geologists are locating and defining the ancient boundaries between the building blocks of Antarctica to piece together the continent's history. At the same time, a map of Antarctic bedrock is being compiled, complete with data on the history of the magnetic changes that have accompanied it.

The constant erosion of the ice surface by wind and sun allows meteorites that have landed on Antarctica to be uncovered and collected. Japan maintains a huge collection of these, which is available to scientists from around the world, and they are revealing much about the earth's neighboring planets. Although this is not strictly Antarctic science, the unique environmental conditions on the Antarctic ice cap provide a convenient collection opportunity for these fascinating fragments.

Studying space

Antarctica has been used for observing astronomical and physical phenomena since exploration began. It is a productive research site for astronomers because the mean elevation is high above the bedrock, and the cold, dry air provides outstanding optical conditions. For upper atmosphere studies, the downward projection of the earth's protective magnetic field lines at high latitudes provides a window for observation undistorted by the deflection induced by the magnetic shield; cosmic rays, for example, can be far more easily observed in Antarctica than in more temperate latitudes. The international Space Ship Earth project is examining how the planet is reacting to the constant bombardment of radiation from outer space. Much international effort is being expended on a better understanding of space weather—the environment in which satellites, astronauts, and some aircraft must operate—and the practical value of these studies is rising steadily with increasingly sophisticated telecommunications and other satellite applications.

International cooperation

Antarctic science is remarkable in the degree to which nations cooperate in planning and executing research and publishing the results. Geophysical findings on a range of phenomena from many Antarctic sites are now available on a web database. Antarctica is a laboratory that has no room for those who want to go it alone. This is partly because costs of travel and field support in Antarctica are extremely high. But most of all, the harsh environment and the threat that it poses unite scientists in their common cause, and compel them to support each other in their arduous and often dangerous work.　MS

Meteorites

By 1999, some 16,000 fragments from space had been found in Antarctica, surpassing the total number of extraterrestrial specimens recovered since antiquity from all other regions of earth. Most of these fragments originated between Mars and Jupiter from the asteroid belt of primitive rock debris, as old as the solar system itself. A few of exceptional interest came from the moon and even Mars. They are believed to have formed by "splash-off" from the lunar or Martian crust, perhaps propelled by an asteroidal impact out of the parent gravity field into an earth-crossing orbit.

No one has ever seen a meteorite fall in Antarctica, probably because during the months-long polar winter night, researchers generally stay inside their warm and comfortable huts. Antarctica's first meteorite, a 1-kilogram (2-lb) stony type, was discovered in 1912 near Douglas Mawson's "Home of the Blizzard" in Adélie Land. In 1961, two more were found: a 7-kilogram (15½-lb) iron meteorite near Lazarev Station in Queen Maud Land, and the 29-kilogram (64-lb) Thiel Mountains pallasite, a rare stony-iron type of meteorite, at the edge of the polar plateau, about 560 kilometers (350 miles) from the Pole. Another small iron meteorite was found in 1964. In 1969, the Yamato Mountains of eastern Queen Maud Land yielded nine meteorites, and since then, many thousands more have been located in "blue-ice fields." Specialists study aerial photographs of such areas before searching them by helicopter and Ski-Doo.

The significance of the 1969 Yamato Mountains' find was recognizing that the nine fragments represented different kinds of meteorites. The statistical impossibility that several different meteorites would land at a single place obviously meant that they had been transported there by ice. Glaciologists and planetologists suggested that extraterrestrial debris, landing mostly on ice sheets, became buried deeply under accumulating snow and ice. At deep levels the ice becomes plastic under pressure and flows laterally, carrying meteorites to other places. Ice that flows to the sea breaks off as bergs, and any entrapped meteorites eventually fall to the sea floor. In places ice moves against nunataks and mountains, where it is forced to the surface and melts. Thus any entrapped meteorites accumulate at the surface and are easily found, often in large numbers.

Antarctic meteorites, protected from weathering by the ice and cold, are collected by sterile techniques in order to preserve chemical features that can date the time of fall. They are sent for examination to the Lunar Receiving Laboratory of the NASA Johnson Space Center in Houston, Texas. One, unlike any other found on earth before, was discovered at Allan Hills in 1982. It closely resembled some lunar rocks called anorthositic breccia—a rock made up of broken fragments and rich in the mineral feldspar. Since then, several more lunar rocks have been found, as well as several of a type called shergottite, believed to have come from the crust of Mars.

The search for meteorites in Antarctica continues in the hopes of finding additional space samples of unusual type, especially any that might carry evidence of life. AF

▲ LIFE ON MARS?
Compared with meteorites collected elsewhere on earth those found in Antarctica are relatively well preserved. Meteorites thought to originate from Mars (such as the one pictured here) are of particular interest because some NASA scientists believe they have detected some mineral features characteristic of biological activity.

◄ MOBILE LABORATORY
These tractor trains are parked up at an overnight GPS survey stop to measure ice sheet surface velocities in the Lambert Glacier Basin, data vital to calculations of the mass balance of the continent. The spidery web of a radar antenna, used to determine ice thickness, appears in silhouette on the right.

▲ SIGNIFICANT SITE
Allan Hills in the Transantarctic Mountains is a site rich in fossil remains. The nearby Allans Hill ice field is the site of a number of significant meteorite finds, including a 4½ billion year old rock believed to have once been a part of Mars.

Medical research at the poles

▼ PIONEERING RESEARCH
While co-expeditioners
Mertz, Madigan, Hunter,
and Hodgeman quiety occupy
themselves, Douglas Mawson's
1911–14 bacteriologist Dr
Archibald McLean (second
from left) studies a specimen.
McLean pioneered studies on
microbes that live on humans.

The supremely self-confident physicist Ernest Rutherford once remarked that all science except his own was "stamp collecting." It is true that medicine, like geology and meteorology, depends strongly on observation, but it is equally true that these subjects, like physics, extend their boundaries by experimentation. The difference is that for these subjects it is nature that provides their laboratories—and for their practitioners the attraction of Antarctica is that it is a magnificent natural laboratory.

Physician Frederick Cook's account of the problems he encountered on the ice-bound vessel *Belgica* in 1898 and 1899 exemplifies purely observational and opportunistic medical science, and it enlivens Antarctic literature to this day. Around the beginning of the twentieth century, however, expeditions to Antartica began planning to carry out medical research deliberately. Erik Ekelöf, medical officer and bacteriologist on Otto Nordenskjöld's 1901 expedition to the Antarctic Peninsula, was the first to conduct bacteriological studies in the Antarctic using the laboratory methods pioneered by Robert Koch and his followers over the preceding two decades. He demonstrated the presence of bacteria and other microbes in the soil of Snow Hill Island, and continued a medical tradition of looking for bacteria in the air, following Joseph Lister's discovery of antiseptic techniques in the 1860s and 1870s, with its emphasis on the prevention of airborne infection. Not surprisingly, Ekelöf found the air of Antarctica very clean, and his findings were confirmed by geologist and bacteriologist Harvey Pirie during the 1902–04 *Scotia* expedition to the South Orkneys, and by Antarctic explorer Jean-Baptiste Charcot on his 1903–05 survey of the Peninsula.

▲ PLATE OF CULTURES
The Australasian Antarctic expeditioners of 1911–14 spent most of their first Antarctic winter confined to their Commonwealth Bay hut. Dr McClean made good use of the isolation—growing cultures from microorganisms within the hut. Notably, he found that *Staphylococcus pyogenes aureus*, a common germ outside Antarctica, could not be cultivated artificially.

▶ THE PRICE OF LIVING IN THE COLD
The harsh conditions of Antarctica can age a person and strip away sophistication. "Before and after" photographs from *Through the First Antarctic Night*, Frederick Cook's account of the *Belgica* expedition illustrate this. From the top, Cook, Amundsen and Racovitza before the journey (left) and upon their return (right).

Work on environmental microbiology continues in the Antarctic today, although it has long since flowered into a subject in its own right. Nevertheless, a medical microbiological strand continues in the investigation of the microbes that live on and in humans. Archie L. McLean, medical officer and bacteriologist on Douglas Mawson's aggressively scientific Australian expedition to the Adélie Coast in 1911–14, pioneered these studies. His most interesting finding was "that after a few months in Adélie Land, *Staphylococcus pyogenes aureus*—a common germ in civilization—could not be cultivated artificially from the throat, nose, or skin of six individuals from whom monthly bacteriological cultures were made." This result was confirmed in studies carried out in the 1960s and the late 1980s, and it is significant because this organism, and particularly its antibiotic-resistant variants, are common causes of iatrogenic infection. The polar findings are as yet unexplained, although it seems likely that prolonged isolation rather than climate is responsible.

Antarctica as a research laboratory

The effect of extreme cold on human physiology has been a consistent theme in polar medical research since Ekelöf's observations on body weight and components of the blood at the end of the nineteenth century. Studies on acclimatization peaked in the 1960s and 1970s, and a conclusion published in the *Lancet* in 1963 still holds good: that "the effects of cold are small,

Before.   After.

Frederick A. Cook.   Frederick A. Cook.

Roald Amundsen.   Roald Amundsen.

Emile Racovitza.   Emile Racovitza.

(We were all reasonally good-looking when we embarked, but we were otherwise when we returned.
The long night effected a radical transformation in our physiognomies.)

usually identifiable and not of great significance on a successful polar expedition." Using the long polar days and nights to study circadian rhythms goes back to research that began in Greenland at the beginning of the twentieth century, and it continues. Paradoxically, it is complicated by the excellent living conditions in modern Antarctic bases: many contemporary medical researchers study people based in Antarctica not because they must survive in a hostile environment but because they are isolated. They form a human laboratory, a scenario almost impossible to replicate anywhere else in the world, where microbiologists can study the transmission of bacteria and psychologists can study group behavior in isolated "captive" groups.

Polar medical research resembles science in general, in the sense that, with rare exceptions such as the much-publicized ozone hole, its results are so specialized that the public never hears about them. But everyone knows about the wind-chill factor, which was first identified in the early 1940s during an Antarctic physiology research program. It is even possible that manned space travel will benefit from Antarctic research: the

European Space Agency has funded a study of Antarctic bases as models for a mission to Mars. Very much closer to home is the remarkable work of the pathologist Johan Hultin; in 1997 Hultin analyzed lung tissue from the body of an Inuit entombed in 1918 in Alaskan permafrost—tissue from which he isolated genes of the 1918 influenza virus that killed tens of millions, and which are currently being studied to find out why this virus was so lethal.   HP

▲ HUMAN GUINEA PIGS
Medical research in Antarctica follows various distinct strands including obvious ones such as the effects of isolation and extended periods of darkness (in winter) and sunlight (in summer). Life for these isolated groups of people is psychologically and physiologically very different from busy and interactive lives back in the "civilized" world.

◄ LIVING APART
Winter-over personnel, such as this scientist taking the daily weather readings at Scott Base, make ideal subjects for various kinds of medical research. Clinical psychologist A.J.W. Taylor's *Antarctic Psychology* (1987) was "the first comprehensive book on isolation of its kind."

Part VI
RESOURCES

Alone among continents, Antarctica belongs to the world, and free exchange of information is a condition of the Antarctic Treaty. Interest in Antarctica as a record of the story of the earth and a predictor of its future has ensured that its geological and biological history is meticulously documented in libraries, in video recordings, and in websites packed with textual and visual information.

Wildlife conservation status

Conservation status information from two internationally recognized sources, the World Conservation Union (IUCN) and the Convention on International Trade in Endangered Species of Wild Fauna and Flora (CITES), are included with other statistics at the top of species entries in the wildlife chapters of this book. An explanation of the status codes and the organizations which generate them are outlined below.

IUCN

The World Conservation Monitoring Center (WCMC) is part of the United Nations Environment Program (UNEP), and provides worldwide biodiversity information for policy and action to conserve the living world. That role includes publication of the *IUCN Red List of Threatened Plants and Animals*, which is an internationally recognized series of lists that categorize the status of globally threatened species. The WCMC determines the relative risk of extinction of species, and the main purpose of the Red List is to catalogue the species that are regarded as threatened at a global level and are at risk of overall extinction.

Red List Categories and Criteria

EXTINCT

A taxon is Extinct when there is no reasonable doubt that the last individual has died.

EXTINCT IN THE WILD

A taxon is Extinct in the wild when it is known only to survive in cultivation, in captivity, or as a naturalized population (or populations) well outside its past range.

CRITICALLY ENDANGERED

A taxon is Critically Endangered when it is facing an extremely high risk of extinction in the wild in the immediate future: an 80 percent reduction over 10 years or three generations; at least 50 percent probability of extinction in the wild within 10 years.

ENDANGERED

A taxon is Endangered when it is facing a very high risk of extinction in the wild in the near future: a reduction of at least 50 percent over 10 years or three generations; severely fragmented or reduced populations; at least 20 percent probability of extinction in the wild within 20 years.

VULNERABLE

A taxon is Vulnerable when it is facing a high risk of extinction in the wild in the medium-term future: 20 percent probability of extinction in the wild within 10 years or three generations; very small or restricted population; at least 10 percent probability of extinction in the wild within 100 years.

LOWER RISK

The Lower Risk category is separated into three subcategories:-

Conservation Dependent: Taxa which are the focus of a continuing conservation program, the cessation of which would result in the taxon qualifying for one of the threatened categories above within a period of five years.

Near Threatened: Taxa which do not qualify for Conservation Dependent, but which are close to qualifying for Vulnerable.

Least Concern: Taxa which do not qualify for Conservation Dependent or Near Threatened.

DATA DEFICIENT

A taxon is Data Deficient when there is inadequate information to make an assessment of its risk of extinction based on its distribution and/or population status. Data Deficient is not a category of threat or Lower Risk, and acknowledges the possibility that future research will show that threatened classification is appropriate.

NOT EVALUATED

A taxon is Not Evaluated when it is has not yet been assessed against the criteria.

CITES

The international wildlife trade has caused massive declines in the numbers of many species of animals and plants. An international treaty, known as CITES, to which 152 nations subscribe, was introduced in 1975 to ban commercial international trade in endangered species and to regulate and monitor trade in other species that might become endangered.

CITES APPENDIX I

The most endangered species are listed in Appendix I, which includes all species threatened with extinction that are or may be affected by trade.

CITES APPENDIX II

Other species at serious risk are listed in Appendix II, which includes all species that although not necessarily currently threatened with extinction may become so unless trade is subject to strict regulation. Appendix II includes other species which must be subject to regulation in order that trade in endangered species listed in Appendix I may be brought under effective control, for example, species similar in appearance to endangered species.

Additional information can be found at the following websites:

http://www.iucn.org/themes/ssc/redlists/ssc-rl-c.htm

http://www.cites.org/CITES/eng/index.shtm

◄ Adélie penguins, Mt Herschel.

The Antarctic Treaty (Full text, 1961)

The Governments of Argentina, Australia, Belgium, Chile, the French Republic, Japan, New Zealand, Norway, the Union of South Africa, the Union of Soviet Socialist Republics, the United Kingdom of Great Britain and Northern Ireland, and the United States of America,

Recognizing that it is in the interest of all mankind that Antarctica shall continue for ever to be used exclusively for peaceful purposes and shall not become the scene or object of international discord;

Acknowledging the substantial contributions to scientific knowledge resulting from international cooperation in scientific investigation in Antarctica;

Convinced that the establishment of a firm foundation for the continuation and development of such cooperation on the basis of freedom of scientific investigation in Antarctica as applied during the International Geophysical Year accords with the interests of science and the progress of all mankind;

Convinced also that a treaty ensuring the use of Antarctica for peaceful purposes only and the continuance of international harmony in Antarctica will further the purposes and principles embodied in the Charter of the United Nations;

Have agreed as follows:

Article I

1. Antarctica shall be used for peaceful purposes only. There shall be prohibited, inter alia, any measure of a military nature, such as the establishment of military bases and fortifications, the carrying out of military manoeuvres, as well as the testing of any type of weapon.

2. The present Treaty shall not prevent the use of military personnel or equipment for scientific research or for any other peaceful purpose.

Article II

Freedom of scientific investigation in Antarctica and cooperation toward that end, as applied during the International Geophysical Year, shall continue, subject to the provisions of the present Treaty.

Article III

1. In order to promote international cooperation in scientific investigation in Antarctica, as provided for in Article II of the present Treaty, the Contracting Parties agree that, to the greatest extent feasible and practicable:

   a. information regarding plans for scientific programs in Antarctica shall be exchanged to permit maximum economy of and efficiency of operations;

   b. scientific personnel shall be exchanged in Antarctica between expeditions and stations;

   c. scientific observations and results from Antarctica shall be exchanged and made freely available.

Article IV

1. Nothing contained in the present Treaty shall be interpreted as:

   a. a renunciation by any Contracting Party of previously asserted rights of or claims to territorial sovereignty in Antarctica;

   b. a renunciation or diminution by any Contracting Party of any basis of claim to territorial sovereignty in Antarctica which it may have whether as a result of its activities or those of its nationals in Antarctica, or otherwise;

   c. prejudicing the position of any Contracting Party as regards its recognition or non-recognition of any other State's rights of or claim or basis of claim to territorial sovereignty in Antarctica.

No acts or activities taking place while the present Treaty is in force shall constitute a basis for asserting, supporting or denying a claim to territorial sovereignty in Antarctica or create any rights of sovereignty in Antarctica. No new claim, or enlargement of an existing claim, to territorial sovereignty in Antarctica shall be asserted while the present Treaty is in force.

Article V

1. Any nuclear explosions in Antarctica and the disposal there of radioactive waste material shall be prohibited.

2. In the event of the conclusion of international agreements concerning the use of nuclear energy, including nuclear explosions and the disposal of radioactive waste material, to which all of the Contracting Parties whose representatives are entitled to participate in the meetings provided for under Article IX are parties, the rules established under such agreements shall apply in Antarctica.

Article VI

The provisions of the present Treaty shall apply to the area south of 60° South Latitude, including all ice shelves, but nothing in the present Treaty shall prejudice or in any way affect the rights, or the exercise of the rights, of any State under international law with regard to the high seas within that area.

Article VII

1. In order to promote the objectives and ensure the observance of the provisions of the present Treaty, each Contracting Party whose representatives are entitled to participate in the meetings referred to in Article IX of the Treaty shall have the right to designate observers to carry out any inspection provided for by the present Article. Observers shall be nationals of the Contracting Parties which designate them. The names of observers shall be communicated to every other Contracting Party having the right to designate observers, and like notice shall be given of the termination of their appointment.

2. Each observer designated in accordance with the provisions of paragraph 1 of this Article shall have complete freedom of access at any time to any or all areas of Antarctica.

3. All areas of Antarctica, including all stations, installations and equipment within those areas, and all ships and aircraft at points of discharging or embarking cargoes or personnel in Antarctica, shall be open at all times to inspection by any observers designated in accordance with paragraph 1 of this Article.

4. Aerial observation may be carried out at any time over any or all areas of Antarctica by any of the Contracting Parties having the right to designate observers.

5. Each Contracting Party shall, at the time when the present Treaty enters into force for it, inform the other Contracting Parties, and thereafter shall give them notice in advance, of

   a. all expeditions to and within Antarctica, on the part of its ships or nationals, and all expeditions to Antarctica organized in or proceeding from its territory;

b. all stations in Antarctica occupied by its nationals; and

c. any military personnel or equipment intended to be introduced by it into Antarctica subject to the conditions prescribed in paragraph 2 of Article I of the present Treaty.

Article VIII

1. In order to facilitate the exercise of their functions under the present Treaty, and without prejudice to the respective positions of the Contracting Parties relating to jurisdiction over all other persons in Antarctica, observers designated under paragraph 1 of Article VII and scientific personnel exchanged under sub-paragraph 1(b) of Article III of the Treaty, and members of the staffs accompanying any such persons, shall be subject only to the jurisdiction of the Contracting Party of which they are nationals in respect of all acts or omissions occurring while they are in Antarctica for the purpose of exercising their functions.

2. Without prejudice to the provisions of paragraph 1 of this Article, and pending the adoption of measures in pursuance of subparagraph 1(e) of Article IX, the Contracting Parties concerned in any case of dispute with regard to the exercise of jurisdiction in Antarctica shall immediately consult together with a view to reaching a mutually acceptable solution.

Article IX

1. Representatives of the Contracting Parties named in the preamble to the present Treaty shall meet at the City of Canberra within two months after the date of entry into force of the Treaty, and thereafter at suitable intervals and places, for the purpose of exchanging information, consulting together on matters of common interest pertaining to Antarctica, and formulating and considering, and recommending to their Governments, measures in furtherance of the principles and objectives of the Treaty, including measures regarding:

a. use of Antarctica for peaceful purposes only;

b. facilitation of scientific research in Antarctica;

c. facilitation of international scientific cooperation in Antarctica;

d. facilitation of the exercise of the rights of inspection provided for in Article VII of the Treaty;

e. questions relating to the exercise of jurisdiction in Antarctica;

f. preservation and conservation of living resources in Antarctica.

2. Each Contracting Party which has become a party to the present Treaty by accession under Article XIII shall be entitled to appoint representatives to participate in the meetings referred to in paragraph 1 of the present Article, during such times as that Contracting Party demonstrates its interest in Antarctica by conducting substantial research activity there, such as the establishment of a scientific station or the despatch of a scientific expedition.

3. Reports from the observers referred to in Article VII of the present Treaty shall be transmitted to the representatives of the Contracting Parties participating in the meetings referred to in paragraph 1 of the present Article.

4. The measures referred to in paragraph 1 of this Article shall become effective when approved by all the Contracting Parties whose representatives were entitled to participate in the meetings held to consider those measures.

5. Any or all of the rights established in the present Treaty may be exercised as from the date of entry into force of the Treaty whether or not any measures facilitating the exercise of such rights have been proposed, considered or approved as provided in this Article.

Article X

Each of the Contracting Parties undertakes to exert appropriate efforts, consistent with the Charter of the United Nations, to the end that no one engages in any activity in Antarctica contrary to the principles or purposes of the present Treaty.

Article XI

1. If any dispute arises between two or more of the Contracting Parties concerning the interpretation or application of the present Treaty, those Contracting Parties shall consult among themselves with a view to having the dispute resolved by negotiation, inquiry, mediation, conciliation, arbitration, judicial settlement or other peaceful means of their own choice.

2. Any dispute of this character not so resolved shall, with the consent, in each case, of all parties to the dispute, be referred to the International Court of Justice for settlement;

but failure to reach agreement on reference to the International Court shall not absolve parties to the dispute from the responsibility of continuing to seek to resolve it by any of the various peaceful means referred to in paragraph 1 of this Article.

Article XII

1.

a. The present Treaty may be modified or amended at any time by unanimous agreement of the Contracting Parties whose representatives are entitled to participate in the meetings provided for under Article IX. Any such modification or amendment shall enter into force when the depositary Government has received notice from all such Contracting Parties that they have ratified it.

b. Such modification or amendment shall thereafter enter into force as to any other Contracting Party when notice of ratification by it has been received by the depositary Government. Any such Contracting Party from which no notice of ratification is received within a period of two years from the date of entry into force of the modification or amendment in accordance with the provision of subparagraph 1(a) of this Article shall be deemed to have withdrawn from the present Treaty on the date of the expiration of such period.

2. If after the expiration of thirty years from the date of entry into force of the present Treaty, any of the Contracting Parties whose representatives are entitled to participate in the meetings provided for under Article IX so requests by a communication addressed to the depositary Government, a Conference of all the Contracting Parties shall be held as soon as practicable to review the operation of the Treaty.

a. Any modification or amendment to the present Treaty which is approved at such a Conference by a majority of the Contracting Parties there represented, including a majority of those whose representatives are entitled to participate in the meetings provided for under Article IX, shall be communicated by the depositary Government to all Contracting Parties immediately after the termination of the Conference and shall enter into force in accordance with the provisions of paragraph 1 of the present Article

b. If any such modification or amendment has not entered into force in accordance with the provisions of subparagraph 1(a) of this Article within a period of two years after the date of its communication to all the Contracting Parties, any Contracting Party may at any time after the expiration of that period give notice to the depositary Government of its withdrawal from the present Treaty; and such withdrawal shall take effect two years after the receipt of the notice by the depositary Government.

Article XIII

1. The present Treaty shall be subject to ratification by the signatory States. It shall be open for accession by any State which is a Member of the United Nations, or by any other State which may be invited to accede to the Treaty with the consent of all the Contracting Parties whose representatives are entitled to participate in the meetings provided for under Article IX of the Treaty.

2. Ratification of or accession to the present Treaty shall be effected by each State in accordance with its constitutional processes.

3. Instruments of ratification and instruments of accession shall be deposited with the Government of the United States of America, hereby designated as the depositary Government.

4. The depositary Government shall inform all signatory and acceding States of the date of each deposit of an instrument of ratification or accession, and the date of entry into force of the Treaty and of any modification or amendment thereto.

5. Upon the deposit of instruments of ratification by all the signatory States, the present Treaty shall enter into force for those States and for States which have deposited instruments of accession. Thereafter the Treaty shall enter into force for any acceding State upon the deposit of its instruments of accession.

6. The present Treaty shall be registered by the depositary Government pursuant to Article 102 of the Charter of the United Nations.

Article XIV

The present Treaty, done in the English, French, Russian and Spanish languages, each version being equally authentic, shall be deposited in the archives of the Government of the United States of America, which shall transmit duly certified copies thereof to the Governments of the signatory and acceding States.

Words from the ice

Specialized English words have been used about Antarctica ever since humans began theorizing that the southern continent existed. But most of the words we associate with it have appeared during the last 100 years—understandably, because this has been the time in which our knowledge of the region itself has largely grown. In 1957, American journalist Walter Sullivan wrote:

> Until the moon or other planets are attained, Antarctica will remain the most unearthly region within the reach of man. The landscape is so alien that a completely specialized vocabulary is needed to describe it.

Much of the Antarctic vocabulary is used by the English-speaking nations: Australia, New Zealand, Britain, South Africa, and the United States. On English-speaking bases in Antarctica, you can hear words and phrases unknown elsewhere, such as **apple huts, googies** (a fiberglass field hut), **getting slotted** (falling into a crevasse), **winfly** (a winter fly-in of supplies), **greenout** (an overwhelming of the senses when returning from Antarctica to a green, vegetated landscape), and **jade** (green).

Many of the early exploring expeditions to the Antarctic represented diverse non-English-speaking nations—the *Belgica* expedition (1897–99), for example, included Roald Amundsen from Norway, Henryk Arctowski from Poland, and Emile Racovitza from Romania. Many of these travelers used words in Antarctica that they had previously encountered in their own countries or—more commonly—in previous Arctic travel. Such words include **anorak** (parka), **finneskoe** (a Lapp or Saami shoe), **krill** (a Norwegian word, applied there and in Antarctic English to small crustaceans), **sastrugi** (a Russian word for hardened icy ripples formed from snow), and **nunataks** (peaks that jut out of the surrounding ice sheet). The word nunatak occurs in more or less the same form (and with the same meaning) in Danish, Finnish, Swedish, Norwegian, and Russian, and is also in use in modern Canadian English.

Snow and ice

Antarctica is surrounded by ice and covered in it, and many appropriate discriminative and descriptive words have been generated for it: **tabular bergs** (vast, flat-topped icebergs which come from Antarctica's extensive floating ice sheets), **growlers** and **bergy bits** (larger or smaller chunks of decaying icebergs), **frazil** (the first stage of sea ice formation), **grease ice** (the next stage, resembling an oil slick on the sea's surface), **blue ice** (blue glacier ice, usually snow-free), **anchor ice** (ice formed on the bottom of a body of water, or on the seabed), and **pancake ice**. Writing of his time at Cape Adare, an English explorer elegantly described the formation of pancake ice—large circles of sea ice looking like waterlily pads, which form by bumping together in the sea:

> As the morning proceeded the swell decreased more and more, until movement was scarcely perceptible; and then, as the coating of ice became more rigid with increased thickness, it was unable to adapt itself even to this small movement, and the sheet broke up into small angular pieces a foot or two broad. As these rubbed gently against each other the corners were removed and the edges were upturned, and before our eyes there had taken place the formation of a field of the pancake ice which has from the earliest times been one of the marvels of the Polar seas.

Priestley, Raymond E. (1914, repr. 1974) *Antarctic Adventure: Scott's Northern Party*, Melbourne University Press, p. 85.

Not all ice terms are elegant—there is nothing poetic about a **snotsicle** (a thread of frozen mucus suspended from the nose of the owner), or an **icequake** (a tremor felt on an ice surface). And most are practical—sea ice concentrations are measured in **eighths**, which tell the **ice master** of a ship what proportion of the sea's surface is covered.

The creatures

Antarctic English includes the common names for its creatures: the huge **Sea elephants,** the **Leopards,** the **Wedds** (Weddell seals), **Icefish, Crabbies** (Crabeater seals), and **Patagonian toothfish.**

Familiar names for albatrosses and other seabirds—**GPs** (giant petrels), **Nellies, Sooties, Stinkers, Wanderers,** and **Yellownoses**—contribute to the language. Many of these animals were a source of food as well as objects of curiosity to the biologists. Photographer Frank Hurley wrote of the large, meaty, and blubbery sea elephants on Macquarie Island:

> Sea-elephant for food and a cave for shelter—what more could any man desire?

Hurley, Frank (1979) *Shackleton's Argonauts: the Epic Tale of Shackleton's Voyage to Antarctica in 1915*, McGraw-Hill Book Co., p. 124.

Penguins are not restricted to Antarctica: they occur as far north as the equator. But they are the main icon of the continent, and there are plenty of penguins in the Antarctic lexicon: **Chinstraps, Gentoos, Jackasses, Kings, Macaronis,** and the bad-tempered Rockhoppers:

> About half a million **Rockhoppers**—crimson-eyed penguins of vitriolic temperament—still nest on Nightingale Island.

Stonehouse, Bernard (1968) *Penguins: the World of Animals series*, Arthur Barker, London/Golden Press, NY, p 10.

Despite efforts by the more particular biologists, who prefer the term **colony,** penguins usually live in **rookeries.** And so do albatrosses.

Whaling and sealing

Hunting whales and seals has been a highly significant activity in the history of the exploration and exploitation of the Antarctic. The whales themselves, hunted first for their oil, and only in the twentieth century for their meat, were named for their colors (the **Blue whales** or **Sulphur-bottoms**), shapes (**Finners, Humpbacks**), for what they yielded (**Baleen** or **Whalebone whales**) or their desirability (**Right whales**), or known by names already established in Arctic waters (**Minke**). The early twentieth century breakthrough of **floating factories** enabled Antarctic whalers to rapidly process their catch near its point of capture in vast ships equipped for **boiling down,** without needing to return to the **shore stations.**

Workers on whaling and sealing ships the **flensers** (who removed the blubber) and **lemmers** (who dismembered whales and cut them into pieces for treatment)—brought their words with them from the far end of the other side of the world. In the early twentieth century, almost all whalers in Antarctica were Scandinavian—on South Georgia in the first decade of the century most (80 percent) were Norwegian, a few (8 percent) Swedish, and smaller numbers Finnish or Danish (2 percent each) (R. Headland, *The Island of South Georgia,* 1984).

Gradually some of their words became part of the English spoken in the region, often using direct borrowings such as the Norwegian **lemmer.** Another Norwegian word, **plan,** means an expansive, flat plane. A **bone plan** was a raised platform, often sloping, where the whale's huge bones were sawn up to extract their oil:

> There was also a second raised plan for sawing up the enormous heads and jawbones and separating the segments of the tremendous backbones, which were then treated to steam pressure boilers to extract the oil ... Sometimes a whaler missed his footing on the 'bone plan', as it was called, and came slithering down into the quagmire of blood, grease and flesh at the bottom.
>
> Saunders, Alfred (1950) *A Camera in Antarctica*, Winchester Publications, London, p 10.

Clothing, shelter, traveling, and food

Arctic influences were not restricted to words associated with whaling and sealing. To keep their feet warm and dry, explorers brought **mukluks** and **finneskoe** (seal or reindeer skin boots) and **sennegrass** (which they stuffed into the soles of their boots to absorb moisture), **anoraks** and **ventiles** (windproof clothing) to protect their bodies from bitter winds, and **nose-wiper mitts** to guard against the formation of **snotsicles**. Even the names of modern materials (such as **Polartec**) reflect their polar origins, or at least the tough images associated with these places.

Polar pyramids—tents designed specially for Antarctic conditions—and **apple huts**, which are round and red and come from Tasmania (the Apple Isle) are specialized types of dwelling made to protect people from the extreme cold. Moving around in Antarctica involves using unusual transport, for example **dog sleds**, **Haggs**, or **weasels**. Haggs (short for Hagglunds) and weasels are oversnow tracked vehicles, the latter designed for the United States army and used in Antarctica until the 1960s. In emergencies, they had other value:

> An operating table was fashioned from sledge boxes and an oxygen mask was made from weasel spare parts.
>
> Sullivan, Walter (1957) *Quest for a Continent*, Secker & Warburg, London, p. 131.

Food assumes gigantic importance in Antarctic life, and has plenty of words of its own there. People have eaten, and relished, delicacies such as **penguin eggs**, **Emperor penguin breast**, **hoosh**, **sledgies** and **upland goose** (*Chloephaga picta leucoptera*), a goose still hunted and eaten in the Falkland Islands.

Hoosh, the staple of early polar expeditions, was a singularly ordinary dull stew, usually made from two other ingredients that feature in Antarctic English: **pemmican** and **sledgies**. **Pemmican** is a mixture of dried

meat with fat, and sometimes, some cereal—its name comes from the language of an indigenous tribe of northern North America. **Sledgies**—also called **sledge biscuits**, **sledging biscuits**, or **Eskimo biscuits**—were famous for their tooth-cracking hardness, as well as their durability on sledging trips. An Australian expeditioner recalling the 1950s wrote:

> Breakfast follows the same routine: melt snow, make coffee, cook the porridge (with a good helping of butter concentrate in it), plaster chunks of butter concentrate onto sledge biscuits.
>
> McLeod, Ian in Robinson, Shelagh, ed. (1995) *Huskies in Harness: a Love Story in Antarctica*, Kangaroo Press, Sydney, p. 51.

These days the talk is more likely to be of **freshies**—vegetables and fruit flown or shipped in. Sometimes it can be months between deliveries, and people can pine badly for a slice of tomato or a green salad at places such as South Pole station:

> The winter airdrop was scheduled for Saturday the thirteenth of June, the only delivery of mail and freshies between February and October.
>
> Warren, Stephen in *Aurora*: the official journal of the ANARE club [Melbourne], 16(2) December 1996, p.32.

The distinctive vocabulary of the Antarctic region extends into the sub-Antarctic: the Antipodes Islands, Auckland Islands, Campbell Island, the Snares and the Bounty Islands, Macquarie, Heard, Kerguelen, the Crozets, Marion, and Prince Edward Islands. The **Macquarie cabbage** (*Stilbocarpa polaris*) occurs on New Zealand's sub-Antarctic islands, as well as on Macquarie Island. In the days when sealers and whalers visited these islands, the plant's fleshy and mustardy leaves were eaten as a green, and regarded as an excellent antiscorbutic or preventer of scurvy. The leaves sound less than appetizing: in 1955, a conscientious investigator of Macquarie Island's plants wrote:

> *Stilbocarpa polaris* is the Macquarie Island Cabbage used by the sealers of the nineteenth century as an anti-scorbutic ...

and added in a footnote:

> The petioles taste like celery when cooked; pickled rhizomes like turnips; and leaves when cooked like wet blotting paper.
>
> Taylor, B.W. (1955) The flora, vegetation and soils of Macquarie Island, *ANARE reports series B, vol. II, Botany* Antarctic Division, Melbourne, p. 131.

Bloody cold!

The weather provides Antarctic words, too: **katabatic winds**, for example, are literally winds falling down from a higher place. Such winds are not exclusively Antarctic, but Antarctica is the king of the katabatic blizzard, and this is the reason why Douglas Mawson called his base at Commonwealth Bay—and his 1915 book about his time there—*The Home of the Blizzard*. Nowhere are katabatic winds fiercer than in Antarctica, and Mawson happened to choose the windiest place on earth to live in when he was there. Antarctica produces the best katabatic winds because it is the highest and coldest continent, and is covered in ice to a depth of more than 4 kilometers (2 ½ miles), so that the air reaching the sea at Commonwealth Bay falls rapidly down from the great height of the polar plateau, and rushes out to sea.

Other Antarctic English terms relate to the sky and the atmosphere: **southern aurora**, **noctilucent clouds** (luminous night time clouds visible at high altitude and high latitude), and the **ozone hole**, first noted by British Antarctic Survey scientists, a truly Antarctic term whose use is no longer restricted to those who have worked in Antarctica. The phrase **wind-chill factor** (a way of calculating how cold it *feels* to stand in a cold wind) was first coined in Antarctica. An American geographer developed the scale after he had lived and worked down south:

> It becomes apparent to anyone subjected to cold that a windy day feels much colder than a calm day on which the thermometer may actually register a considerably lower temperature. I adopted the word wind-chill to express this factor, recognizing that it was in reality a rate at which the body was cooling.
>
> Siple, Paul (1959) *90° South: the Story of the American South Pole Conquest*, GP Putnam's Sons, NY, p. 71.

Like English anywhere, new Antarctic words are constantly being born, while others are quietly dying. Antarctica, or the **Big Pav** (from the Australian dessert, the pavlova), will always have its special words. BH

Antarctica around the world

Many sites around the world outside Antarctica have an association with the south polar regions, some very directly so, others rather tenuously. They offer insights into the history of the region and an affordable alternative to going to the Antarctic as a tourist. Some of these "low-latitude" Antarctic locations are listed below. Visit http://www.antarctic-circle.org/llag.htm for more.

England

London and its environs

Among the museums and public collections worth visiting, the recently enlarged **National Maritime Museum** at **Greenwich, S.E.10**, has also expanded its Antarctic collection. Along with the Scott Polar Research Institute in Cambridge, this is Britain's largest accumulation of Antarctic memorabilia. It contains numerous paintings from the Cook era onward, the autographed banjo that L.D.A Hussey played on Elephant Island, a Scott sledge, his sledge flag flown at the South Pole, and much more.

The **Royal Geographical Society, 1 Kensington Gore, S.W.7**, is a treasure trove of artefacts, paintings, and manuscripts, though most are stored away. A large model of the *Discovery* sits just inside the entrance, and paintings of James Cook, Robert Scott, and his mentor, Sir Clements Markham, hang nearby (a bust of Markham is set outside, next to the doorway). A powerfully evocative bronze statue of Ernest Shackleton occupies a niche on the building's Exhibition Road façade.

Shackleton's family lived at **12 Westwood Hill, S.E.26**—this house in which he grew up still stands and is marked with a historic blue plaque. As a youngster, Shackleton attended nearby **Dulwich College, S.E.21**. The college has a large collection of "Shackletonia" in its archives, but the star attraction, on show in the North Cloisters, is the *James Caird*, the 7-meter (23-foot) boat that made the incredible 1,287-kilometer (800-mile) voyage from Elephant Island to South Georgia. With raised sails it sits on a bed of stones brought from South Georgia.

The **British Library, 96 Euston Road, N.W.1**, has just one Antarctic item on public display, but it is one that should not be missed: Scott's journal opened to his final entry, written on 17 March 1912.

Houses of Antarctic explorers are scattered about London. Scott rented **56 Oakley Street, S.W.3** (marked by a blue plaque), from 1904 to 1908, and lived with his mother and sisters at **80 Hospital Road, S.W.3**, while organizing the *Discovery* expedition. Shackleton had many addresses over the years, one of the more interesting was **11 Vicarage Gate, Kensington, W.8**, where he moved in 1913; it is now the Abbey House Hotel. Two other blue-plaque Antarctic house are those of Sir James Clark Ross at **2 Eliot Place, Blackheath, S.E.3**, and Dr Edward A. Wilson at **42 Vicarage Crescent, Battersea, S.W.11**.

Statuary and memorials also abound in London and its environs. Some notable ones are the bronze statue of Robert Scott in sledging gear sculpted by his artist wife Kathleen— she did a second version in marble, which may be found in Christchurch, New Zealand. The bronze version, unveiled in 1915, stands in **Waterloo Place, S.W.1**, opposite the statue of another tragic polar hero: Arctic explorer Sir John Franklin. At **St Paul's Cathedral, E.C.4**, on 5 May 1916, Prime Minister Asquith unveiled a large bronze tablet in the south transept in memory of Scott and his four polar companions—the men are depicted hauling a sledge. Above Scott's head is the inscription: "Death is swallowed up in victory."

The bronze plaque in the school library at **Eton College**, near **Windsor**, is another Kathleen Scott work. It memorializes L.E.G. Oates, the "very gallant gentleman" who willingly went to his death to save his comrades.

Elsewhere in England

The **Scott Polar Research Institute (SPRI), Lensfield Road, Cambridge**, is with little doubt the ultimate destination for polar pilgrims. It has long been the world's preeminent polar educational and research institution, but it is the museum, archives, and library that draw enthusiasts, writers, and collectors from near and far. The relics, journals, paintings, and expedition equipment on display are numerous and varied. Some favorite items include five framed Antarctic sledging flags, Captain Scott's "housewife" or repair kit, and Edward Wilson watercolors (only a tiny selection of the SPRI's collection are on show), the wooden box used for anonymous submissions to the *South Polar Times*, Oates' sleeping bag, and his last letter to his mother. Elsewhere in the building are the brass ship's bell from the *Terra Nova* (rung twice a day) and a spar from the *Endurance*.

The **Oates Memorial Museum** in the lovely village of **Selborne, Hampshire**, shares space with the Gilbert White Museum in "The Wakes", where White wrote his famous natural history book. The collection focuses on Captain Oates who died on the return from the South Pole and includes artefacts, paintings, a life ring, and a pennant from the *Terra Nova*, together with various pieces of clothing and equipment.

In a similar vein, the exhibit memorializing Dr Edward Wilson at **the Cheltenham Art Gallery & Museum, Clarence Street, Cheltenham, Gloucestershire**, is a popular one. Among special items there is the prayer book from which, on 12 November 1912, surgeon Edward Atkinson read the service for Scott, Wilson, and "Birdie" Bowers beside the tent in which they died. A short walk away, on The Promenade, is a bronze statue of Dr Wilson, sculpted by Kathleen Scott and reminiscent of the one she did of her husband in London. Wilson was born in Cheltenham at **91 Montpellier Terrace**; the fact is noted in a large inscription on the stone façade of the building, which now offers bed and breakfast accommodation.

Plymouth was Scott's hometown, and his skis and a model of the *Terra Nova* are preserved at the **Plymouth City Museum & Art Gallery, Drake Circus**. A few miles away on **Mount Wise** (really a very small hill) in **Devonport** stands the national memorial to Scott and his companions, a mammoth affair of bronze and granite that incorporates some very powerfully executed scenes and bas-relief portraits. The Scott tragedy gave rise to a tremendous number of memorial tributes throughout the Empire. The one at Mount Wise is notable for its size; others have attributes equally as compelling. The four Scott Party Memorial Windows **at St Peter's Church, Binton, Warwickshire**, are admired for their skilful depiction of the polar journey. Scott's brother-in-law was rector of the church and is buried in the churchyard, his grave marker designed by his sister, Kathleen.

On a wall in **St Mary the Virgin Church at Gestingthorpe, Essex**, hangs a brass plaque in memory of Captain Oates who grew up across the road in the local manor house, **Gestingthorpe Hall**. Placed there a year after his death by his brother officers of the Sixth Inniskilling Dragoons, it still shines brightly; his mother used to polish it every week throughout her long life.

In the pleasant village of **Aston Abbots, Buckinghamshire**, behind **St James the Great Church**, is the tomb of Sir James Clark

Ross who gave his name to the sea, the island, and the ice shelf. Inside the church is a memorial window to him and his wife Ann.

Another grave, earlier still, is that of Edward Bransfield in the **Brighton Extra Mural Cemetery, Brighton, East Sussex**. The recently refurbished raised tomb includes the inscription "the first man to see mainland Antarctica in January 1820" (and if Bransfield was not the first, he was nearly so). Bransfield, who died in obscurity, lived at **11 Clifton Road**, later moving to **61 London Road**; both houses are still standing in Brighton.

Wales
Cardiff

Cardiff has strong connections with the Antarctic. It was the last British port to see *Terra Nova* off and the first to welcome her back. The **Royal Hotel** in **St Mary Street** hosted a farewell banquet for the officers and scientists of Scott's expedition on 13 June 1910. The *Terra Nova* departed two days later. On 17 June 1913, a welcome-back dinner was also held there, though surely a more somber gathering. At the **Welsh Industrial and Maritime Museum, 126 Bute Street, Cardiff Bay**, the *Terra Nova*'s figurehead, a lovely blonde lady, holds reign. In 1916 a large bronze memorial tablet to commemorate Scott's sacrifice was installed by the grand staircase in the **Cardiff City Hall**. Its pictorial detail includes penguins, seals, skis, a sledge, and dogs. The **Clock Tower** rising from the lake in nearby **Roath Park** is a memorial to Scott's polar party in the form of a lighthouse.

Scotland

The *Discovery*, Scott's ship on his first Antarctic expedition, was built in **Dundee**. For years it languished in London but has now returned to its birthplace where it is the centerpiece of the harborside development: **Discovery Point**. The ship is open for visits and the wardroom can be hired for functions. The on-shore exhibits are extensive and very well presented.

Sir Ernest Shackleton lived at **14 South Learmouth Gardens, Edinburgh**, from April 1904 while he was secretary of the Royal Scottish Geographical Society. This building is now Channings, an upmarket hotel. Brother officers of "Birdie" Bowers, who died with Scott, erected a memorial to him at **St Ninian's Church** in **Port Bannatyne** on the **Isle of Bute** about a mile north of the center of **Rothesay**. (An identical one was also installed in **St Thomas Cathedral, Bombay, India**—the oldest English building in Bombay and "since 1718 the center of Bombay's Christian worship.")

Ireland

Most Antarctic sites in Ireland have to do with Shackleton in one way or another. Before the family moved to Croydon, south of London in 1884, they spent four years at **35 Marlborough Road, Dublin**, a handsome brick house that recently received a plaque noting this fact. Sir Ernest was born in 1874 at **Kilkea House** in **County Kildare**, and in nearby **Athy**, at the **Athy Heritage Centre** in the Town Hall, an exhibit on the famous native son includes a sledge used during the *Nimrod* expedition.

To the west, at **Anascaul** on the **Dingle Peninsula, County Kerry**, is Tom Crean's **South Pole Inn**, the pub he opened after his Antarctic exploring days with Scott and Shackleton ended. It has some appropriate Antarctic decorations and a guest book that records the names of hundreds of thirsty Antarcticans. Crean's grave is about a mile away in the very wild though picturesque **Ballinacorty Cemetery**. A monument in the **Town Park, Kinsale, County Cork**, has recently been dedicated to the McCarthy brothers: Mortimer, a member of Scott's *Terra Nova* expedition, and Timothy who accompanied Shackleton in the *James Caird*.

France
Paris

France's two most famous Antarctic explorers are buried in **Paris**. Jules-Sébastian-César Dumont d'Urville, who led the great 1838-40 expedition to the South lies in the **Cimetière Montparnasse**. He, his wife, and son were killed in a fiery railroad crash in 1842. His monument is a large rounded obelisk. North of the Seine, in the **Cimetière Montmartre**, is the mausoleum of Jean-Baptiste Charcot, who also died tragically when he went down with his famous ship *Pourquoi-Pas?* during a gale off Iceland. Both cemeteries are known for their celebrity residents.

Norway
Oslo and its environs

Oslo's best known Antarctic site is the **Fram Museum**, where Nansen and Amundsen's famous ship *Fram* has been enclosed in a large A-frame structure since 1936. There are various displays on the ship, not only relating to Nansen and Amundsen but also to Carsten Borchegrevink. Oscar Wisting, who was at the South Pole with Amundsen, asked to spend a night in his old cabin soon after the museum opened; he was found dead there the next morning. The **Ski Museum**, at the **Holmenkollen ski jump**, north of Oslo, has Amundsen artefacts including a tent, skis, clothing, provision boxes, even a stuffed sledge dog. Amundsen's house, "**Uranienborg,**" is located south of the city, at **Svartskog** on the east side of the fjord and now open as a museum. His bedroom features portholes, and everything is just as he left it when he set out on his ill-fated rescue attempt of Umberto Nobile. Down by the fjord, is a bronze statue of Amundsen, accompanied by a sledge dog.

Elsewhere

The **Bjaaland Museum** in **Morgedal, Telemark**, remembers Olav Bjaaland, the last of the five in Amundsen's polar party to die, in 1961. Oscar Wisting was born in **Larvik** and a statue of him stands outside the **Sjøfartsmuseum**; close by is a bust of Colin Archer, the designer of the *Fram* who also lived in Larvik. A statue of Amundsen graces the public square in **Tromsø** and some Amundsen artefacts are preserved in the **Polar Museum** there. Although there is no monument to Amundsen in Britain, Scott is commemorated in Norway, by a rough-hewn granite obelisk near the railroad station at **Finse** on the line between Oslo and Bergen.

Japan

Lieutenant Nobu Shirase was Japan's participant in the Heroic Age; his team actually met some of Amundsen's party in the Bay of Whales. The **Shirase Antarctic Expedition Memorial Museum** is located in his hometown of **Konoura**, 500 kilometers (310 miles) north of Tokyo. Among the exhibits is a full-scale model of the stern section of his ship, *Kainan Maru*, and equipment and clothing from his expedition.

South Africa

British ships bound for the Antarctic generally re-provisioned in **Cape Town**, so it is not surprising to find a memorial to Scott in **Adderly Street**. The **Johannesburg Municipal Gallery** has a lovely oil portrait of Scott's wife Kathleen by Charles Shannon. Dr Reginald Koettlitz, physician on the *Discovery* expedition, is buried at **Somerset East**.

United States
Washington and its environs

The **Navy Museum** in the **Washington Navy Yard** is a cavernous building filled with exhibits including a whole section on polar exploration, the largest of its kind in the United States. Byrd, Ronne, Wilkes, and other Americans are highlighted, but attention is also paid to Scott, Shackleton, and Dumont D'Urville. Lincoln Ellsworth's *Polar Star*, the plane in which he and Hollick-Kenyon made the first traverse of Antarctica, can be found at the Smithsonian's **National Air and Space**

Museum on the **Mall**. A plaque beside a footpath just **north of the Capitol** notes where once Lieutenant Charles Wilkes' house stood. Not too far away, on **Pennsylvania Avenue** at **Eighth Street**, is the elaborate **Navy Memorial**. Among its many bas-reliefs is one devoted to Wilkes. His grave is in **Arlington National Cemetery** across the **Potomac**, and the marker states, incorrectly, that "he discovered the Antarctic Continent." Three other Antarctic graves are there: that of Admiral Richard Byrd with a simple military issue headstone, behind it the far larger marker for Bernt Balchen, and nearby that of Finn Ronne. A large bronze statue of Byrd stands on the **Avenue of Heroes** across from the **Visitors Center**.

New England

Although Byrd was from an old Virginia family, he married a Boston woman and when not exploring in the north or south was at home at **9 Brimmer Street**, a lovely brick townhouse at the foot of **Boston's Beacon Hill**. Igloo, the fox terrier that accompanied Byrd on his early Antarctic travels, is buried in the **Pine Ridge Cemetery for Small Animals** in **Dedham**, just outside Boston. The iceberg-shaped, pinkish gravestone is inscribed "Igloo. He was more than a friend."

The splendid **Peabody Essex Museum** in **Salem, Massachusetts**, has a number of Antarctic artefacts though few are displayed. However, Charles Wilkes' superb oil painting of his flagship *Vincennes* hangs in the East India Marine Hall.

Many of the early sealers in the South Shetlands called **Stonington, Connecticut**, their home port. Nathaniel Brown Palmer—who along with Bellingshausen, Smith, and Bransfield, claimed to be the first to sight the Antarctic continent in 1820—is buried in the tranquil **Evergreen Cemetery**. The grave of Phineas Wilcox, the first mate of Palmer's *Hero*, is in the **Miner Burying Ground** nearby. A sign marks Captain Palmer's birthplace on **Water Street** and the **Stonington Historical Society** headquarters now occupies his ornate mansion on **Palmer Street**. Within are various paintings, relics, and displays relating to Palmer, the Antarctic, and sealing.

Elsewhere

The enormous **American Museum of Natural History** in **New York City** contains a ground-floor alcove display that focuses on Lincoln Ellsworth—a Museum trustee—but also includes many interesting items associated with Amundsen and Nordenskjöld. In **Annapolis, Maryland**, the excellent **Naval Academy Museum** has, among other things,

Wilkes' dress cocked hat, and sledges associated with both Byrd and Scott. The **Virginia Aviation Museum** at the **Richmond International Airport** is home to Byrd's aircraft, *Stars & Stripes*. Another famous Byrd aircraft, the Ford tri-motor *Floyd Bennett*—first to *fly over* the South Pole—is at the **Henry Ford Museum** in Dearborn, Michigan. The *Que Sera Sera*—first plane to *land* at the South Pole—is at the **National Museum of Naval Aviation** in **Pensacola, Florida**.

Ohio State University is home to the **Byrd Polar Research Center**. A few items are on display, but this is primarily a research and educational center, the closest thing to an American SPRI. The **University Archives** nearby house the papers of Byrd, Sir Hubert Wilkins, and Dr Frederick Cook. Ohio is also the location of a remarkable monument to John Cleves Symmes. He originated the Theory of Concentric Spheres and Polar Voids, or simply the Hollow Earth Theory. Appropriately, his monument on **Third Street** in **Hamilton** is a stone obelisk, atop which sits a round stone sphere with a hole drilled through it.

Argentina

Argentina is a likely jumping-off place for those traveling to the Antarctic Peninsula, and there are several sites to visit on the way. In **Buenos Aires**, the *Uruguay*, a famous Antarctic ship, is now a museum moored at **Dock One** along the lengthy **Perto Madero**. Far to the south, in **Ushuaia**, the **Maritime Museum** is located in the **Presidio**. It has the largest collection of same-scale Antarctic ship models anywhere, numerous artefacts, particularly ones associated with Nordenskjöld's expedition, and an extensive philatelic display.

Chile

Santiago's Natural History Museum has dioramas of Antarctic bird life, archeological items and relics from the South Shetlands, and displays of polar equipment.

Australia

Tasmania: Hobart

Hobart has traditionally been the departing point for voyages to the Antarctic from Australia; among the earliest were Sir James Clark Ross and Dumont d'Urville. At the **Hobart Crematorium and Cemetery, Cornellian Bay,** is a memorial to the sailors on Dumont d'Urville's expedition who died at sea or in hospital in Hobart and a commemorative rose garden. There is some excellent material on sealing, whaling, and Antarctic expeditions leaving from Tasmania at the **State Library of Tasmania** in **Murray Street**; early

whaling gear and Antarctic expedition equipment may be seen at the **Tasmanian Museum and Art Gallery** in **Macquarie Street**. In **St Albans Anglican Church, Main Road, Claremont**, are stained glass windows memorializing the Scott expedition.

There is a small museum of Heroic Age artefacts in the foyer of the **Australian Antarctic Division, Channel Highway, Kingston**, and the Division also boasts world-class collections of polar library materials and Antarctic image resources. A bust of Amundsen may be found in the **Centenary Building** at the **University of Tasmania Sandy Bay** campus. The same building is home to two special Antarctic-related organizations: the **Institute of Antarctic and Southern Ocean Studies** (IASOS) and the prestigious **Antarctic Cooperative Research Centre** (CRC).

One can spend a night in the Amundsen Suite at **Hadley's Hotel** in **Murray Street**—Amundsen was a guest there in March of 1912 on his return from the Pole. During his short stay he sent a telegram from the **Hobart General Post Office** on the corner of **Macquarie** and **Elizabeth Streets**, informing the world of his triumph. "**Antarctic Adventure**", **Salamanca Square**, presents hands-on exhibits. Hobart's Maritime Place was recently renamed **Mawson Place** in honor of the great explorer; the associated **Waterside Pavilion** features an Antarctic display."

New South Wales: Sydney and Newcastle

Among its vast collection of Antarctic documents and photographs, **Sydney's Mitchell Library** holds Apsley Cherry-Garrard's Antarctic sketchbooks and Frank Hurley's 20 surviving Paget color transparencies from the *Endurance* expedition. The **Powerhouse Museum, Darling Harbour,** Sydney, has several items associated with Mawson including a sledge, clothing, snow goggles, and Charles Laseron's polar medal. "In Blizzard Bound," an exhibit at the **Newcastle Regional Museum**, also features items from Mawson's expeditions.

ACT: Canberra

Mawson, a suburb of Canberra, was established in 1966. The streets were named for persons and ships associated with the Antarctic, both Australian and otherwise, including Colbeck, Hurley, Joyce, Markham, Shackleton, and Wilkins.

Victoria: Melbourne

Captain Cook's cottage is a popular tourist spot in a lovely setting in **Fitzroy Gardens, Melbourne**. It was moved there from Great

Ayton, North Yorkshire, in 1934. (The world is full of Cook sites—**North Yorkshire, England,** being particularly rich in them.) In **St Paul's Cathedral** a wood plaque was recently installed honoring Australian Antarctic expeditioners who made the ultimate sacrifice. Outside the **Polly Woodside Maritime Museum, Southbank,** is a memorial to the Antarctic ship *Nella Dan,* while inside the Museum is a large model of the ship and a "Secrets of the Frozen World" exhibit.

South Australia: Adelaide

A bronze bust of Australia's great Antarctic explorer, Sir Douglas Mawson, stands in an outdoor setting at the **University of Adelaide** where he taught for nearly 50 years. Although the University's Mawson Institute closed in 1990, there is a display of Mawson memorabilia—including the famous "half-sledge"—at the **Tate Geological Museum.** At the **Waite Campus,** hundreds of Mawson objects are in storage, awaiting a better and more accessible venue, an effort also involving the **South Australian Museum, North Terrace,** which also has many Mawson items.

Western Australia: Esperance

Charles Sandell brought back souvenirs and relics from Mawson's Australasian Antarctic Expedition, and donated them to the **Esperance Municipal Museum** in his hometown: among them is a propeller from the "air tractor" and a 1911 Christmas bottle of port.

New Zealand

New Zealand has had a particularly close association with Antarctic expeditions from the Heroic Age to the present, and many sites there are worth visiting.

Christchurch and its environs

The **Visitor Centre** at the **International Antarctic Centre** beside **Christchurch Airport** provides an interactive experience focusing on Antarctic science and the polar environment, a variety of displays including some Heroic Age artefacts, and a gift shop and café. The **Canterbury Museum, Rolleston Avenue, central Christchurch,** shares with the SPRI in England the title of world's premier Antarctic museum. Some favorite items include Frank Hurley's Model B Kodak camera, the stove that accompanied Shackleton on the *James Caird* and the silver communion service from Scott's Cape Evans hut. Close by in the port of **Lyttelton** is the **Lyttelton Museum** on **Gladstone Quay.** This eclectic local history museum has an Antarctic component including a stuffed Emperor penguin, polar medals, a model of the *Discovery,* and a Shackleton sledge. In the center of **Christchurch** on **Worcester Street** stands the statue of Captain Scott sculpted by his wife Kathleen, a copy of the bronze one in London. Because of wartime metal shortages Italian marble was used. Two plaques with Antarctic connections can be found at **Christchurch Cathedral.**

Auckland and its environs

The **Auckland Museum** in the **Domain** has a small display of Heroic Age memorabilia. **Kelly Tarlton's Antarctic Encounter** is a theme park of sorts (offering rides in Snow Cats) about four miles from the center of town. Among other things, particularly penguins, it features a replica of the wardroom in Scott's hut at Cape Evans.

Wellington and its environs

The **Alexander Turnbull Library** is New Zealand's national library and holds many Antarctic-related manuscripts and journals, as well as two copies of the very rare *Aurora Australis,* issued at Cape Royds during Shackleton's *Nimrod* expedition. **Mount Victoria** rises above the city and on top is a memorial to **Richard Byrd.** Shaped like a polar tent, facing south, it was dedicated in 1962 and extensively repaired a few years ago. The graves of two important Antarcticans associated with Shackleton's *Endurance* expedition are in **Karori Cemetery:** Harry "Chips" McNeish, the carpenter who kept the *James Caird* together, and Colonel Thomas Orde-Lees.

Port Chalmers and Dunedin

The departure of Scott's *Discovery* from Lyttelton in 1901 was marred by the death of Charles Bonner who fell from the top of the mainmast when the ship encountered an ocean swell. He was buried at the next port of call in the **Port Chalmers Cemetery** where a stone obelisk marks the grave. The most impressive monument in **Port Chalmers**—given size and setting—is the **Scott Memorial,** a circular 9-meter (30-foot) tall stone column with an anchor on top, set high on a hilltop overlooking the harbor. A bronze bust of Richard Byrd can be found in **Unity Park, Dunedin;** it was presented to the city by the National Geographic Society.

Elsewhere

A memorial to Scott and his companions is situated in the park beside **Lake Wakatipu** in **Queenstown.** It consists of two plaques—one with Scott's last message, the other with the names of the polar party—set on a huge glacial boulder. At **Waitaki Boys' High School** in **Oamaru** "Birdie" Bowers's sledge flag, made by his sister, is in the Hall of Memories, which also contains a tablet in Scott's memory. RS

Antarctic links

Every imaginable Antarctic resource can be found on the Internet with a little patience—a recent search under "Antarctica" scored over 150,000 hits. The 50 sites listed below cover the gamut, from history to glaciology to current events, providing something for every Antarctic taste. Websites have a way of becoming inactive or disappearing outright, but those included here should be around for a while.

Principal websites

Four of the sites listed stand out because of the quality and depth of their content, their navigational ease, and the fact that they complement one another.

The first is the website of the **Scott Polar Research Institute** (SPRI), the world's leading academic polar organization <http://www.spri.cam.ac.uk>. This site is very detailed and comprehensive, and even has versions of its homepage in eight languages other than English. SPRILIB, a searchable database with over 36,000 bibliographic records, and the Picture Library Database are excellent. The Index to Antarctic Expeditions (with links, summaries and related literature) is useful, as are the worldwide Polar Museums Directory and the Directory of Polar and Cold Regions Organizations. There are even Kid's Pages including Polar Jokes.

For those with a passion for Antarctic history, the **Antarctic Philately** site <http://www.south-pole.com> is the place to go. Besides copious information on stamps and postal history, there is voluminous coverage of south polar exploration from Cook to nearly the present. The highpoints are the numerous biographical entries and a well-done time line stretching from 1519 to 1959. Some portions are nearly book-length. This elegantly designed site also features many seldom-seen photo illustrations.

For keeping up with the news, go to **The Antarctican** <http://www.antarctican.com> Produced in Tasmania, a site describing itself as "delivering the latest news and comment on Antarctic life, South Polar endeavor, the world of the ice, and the Southern Ocean around it." The individual pieces, covering all Antarctic subjects, are original and in a common format, not reprints, or links to newspapers. The Polemic section lets you make your views known on things Antarctic and you can post your Antarctic photos on Ice Picks.

Gateway Antarctica <http://www.icair.iac.org.nz> originates at the University of Canterbury in Christchurch, New Zealand. It strives for "increased understanding and more effective management of the Antarctic and the Southern Ocean by being a focal point and a catalyst for Antarctic scholarship, attracting national and international participation in collaborative research, analysis, learning and networking." Gateway to Antarctica is one portion of the Gateway Antarctica site. Once through the Gateway, there are sections on History (including the very lengthy "Flight of the Puckered Penguins", focusing on Antarctic aviation in the 1950s), Tourism (the "Visitor's Introduction to Antarctica and its Environment" is praiseworthy), Logistics, Environment (with lots on ozone depletion), Education, Images, Science, News, and Treaty (a fine background on the Antarctic Treaty).

Other websites
Antarctic Centers

Although SPRI has the best site of the academic centers, others worth going to include that of the **Byrd Polar Research Center** at the Ohio State University <http://www-bprc.mps.ohio-state.edu>, which among other things has a biography of Richard Byrd and a notable collections of links entitled Polar Pointers.

The Antarctic Co-operative Research Centre at the University of Tasmania was "established in 1991 and is now one of the largest research organizations in the world concerned with polar regions." Its site, **Antarctic CRC** at <http://www.antcrc.utas.edu.au/antcrc>, is thoroughly scientific in focus. It gives details on conferences, media releases, and on-line newsletters, and has a listing of commonly used Antarctic acronyms and abbreviations.

Glacier <http://www.glacier.rice.edu> administered by Rice University in Texas, has many short sections on all kinds of Antarctic subjects, making it particularly suitable for students. Much of it is presented in question and answer format. The coverage of ice and glaciers, life at the stations and polar clothing is first rate.

The Library at **CRREL** (Cold Regions Research and Engineering Laboratory) in Hanover, New Hampshire, "is recognized as the world's foremost collection of cold regions scientific and technical literature." Its site <http://www.crrel.usace.army.mil/library> has several searchable databases and many links.

National programs

Each country with an Antarctic presence has a website for its program but most are either sparse in content or of limited interest. Several do stand out, however.

The best starting point is the site maintained by **COMNAP** (Council of Managers of National Antarctic Programs), <http://www.comnap.aq>, which "was established in 1988 to bring together those managers of national agencies responsible for the conduct of Antarctic operations in support of science." It lists all the stations, has details on the facilities and activities of each, and links to all the national programs (the 'aq' in the web address is Antarctica's own internet domain.).

Another helpful source on national programs is the **SCAR** (Scientific Committee on Antarctic Research) site <http://www.scar.org>. "SCAR is charged with the initiation, promotion and coordination of scientific research in Antarctica. It also provides scientific advice to the Antarctic Treaty System." The site has an excellent listing of facts (some Antarctic statistics) and information on various scientific working groups and specialists.

National sites of note include:
- the **New Zealand Antarctic Institute** <http://www.antarcticanz.govt.nz> (some good school resources—frequently asked questions, information sheets, and education database);
- the **British Antarctic Survey** <http://www.antarctica.ac.uk> (all about British bases, copyright-free photos, news stories);
- the United States National Science Foundation's **Office of Polar Programs** <http://www.nsf.gov/od/opp> (sections on scientific programs by discipline, journal articles, lots of photographic and satellite images); and
- the Australian Antarctic Division's site **Antarctica Online** <http://www.antdiv.gov.au> (maps of station areas, searchable catalogue of library holdings, many copyright-free photos, database of polar words and phrases, station webcams, and the Antarctic Artefacts Register).

Also interesting is **The New South Polar Times** <http://205.174.118.254/nspt/home.htm>, a newsletter written by the staff of the

Amundsen-Scott South Pole Station. Along similar lines, **The Antarctic Sun** <http://www.polar.org/AntSun/index.htm> is published during the austral summer at McMurdo Station. The contents are informal and chatty; and there are even cartoons.

For a virtual tour of either South Pole Station or McMurdo Station, go to the **CARA** sites (Center for Astrophysical Research in Antarctica) at:

- <http://astro.uchicago.edu/cara/vtour/pole>; and
- <http://astro.uchicago.edu/cara/vtour/mcmurdo>.

News

In addition to the highly-recommended The Antarctican, described earlier, Antarctic news may be found at **70South** <http://www.70south.com> (links to news sources, contributed items, and interesting or important Antarctic anniversaries for the current month).

History

The Antarctic Philately site, noted earlier, is *the* site when it comes to history. Many webpages will have some coverage of Antarctic history and the better-known explorers. Gateway to Antarctica, also noted earlier, gives good coverage.

Virtual Antarctica <http://www.terraquest.com/antarctica/index.html> is maintained by Mountain Travel-Sobek and World Travel Partners includes a chronology of Antarctic exploration from 1772 to 1995, as well as other information aimed at the Antarctic tourist. **Adventure Network**, another tour operator, has a concise history section <http://www.adventure-network.com>.

Those interested in Antarctic historic sites, especially the huts of the explorers, should check out **Heritage-Antarctica.org** <http://www.heritage-antarctica.org>, which highlights the activities of New Zealand's Antarctic Heritage Trust in maintaining the historic huts in the Ross Sea sector. Included is a listing of historic sites with accompanying maps. This site also serves the United Kingdom Antarctic Heritage Trust. Efforts at restoring Mawson's huts at Commonwealth Bay are the focus of **The AAP Mawson's Huts Foundation** site <http://203.63.165.141>.

Sites on individual explorers are becoming common. At the moment Shackleton seems to attract the most attention. The most enthusiastic is **Welcome to the Shack!** <http://home.nycap.rr.com/gn>. Among its contents are frequently asked questions, Shack Books, Book Review, Shack Links, and News

Archive. The Message Board features queries posted by readers with responses from others.

Boston's WGBH public television station has an informative site entitled **Shackleton's Antarctic Odyssey** <http://www.pbs.org/wgbh/nova/shackleton>. There are useful biographies of all Shackleton's men, details on the lesser known Ross Sea party, historic maps, short accounts of other heroic age expeditions, educational resources such as lesson plans, and information on clothing, food, etc.

Shackleton's old school, London's Dulwich College, maintains a collection of **Shackleton Links** at <http://www.dulwich.org.uk/history/shacklinks.htm>.

The **James Caird Society** was established in 1994 to honor and perpetuate Shackleton's life and deeds. This very active group has placed a considerable amount material on its website at <http://www.jamescairdsociety.com> including some good photographs, quite a bit of history on Shackleton's *Endurance* expedition, the latest Shackleton news and details on the Society itself.

The explorer's cousin and family historian, **Jonathan Shackleton**, has a site <http://indigo.ie/~jshack/ernest.html> that notes recent books, videos and exhibitions, and recounts Shackleton's four Antarctic expeditions.

Frank Hurley, Shackleton's Australian photographer, is the subject of Shane Murphy's site <http://frankhurley.com>.

At present **Captain Scott** receives less web attention than Shackleton. His ship *Discovery* is beautifully preserved in Dundee, and the associated web site <http://www.rrs-discovery.co.uk> has lots on Scott, his men, the ship and the extensive exhibits on shore at Discovery Point.

A wealth of material on **Captain Cook** including a very detailed chronology of his second (Antarctic) voyage may be found at <http://www.CaptainCookStudyUnit.com>. This also offers information on stamps, coins, medals, and Cook's ships, and has a long bibliography as well.

Education

In addition to those already mentioned, sites that should appeal to students and teachers include **Share the Journey** <http://www.sofweb.vic.edu.au/claypoles>, hosted by the Victoria Department of Education in Australia; it follows the year spent at Commonwealth Bay by Yvonne and Jim Claypole and includes diary entries and questions for students.

Antarctica as an Educational Resource is a site maintained by The School of Biological Sciences at the University of Auckland

<http://www.sbs.auckland.ac.nz/biology_web_pages/antarctica/index.htm>. Its aim "is to provide a focus for teachers interested in using Antarctica as an educational resource." There are specific sections (with questions posed) covering History, Environment, Biology, Adaptive Radiation, and Human Involvement.

TEA (Teachers Experiencing Antarctica and the Arctic) <http://tea.rice.edu> is a terrific educational site sponsored by the National Science Foundation and facilitated by Rice University, the American Museum of Natural History and CRREL. School-teachers participate in research projects in Antarctica and post their journals here, or can e-mail them and learn more. The content is voluminous.

Tourism

Two useful sites for Antarctic tourists stand out.

IAATO (International Association of Antarctica Tour Operators) is "a member organization founded in 1991 to advocate, promote and practice safe and environmentally responsible private-sector travel to the Antarctic." Its site <http://www.iaato.org> has a definitive listing of tour operators with contacts and web addresses, a variety of tourism statistics, a book list, and a section on Guidance for Visitors to the Antarctic.

Lonely Planet, the guidebook publisher, has helpful information such as Facts for the Traveler, When to Go, and Getting There & Away on <http://www.lonelyplanet.com/destinations/antarctica/antarctica>.

Webcams and images

Webcams are scattered about Antarctica at various research stations. **WebCam Central** <http://www.camcentral.com/Antarctica.html> is the best place to go for links to those in operation.

AASTO (Automated Astrophysical Site-Testing Observatory) presents from the South Pole live images updated every 10 minutes <http://bat.phys.unsw.edu.au/~aasto>. Also here are researcher diaries and an extensive Photogallery.

For icebergs, the Antarctic Meteorology Research Center's site, Iceberg Images at the **AMRC** <http://uwamrc.ssec.wisc.edu/amrc/iceberg.html>, has a vast number of images and gives the current status and positions of the larger ones.

For copyright-free satellite aerials taken in 1997, visit **NASA**'s website <http://svs.gsfc.nasa.gov/imagewall/Antarctica.html>.

Links

Nearly every website has links to other sites. Those of the Byrd Polar Research Center and CRREL, mentioned above, are particularly extensive. Some other choices such as the **Arctic and Antarctic Advice Agency Austria** (yes, Austria!) <http://www.arctic.at> features links by category and country and quite a bit more.

Stephanie Bianchi's **Some Polar Websites** <http://www.acs.ucalgary.ca/~tull/polar/polar.htm> is very comprehensive, and arranged by subject.

The Polar Web <http://www.urova.fi:80/home/arktinen/polarweb/polarweb.htm> is a collaborative project of the Polar Libraries Colloquy and is managed by the Arctic Centre at the University of Lapland in Finland. Although the emphasis is more Arctic than Antarctic, it is nonetheless a worthwhile compilation with helpful descriptions of each link.

Discoverers Web <http://www.win.tue.nl/~engels/discovery> is a massive collection of links covering all aspects of travel, discovery, and exploration.

Arranged mostly by region and era it has a more than adequate polar section.

Miscellaneous

Some Antarctic websites defy easy categorization. Here are a few.

A searchable database at <http://webhost.nvi.net/aspire> provides probably more than anyone will ever want to know about the **Antarctic Treaty**.

Welcome to the Ice <http://www.theice.org> is the work of Robert Holmes and has a bit of everything. Its Discussion Board includes many e-mail queries with comments and responses posted by others. It has an informative Did You Know? section and frequently asked questions with answers. Charles Neider's "Historic Guide to Ross Island" is among the offerings.

The CIA's **Factbook on Antarctica** may be found at <http://www.cia.gov/cia/publications/factbook/geos/ay.html>. As it employs the standard format used for all countries, the categories are not always relevant—percentage of land in permanent crops, for example—but it is still worth a look.

The **Tasmanian Polar Network (TPN)** <http://www.tpn.aq> gives Antarctic shipping schedules, information about polar ships calling at Hobart, some news on polar events, and dates of conferences.

Raytheon Polar Services is the logistical arm of the United States Antarctic Program and its site <http://www.polar.org> includes lots on American activities in the Antarctic including the useful Participant Guide and the cruise histories and operating schedules of American polar vessels.

The Antarctic Circle <http://www.antarctic-circle.org> is a "forum and resource on historical, literary, bibliographical, artistic and cultural aspects of Antarctica and the South Polar regions." There are individual pages covering Antarctic events, books and booksellers, organizations, and historic time lines. Two interesting and lengthy features are The Low-Latitude Antarctic Gazetteer, a compilation of Antarctic sites outside the Antarctic (statues, explorers' houses, etc.), and "Tekeli-li," Fauno Cordes' annotated bibliography of Antarctic fiction. RS

Gazetteer

- The gazetteer contains the names on the following maps: Antarctica (pages 14–15), Antarctic Peninsula (page 89), Gerlache Strait (page 95), Ross Sea and Ross Island (page 113), Antarctica and the Sub-Antarctic Islands (pages 146–147) and The Arctic (pages 176–177).
- The country or ocean is listed for all names outside Antarctica and the Sub-Antarctic islands.
- Names that have a symbol (stations and summits) are given a grid reference in the gazetteer according to the location of the symbol. Names without a symbol are referenced in the gazetteer according to the start or center of the name.

A

Abbot Ice Shelf 14 D27
Abbott Peak 113 D8
Academician Vernadsky station, Ukraine 14 E30, 89 C3, 95 A3
Access Point 95 B2
Adams Glacier 15 E12, 147 C4
Adare Peninsula 113 D7
Adelaide, Australia 147 F5
Adelaide Island 14 E30, 89 A4, 147 C10
Adie Inlet 89 D4
Admiralty Inlet, Canada 177 D18
Admiralty Mountains 113 D6
Admiralty Sound 89 G3
Åland, Finland 177 G7
Alaska, USA 176 F24
Alaska Peninsula, USA 176 G25
Alaska Range, USA 176 F25
Alberta, Canada 176 H21
Alcock Island 95 D2
Aleutian Islands, USA 176 H27
Alex Point, USA 176 G24
Alexander Archipelago, USA 176 G23
Alexander Island 14 D29, 147 C10
Alexandra Mountains 113 C10
Alfred Faure station, France 147 E2
Allison Peninsula 14 D28
Almirante Brown station, Argentina 14 F30, 89 D3, 95 C2
Ambarchik, Russian Federation 176 E29
Amery Ice Shelf 15 E8, 147 C3
Ammassalik, Greenland 177 E13
Amundsen Glacier 113 A10
Amundsen Gulf, Canada 176 D22
Amundsen-Scott station, USA 15 A10
Amundsen Sea 14 E26, 147 C9
Anadyr', Russian Federation 176 F28
Anadyrskiy Zaliv, Russian Federation 176 F27
Anare Mountains 15 D17, 113 D6
Anchorage, USA 176 F24
Andersson Island 89 G2
Andes, South America 14 J29, 146 F10
Andvord Bay 95 C2
Angot Point 95 D1
Antarctic Peninsula 14 E30, 89 C4, 95 C3, 147 C11
Antarctic Sound 89 G2
Antipodes Islands, NZ 147 E7
Anvers Island 14 F30, 89 D3, 95 B2, 147 C10
Arago Glacier 95 C2
Arctic Bay, Canada 177 D18
Arctic Ocean 176 C27

B

Arctowski Peninsula 95 C2
Arctowski station, Poland 14 F31, 89 F2
Argentina 14 J30, 146 F10
Argentine Islands 89 C3, 95 A3
Argo Point 89 E4
Arkhangel'sk, Russian Federation 177 F5
Arrowsmith Peninsula 89 B4
Arthur Harbor 95 A2
Artigas station, Uruguay 14 F31, 89 F2
Aspland Island 89 G1
Astrolabe Island 89 F2
Asunción, Paraguay 146 G11
Athabasca River, Canada 176 G21
Auckland, NZ 147 F6
Auckland Islands, NZ 147 D6
Aurora Glacier 113 D9
Australia 147 G5
Avery Plateau 89 C4
Axel Heiberg Island, Canada 177 C19

Back River, Canada 177 E20
Backdoor Bay 113 C8
Baffin Bay, Canada 177 D16
Baffin Island, Canada 177 E17
Bagshawe Glacier 95 C2
Bahía Grande, Argentina 14 H30
Bailey Peninsula 15 E12
Balleny Islands 15 E17, 147 C6
Baltic Sea, Europe 177 G7
Bancroft Bay 95 D2
Banks Island, Canada 176 D22
Banzarre Coast 15 E13
Barents Sea, Arctic Ocean 177 D5
Barilari Bay 89 C3
Barne Glacier 113 D8
Barne Inlet 113 B6
Barrow, USA 176 D25
Bart Bay 95 C2
Bartlett Inlet 113 C10
Bathurst Island, Canada 177 C19
Bay of Whales 15 C20, 113 C9
Bear Peninsula 14 D25
Beardmore Glacier 15 B17, 113 B6
Beaufort Island 113 C6
Beaufort Sea, Arctic Ocean 176 D23
Beaumont Bay 113 B6
Beerenberg, Jan Mayen 177 D10
Belarus 177 H7
Belcher Islands, Canada 177 G18
Belfast, UK 177 H10
Belgicafjella 15 D3
Bellingshausen Sea 14 E28, 147 C10
Bellingshausen station, Russian Federation 14 F31, 89 F2
Beloye More (White Sea), Russian Federation 177 E6
Beneden Head 95 C2
Bergen, Norway 177 F9
Bering Sea, Pacific Ocean 176 G27
Bering Strait, Russian Federation/USA 176 F26
Berkner Island 14 C32, 147 B11
Berlin, Germany 177 H8
Bermel Peninsula 89 C5
Berthelot Islands 95 A3
Bird Island 147 D11
Bird Island station, UK 147 D11
Biscoe Bay 95 B2
Biscoe Islands 14 F30, 89 B3, 147 C10
Bismarck Strait 89 C3, 95 B2
Bjørnøya (Bear Island), Svalbard 177 D8
Black Island 113 C6
Blodgett Iceberg Tongue 15 E13
Bluff Island 95 D2

Bob Island 95 B2
Bodø, Norway 177 E8
Bolivia 146 H10
Bone Bay 89 F2
Booth Island 95 A3
Boothia Peninsula, Canada 177 D19
Borden Island, Canada 177 C21
Borden Peninsula, Canada 177 D18
Börgen Bay 95 B2
Botswana 147 G1
Bounty Islands, NZ 147 E7
Bouquet Bay 95 C2
Bourgeois Fjord 89 B4
Bouvetøya, Norway 147 D1
Bowers Mountains 113 D6
Bowman Island 15 E11, 147 C4
Boyle Mountains 89 B4
Brabant Island 14 F30, 89 D3, 95 C2
Bransfield Island 89 G2
Bransfield Strait 14 F30, 89 E2
Brazil 146 H11
Brialmont Cove 95 D2
Bristol Bay, USA 176 G25
Britannia Range 113 C5
British Columbia, Canada 176 G22
Brodeur Peninsula, Canada 177 D18
Brooklyn Island 95 C2
Brooks Range, USA 176 E25
Brown Bluff 89 G2
Bruce Plateau 89 C4
Brunt Ice Shelf 15 D34
Bryde Island 95 B2
Buckle Island 15 E17
Buenos Aires, Argentina 146 F11
Buls Bay 95 C2
Bunger Hills 15 E11
Bunger Island 14 D26
Bush Mountains 113 B8
Bussey Glacier 95 A3
Butler Passage 95 B2
Byers Peninsula 89 E2
Bylot Island, Canada 177 D17
Byrd Glacier 15 C16, 113 B5
Byth Point 89 E2

C

Cabinet Inlet 89 D4
Cabo de Hornos (Cape Horn), Chile 14 G30
Cabo Vírgenes 14 H30
Caird Coast 15 C34
Campbell Island, NZ 147 D6
Canada 176 F20
Canada Basin, Arctic Ocean 176 C24
Canberra, Australia 147 F5
Cape Adare 15 D18, 113 D7, 147 B6
Cape Agassiz 89 D5
Cape Alexander 14 E30, 89 D4
Cape Andreas 95 E1
Cape Ann 15 E6, 147 C2
Cape Anna 95 C2
Cape Armitage 113 C8
Cape Barrow, Canada 176 E21
Cape Bellue 89 B4
Cape Bird 113 D8
Cape Boothby 15 E6
Cape Borley 15 E6, 147 C2
Cape Carr 15 E14, 147 C5
Cape Casey 89 D4
Cape Cheetham 15 E17, 113 D6
Cape Chidley, Canada 177 F16
Cape Cloos 89 C3, 95 A3
Cape Cockburn 95 C2
Cape Colbeck 113 C10
Cape Columbia, Canada 177 B18
Cape Crozier 113 D10
Cape Darnley 15 E7, 147 C3
Cape Denison 15 E15
Cape Disappointment 89 E3, 95 D3
Cape Ducorps 89 F2
Cape Elliott 15 E11

Cape Errera 95 B2
Cape Evans 113 C8
Cape Fairweather 89 E3, 95 D2
Cape Filchner 15 E10
Cape Flying Fish 14 D26
Cape Framnes 14 E30, 89 E3
Cape Freshfield 15 E16
Cape Garry 89 D2, 95 C1
Cape Goodenough 15 E13, 147 C5
Cape Hallett 113 D7
Cape Herlacher 14 D25, 147 B9
Cape Herschel 89 E3, 95 D2
Cape Hooker 113 D6
Cape Hudson 15 E16, 147 C6
Cape James 89 D2
Cape Juncal 89 G2
Cape Kjellman 89 F2
Cape Knowles 14 D30
Cape Lancaster 95 B2
Cape Longing 89 F3
Cape Lookout 89 H1
Cape MacKay 113 C9
Cape Marsh 89 F3
Cape Mascart 89 A4
Cape Maude 113 B7
Cape Melville 89 G2
Cape Mikhaylov 15 E12
Cape Monaco 95 A2
Cape Moore 15 D17, 113 D6
Cape Murray 113 C6
Cape North 15 D17, 113 D6
Cape Northrop 89 C4
Cape of Good Hope, South Africa 147 F1
Cape Peremennyy 15 E11
Cape Poinsett 15 E12, 147 C4
Cape Renard 89 C3, 95 B3
Cape Robert 15 E14
Cape Robinson 89 D4
Cape Roquemaurel 89 F2
Cape Roux 95 C2
Cape Royds 113 D8
Cape Selborne 15 B17, 113 B6
Cape Shirreff 89 E2
Cape Tennyson 113 D9
Cape Town, South Africa 147 F1
Cape Tuxen 95 A3
Cape Valentine 89 H1
Cape Waldron 15 E12, 147 C4
Cape Washington 15 D17, 113 D6
Cape Willems 95 B2
Cape Wollaston 95 E1
Capitán Arturo Prat station, Chile 14 F31, 89 F2
Carney Island 14 D24, 147 B9
Casey station, Australia 15 E12
Chapman Point 89 E3
Charcot Cove 113 C6
Charcot Island 14 E29
Charlotte Bay 95 D2
Chatham Islands, NZ 147 E7
Chaunskaya Guba, Russian Federation 176 E29
Chile 14 J29, 146 F10
Christchurch, NZ 147 E6
Christiania Islands 89 E2, 95 D1
Chukchi Sea, Arctic Ocean 176 D27
Chukotskiy Khrebet, Russian Federation 176 E27
Churchill, Canada 177 G19
Churchill Mountains 113 B6
Churchill Peninsula 89 D4
Churchill River, Canada 176 G20
Cierva Cove 95 D2
Clarence Island 14 F31
Clark Peninsula 15 E12
Claude Point 95 C2
Coast Mountains, Canada/USA 176 H22
Coats Island, Canada 177 F18
Coats Land 15 C34, 147 B12

Kong Frederik VIII Land, Greenland 177 C12
Konsfjorden, Svalbard 177 C9
Korff Ice Rise 14 C29
Koryakskiy Khrebet, Russian Federation 176 F28
Kotzebue Sound, USA 176 E26
Krause Point 15 E10
Krogh Island 89 B4
Kronprins Olav Kyst 15 E5
Kronprinsesse Martha Kyst 15 D35
Kuril'skiye Ostrova (Kuril Islands), Russian Federation 176 J30
Kyyiv, Ukraine 177 H7

L

La Gorce Mountains 113 A11
Labrador Sea, Canada/Greenland 177 G15
Ladozhskoye Ozero, Russian Federation 177 G6
Lady Newnes Bay 15 D18, 113 D6
Laënnec Glacier 95 C2
Lake Athabasca, Canada 176 G21
Lake Vanda 113 C6
Lake Vostok 15 C11
Lambert Glacier 15 D7, 147 B3
Land Glacier 14 C22
Lanusse Bay 95 C2
Lapland, Europe 177 E7
Larrouy Island 89 C3
Larsemann Hills 15 D8
Larsen Ice Shelf 14 E30, 89 D4, 95 D3
Larsen Inlet 89 F3
Latady Island 14 D29, 147 C10
Latvia 177 G7
Laurie Island 14 F32
Lauritzen Bay 15 E16
Laussedat Heights 95 C2
Lavoisier Island 14 E30, 89 B4
Law Dome 15 E12
Law Plateau 15 D8
Lazarevisen 15 E2
Leay Glacier 95 B3
Lecointe Island 95 C2
Lemaire Channel 95 B3
Lemaire Island 95 C2
Lena, Russian Federation 176 E33
Lennox-King Glacier 113 B6
Leonardo Glacier 95 D2
Leppard Glacier 89 D3
Lesotho 147 G1
Leverett Glacier 113 A11
Lewis Bay 113 D9
Liard Island 89 B4
Liège Island 89 D2, 95 C1
Lillie Glacier 113 D6
Lincoln Sea, Canada/Greenland 177 B15
Lion Island 95 B2
Lister Glacier 95 C2
Lithuania 177 G7
Little Diomede Island, USA 176 F26
Liv Glacier 113 A9
Livingston Island 14 F30, 89 E2
Lomonosov Ridge, Arctic Ocean 177 A22
London, UK 177 H10
Longyearbyen, Svalbard 177 C8
Loper Channel 89 H1
Low Island 89 D2, 95 C1
Luitpold Coast 15 C33
Lützow Holmbukta 15 E4
Lyddan Island 15 D34

M

McClure Strait, Canada 176 C22
Mackay Glacier 15 C16, 113 C6
Mackenzie Bay, Antarctica 15 E8
Mackenzie Bay, Canada 176 E23

Mackenzie Glacier 95 C2
Mackenzie King Island, Canada 177 C21
Mackenzie Mountains, Canada 176 F22
Mackenzie River, Canada 176 E22
Macleod Point 95 D2
McMurdo Sound 113 C6
McMurdo Station, USA 15 C17, 113 C8
Macquarie Island, Australia 147 D6
Macquarie Island station, Australia 147 D6
Mac.Robertson Land 15 D7, 147 B3
Madagascar 147 G2
Magadan, Russian Federation 176 G30
Maitri station, India 15 D2
Makarov Basin, Arctic Ocean 177 A28
Mamelon Point 89 C4
Manitoba, Canada 177 G19
Mansel Island, Canada 177 F18
Marguerite Bay 14 E30, 89 A5
Marie Byrd Land 14 C25, 147 B9
Mariner Glacier 113 D6
Marion Island 147 E2
Marion Island station, South Africa 147 E2
Marshall Mountains 113 B6
Martin Peninsula 14 D25
Masson Island 15 E10, 147 C4
Matha Strait 89 B4
Matochin Shar, Russian Federation 177 D4
Mawson Coast 15 E7
Mawson Escarpment 15 D7
Mawson Glacier 15 C16, 113 C6
Mawson Peninsula 15 E16
Mawson station, Australia 15 E7
Melbourne, Australia 147 F5
Melchior Islands 95 B2
Melville Island, Canada 177 C21
Melville Peninsula, Canada 177 E18
Mendeleyev Ridge, Arctic Ocean 176 B27
Mertz Glacier 15 E15
Mertz Glacier Tongue 15 E15
Mikhaylov Island 15 E9, 147 C3
Mikkelsen Harbor 95 E1
Milburn Bay 95 E1
Mill Inlet 89 C4
Mill Island 15 E11, 147 C4
Millerand Island 89 B5
Minna Bluff 15 C18, 113 C6
Minsk, Belarus 177 H7
Mirny station, Russian Federation 15 E10
Mitchell Point 95 C2
Mobiloil Inlet 89 C5
Molodezhnaya station, Russian Federation 15 E5
Montagu Island 147 D12
Montevideo, Uruguay 146 F11
Moody Point 14 F31, 89 H2, 147 C11
More Laptevykh, Russian Federation 176 D32
Moser Glacier 95 C2
Moskva, Russian Federation 177 G6
Moubray Bay 15 D18, 113 D7
Mt Albert Markham 113 B5
Mt Bird 113 D8
Mt Borland 15 D7
Mt Bulcke 95 C2
Mt Coman 14 D30
Mt Dudley 89 B5
Mt Erebus 15 C17, 113 D9
Mt Ford 113 D6

Mt Frakes 14 C25
Mt Français 89 D3, 95 B2
Mt Frazier 113 C10
Mt Gaudry 89 A4
Mt Haddington 89 G3
Mt Joyce 113 C6
Mt Larsen 113 D6
Mt Logan, Canada 176 F24
Mt Longhurst 113 C5
Mt McClintock 15 B16, 113 B5
Mt McKinley, USA 176 F25
Mt Markham 15 B17, 113 B6, 147 A6
Mt Melbourne 15 D17, 113 D6
Mt Menzies 15 D7, 147 B3
Mt Miller 113 B6
Mt Minto 15 D17, 113 D6, 147 B6
Mt Moberly 95 B2
Mt Murchison 113 D6
Mt Northampton 113 D6
Mt Reeves 89 A4
Mt Sandow 15 E11
Mt Saunders 113 A6
Mt Scott 95 A3
Mt Shackleton 95 B3
Mt Sidley 14 C24, 147 B8
Mt Siple 14 D24
Mt Southard 15 D16, 113 D5
Mt Takahe 14 C25
Mt Tennant 95 C2
Mt Terra Nova 113 D9
Mt Terror 113 D9
Mt Walker 95 C2
Mt William 95 B2
Mt Zeppelin 95 D2
Mozambique 147 G2
Mozambique Channel, Indian Ocean 147 G2
Mulock Glacier 113 C6
Mulock Inlet 15 C17, 113 C6
Murmansk, Russian Federation 177 E6
Mys Chelyuskin, Russian Federation 177 C34
Mys Dezhneva, Russian Federation 176 E27
Mys Navarin, Russian Federation 176 F28

N

Namibia 147 G1
Nanisivik, Canada 177 D18
Nansen Basin, Arctic Ocean 177 B2
Nansen Island 95 C2
Napier Mountains 15 E6
Narsarsuaq, Greenland 177 F14
Neko Harbor 95 C2
Nelson Island 89 F2
Nelson River, Canada 177 G19
Nemo Peak 95 B2
Neny Fjord 89 B5
Neptunes Bellows 89 E2
Neumayer Channel 95 B2
New Bedford Inlet 14 D30
New Zealand 147 E6
Newfoundland, Canada 177 H16
Newman Island 14 C22
Newtontoppen, Svalbard 177 C8
Neyt Point 95 D1
Nilsen Plateau 113 A9
Nimrod Glacier 113 B6
Nimrod Passage 95 A2
Ninnis Glacier 15 E15
Ninnis Glacier Tongue 15 E15
Nizhniy Novgorod, Russian Federation 177 G5
Nobile Glacier 95 D2
Nome, USA 176 F26
Nord, Greenland 177 B11
Nord-Jan, Jan Mayen 177 D10
Nordaustlandet, Svalbard 177 B7

Nordenskjöld Coast 95 E2
Nordkapp, Norway 177 D7
Nordvik, Russian Federation 177 D34
North Atlantic Ocean 177 G12
North Pacific Ocean 176 J27
North Pole, Arctic Ocean 177 A9
North Sea, Europe 177 G9
Northcliffe Glacier 15 E10
Northwest Territories, Canada 176 F21
Norton Sound, USA 176 F26
Norway 177 F9
Norwegian Sea, Atlantic Ocean 177 D9
Novaya Zemlya, Russian Federation 177 D4
Noville Peninsula 14 D27
Novolazarevskaya station, Russian Federation 15 D2
Novosibirskiye Ostrova, Russian Federation 176 C31
Nunavut, Canada 177 E19
Nunivak Island, USA 176 G26
Nuuk, Greenland 177 F15
Ny Ålesund, Svalbard 177 C8

O

Oates Land 15 D16, 113 D6, 147 B6
Ob' Bay 15 D17, 113 D6
Observation Hill 113 C8
Okhotskoye More (Sea of Okhotsk), Russian Federation 176 H31
Okuma Bay 113 C10
Old Crow, Canada 176 E23
Omega Island 95 C2
Omolon, Russian Federation 176 F30
Ontario, Canada 177 H18
Orcadas station, Argentina 14 F32, 147 C11
Orkney Islands, UK 177 G10
Orléans Strait 95 E1
Orne Harbor 95 C2
Oscar II Coast 89 E3, 95 D3
Oslo, Norway 177 G8
Ostrov Dzheksona, Russian Federation 177 B4
Ostrov Greem Bell, Russian Federation 177 B1
Ostrov Kotel'nyy, Russian Federation 176 C32
Ostrov Novaya Sibir', Russian Federation 176 D30
Ostrov Ratmanova, Russian Federation 176 E26
Ostrov Vrangelya (Wrangel Island), Russian Federation 176 D28
Outer Hebrides, UK 177 G10
Ozero Taymyr, Russian Federation 177 C35

P

Palmer Archipelago 14 F30, 89 D3, 95 A2
Palmer Land 14 D30, 147 B10
Palmer station, USA 14 F30, 89 C3, 95 A2
Paradise Harbor 95 C2
Paraguay 146 G11
Parry Islands, Canada 177 C20
Pasteur Peninsula 95 C2
Patagonia, Argentina/Chile 146 E10
Paulding Bay 15 E13
Paulet Island 89 H2
Pavlov Peak 95 D2
Peace River, Canada 176 G21
Pechora, Russian Federation 177 E4
Pelseneer Island 95 C2
Pendleton Strait 89 B3

Index

Page numbers in **bold** print refer to main entries. Page numbers in *italics* refer to photographs.

elephant seals, 91, 105, 152, *153*, 156, 157, 158, *159*, 160, *160*, 170, *170–1*, 173, 218, *218*, 219, 220, *220*, 221, *221*, 232, *233*, *236*, **246–9**, *246*, *247*, *248–9*, 351, **366–7**, *400*, 401, 532, 533, *533*, 535
Eliza Scott (ship), 401
Elizabethides, 543
Elk *see* Moose
Ellesmere Island, *38–9*, 180, 374, 510, 512, 513
Ellsworth, Lincoln, 35, 91, 481, 483, **492–3**, *492*, *493*, 513, 518, 551, 583, 584
Ellsworth base, 501
Ellsworth Land, 28, *56*
Ellsworth Mountains, 28, 29, 35, 58, 77
Elmerson Peninsula, *38–9*
Elseshul, 152
Emma Island, 93
Emperor penguin (*Aptenodytes forsteri*), *2–3*, *4–5*, 109, *109*, 122, *122*, 130, *132–3*, 141, 326, *326–7*, 329, 334, *335*, **336–7**, *336*, *337*, *338*, **339**, *339*, 342, 394, *417*, 426, *437*, 448, 454, 558, 559, *559*, 581
endangered species, 576
Endeavour (ship), 388
Enderby, Charles, 166
Enderby Brothers, **399**, 539, 540
Enderby Island, 166, *166*
Enderby Land, 132, 399, 436–7, 486, 498, 540
Endurance (ship), 68, 69, 73, *378–9*, 466, **468–70**, *468–9*, *470–1*, 472, 473, 526, 545, *545*, 548, 562, 582
energy sources, alternative, 528, *528*
England *see* Britain
ENSO, 234
Enterprise Island, 93
environmental impact *see* human activity/impact; tourism
environmental microbiology, 572
Eocene Epoch, *30*
ephiphytes, 213
Epilobium latifolium, 365
equator, 22, 26, 52
equinoxes, **23**
Eratosthenes of Cyrene, 382
Erebus (ship), 130, 408, **410–11**, *410*, *504–5*, 509, 544, *544*
Erebus Bay, *61*, *75*
Erebus Gulf, 90, 434, 435
Erebus Ice Tongue, *8–9*
Erect-crested penguin (*Eudyptes sclateri*), *303*, 358
Eric the Red, 180
Erignathus barbatus, *219*, 366, *366*, 376
Eriophorum species, 365
Errera Channel, **94**, 96, *96*, 97
Escudero, Julio, 519
Esperanza station, **88**
Eubalaena australis, 152, *250–1*, *253*, 258, *258*, **259–61**, *259*, *260–1*, 263, 401, 414, 415
Eubalaena glacialis, 259
Eudyptes, 352, **358**
    E. chrysocome, 155, 170, *170*, 316, *352*, **354–5**, *354*, *355*, 580
    E. c. chrysocome, 355
    E. c. filholi, 355
    E. c. moseleyi, 355
    E. chrysolophus, 148, 152, 156, 160, 171, 316, 327, 328, *353*, **356**, *356*, 580
    E. pachyrhynchus, 352, 358, *358*
    E. robustus, 358
    E. schlegeli, 170, *193*, 330, **357**, *357*

E. sclateri, *303*, 358
Eudyptula minor, 326, 330, **358**, *359*
Eumetopias jubatus, **366**
Euphausia crystallorophias, 203, 207
Euphausia superba, 202, 203, **207–8**, *207*, *208–9*, **210**, **534** *see also* krill
euphausiids, 207, 210, 211 *see also* krill
Eurasia Basin, 38, 39
Europe
    formation of, 30, 38
    global warming &, 76
    mapping &, **538–9**
    see also under country e.g. Germany
European Arctic region, **178–80**, *178*, *179*, *180*
European reindeer, 178, 182, *183*, 374
European Space Agency, 573
European Union, 143, 180
Evans, Edgar, 456, *457*, 562, *562*
Evans, Edward "Teddy", 123, 126–7, 452, 455, *455*
Evans, Frederick, 442, 446
Evans, Hugh Blackwell, 421
evaporation, 41, 128, 197
evolution, **40–7**
Exasperation Inlet, 97
Exclusive Economic Zones, 534
expeditions and exploration
    Arctic &, *504–5*, **506–13**, *506*, *507*, *508–9*, *510*, *511*, *512*, *513*, 560
    Australasia &, 54, 134, 451, **460–5**, *460–1*, *462–3*, *464–5*, **466**, 484, 546, *548*, 585, 555, *572*
    Australia &, 134, **414–16**, **482–3**, *482*, *483*, **486**, *486*, *487*, **488**, *488–9*, 490, *491*, 492, 493, **498–9**, *498–9*
    Austria-Hungary &, 510
    Belgium &, 93, 94, **417–18**, *417*, *418*, *419*, *420*, 545–6, *545*, 580
    Britain &, 34, 62, 90, 107, 108–9, 116, 118, 123, 124, **149**, 150, 156, 164, 166, 170, 172, 173, 382, **386–7**, *386*, **388–90**, *388*, *390–1*, **391**, 395, **396–7**, *396*, 399, **401**, **408–11**, *408*, *410*, *411*, **412–13**, *412*, *413*, **421–3**, *422*, *423*, *424–5*, **426–9**, *426*, *427*, *428*, *429*, *436–7*, *437*, **442–7**, *442–3*, *444–5*, *446–7*, **452–7**, *452–3*, *454*, *455*, *456*, *457*, **468–75**, *468–9*, *470–1*, *472*, *473*, *474*, *475*, **478–9**, *478*, 480, **482–3**, *486*, *486*, *487*, **488**, *488–9*, **490**, *490–1*, *504–5*, **507–10**, *510*, 513, 531, 541, *541*, **544**, 545, 550, *550*, 554, *554*, 560, **562**, *562*, 567
    clothing &, **560**, *560*, *561*, *581*
    Commonwealth &, 116, **502–3**, *502*, *503*, 567
    early years (1487–1900) of, **380–423**, 528
    food &, **581**
    France &, 91, 103, 107, 108, 166, **172**, **173**, **402**, *402*, *403*, **404**, *404*, **438–41**, *438–9*, *440–1*
    Germany &, 62, 104, 107, 149, 173, 426, **430–1**, *430–1*, **458**, *458*, 465, **494**, *494*, 510, 550, 551
    Heroic Age &, 132, 134, **426–75**, 540, 545, **562**, 567, 584, 585
    Italy &, 512
    Japan &, 139, **459**, *459*
    modern age (1921–59) &, **476–503**
    Netherlands &, **387**
    Norway &, 38, 140, 164, **414–15**, **448–51**, *448–9*, *450*, *451*, 452, 455, 486, 494, 500, **511**, *511*, 544, 545, 555

Portugal &, 382, **384–5**, 544
Russia &, 149, 164, **392–4**, *392*, 508, 561
Spain &, 384, **385**
Sweden &, 88, 90, 91, 108, **432–5**, *432–3*, *434*, *435*, 438, 511, 554
United States &, 35, 404, **405–7**, *405*, *406*, 482, 483, **484–5**, *484*, *485*, **492–3**, *492*, *493*, **495–7**, *495*, *497*, **510–11**, **512–13**, **551–2**
    world sites commemorating, **582–5**
    see also aviation; shipping
Explorer (ship), 527
extinctions, **45–6**, 576

F
Fairy penguin *see* Little penguin
Fairy prion (*Pachyptila turtur*), **301**, *301*
Fairbanks, 185
Falkland fur seal *see* South American fur seal
Falkland Islands, 35, 93, 100, 146, **155**, *155*, 212, 226, 230, 246, 300, 306, 312, 315, 316, 340, 355, 358, 359, 398, *399*, 401, 411, 432, 480, 486, 553, 581
Falklands Islands Dependency Survey (FIDS), 149, 480, 540, 563
Falklands War, 148, **152–3**, **155**
False Cape Renard, 104, *104–5*
Fanning, Edmund, 149
Faraday station, 108–9
fast ice, *64–5*, 72, *75*, 203, 204, 242, *243*, 334, 366
faulting, 30, 31, 37, 39
feld mark, *195*
fellfields, 365
ferns, 42, 194, 197
field ice, 389
Filchner, Wilhelm, 62, **458**
Filchner Ice Shelf, 62, 69, 141, 197
Fimbul Ice Shelf, 69, 140
Fin whale (*Balaenoptera physalus*), 150, *252*, 257, 262, **265–6**, 530, 534, 580
Finback *see* Fin whale
Finch Creek, *193*
Finé, Orance, 538
Finland, 178, 546, 580
Finner whale *see* Fin whale
Fiordland penguin (*Eudyptes pachyrhynchus*), 352, 358, *358*
fiords, 137, 173, 179, 186, **458**, 490
Fire oak (*Casuarina cunninghamiana*), 47
fireweed, dwarf, 365
fish, 34
    Antarctica &, 200, 204, *207*, **210–12**, *211*, 213
    Arctic &, **362–3**
Fish Islands, 109
fishing, 180, 182, **208**, 211–12, *212*, 528, **530–1**, **534**, 535, *535*, 570
    see also seabirds
Flat-headed whale *see* Southern bottlenose whale
flat ice, *62–3*, *64–5*
Fletcher, Francis, 386, 387
flies, 192, 194
floating ice, 56, 77, 198
flowering plants, 40, 43, **45**, *45*, **46–7**, *47*, 97, 160, 166, *167*, *169*, *179*, 181, *187*, 192, 194, *194*, 195, **196–7**, 362, 364, 365, *365*
Flower's whale *see* Southern bottlenose whale
Floyd Bennett (aircraft), 584
flukes *see* whales
Flying Fish (ship), **405**, 406, 407
fog, 178, 390, 395, 542

folding, 30, **35–6**
footwear, 560, 561
Forbidden Plateau, 93
Ford, Edsel, 484, 550
forests, 36, 41, **42**, *42*, 44, 186, 166, *168–9*, 362, *362* *see also* trees
Forster, John Reinhold, 388
Fortuna Bay, *416*
fossils, 24, 30, 34, *34*, 36, 37, **41**, *41*, **42**, *42*, *43*, 44, **45–7**, *45*, 47, 91, 202, 284, 294, 320, 414, 432
Foster, Henry, 160
foxes, 178, 182, 370, **374**, 376
foxtails, 365
Foyn, Svend, **414**, *414*, 415, 416
Foyn Land, 414
Fram (ship), 38, **448**, *448*, 451, 452, *459*, **511**, 544, 545, *545*, 583
Fram Basin, 39
"Framheim", 448, *449*, 450
Framnes Mountains, *28*
Français (ship), **438**, *439*
France
    Antarctic "locations" in, **583**
    bases of, 54, **135**, **143**, 500
    expeditions and explorers of, 91, 103, 107, 108, 166, **172**, **173**, **402**, *402*, *403*, **404**, *404*, **438–41**, *438–9*, *440–1*
    mapping by, 538, **539**, *539*
    territorial claims of, 518
Franklin, Sir John, 130, 186, 508, **509–10**, *509*, 544, 582
Franklin expedition, 186, *504–5*, 508, **509–10**, 544
Franklin Island, 130
Franz Josef Land, *38*, **181–2**, *182*, 510, 511
Fratercula arctica, 372, *372*
Fratercula corniculata, 372
frazil, 72, 580
Freeman, Thomas, 401
freezing point, **192**
Fregetta tropica, **309**, *309*
frigate birds, 283
Frobisher, Sir Martin, 507, *507*
Frobisher Bay, 186
frost flowers, *61*
frostbite, 560, 562, *562*
Fuchs, Vivian, 62, 502, *502*, *503*, 567
fuel spills, 100, 199, 528–9
Fulmar prion (*Pachyptila crassirostris*), **302**, *302*, 303
fulmarine petrels, **294**, *294*
Fulmarus glacialis, 283, 370
Fulmarus glacialoides, 283, *295*, **297–8**, *297*
fulmars, 91, 135, 141, 152, 194, **294–307**, *294*, *295*, *296*, *297*, *298*, *299*, 300, *301*, *302*, *303*, *304*, *305*, *306–7*, 370
fungi, 128, 198, 199
fur seals, 91, 100, **216–17**, 218, *219*, **222–5**, *222*, **226–9**, *226*, *227*, *228*, *229*, 230, 398–9, *398–9*, 530, *530*, 532, 533, 535
Furious Fifties, 50, 52
Fury (ship), 508

G
Gabriel de Castilla station, 162
gadfly petrels, **294–5**
Galapagos fur seal, *219*
Galapagos Islands, 287
Galapagos penguin (*Spheniscus mendiculus*), 326, 359
Galindez Island, 108
Gama, Vasco da, 384, *384*
Gamburtsev Subglacial Mountains, 28, 77
Gammle Metten, 179

Acknowledgments

The authors would like to thank the many people who have contributed to this remarkable project. Most thanks go to all the contributors named within the author list who must be acknowledged here for their exceptional assistance far beyond their written work. Thanks, too, to the dedicated production team and all at Global Publishing. We owe a lot to those we have traveled with, and those who have taken us to the polar regions. Among the many Antarcticans we would like to thank, in no particular order, are Chris Downes, Margot Morrell, Andrew Prossin, the Wikander family, Carla Santos, Greg Mortimer, Andie Smithies, Graeme Watt, Emily Slatten, Ingrid Visser, Dava Sobel, Ben Hodges, Lee Belbin, Peter Gill, Nigel de Winser, Francis Herbert, Sir Ed Hillary, Phil Law, Glenn A. Baker, Dean Peterson, Rob McNaught, Rinie Van Meurs, Chris McKay, Esteban de Salas, John Borthwick, Dennis Collaton, Chris Doughty, Tony Press, Kerry Lorimer, Bernhard Lettau, Glenn Browning, Pat Quilty, Aaron & Cathy Lawton, Rob Harcourt, Austin Simpson, Pamela Wright, Scott McPhail, Andrew Fountain, Michael Bryden, Zaz Shackleton, Jayne Paramor, Bill Davis, Des Cooper, Jack Sayers, Darrel Schoeling, Barry Griffiths, Dave Briscoe, Denise Landau, Heather Jeffery, Dick Filby, Mark Eldridge, Tim Bowden, Ken Morton, Andy Beatty, Lincoln Hall, Stephen Martin, Warren Papworth, Adelie Hurley, Toni Hurley-Mooy, Craig Bowen, Barry Boyce, Bob Finch, Tony Harrington, John Palmer and Ross Brewer. There are many others in the Antarctic community who helped a lot, not least of all our Russian captains and their crews. Several institutes were of inestimable assistance including the Scott Polar Research Institute, Oficina Antártida Ushuaia, Australian Geographic Society, Royal Geographical Society, Cambridge University Library, State Library of NSW, Library of the Australian Antarctic Division, US Naval Library.

The publishers would like to thank the following people for their assistance in the production of this book: Louise Buchanan, Robert Clancy, Vanessa Finney, Nick Gales, Peter Gill, Mark Hindell, David Howard, Harvey Marchant, Gary Miller, Therese Potma, Patrick Quilty, Patrick Toomey.

Photographers: Kim Westerskov, David McGonigal, Grant Dixon, Mike Craven, Tony Palliser, Luke Saffigna, Malcolm Ludgate, Craig Potton, Stewart Campbell, Kelvin Aitken, Kevin Deacon, Debra Glasgow, Peter Gill, Albert Kuhnigk, Robyn Stewart, Garry Phillips, Harvey Marchant, Steve Nicol, André Martin, Graham Robertson, Patrick Toomey, Jane Francis, J. Howe, R. Hunt, Geoff Longford, Glen A. Baker, David Keith Jones, and Barry Griffiths, Quest Nature Tours.
Cover photograph: Kim Westerskov

Global Book Publishing would be pleased to hear from photographers wishing to supply photographs.

The Publisher believes that permission for use of the historical and satellite images in this publication, listed below, has been correctly obtained, however if any errors or omissions have occurred, Global Book Publishing would be pleased to hear from copyright owners.

Kelvin Aitken: 250–51, 257, 263, 267, 269, 277.
The Art Archive: 380–81, 424–25, 507 top, 511 bottom.
Australian Picture Library/Corbis: 180, 182, 183 bottom, 184, 185 top, 368, 377 bottom, 382 bottom, 384 bottom, 385 top, 386 top, 476–77, 483 top, 483 center, 485 bottom, 494, 495 top, 497 top, 501 top, 502, 503 top, 513 top, 550 bottom.
Hakluyt Society: 394.
Mitchell Library, State Library of New South Wales: 378–79, 390.
NASA: 22 top left, 69 bottom, 79, 541 bottom, 571 top.
National Library of Australia: 384 top, 385 center, 400 top, 400 bottom, 403 bottom, 404 top, 413 bottom, 468 bottom, 470 left, 470–71, 471, 472 top left, 472 center, 485 top, 487 top right, 487 bottom, 488 top, 490, 491 bottom left, 491 center right, 491 bottom right, 492 bottom, 493 top left, 504–505, 508 top, 509 top, 509 center left, 510 top, 541 center right.
Royal Geographical Society: 382 top, 412.
Pat Vickers-Rich: 43 bottom.

Satellite image, page 548 bottom, received and processed by the Australian Bureau of Meteorology from USA's NOAA series of polar orbiting meteorological satellites.

Main cover photograph:
Barne Glacier, Ross Island.

Front cover inset photographs, left to right:
Zodiac under ice arch, Antarctic Peninsula.
King penguin.
Amundsen–Scott station, South Pole.
Breaching, young Humpback whale.
Ice mask.
Adélie penguins take the plunge.
Spigot Peak, Antarctic Peninsula.
Weddell seal in breathing hole.

Back cover inset photographs, left to right:
Courtship dance, Wandering albatrosses.
Cape Renard, Antarctic Peninsula.
Emperor penguin chick.
Southern right whale.
Sea ice survey, McMurdo Sound.
Preening, White-capped albatrosses.
King penguins, South Georgia.
Young Southern elephant seals.

pp 18–19: Cape Royds, Ross Sea.
pp 84–85: Ross Ice Shelf
pp 188–89: Emperor penguins in blizzard.
pp 378–79: *Endurance* in ice. Photographer, Frank Hurley.
pp 514–15: McMurdo Sound, near Cape Bernacchi.
pp 536–37: *Kapitan Khlebnikov* breaks through sea ice.
pp 574–75: Stranded iceberg, Ross Sea.

© Global Book Publishing Pty Ltd 2001
Text © Global Book Publishing Pty Ltd 2001
Maps © Global Book Publishing Pty Ltd 2001
Illustrations from the Global Illustration Archives
© Global Book Publishing Pty Ltd 2001
Photographs from the Global Photo Archive (except where credited above)
© Global Book Publishing Pty Ltd 2001